Flexible and Wearable Sensors

With rapid technological developments and lifestyle advancements, electronic sensors are being seamlessly integrated into many devices. This comprehensive handbook explores current, state-of-the-art developments in flexible and wearable sensor technology and its future challenges.

Numerous recent efforts have improved the sensing capability and functionality of flexible and wearable sensors. However, there are still many challenges in making them super-smart by incorporating features such as self-power, self-healing, and multifunctionality. These features can be developed with the use of multifunctional nanostructured materials, unique architectural designs, and other advanced technologies. This book provides details about the recent advancements, materials, and technologies used for flexible and wearable sensors. Its wide range of topics addresses the fundamentals of flexible and wearable sensors, their working principles, and their advanced applications.

This handbook provides new directions to scientists, researchers, and students to better understand the principles, technologies, and applications of sensors in healthcare, energy, and the environment.

Dr. Ram K. Gupta is an Associate Professor at Pittsburg State University. Dr. Gupta's research focuses on nanomagnetism, nanomaterials, green energy production and storage using conducting polymers and composites, electrocatalysts for fuel cells, optoelectronics and photovoltaics devices, organic-inorganic hetero-junctions for sensors, bio-based polymers, bio-compatible nanofibers for tissue regeneration, scaffold and antibacterial applications, and biodegradable metallic implants. Dr. Gupta has published over 250 peer-reviewed articles, made over 350 national/international/regional presentations, chaired many sessions at national/international meetings, and edited/written several books/chapters for leading publishers. He has received several million dollars for research and educational activities from external agencies. He is currently serving as Associate Editor, Guest Editor, and Editorial Board Member for various journals.

Flexible and Wearable Sensors

Materials, Technologies, and Challenges

Edited by
Ram K. Gupta

CRC Press
Taylor & Francis Group
Boca Raton London New York

CRC Press is an imprint of the
Taylor & Francis Group, an **informa** business

Front cover image: Andrew Berezovsky/Shutterstock

First edition published 2023
by CRC Press
6000 Broken Sound Parkway NW, Suite 300, Boca Raton, FL 33487-2742

and by CRC Press
4 Park Square, Milton Park, Abingdon, Oxon, OX14 4RN

CRC Press is an imprint of Taylor & Francis Group, LLC

© 2023 selection and editorial matter, Ram K. Gupta; individual chapters, the contributors

Library of Congress Cataloging-in-Publication Data

Names: Gupta, Ram K., editor.
Title: Flexible and wearable sensors : materials, technologies, and challenges / edited by Ram K. Gupta.
Description: First edition. | Boca Raton : CRC Press, [2023] | Includes bibliographical references and index.
Identifiers: LCCN 2022041685 (print) | LCCN 2022041686 (ebook) | ISBN 9781032288178 | ISBN 9781032289809 | ISBN 9781003299455
Subjects: LCSH: Flexible electronics. | Wearable technology. | Detectors.
Classification: LCC TK7872.F54 F47 2023 (print) | LCC TK7872.F54 (ebook) | DDC 621.381--dc23/eng/20221122
LC record available at https://lccn.loc.gov/2022041685
LC ebook record available at https://lccn.loc.gov/2022041686

ISBN: 978-1-032-28817-8 (hbk)
ISBN: 978-1-032-28980-9 (pbk)
ISBN: 978-1-003-29945-5 (ebk)

DOI: 10.1201/9781003299455

Typeset in Palatino
by Deanta Global Publishing Services, Chennai, India

Access the Support Material: https://www.routledge.com/9781032288178

Contents

List of Contributors

Rameshwar Adhikari
Central Department of Chemistry
and
Research Centre for Applied Science and
Technology
Tribhuvan University (TU)
Kirtipur, Nepal

Arpana Agrawal
Department of Physics
Shri Neelkantheshwar Government Post-
Graduate College
Khandwa, India

Eider Aparicio-Martínez
Centro de Investigación en Materiales
Avanzados, S. C.
Chihuahua, Mexico

Mannix P. Balanay
Department of Chemistry
Nazarbayev University
Nur-Sultan, Kazakhstan

Bakhytzhan Baptayev
National Laboratory of Astana
Nur-Sultan, Kazakhstan

Ayşe Çelik Bedeloğlu
Department of Polymer Materials
Engineering
Bursa Technical University
Bursa, Turkey

Yuezhen Bin
Department of Polymer Science and
Engineering
School of Chemical Engineering
Dalian University of Technology
Dalian, China

Muhammad A. Butt
Samara National Research University
Samara, Russia
and
Warsaw University of Technology
Institute of Microelectronics and
Optoelectronics
Warsaw, Poland

Prabhat K. Dwivedi
Centre for Nanosciences
Indian Institute of Technology
Kanpur
Kanpur, India

Rocio B. Dominguez
Centro de Investigación en Materiales
Avanzados, S. C.
Chihuahua, Mexico

Hamide Ehtesabi
Faculty of Life Sciences and
Biotechnology
Shahid Beheshti University
Tehran, Iran

Fatma S. Erdonmez
Department of Polymer Materials
Engineering
Bursa Technical University
Bursa, Turkey

Mustafa Erol
Department of Metallurgical and
Materials Engineering and Center for
Fabrication and Application of Electronic
Materials
Dokuz Eylul University
Izmir, Turkey

Shadi Fathi
Department of Chemical Engineering
University of Mohaghegh Ardabili
Ardabil, Iran

Sahar Foroughirad
Faculty of Polymer Engineering
Sahand University of Technology
Tabriz, Iran
and
Borna Chemi Arya Knowledge Based Co.
Tabriz, Iran

Chaudhery Mustansar Hussain
Department of Chemistry and
Environmental Science
New Jersey Institute of Technology
Newark, NJ, USA

Soheil Jalali
Department of Chemical Engineering
University of Mohaghegh Ardabili
Ardabil, Iran

Abolghasem Jouyban
Pharmaceutical Analysis Research Center
Tabriz University of Medical Sciences
Tabriz, Iran

Bhim P. Kafle
School of Engineering
Kathmandu University
Dhulikhel, Nepal

Chun-Won Kang
Department of Housing Environmental
Design, and Research Institute of Human
Ecology
Jeonbuk National University
Jeonju, Republic of Korea

Selcan Karakuş
Department of Chemistry, Faculty of
Engineering
Istanbul University-Cerrahpasa
Istanbul, Turkey

Zahra Karimzadeh
Department of Pharmaceutical
Chemistry
Tabriz University of Medical Sciences
Tabriz, Iran

Nikolay L. Kazanskiy
Samara National Research University
Samara, Russia
and
IPSI RAS-Branch of the FSRC
"Crystallography and Photonics" RAS
Samara, Russia

Özgür Yasin Keskin
Department of Metallurgical and
Materials Engineering
Dokuz Eylul University
Izmir, Turkey

Svetlana N. Khonina
Samara National Research University
Samara, Russia
and
IPSI RAS-Branch of the FSRC
"Crystallography and Photonics" RAS
Samara, Russia

Haradhan Kolya
Department of Housing Environmental
Design, and Research Institute of Human
Ecology
Jeonbuk National University
Jeonju, Republic of Korea

Naveen Kumar
Department of Physics
Maulana Azad National Institute of
Technology
Bhopal, India

Rajnish Kurchania
Department of Physics
Maulana Azad National Institute of
Technology
Bhopal, India

Aman Mahajan
Department of Physics
Guru Nanak Dev University
Amritsar, India

Mansour Mahmoudpour
Department of Food Science and
Technology
Nutrition Research Center
Tabriz University of Medical Sciences
Tabriz, Iran

Sadaf Mehrasa
Department of Chemical Engineering
University of Mohaghegh Ardabili
Ardabil, Iran

Shubham Mishra
Centre for Nanosciences
Indian Institute of Technology Kanpur
Kanpur, India

Changwoon Nah
Department of Polymer-Nano Science and
Technology
Jeonbuk National University
Jeonju, Republic of Korea

Tania K. Naqvi
Centre for Nanosciences
Indian Institute of Technology Kanpur
Kanpur, India

Samia Nawaz
Department of Chemistry
University of Sahiwal
Sahiwal, Pakistan

C. Nithya
Department of Chemistry
PSGR Krishnammal College for Women
Coimbatore, India

Sarbaranjan Paria
Department of Polymer-Nano Science and
Technology
Jeonbuk National University
Jeonju, Republic of Korea

Erdem Tevfik Özdemir
Department of Metallurgical and
Materials Engineering
Dokuz Eylul University
Izmir, Turkey

Meijie Qu
Department of Polymer Science and
Engineering
School of Chemical Engineering
Dalian University of Technology
Dalian, China

Jyoti Rani
Department of Physics
Maulana Azad National Institute of
Technology
Bhopal, India

Behnaz Ranjbar
Radsys Pooshesh Knowledge Based Co.
Tehran, Iran

Zahra Ranjbar
Institute for Color Science and Technology
and
Center of Excellence for Color Science and
Technology
Tehran, Iran

Komal Rizwan
Department of Chemistry
University of Sahiwal
Sahiwal, Pakistan

Mahmoud Roushani
Department of Chemistry
Ilam University
Ilam, Iran

Fahimeh Hooriabad Saboor
Department of Chemical Engineering
University of Mohaghegh Ardabili
Ardabil, Iran

Sagar Sardana
Department of Physics
Guru Nanak Dev University
Amritsar, India

Uswa Shafqat
Department of Chemistry
University of Sahiwal
Sahiwal, Pakistan

Jafar Soleymani
Pharmaceutical Analysis Research Center
Tabriz University of Medical Sciences
Tabriz, Iran

G. Subashini
Department of Chemistry
PSGR Krishnammal College for Women
Coimbatore, India

Aylin Altınış ık Tağaç
Center for Fabrication and Application of
Electronic Materials
and
Department of Chemistry
Dokuz Eylul University
Izmir, Turkey

Yerbolat Tashenov
Department of Chemistry
L. N. Gumilyov Eurasian National
University
Nur-Sultan, Kazakhstan
and
Department of Chemistry
Nazarbayev University
Nur-Sultan, Kazakhstan

Yub N. Thapa
Central Department of Chemistry
Tribhuvan University (TU)
Kirtipur, Nepal
and
Research Centre for Applied Science and
Technology
Tribhuvan University (TU)
Kirtipur, Nepal

and
Tribhuvan Multiple Campus
Tribhuvan University (TU)
Tansen, Nepal

Hatice Aylin Karahan Toprakci
Department of Polymer Materials
Engineering
Yalova University
Yalova, Turkey

Ozan Toprakci
Department of Polymer Materials
Engineering
Yalova University
Yalova, Turkey

Ömer F. Ünsal
Department of Polymer Materials
Engineering
Bursa Technical University
Bursa, Turkey

Hai Wang
Department of Polymer Science and
Engineering
School of Chemical Engineering
Dalian University of Technology
Dalian, China

Neda Zalpour
Department of Chemistry
Ilam University
Ilam, Iran

Rui Zhang
Department of Polymer Science and
Engineering
School of Chemical Engineering
Dalian University of Technology
Dalian, China

1

Flexible and Wearable Sensing Devices: An Introduction

Muhammad A. Butt, Nikolay L. Kazanskiy, and Svetlana N. Khonina

CONTENTS

1.1 Introduction

Wearable sensing devices, as the name suggests, are merged into wearable items or directly with the body to aid in the surveillance of health and/or provide clinically valuable data for therapy. Because many illnesses and impairments require constant monitoring in today's world, patient monitoring for prompt intervention is critical [1]. As a result, using wireless body sensor networks to monitor patients is one of the most significant fields of applications of wearable technology in the healthcare profession. Patient monitoring is extremely sensitive and vital during epidemic occurrences or emergency medical service situations. In these cases, immediate patient monitoring allows the medical team to take the appropriate actions as soon as possible. Patient monitoring alerts caregivers to potentially dangerous situations, and most of these systems use physiological input data to control support items directly. In any case, given the current global situation, smart wearable sensing devices for patient checking can be a revolution in the fight against diseases. The Centers for Disease Control and Prevention (CDC) describe an epidemic as a rapid upsurge in the number of cases of a transmittable disease within a community or topographical region over some time.

Most studies in this field relied on rigid electrical devices produced on semiconductor electronics platforms until about a decade ago. Wearable sensing devices that utilize stretchable and flexible electronics have recently gained a lot of interest [2]. Unlike rigid electronics, these flexible sensors have an extensive range of mechanical features, making production more complex. The most prevalent way for creating soft sensors has been to integrate deformable conducting material designs onto an elastic substrate via transfer printing, screen printing, photolithography, microchannel molding, and lamination. However, there are some disadvantages to these technologies including high costs, multi-step production procedures, limited durability, and prototyping and grading challenges. Consequently, 3D printing is a viable solution that can be utilized with other techniques. Its main advantages are extraordinary-resolution quick patterning and the capacity to unswervingly produce biological devices.

There is a rising awareness of elastic and wearable medical devices for consistent and uninterrupted surveillance of human well-being data [3]. Novel devices are being developed to make uninterrupted surveillance of vital signs as pleasant as possible. They may monitor diverse health pointers such as blood pressure (BP), pulse, body temperature, blood glucose, and others non-invasively and in real time by simply attaching these wearables to the human body surface [4–7]. In medicine, taking a person's body temperature is crucial. A fluctuation in body temperature is a symptom of several disorders. The progression of some ailments can be tracked by taking a body temperature reading. This enables the physician to evaluate the efficacy of therapies based on body temperature. The response to a disease-specific stimulus is a fever. To sustain the body's natural defenses, the body's usual temperature fluctuates. The most typical type of disease-related (pathological) rise in body temperature is fever.

When a person's physical health signs are irregular, real-time scrutiny of critical signals can alert users and healthcare workers to pursue extra medical help, evading the condition when the superlative treatment time is lost. Due to the numerous advantages they offer consumers, flexible electronics are becoming increasingly widespread. Flex circuits will be employed for a range of applications as more businesses realize their potential for increased customizability, scalability, and mobility. Due to their affordability, flexible electronics are being used by many organizations and sectors in systems and applications. Flex circuit production has traditionally been linked with high production costs. A low-cost substitute for more conventional electronics is now available because of advances brought about by the ongoing rise of these items. Although the field of flexible electronics is still in its infancy, the fact that it can function with less expensive substrates immediately sets it apart from more conventional electronics, which depend on pricey substrates like glass.

Flex circuits enable businesses to give customers more individualized experiences. As an illustration, consider goods like smart bracelets whose screens use flexible electronics. Systems, sensors, and displays are combined in a smart band to monitor and transmit data to users. These gadgets frequently enable users to monitor their well-being and communicate with their peers. These complicated gadgets may be worn on the wrist thanks to flexible circuits. One of the key advantages of flexible electronics is mobility. Conventional electronics sometimes include glass-based displays constructed of parts that are more likely to break, making them difficult to move. The organic thin-film transistors used in flex circuits, which enable goods to be substantially smaller and lighter, eliminate this issue. These modifications have made it such that devices using flexible electronics frequently have the strength and dimensions required for simple transportation.

There have been some exciting developments in wearable electronics with new materials, methods, and sensing mechanisms [8, 9]. Wireless heart rate monitors, fitness trackers, and wearable body sensors are all experiencing a surge in popularity. Patients use these devices for diverse reasons, including monitoring their daily walks and sleep quality, as well as monitoring their heart rates. Patients are asking doctors to analyze their wearable technology's heart rate data. Questions concerning the data collected by these devices are frequently rejected by physicians because they are not medical grade. We feel that this is a chance to interact and cooperate with the patient that has been too often squandered. Figure 1.1 illustrates the newly established flexible electronics for personal healthcare.

Various substrates have been employed to develop optical sensing devices for refractive index sensing in recent years [10–14]. Optical sensing devices are unique in that they are resilient to EM radiation, can probe nanoscale volumes, allow for non-intrusive evaluation of biological substances at quite deep penetration depths, and usually employ low-cost, water and corrosion-resilient sensing constituents. Ion, protein, and virus concentration, along with pulse, BP, blood oxygenation, abdominal and thoracic breathing rates, targeted localized bending, and movement, have all been detected and quantified using these resources. Flexible wearable sensing devices could be used to measure signs of spasticity, balance, and other motor-related conditions. More digital biomarker systems that can regularly check patient movement and enhance patient care can be developed by designing patient-friendly sensors that are simple to attach to the hips, knees, and legs. In the environment of wearables, optical sensors, like all other monitoring equipment, must deal with the increased concerns of correct signal-to-noise ratio, inadequate dynamic range, signal specificity, and human variability.

In addition, optical sensing equipment has a problem with adjacent light intrusive with signal readings, in addition to poor light diffusion into the skin and other biofluids. Photonic fabrics [15], exceptional colorimetric [16] and fluorometric materials [17], and flexible photonics [18] are among the new optical sensing constituents and integration approaches now being explored to address these issues. One of the primary factors considered while building wearable sensing devices is power usage. The optimal power source for wearable sensing devices used to check the status of neonates is compact, lightweight,

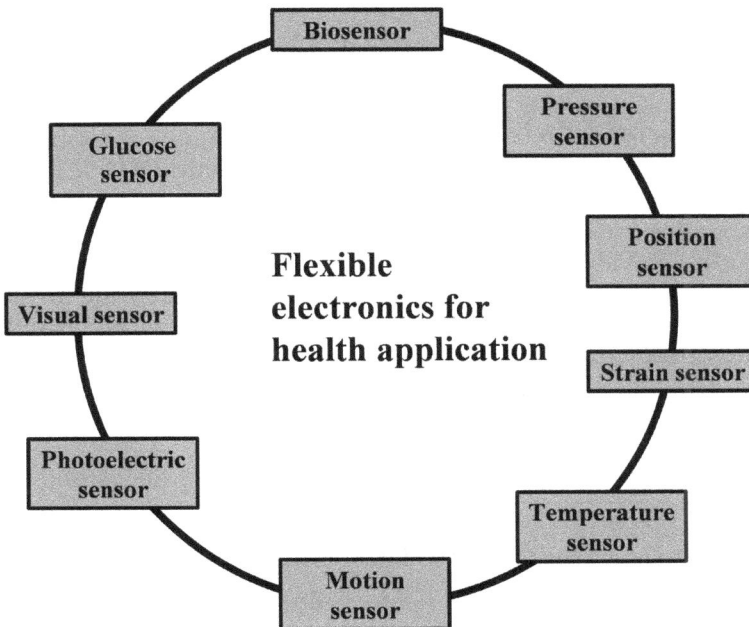

FIGURE 1.1
Flexible electronics for healthcare surveillance.

and long-lasting. Wearable sensing devices can be powered in a variety of ways, including wireless power transmission and rechargeable batteries.

1.2 The Global Wearable Sensing Devices Market

The electronic sector has exploded because of digitization. Consequently, the wearable technology business is widely used for self-health surveillance purposes and is growing rapidly. Public well-being and safety mindfulness has increased owing to increased consumer electronics expenditure, urbanization, and better lifestyles among the expanding population, all of which are supporting the development of wearable equipment. As stated by the Consumer Technology Association [19], smartwatch unit trades escalated from 5 million to 141 million from the year 2014 to 2018. People are becoming more conscious of their well-being, which is driving this development. Additionally, the number of wearable equipment units sold has increased substantially in the past few years. The use of sensors in these wearables has augmented because of this trend. As said by Cisco Systems, the rise in connected wearables from 325 million in 2016 to 593 million in 2018 has a lot of promise, mainly in healthcare. Wearable device demand is expected to grow across different industries because of several organizations spending heavily on Internet of Things (IoT) equipment and IoT sensor utilization declining [19].

Jersey City, December 14, 2021 (Globe newswire)-Verified Market Research just published a report on the "Wearable Sensing Devices Market" by type, device, and geography. The worldwide wearable sensing devices market was valued at US$660.89 million in 2020, according to Verified Market Research, and is predicted to reach US$5208.05 million by 2028, with a CAGR of 29.3% from 2021 to 2028 [20] (Figure 1.2). By 2020, optical and optoelectronic technologies are estimated to make up 13% of the wearable market as they are taking a part in other market categories such as chemical, elastic, and pressure measuring devices.

For instance, a specialized fitness scrutiny device, which is a wearable sensor, is currently owned by over a tenth of Americans, up from a third in 2012 [21]. By acquiring information from blood oxygen level, pulse, movement, speed, step count, and even eating and sleeping schedules utilizing mobile phones and web connections, fitness trackers and smartwatches may provide customized health summaries [22]. For the increasing number of older people who live alone, such tools are particularly alluring for at-home health monitoring. By permitting elderly users, their loved ones, carers, and healthcare experts to review virtual patient data, hospital resource use and emergency response times are reduced.

1.3 Wearable Sensing Devices and Their Implementations

Optical sensors are currently accessible on the market. Detecting blood oxygenation and tracking pulse are the most usual optical sensing technologies employed in wearable electronics. When diagnosing a clinical or biological occurrence, the absorption and scattering features of light regarding the place of the body were utilized to characterize it. The most

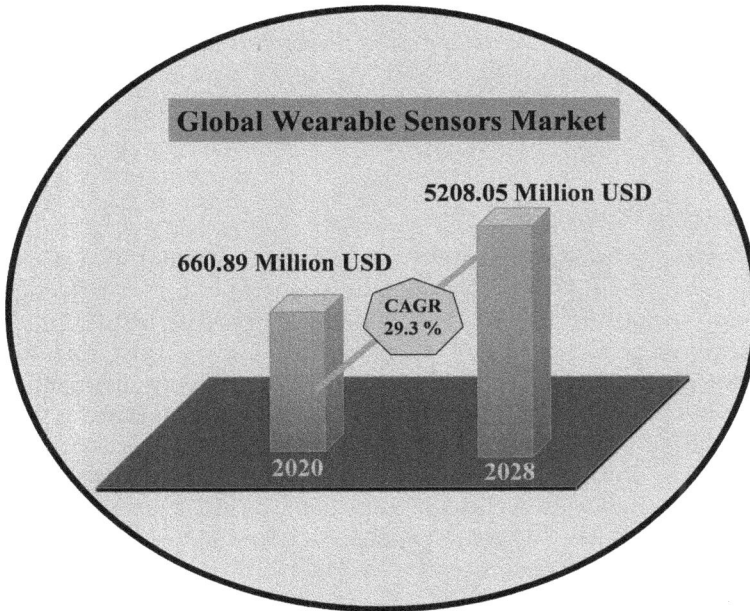

FIGURE 1.2
Global wearable sensing devices market. Encouraged by reference [20].

prominent example is the oximetry technique, which uses differences in the optical features of hemoglobin in its oxygenated and deoxygenated statuses. Using an examination of the pulsatile constituent of the bloodstream, it is feasible to estimate essential physiological parameters, for example, pulse infidelity or unpredictability utilizing photoplethysmography and oxygen saturation in arterial blood via pulse oximetry.

A light source of a specific wavelength is shone on the area of the skin where the examination is to be conducted to undertake non-invasive biomedical testing. The sensor finds light that has been absorbed, refracted, or reflected, describes the data and analyzes it. Since the wavelength controls how far the light may go through the skin, it is the most important component for transmitting an optical signal.

A lot of health-related data is included in basic physiological signals. Users are becoming more accustomed to the idea of having every step, pulse, and breath recorded and scrutinized to offer an uninterrupted response to their well-being and daily events. Heart tracking has been the primary application of wearable optical sensors so far. Several wearable gadgets have focused on photoplethysmography signals. Photoplethysmography signals are volumetric assessments of an organ, most often blood vessels under the skin. Variations in optical absorption can be utilized to examine the oscillation of arterial volume caused by blood pumping by illuminating a perfuse patch of the skin and recording the reflected or transmitted signal. The frequency of fluctuating optical absorption determines the pulse rate, and the amplitude of the AC constituent corresponds to BP in both the systolic and diastolic stages of the cardiac cycle. Using wavelengths of light with different relative absorption by oxygen-loaded hemoglobin may also be utilized to measure blood oxygenation. Photoplethysmography sensors generally offer a stronger signal with less motion artifact when employed in transmission mode rather than reflection mode. Transmission mode photoplethysmography sensors, on the other hand, have a limited signal site because they must be employed in places with strong blood profusion, where optical signs may pass to the other side of the tissue and be verified by receiving photo-diodes (PDs). In this section, the four most widely utilized implementations of flexible wearable sensing devices, i.e., temperature, pressure, glucose, and pulse rate surveillance are discussed.

1.3.1 Temperature Measuring Devices

Temperature sensing is among the most significant functional aspects for measuring human body temperature (HBT), with an emphasis on individuals with long-term chronic situations, usually healthy, drowsy, and wounded surgical patients, including the well-being of medicare employees. Wearable sensing devices are intriguing not only in the well-being sector but also in measuring and analyzing the body temperature of healthy people engaged in strenuous outdoor pursuits. Wearable temperature measuring devices (TMDs) are excellent for employees working in severe environments, in addition to ath-letes' tireless activities for fitness. Dehydration, tiredness, and other serious health impli-cations result from rising levels of climatic conditions, particularly warmth and humidity. As a result, it is critical to develop wearable TMDs not only for the surveillance of human well-being but also for assessing the nearby environment.

Depending on whatever area of the body was used for the measurement, the body temperature is always fluctuating. There is no universally accepted definition of normal temperature because of this. A healthy person's body temperature fluctuates through-out the day and varies according to their activity. Rectal temperature measurements often show that the body temperature is 0.5°C higher in the evening than it is at other moments of the day for physiological reasons. Any kind of physical effort also raises body temperature.

Traditional sensors are constructed in a range of geometric forms dependent on the uses, medium processability, and manufacturability in the desired shape. Through variations and physical interaction with hot surfaces, the temperature sensing system identifies these various structures. Contact and non-contact sensors are the two basic categories of TMDs. Contact sensors are utilized to keep track of a wide range of surfaces, including solids, liquids, and gaseous phases. Non-contact sensors, on the other hand, can detect heat irra-diation emitted by hot surfaces from distance. Some of the most popular TMDs are ther-mocouples, thermostats, thermistors, negative temperature coefficient thermistors (NTCs), resistive temperature detectors (RTDs), and silicon (Si)-based sensors.

Printed wearable sensing devices for monitoring critical bio-signals, for example, HBT, respiration rate, BP, glucose, and electrophysiology have lately piqued the interest of the biomedical research community. These sensors are constructed from biocompatible mate-rials that stick to the desired surfaces in an amenable way. These sensor devices are fre-quently reproduced as e-tattoos that are placed directly on the skin or are integrated into deformable textiles. Among the many essential indications, HBT is given special atten-tion, and it is employed as a timely pointer for a variety of illnesses. For long-term in situ monitoring, TMDs should be flexible and elastic, enabling comfortable integration into human skin. Commercially manageable sensing devices are based on rigid planar substrates that cannot be employed for wearable sensing implementations on non-planar surfaces. Standard cleanroom processes cannot be utilized to construct these sensors due to the lower glass transition temperature of polymeric substrates. Consequently, the new printed electronics automation enables the production of electronic gadgets, circuits, and systems on a wide range of substrates in ambient conditions.

A new generation of flexible electronic applications is now possible thanks to printing methods that are very effective and compatible with polymeric materials (both inks and substrates). A novel family of materials known as conductive flexible polymers has been designed for a variety of uses, including photovoltaic solar cells, transistors, molecular devices, sensors, and actuators. There are a variety of printing methods available. Flexible polymers play a crucial role in the creation of novel conductive circuits because they offer several benefits over conventional rigid substrates, such as greater contact areas, the potential to fold or roll, and weight reduction. There are two types of printed electronics technologies: contact and non-contact. Methods where the printing plate comes into direct touch with the substrate, such as screen printing, flexography, gravure printing, and soft lithography, comes under the category of contact-printing techniques. Whereas the methods that do not make direct contact with the substrate, such as inkjet printing, aerosol printing, and laser direct writing, are known as non-contact printing techniques.

The benefit of non-contact printing techniques over contact printing techniques is that the substrate only touches the deposition medium. As a result, there are fewer chances of contamination and substrate damage, and the alignment of the patterns is more precise. This last problem is a crucial capability for designing multilayered devices. Non-contact printing methods just require a digital picture rather than a physical mask of the image that will be printed, simplifying the switching process, and incurring no extra costs. However, non-contact technologies also have certain challenges when creating multilayered devices, which is necessary. They operate with a variety of substrates, including wood, glass, metals, and – most intriguingly – rubbers and polymers. These materials need relatively low temperatures since high processing temperatures and thermal pressures run the danger of causing damage and deformation.

Ali et al. suggest a two-step deposition method for constructing an HBT measuring device at room temperature using an inkjet material printer [23]. A silver-based interdigital electrode and a carbon black sheet receptive to HBT make up the sensor. The device has two terminals that keep track of resistance variations as a function of temperature. At temperatures ranging from 28 to 50°C, the spacing of the interdigital electrodes was set for high receptivity and linear resistance behavior was found. The reaction to HBT data is highly linear. Figure 1.3(a–c) shows how Yan et al. employed a lithographic filtration technique to generate stretchable graphene thermistors with fundamentally high elasticity [24].

The electrodes were silver nanowire (Ag NW), while the heat-detecting channels were built of 3D crumpled graphene. The detecting channel and electrodes were completely

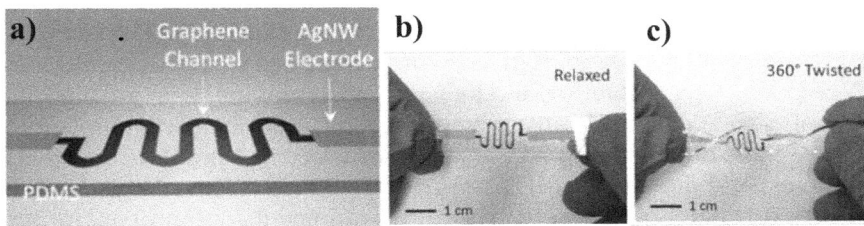

FIGURE 1.3
a) Graphical illustration of the flexible graphene thermistor. Adapted with permission [24]. Copyright (2015), American Chemical Society. b) Fabricated sensor in a relaxed state. Adapted with permission [24]. Copyright (2015), American Chemical Society. c) Fabricated sensor in a 360° twisted state. Adapted with permission [24]. Copyright (2015), American Chemical Society.

encased in an elastomer matrix to attain high elasticity. Temperature sensing features were observed at various strains ranging from 10% to 50%. The sensors can keep working although pushed to their limits. Strain-dependent thermal indices were seen in the devices, suggesting that strain may be employed to effectively change the thermistors' receptivity. The unique variable thermal index surpasses typical ceramic thermistors for varied utilization in wearable electronics.

1.3.2 Pressure Measuring Devices

Several studies have been conducted on the creation of flexible pressure measuring devices (PMDs). To create flexible PMDs, two main parts must be utilized: the first is to utilize intrinsically flexible materials, and the second is to generate elasticity by using specifically constructed structures. The wearables comprise functional polymers, for instance, silicone rubber has piqued curiosity among researchers. Silicone rubber, particularly polydimethylsiloxane (PDMS), is exceptionally supple and, due to its biocompatibility, is ideal for wearable electronic devices. Composite materials were developed to advance the electrical functioning of silicone rubber [25]. These materials have better electrical features than bulk materials, which can considerably advance the electrical functioning of the sensing device. Another important way for providing elasticity to the sensor is to make the sensing layer porous or in the shape of a nanowire. Although the materials are essentially breakable, their nano-or micro-scale structural form can significantly offer flexibility. When a naturally flexible material is formed with a variety of structures, its elasticity may be substantially varied. The use of flexible materials and a sensing film with a particular assembly not only improves mechanical strength but also progresses electrical functioning.

Flexible PMDs can give crucial information regarding specific needs within the human body as well as during contact with the outside world. The devices based on Si have been around for approximately 50 years. Most Si-based devices employ the piezoresistive or capacitive sensing mechanisms, which are manufactured utilizing well-recognized Si micromachining processes. Although the lateral dimensions and functionality of the sensors have been enhanced, the fundamental limitations of Si-based pressure sensing, particularly the minimum height and fragility, have yet to be overcome. A few attempts to develop thin Si-based pressure sensing devices have been made, but they all need considerable changes in manufacturing techniques. Traditional semiconductor and metal-based PMDs are constrained by rigidity, fragility, poor receptivity, narrow sensing spans, limited tensile capacity, and low resolution, making them hard to utilize in implementations demanding flexible touch or wearables. As a result, flexible sensors suitable for arbitrarily curved surfaces have been studied, with important consequences in the fields of human-machine interfaces, wearable electronic devices, electronic skin, and treatment [26]. Flexible sensors, on the other hand, have trials in terms of extraordinary receptivity, elevated resolution, quick response, decent stability, and long resilience, making them unsuitable for these emerging implementations. The transduction methods of flexible PMDs are allocated into three categories: piezoresistive, capacitive, and piezoelectric. Countless efforts have been undertaken to expand the elastic PMD's features based on these mechanisms. The for and against of three distinct types of pressure sensing equipment are compared in Table 1.1.

A variety of joint movement trials were utilized to evaluate the functioning of the piezoelectric PMD. The sensing device was placed on the knuckle in the manner shown in Figure 1.4a, and the output voltage (V_{out}) feedback was relative to the level of bending of the finger, attaining a maximum of 1.8 V. The sensor was mounted to the wrist in Figure 1.4(b–d) to discriminate between diverse wrist actions, such as upward turn,

TABLE 1.1

For and against Three Types of PMDs

Transduction principles	For	Against
Capacitance	Exceptional responsiveness, simple setup, large dynamic span, good temperature consistency, quick reaction, and suitable for minor force tests.	Poor linearity, EM interference vulnerability, and parasitic capacitance vulnerability.
Piezoresistivity	Simple manufacturing, low cost, large deformation, excellent anti-interference capability, and compact size.	Poor temperature endurance, low responsiveness, and poor stability.
Piezoelectricity	Extremely receptive, fast response, low power utilization.	Poor stretchability and low spatial resolution are only valid for dynamic testing.

Adapted from reference [26].

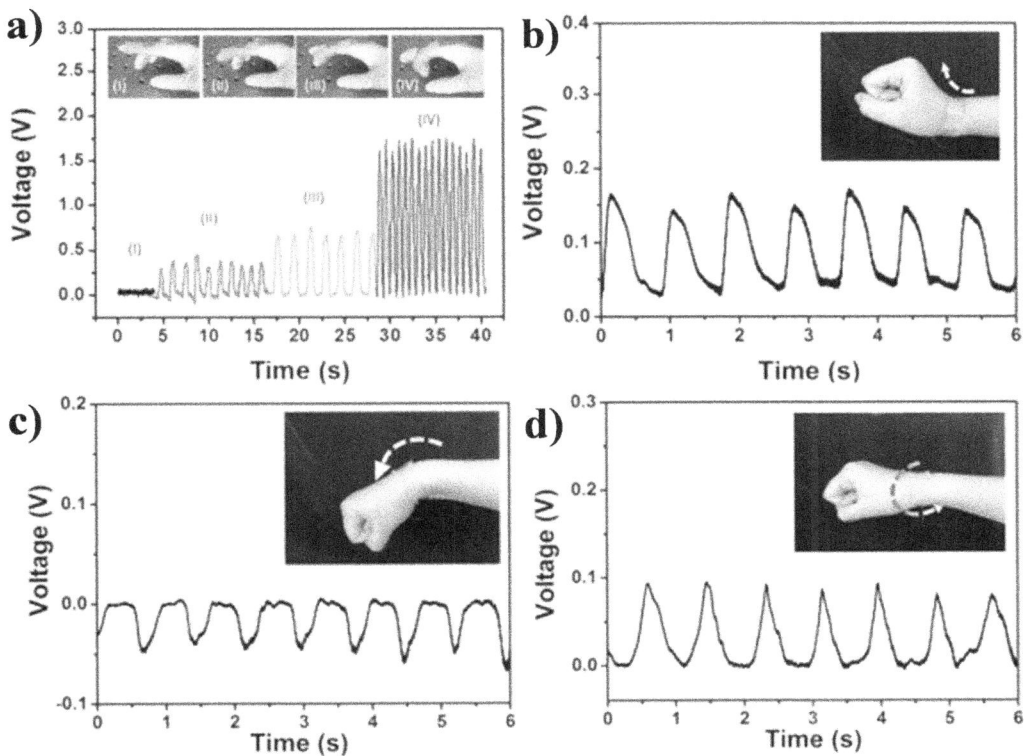

FIGURE 1.4
Piezoelectric PMD employed for perceiving joint motion states: a) finger movement states; b) variation in generated voltage under upward tuning wrist motion, c) variation in produced voltage under downward turning wrist motion; d) variation in produced voltage under torsion wrist motion. Adapted with permission [27]. Copyright The Authors, some rights reserved; exclusive licensee MDPI. Distributed under a Creative Commons Attribution License 4.0 (CC BY).

downward turn, and twisting [27]. As the wrist moved up and down, the sensor bent, resulting in maximum and minimum output voltages of 0.15 V and 0.05 V, respectively. The variance in output voltage might be because of changes in sensing device distortion. When the wrist was twisted, the sensing device was exposed to a combination of upward, shear, and downward forces, and the V_{out} was a positive or negative amalgamation of relevant reactions. Pressure-sensitive organic transistor-based PMDs as thin foil-based screen-printed devices have lately been disclosed [28]. Nevertheless, they do need a lengthy printing process along with appropriate stimulation and reading electronics. Body pressure in the medium pressure range includes BP, pulse, radial artery wave, phonation vibration, and skin modulus. The elevated-pressure span contains the weight of a human or atmospheric pressure at high elevations. By monitoring these distinct kinds of pressure, eye issues, heart illness, and wounded voice cords can be tracked. Wearable sensing devices have therefore been intensively researched for application in medical diagnosis and well-being management. Various sensing methods, including piezoelectric, piezoresistive, and capacitance systems, are utilized to transform physical inputs into electrical signals.

1.3.3 Uninterrupted Glucose Surveillance

Over 415 million people worldwide suffer from diabetes. The most prevalent kind of diabetes, type I, is caused when the pancreas cells are attacked by the body, which prevents it from producing insulin [29]. Type II diabetes is brought on by the body's inability to produce adequate insulin or by the inefficiency of the insulin that is created [30]. Blood sugar (BS) control is a crucial component of treating diabetes, even though diabetics can have healthy, happy, and productive lives. To do this, BS rates must be tightly controlled. Medicare automation has improved rapidly since the first BS test strip was manufactured in 1965 [31]. One of the most significant moments in the history of BS monitoring was the introduction of uninterrupted glucose surveillance (UIGS) as a technique for regulating BS levels in 1999. Since 1999, UIGS has become more sophisticated, with new devices continuously expanding on the accomplishment of those that came before them. Flash uninterrupted glucose surveillance (FUIGS) is one of the most important new kinds of UIGS [32, 33].

Self-monitoring of blood glucose (SMBG) is still a popular method of glucose surveillance [34]. This involves puncturing the patient's finger and taking a little amount of blood to assess the level of BS in their system. The patient then administers insulin or glucose as needed. While this technique is straightforward, it does have some drawbacks. SMBG may be bothersome for patients since testing might be unpleasant. Many individuals characterize it as antisocial, and it is impacted by everyday actions. This can be troublesome, especially for those who are younger. Notably, regular testing is required for this strategy to be effective, which decreases its appeal to people looking for discreet and simple solutions to keep track of their BS levels [35, 36]. UIGS, on the other hand, enables diabetics to constantly keep track of their BS levels by employing a wearable gadget that collects information on the body's BS levels. UIGS devices are accessible around the clock and have warnings that sound when a victim's BS levels are unusually high or low [37]. Finger puncturing is no longer necessary or has been diminished, which is good for the patient's well-being. Individuals can also use UIGS to evaluate data and discover tendencies and patterns, which they can subsequently share with their healthcare practitioners.

While the investigation into new technologies for scaled-down electrochemical biosensors is still underway, other sensing approaches have lately become much more appealing

[38]. Many of the constraints linked with electrochemical sensors, such as sensor receptivity and constancy being reliant on the enzyme used and intrusion with active compounds, can be overwhelmed by employing sensing technologies. Near-infrared recognition and Raman spectroscopy for non-intrusive recognition, as well as fluorescence-based sensors for implanted systems, are among the optical identification technologies reported in the literature [39]. The Eversense sensor [40], a fully entrenched UIGS arrangement that offers real-time BS assessments through an external coupled transmitter for an anticipated lifespan of 180 days, was recently developed by Senseonics (Senseonics, Inc., Germantown, MD, USA) by employing this last group of optical detection technologies [41]. The Eversense UIGS is presently only certified for use in European countries (CE mark established in 2016) [42]. It has a lifespan of three months and the precision of the entire integrated automation remains a key problem; the method's main advantages are its endurance and ease of use.

A BS sensor that is both economical and exceedingly wearable has been developed with a rapid data collection time that enables an unobstructed, durable UIGS system [43]. The new biosensor indicates pulse information to uninterrupted constituents of arterial blood volume pulsation during the fluctuation of BS concentration at the wrist tissue. Combined visible-near-infrared spectroscopy is utilized to evaluate the reflected optical power. Near-infrared light, as seen in Figure 1.5a, may penetrate epidermises and touch the arteries in the subcutaneous tissue, but visible light, for example, green and red light, can solely hit the capillaries and arterioles in the dermis tissue. Similarly, several spectra data may be generated for a multivariate model by gathering reflected light at various visible wavelengths at different levels of BS content. Figure 1.6a illustrates how near-infrared light may approach the arteries in the subcutaneous tissue via many layers of skin, whereas visible light, such as green and red light, can only enter the capillaries and arterioles in the dermis [43]. Thus, a diversity of spectral information that can be utilized in a multivariate model is acquired by gathering reflected light at different levels of BS concentration. Figure 1.5b

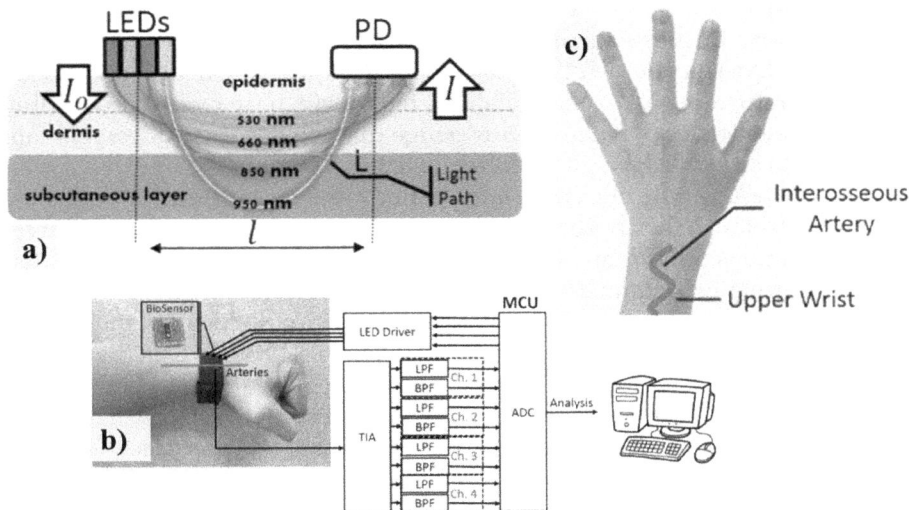

FIGURE 1.5
a) Diffused reflectance photoplethysmography signal in multiple wavelength assessments. Model system: b) block drawing of analog signal processing, c) sensor worn on the wrist. Adapted with permission [43]. Copyright (2019), Elsevier.

shows how the proposed prototype gathers information via analog signal processing. The detector is fastened to the wristband's back and makes straight contact with the skin tissue. It is positioned between the interosseous arteries on the wrist, as shown in Figure 1.5c, and has an overall device size of 15×15 mm^2.

1.3.4 Pulse Rate Monitors

Wearable optical pulse rate monitors (PRMs) based on photoplethysmography have gained in popularity, with a spate of tech companies creating and selling them. Several companies have created optical photoplethysmography sensors that may be placed on the wrist, across the chest, or even in the ear utilizing headphone-based optical sensors that function in reflection mode. Owing to their tiny size and ability to create superior signals via transmission-based assessments owing to their location on the finger, ring photoplethysmography sensors are also of interest. Commercially accessible sensors may currently measure photoplethysmography for pulse observation together with blood oxygen levels, tracking step counts, respiration rate, temperature, and even estimating sleep quality.

Athletes represent a burgeoning market for the application of wearable sensing devices. Scientific advancements have enabled individual endurance athletes, sports teams, and physicians to track player motions, workloads, and physiological indicators to enhance performance while minimizing damage. Tracking these factors may enable the detection of biomechanical exhaustion and timely recognition of harm during training and competition. Surveillance may also aid in the creation of better training routines to increase fitness levels. GPS has been used to track competitors' speed and location in soccer, navigation, cross-country skiing, and field hockey. Football and rugby were the most well-documented uses of GPS in professional sports.

In settings where hyperthyroidism is a problem, such as elevated temperature climates and interior buildings lacking climate control, monitoring core body temperature is critical. An excessive core temperature fluctuation during an athlete's early acclimatization to physical activity is another cause for worry. Precise core body temperature control has been a key difficulty in sports medicine. During sporting activity, the core temperature can be measured. External temperatures have been demonstrated to be a poor predictor of core body temperature. Newer commercialized biosensors have overcome this limitation by employing a telemetric core temperature gauge that relies on a swallowable capsule to communicate data systems through radiofrequency.

Optical PRMs work by flashing a light into the body and measuring how well the light is dispersed by blood flow. The approach is most effective in areas of the body where physiological considerations other than blood flow limit the flow of light that can be dispersed or absorbed. Information obtained from parts of the body including tissues such as bone and muscle may be erroneous. When the body moves, particular parts of the body, namely the wrists and ankles, move more, which has an adverse effect. Figure 1.6 depicts the many areas on the body where sensors may be positioned to acquire the most accurate pulse.

People with a variety of skin tones can trust smartwatches and wrist-worn fitness trackers that measure pulse. However, a new study indicates that their accuracy varies depending on the sort of daily activity. Researchers enlisted 53 participants of various skin tones to text six different devices that use a unique sensor to measure changes in blood flow via the skin [44]. The participants also wore ECG patches while doing various tasks, such as sitting motionless, breathing deeply, walking, and typing. There were no significant changes in heart rate accuracy between those with light, medium, or dark skin tones, according to the study. However, the accuracy varied depending on the activity. The reported heart rate

Forehead

It's an excellent place to keep an eye on your pulse rate. With little relative movement or noise, a clear signal may be created. Long-term wear or continuous monitoring is not advised.

Ear

One of the most important areas on the body for determining biometrics, allowing for measurements which are more hard to achieve in other areas. Prolonged wear or constant monitoring is not advised.

Arm and chest

There is more comparable motion than the head due to the huge muscles in that area, however there is a lot of blood flow. Constant monitoring and patching are likely in this area.

Wrist

This is a common device post, although it is also one of the most difficult to detect attributed to the prevalence of tendons and ligaments that scatter light. Extremely susceptible to movement noise.

Ankle

It is hard to measure this location since it is tightly packed with tendons and ligaments and has a little blood flow.

Calf and Quad

The calf has a lot of blood flow and is a wonderful place to be while you're not walking or jogging.

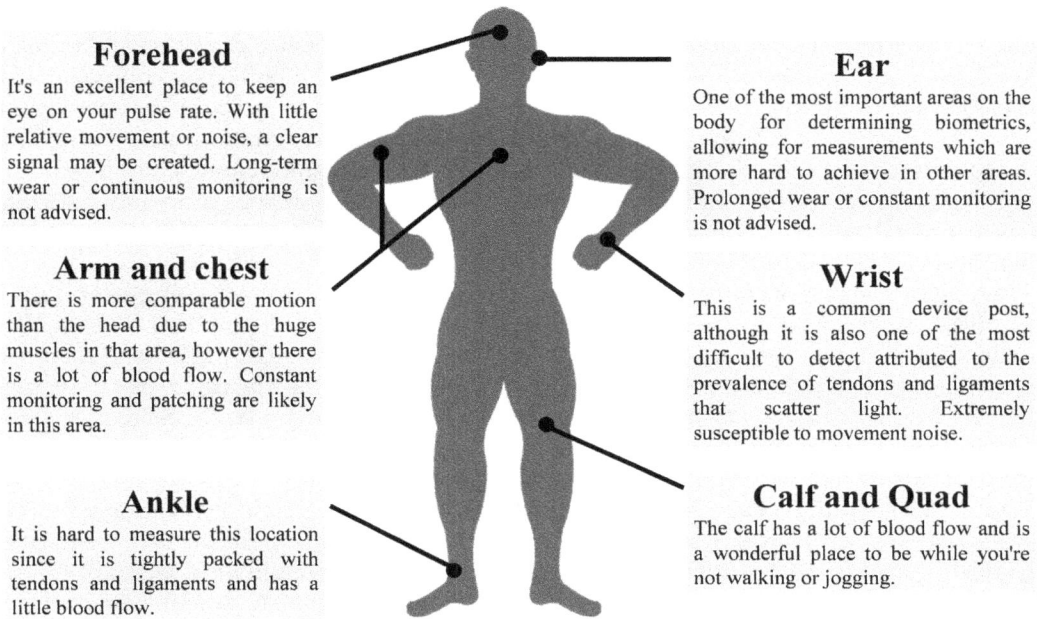

FIGURE 1.6
Diverse probable places in the body to place PRM to acquire the best pulse.

tended to be greater than the genuine heart rate while people were walking. Because of wrist motions that altered the sensors' skin contact during typing, the reported heart rate was lower than the genuine heart rate.

Another research compared the accuracy of two commercially available PRMs based on photoplethysmography during exercise. An exercise program encompassing sitting, lying, walking, running, cycling, and everyday chores including hand motions was performed by 21 healthy participants (15 men and six women) [45]. HR estimate was compared to data from an ECG signal as a control. Mio Alpha 77.83% and Scosche Rhythm 76.29% were the heart rate estimate reliability ratings for a 5% accuracy versus reference. The projected findings showed that device performance is affected by a variety of aspects, including activity, sensor type, and device placement.

1.4 Skin-Like Wearable Sensing Devices

The skin is the biggest organ in the body, and it acts as the body's primary physical, thermal, and hygroscopic barrier between the outside world and its interior organs. Significantly, the skin has the greatest number of sensory receptors for detecting pressure, humidity, temperature, and pain, among other environmental stimuli. Our comprehension of human skin's sensory properties has been produced in remarkable skin-inspired elastic and stretchy gadgets. Wearable electronics and other related industries have been transformed because of these breakthroughs. Existing skin-like wearable gadgets have previously been employed for individual well-being surveillance such as glucose, uric, acid, lactose, BP, and stress level. Next-generation soft robotics, on the other hand, necessitate a diversity of stretchable devices to be worn on soft-bodied robots for detection and perception during communication with their environment, where the perceived competence of skin-like electronics will be extremely valuable in improving soft robot prototypes. There has been a major attempt to produce skin-like stretchy and wearable sensing devices for incorporation with soft robotic systems due to the intrinsic material match and functional complementarity between skin-like devices and soft robots. Soft robots now include biological, optical, strain, and tactile sensing capabilities, allowing them to communicate with their users and surroundings more perceptively.

Two different sorts of challenges are presented by wearable technology: engineering difficulties that require a chemist and chemistry problems that need engineers. Due to how easily skin stretches, wrinkles, and bends as individuals move, it is difficult to establish contact between an electrode and a person. Gels can secure the electrode, but only temporarily because they are watery and eventually dry up. One potential remedy is ionic liquids. Ionic liquids, which are composed of salts that are liquid at normal temperatures and have strong electrical conductivity, are slow to evaporate.

Figure 1.7(a–c) illustrates the usage of PDMS and transparent electrodes for a flexible PMD. The device has an 82% transmittance and an 18-millimeter bending radius. Chen et al. [46] discuss the effect of strengthening temperature on the mechanical features of PDMS. The PDMS film exhibits decent compression characteristics but poor dynamic responsiveness in outcome display at a lower strengthening temperature of 80°C. At higher temperatures, the compression property of PDMS films was drastically reduced. For PDMS film, a strengthening temperature of roughly 110°C was found to be the best negotiation between compression feature and dynamic receptiveness. When cured at 110°C, the PMD exhibits a great receptivity of 0.025 kPa^{-1} and a high response characteristic. In the future, the idea paves the way for intelligent transparent sensing implementations.

1.4.1 Textiles

The initial and primary purpose of textiles was to shield our bodies from harmful natural conditions like ultraviolet radiation, cold winds, rain, and so on. Later, clothes started to have functional and esthetic qualities. Smart and interactive textiles are a recent development in the textile industry. These intelligent fabrics safeguard us from harmful environmental circumstances, keeping us under observation and occasionally even aid in the treatment of illnesses or accidents.

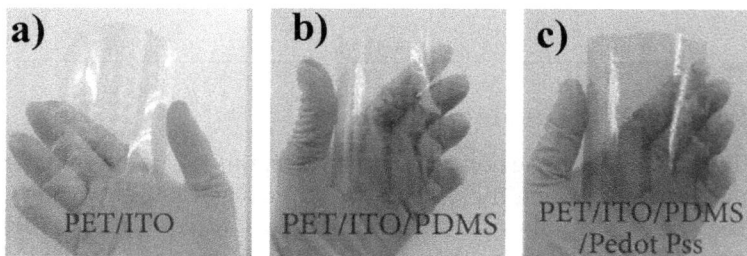

FIGURE 1.7
Flexible substrates for biosensing devices: a–c) PMD based on PDMS and PEDOT: PSS. Adapted with permission [46]. Copyright The Authors, some rights reserved; exclusive licensee Hindawi. Distributed under a Creative Commons Attribution License 4.0 (CC BY).

Smart textiles are gaining popularity as a technology for wearable systems as they allow enhanced assurance between the sensor and the user, i.e., the device is lightweight, compact, and does not block the user's motions. The incorporation of microscopic and embedded sensors into woven fabrics, in addition to their ease of construction and disassembly, increases the system's application. Smart textile solutions are continually pointing to even more downsizing, high energy efficiency, and remote access, the majority of which are under IoT device requirements. Examples of such improvements include resistive sensors incorporated in fabric patches and twin-core microfibers for capacitive readings, as well as various embedding methods and electronics.

Textronics play an important role in IoT, which enables new functionality by securely linking smart clothes for a spectrum of uses. A conductive substance (yarn, fabric, etc.) is required for the creation of flexible and elastic textile-based electronics, and manufacturing procedures play a vital part in influencing the attributes of electronic textiles. Due to their cost-effectiveness, versatility, and simplicity of implantation, textile-based sensing devices, electrodes, and other components appear to be the preferred choice for uninterrupted wearable surveillance. Integrating smart competencies into textiles gives significant benefits in healthcare, sports, automobiles, and defense. These signs of progress have a significant impact on the fourth Industrial Revolution.

1.4.2 Tattoos

An essential category of wearable sensing devices that seeks to give point-of-care diagnostics with a comfortable and secure user experience has been developed because of active research in the field of epidermal electronics. Traditional medical equipment is cumbersome, stiff, and impractical since it does not provide continuous monitoring of important health markers while carrying on with daily activities. The human body's supple and curved structure calls for skin-like sensors that may be transferred easily by tattooing. The discovery of naturally soft electronic materials, bendable designs, and innovative material processing techniques have made it feasible to create flexible and stretchy electronics. Bio-integrated electronics have created an interesting utility in physiological signal and sweat management, peripheral signal detection, and data display, among recognized solutions, for instance, flexible display and e-skin. Flexible and wearable electronics seem to be a feasible option. Among various developments, epidermal electronics, sometimes referred to as electronic tattoos (e-tattoos), appear to be a potential rival to wearable electronics. E-tattoos are extremely thin and ultra-soft electronics, sensors, and actuators that have stiffness and mass density like the human epidermis. E-tattoos can attain maximum conformability to varied skin textures due to their mechanical qualities, producing a non-invasive yet intimate relationship with the skin. Compared with traditional bulky and rigid sensors, the softness and conformability provide numerous distinct benefits. To begin with, conformable contact expands the interaction area between the devices and the skin, lowering the contact impedance and facilitating signal transmission across the sensor-skin interface.

It shows how an easily deployable, untethered, battery-free passive "electronic tattoo" may interact with a person's skin to collect and transmit physiological data. The ultrathin film conforms well to human skin and offers a superior signal-to-noise ratio compared with the industry-reference Ag/AgCl electrodes [47]. A stretchy printed Ag-In-Ga coil and printed biopotential collection electrodes are employed in a wireless power transfer (WPT) device to provide the necessary energy. Electronics for information and data collection are interfaced with the tag. This system is known as "data-by-request". The electrophysiological data of the patient is received and recorded on the caregiver device by bringing the scanning device close to the tattoo that has been placed. If the WPT gadget is placed under the skin, it can deliver 100 mW of measured power and more than 300 mW if it is transmitted across the skin [47].

1.4.3 Bands

Even the most technologically advanced fitness trackers on the market can only monitor heart rate and count steps in most cases. However, scientists have created a tool that is considerably more powerful. It can track compounds in perspiration that can be used to detect drug usage and medical issues and assist trainers and instructors to enhance the productivity of professional athletes. It can be built into a headband or wristband. Other methodologies for on-skin deployments are obligatory for advanced sensing devices that incorporate stiff parts, for example, integrated circuit chips. Healthcare wristbands are commonly used to provide direct skin interaction without producing pain. To maintain track of the primary signal of the radial artery, a polymer transistor-based detector may be conveniently secured on the body using a typical wristband. A seamlessly integrated bracelet with a chemical sensor array and FPCB was successfully formed for in situ evaluations of sweat molecules and electrolytes [48]. Likewise, blood sugar levels in the interstitial fluid were accurately determined using a graphene-based wristband framework [49].

A wristband electrochemical sensor for measuring phenylalanine (PHE) levels addressed the need to manage PHE levels in PHE hydroxylase deficient individuals [50]. To prevent interferences in biofluids, the suggested electrochemical sensor is based on a screen-printed electrode (SPE) modified with a Nafion membrane. Notably, the electrochemical instrument is built onto a wristband to improve operator engagement and interest patients in PHE self-monitoring. A paper-based sampling technique, on the other hand, is intended to alkalinize the real sample even without the requirement for sample preparation, simplifying the analytical procedure. Ultimately, the wearable gadget is put through its paces to detect PHE in saliva and blood serum. The wristband-based sensor is projected to influence phenylketonuria (PKU) self-supervising, easing PKU individuals' day-to-day lives and promoting optimal medication and illness maintenance.

1.5 Perspectives and Concluding Remarks

Remote patient monitoring is a perfect example of how technology is never static. Wearables are poised to play a crucial role in preventative medicine as the usage of linked devices spreads, continually monitoring for illness early warning indicators in at-risk individuals and controlling disease development in those who have already been diagnosed. The average lifespan has recently risen in most countries because of notable advancements in medicine, well-being services, and individual and environmental cleanliness. However, a decrease in mortality rate shared with plummeting fertility rates is likely to cause a huge senior population shortly, putting significant strain on the socioeconomic systems of countries. Consequently, creating cost-efficient, easy-to-use sensor solutions for healthcare and well-being is vital. Remote medical surveillance, which uses non-intrusive and wearable sensing devices, actuators, and modern communication and data automation, is a cost-effective method of allowing the elderly to stay at home rather than in expensive Medicare centers. By using these gadgets, Medicare personnel will be able to follow their patients' key physiological signals in real-time, diagnose health issues, and provide responses distantly.

Wearables can study unusual situations of the physicochemical constituents of the body in real time, owing to non-intrusive monitoring and exceptional precision. The bulk of commonly produced wearables are relatively rigid units linked to wristbands, with design options restricted by power supply battery weight and volume. A few essential devices have been described in this chapter, including temperature, pressure, glucose, and pulse rate monitoring. Among the most essential physiological components for measuring HBT is temperature detection, with an emphasis on patients with long-term chronic illnesses, typically healthy, unconscious, and damaged surgical patients, and the well-being of Medicare workers. Flexible PMDs can offer useful information regarding the exact demands of the human body when in contact with the outside world. UIGS enables patients to constantly keep track of their BS levels by employing a wearable gadget that collects data on the body's BS levels. UIGS devices have alarms that sound when a patient's BS levels are excessively high or low. They are accessible 24 hours a day. Wearable optical pulse rate monitors based on photoplethysmography have become increasingly popular, with a host of tech companies developing and marketing them.

Wearables with skin-like features are a comparatively new sort of system that has just progressed into research labs and pre-commercial models. Skin-like sensors have undertaken numerous tasks formerly unreachable to standard sensors because of their deformability, lightness, agility, and elasticity. Recent years have witnessed a significant surge in both research and commercialization of wearable sensing devices. Wearable sensor effectiveness, unfortunately, has been a combination of advancements and failures. The majority of industrial development has been in the clever adaptation of current mechanical, electrical, and optical body measurement techniques. This adaptation has required improvements in sensor technology miniaturization, conformal and flexible design, and software development that enhances the value of the target data. Nevertheless, commercial acceptance of chemical sensing modalities has faced significant difficulties, particularly for non-invasive chemical sensors. Significant basic obstacles have also been encountered when attempting to increase the accuracy of detection of existing mechanical, electrical, and optical sensing modalities. Understanding that the skin is more of an information barrier than a source may help to understand many of these difficulties. The roadmap for developing the following generation of innovations and technological advances becomes more evident with a fuller understanding of the underlying difficulties encountered by wearable sensing devices and the state-of-the-art in wearable sensor technology.

References

1. M. Butt, N. Kazanskiy and S. Khonina, "Revolution in flexible wearable electronics for temperature and pressure monitoring - A review," *Electronics*, vol. 11, no. 5, p. 716, 2022.
2. N. Kazanskiy, M. Butt and S. Khonina, "Recent advances in wearable optical sensor automation powered by battery versus skin-like battery-free devices for personal healthcare-A review," *Nanomaterials*, vol. 12, no. 3, p. 334, 2022.
3. Y. Liu X. Huang J. Zhou et al., "Bandage based energy generators activated by sweat in wireless skin electronics for continuous physiological monitoring," *Nano Energy*, vol. 92, p. 106755, 2022.

4. H. Zafar, A. Channa, V. Jeoti and G. Stojanovic, "Comprehensive review on wearable sweat-glucose sensors for continuous glucose monitoring," *Sensors*, vol. 22, p. 638, 2022.

5. M. Arif and A. Kattan, "Physical activities monitoring using wearable acceleration sensors attached to the body," *PLoS ONE*, vol. 10, no. 7, p. e0130851, 2015.

6. K. Bayoumy M. Gaber A. Elshafeey et al., "Smart wearable devices in cardiovascular care: Where we are and how to move forward," *Nature Reviews Cardiology*, vol. 18, pp. 581–599, 2021.

7. J. Heikenfeld A. Jajack, J. Rogers, P. Gutruf, L. Tian, T. Pan, R. Li, M. Khine, J. Kim, J. Wang and J. Kim, "Wearable sensors: modalities, challenges, and prospects," *Lab Chip*, vol. 18, pp. 217–248, 2018.

8. J. Wang, T. Dai, Y. Zhou, A. Mohamed, G. Yuan and H. Jia, "Adhesive and high-sensitivity modified Ti3C2Tx (MXene)-based organohydrogels with wide work temperature range for wearable sensors," *Journal of Colloid and Interface Science*, vol. 613, pp. 94–102, 2022.

9. R. Brito-Pereira, C. Ribeiro, N. Pereira, S. Lanceros-Mendez and P. Martins, "Printed multi-functional magnetically activated energy harvester with sensing capabilities," *Nano Energy*, vol. 94, p. 106885, 2022.

10. M. A. Butt, "Numerical investigation of a small footprint plasmonic Bragg grating structure with a high extinction ratio," *Photonics Letters of Poland*, vol. 12, no. 3, pp. 82–84, 2020.

11. M. A. Butt, S. N. Khonina and N. L. Kazanskiy, "Sensitivity enhancement of silicon strip waveguide ring resonator by incorporating a thin metal film," *IEEE Sensors Journal*, vol. 3, pp. 1355–1362, 2020.

12. M. A. Butt, S. N. Khonina and N. L. Kazanskiy, "Highly sensitive refractive index sensor based on hybrid plasmonic waveguide microring resonator," *Waves in Random and Complex Media*, vol. 30, no. 2, pp. 292–299, 2020.

13. S. N. Khonina, N. L. Kazanskiy and M. A. Butt, "Evanescent field ratio enhancement of a modified ridge waveguide structure for methane gas sensing application," *IEEE Sensors Journal*, vol. 20, no. 15, pp. 8469–8476, 2020.

14. M. A. Butt, N. L. Kazanskiy and S. N. Khonina, "Highly sensitive refractive index sensor based on plasmonic Bow Tie configuration," *Photonic Sensors*, vol. 10, no. 3, pp. 223–232, 2020.

15. A. Leal-Junior, L. Avellar, A. Frizera and C. Marques, "Smart textiles for multimodal wearable sensing using highly stretchable multiplexed optical fiber system," *Scientific Reports*, vol. 10, p. 13867, 2020.

16. T.-H. Nguyen, L. Mugherli, C. Rivron and T.-H. Tran-Thi, "Innovative colorimetric sensors for the selective detection of monochloramine in air and in water," *Sensors and Actuators B: Chemical*, vol. 208, pp. 622–627, 2015.

17. M. Beutler, K. Wiltshire, B. Meyer, C. Moldaenke, C. Luring, M. Meyerhofer, U.-P. Hansen and H. Dau, "A fluorometric method for the differentiation of algal populations in vivo and in situ," *Photosynthesis Research*, vol. 72, pp. 39–53, 2002.

18. G. Righini, J. Krzak, A. Lukowiak, G. Macrelli, S. Varas and M. Ferrari, "From flexible electronics to flexible photonics: A brief overview," *Optical Materials*, vol. 115, p. 111011, 2021.

19. https://www.globenewswire.com/en/news-release/2021/12/14/2351923/0/en/Wearable-Sensors-Market-size-worth-5-208-05-Million-Globally-by-2028-at-29-3-CAGR-Verified-Market-Research.html.

20. https://www.verifiedmarketresearch.com/product/wearable-sensors-market/.

21. S. Tang, "Wearable sensors for sports performance," in *Textiles for Sportswear*, 2015, pp. 169–196.

22. D. Seshadri, R. Li, J. Voos, J. Rowbottom, C. Alfes, C. Zorman and C. Drummond, "Wearable sensors for monitoring the physiological and biochemical profile of the athlete," *NPJ Digital Medicine*, vol. 2, p. 72, 2019.

23. S. Ali, S. Khan and A. Bermak, "Inkjet-printed human body temperature sensor for wearable electronics," *IEEE Access*, vol. 7, p. 163981, 2019.

24. C. Yan, J. Wang and P. Lee, "Stretchable graphene thermistor with tunable thermal index," *ACS Nano*, vol. 9, pp. 2130–2137, 2015.

25. H. Kim and Y. Kim, "High performance flexible piezoelectric pressure sensor based on CNTs-doped 0–3 ceramic-epoxy nanocomposites," *Materials and Design*, vol. 151, pp. 133–140, 2018.

26. X. Wang, J. Yu, Y. Cui and W. Li, "Research progress of flexible wearable pressure sensors," *Sensors and Actuators A: Physical*, vol. 330, p. 112838, 2021.
27. A. Wang, M. Hu, L. Zhou and X. Qiang, "Self-powered wearable pressure sensors with enhanced piezoelectric properties of aligned P(VDF–TrFE)/MWCNT composites for monitoring human physiological and muscle motion signs," *Nanomaterials*, vol. 8, p. 1021, 2018.
28. Y. Yeo, S. Park, Y. Yi, D. Kim and J. Lim, "Highly sensitive flexible pressure sensors based on printed organic transistors with centro-apically self-organized organic semiconductor microstructures," *ACS Applied Materials & Interfaces*, vol. 9, no. 49, pp. 42996–43003, 2017.
29. L. A. DiMeglio, C. Evans-Molina and R. Oram, "Type 1 diabetes," *Lancet*, vol. 391, pp. 2449–2462, 2018.
30. A. Olokoba, O. Obateru and L. Olokoba, "Type 2 diabetes mellitus: A review of current trends," *Oman Medical Journal*, vol. 27, no. 4, pp. 269–273, 2012.
31. S. Clarke and J. Foster, "A history of blood glucose meters and their role in self-monitoring of diabetes mellitus," *British Journal of Biomedical Science*, vol. 69, no. 2, pp. 83–93, 2012.
32. I. Hirsch and E. Wright, "Using flash continuous glucose monitoring in primary practice," *Clinical Diabetes*, vol. 37, no. 2, pp. 150–161, 2019.
33. M. Krakauer, J. Botero, F. Lavalle-Gonzalez, A. Proietti and D. Barbieri, "A review of flash glucose monitoring in type 2 diabetes," *Diabetology & Metabolic Syndrome*, vol. 13, p. 42, 2021.
34. O. Schnell, M. Hanefeld and L. Monnier, "Self-monitoring of blood glucose," *Journal of Diabetes Science and Technology*, vol. 8, no. 3, pp. 609–614, 2014.
35. J. Kirk and J. Stegner, "Self-monitoring of blood glucose: Practical aspects," *Journal of Diabetes Science and Technology*, vol. 4, no. 2, pp. 435–439, 2010.
36. H. Lee, Y. Hong, S. Baik, T. Hyeon and D.-H. Kim, "Enzyme-based glucose sensor: From invasive to wearable device," *Advanced Healthcare Materials*, vol. 7, p. 1701150, 2018.
37. G. Cappon, G. Acciaroli, M. Vettoretti, A. Facchinetti and G. Sparacino, "Wearable continuous glucose monitoring sensors: A revolution in diabetes treatment," *Electronics*, vol. 6, no. 3, p. 65, 2017.
38. F. Ribet, G. Stemme and N. Roxhed, "Ultra-miniaturization of a planar amperometric sensor targeting continuous intradermal glucose monitoring," *Biosensors and Bioelectronics*, vol. 90, pp. 577–583, 2017.
39. C. Chen, X. Zhao, Z. Li, Z. Zhu, S. Qian and A. Flewitt, "Current and emerging technology for continuous glucose monitoring," *Sensors*, vol. 17, p. 182, 2017.
40. A. Colvin and H. Jiang, "Increased in vivo stability and functional lifetime of an implantable glucose sensor through platinum catalysis," *Journal of Biomedical Materials Research Part A*, vol. 101, pp. 1274–1282, 2013.
41. J. Kropff, P. Choudhary, S. Neupane, K. Barnard, S. Bain, C. Kapitza, T. Forst, M. Link, A. Dehennis and J. DeVries, "Accuracy and longevity of an implantable continuous glucose sensor in the PRECISE study: A 180-day, prospective, multicenter, pivotal trial," *Diabetes Care*, vol. 40, pp. 63–68, 2017.
42. A. Dehennis, M. Mortellaro and S. Ioacara, "Multisite study of an implanted continuous sensor over 90 days in patients with diabetes mellitus," *Journal of Diabetes Science and Technology*, vol. 9, pp. 951–956, 2015.
43. V. Rachim and W.-Y. Chung, "Wearable-band type visible-near infrared optical biosensor for non-invasive blood glucose monitoring," *Sensors and Actuators B: Chemical*, vol. 286, pp. 173–180, 2019.
44. B. Bent, B. Goldstein, W. Kibbe and J. Dunn, "Investigating sources of inaccuracy in wearable optical heart rate sensors," *NPJ Digital Medicine*, vol. 3, p. 18, 2020.
45. J. Parak and I. Korhonen, "Evaluation of wearable consumer heart rate monitors based on photopletysmography," in *36th Annual International Conference of the IEEE Engineering in Medicine and Biology Society*, pp. 3670–3673, 2014.
46. L. Chen, X. Chen, Z. Zhang, T. Li, T. Zhao, X. Li and J. Zhang, "PDMS-based capacitive pressure sensor for flexible transparent electronics," *Journal of Sensors*, vol. 2019, p. 1418374, 2019.

47. J. Alberto, C. Leal, C. Fernandes, P. Lopes, H. Paisana, A. Almeida and M. Tavakoli, "Fully untethered battery-free biomonitoring electronic tattoo with wireless energy harvesting," *Scientific Reports*, vol. 10, p. 5539, 2020.
48. W. Gao S. Emaminejad H. Y. Y. Nyein S. Challa K. Chen A. Peck H. M. Fahad H. Ota H. Shiraki D. Kiriya et al., "Fully integrated wearable sensor arrays for multiplexed in situ perspiration analysis," *Nature*, vol. 529, pp. 509–514, 2016.
49. L. Lipani, B. Dupont, F. Doungmene, F. Marken, R. Tyrrell, R. Guy and A. Ilie, "Non-invasive, transdermal, path-selective and specific glucose monitoring via a graphene-based platform," *Nature Nanotechnology*, vol. 13, pp. 504–511, 2018.
50. M. Parrilla, A. Vanhooydonck, R. Watts and K. Wael, "Wearable wristband-based electrochemical sensor for the detection of phenylalanine in biofluids," *Biosensors and Bioelectronics*, vol. 197, p. 113764, 2022.

2

Materials and Technologies for Flexible and Wearable Sensors

Arpana Agrawal and Chaudhery Mustansar Hussain

CONTENTS

2.1 Introduction

Sensors act as the sensing organs of data processing systems in a variety of applications in our daily lives, as well as providing accurate and efficient real-time information about health conditions [1, 2]. There is a rapidly growing demand for miniaturized, cost-effective, portable, environmentally friendly, high sensitivity, fast response sensors that have significant adaptability to human skin/body and are capable of measuring a wide range of physiological or metabolism factors [2–4]. Sensors are mainly categorized as flexible sensors and non-flexible sensors. The former are bendable to a certain extent without affecting their qualities while the latter are rigid and constructed from brittle materials. Non-flexible sensors are commonly utilized in a wide range of applications, although they have several drawbacks such as rigidity, intransigency, and so on, which are especially noticeable when the sensors are used to monitor physiological parameters of living beings that require stress or other harsh conditions, potentially harming the sensor. Consequently, flexible sensors are considered the ideal choice due to their superior mechanical and thermal qualities, low fabrication cost, lightweight design, and so on.

Flexible sensors can efficiently perform real-time monitoring of a variety of physiological characteristics and activities. Several physiological markers can reveal the health condition of humans and hence warrants special considerations for assessing and identifying any disease. The noticeable health markers are roughly classified as the motion of body parts including the motion of hands, limbs, feet, face, and throat, and vital indicators such as breath rate or heart rate, wrist pulse rate, blood pressure, skin temperature, and so on, as well as various metabolism factor (glucose, pH, electrolytes, lactic acid, etc.) [1–4]. These factors can be measured by a variety of wearable sensors, which are mainly classified as physical or chemical-based sensors. Physical sensing devices primarily measure physical-parameter-based signs and electrophysiological impulses, whereas chemical sensing devices are based on electrochemical responses. Commonly utilized sensing with wearable, flexible sensors includes electrochemical sensing to monitor glucose, cholesterol, and so on; piezoresistive and piezoelectric sensors for physiological parameter monitoring as wearable, flexible electronic devices; sensors with ion-electron potentiometric transducers for monitoring skin metabolites; and flexible organic electrochemical transistors that convert biochemical signals into electrical signals and hence examine saliva.

So far, various novel materials including polymers, semiconductors, ferroelectric materials, two-dimensional materials, cotton fibers, and various metals for electroding serving as substrates, active elements, or electrodes have been employed for the realization of such commercially available wearable and flexible sensors [5–11]. Moreover, modern fabrication technologies, such as lithography or printing for obtaining patterns for electroding, have eased the fabrication process of various sophisticated wearable sensors with enhanced sensitivity and reliability [12–14]. Such technologies add a further dimension to the utility of such sensors by reducing their skin contact area. Accordingly, the present chapter provides an outline of the utility of various materials and technologies employed for fabricating several flexible wearable sensors along with an illustration of some recently developed flexible and wearable sensors for monitoring various physiological/biological or metabolism parameters.

2.2 Materials for Flexible and Wearable Sensors

The three essential components of flexible and wearable sensors are the substrate, active element, and electrode. Standard inorganic electronic materials are responsive to a wide range of stimuli but, due to their rigidity and frangibility, are incapable of mechanical compliance. As a result, a collaboration between diverse materials can be a solution for combining excellent measurement performance, flexibility, wearability, bendability, and mechanical durability in a single device. The subsequent sections will concentrate on the most commonly utilized materials and their roles in wearable sensors.

It should be noted that the essential building elements of wearable and flexible sensors are flexible materials. In general, polymers are considered the most likely candidate material for such applications owing to their particular molecular architectures, along with their intrinsic mechanically flexible nature, and can be served as a substrate or channel material for fabricating flexible and wearable sensors. Generally, flexible polymers including polyimide (PI), polyethylene (PET), poly(ethylene naphthalate) (PEN), and polydimethylsiloxane (PDMS) serve as substrate materials while polymers such as polyaniline (PANI), poly(3-hexylthiophene) (P3HT), poly(vinylidene fluoride) (PVDF), poly(vinylidene fluoride-trifluoroethylene) (P(VDF-TrFE)), and poly(3,4-ethylenedioxythiophene) polystyrene sulfonate (PEDOT-PSS) are engaged as channel and electrode materials because of their conducting nature [5–7, 15, 16]. PI film/layers can retain their flexibility and tensile strength even at high temperatures (up to 360 °C) and acids/alkalis can be engaged in regular micro-manufacturing processes, giving greater versatility in designing wearable sensors. PDMS, a commercial silicone elastomer, is also an excellent polymer with low Young's modulus, high intrinsic stretchability, non-volatile and non-toxic behavior, hydrophobicity, and ease of usability and processability via spin coating and casting methods, making it particularly suitable for developing flexible sensors.

Yoon et al. [17] have designed a highly flexible human stress monitoring patch using a piezoelectric material (P(VDF-TrFE)). This flexible patch is stamp size (25 mm×15 mm×72 μm) and comprises three sensors – one monitors the skin temperature and two monitor the skin conductance and pulse wave. The small-size patch is a multilayered structure that facilitates reduced contact with the skin and is comfortable and was prepared using the associated micro-fabrication process. Figure 2.1(a) schematically illustrates the multi-layered structure of the fabricated flexible stress monitoring patch that comprises a skin contact layer and an insulating layer, as well as pulse-wave detection. The skin contact layer has aluminum electrodes that sense skin temperature and skin conductance via direct skin connections. The parylene-C insulation layer is employed for isolating the skin contact layer and the pulse-wave sensor layer electrically, while a silver electrode is placed between the piezoelectric and PI support membranes with windows in the pulse-wave detecting layer. The front and back sides of the fabricated stress-examining patch with a comparison to the US penny are shown in Figure 2.1(b). The workings of the fabricated patch, shown attached to the human wrist of the volunteer, are demonstrated in Figure 2.1(c). The constructed stress patch monitors skin temperature and skin conductance with high sensitivity (0.31/°C; 0.28 V/0.02 S, respectively) and a pulse wave with a 70 msec response time in the human physiological range.

Aleeva et al. [5] monitored glucose using an amperometric biosensor and front-end electronics employing crosslinked PEDOT-glucose oxidase. For distant glucose monitoring, a structure consisting of three electrodes was constructed on a PET substrate. Manjunatha et al. [6] developed PVDF film-based nasal sensors to detect human breathing patterns. A piezoelectric sensor based on PVDF polymer for monitoring respiration while walking in a dynamic environment has also been developed [18].

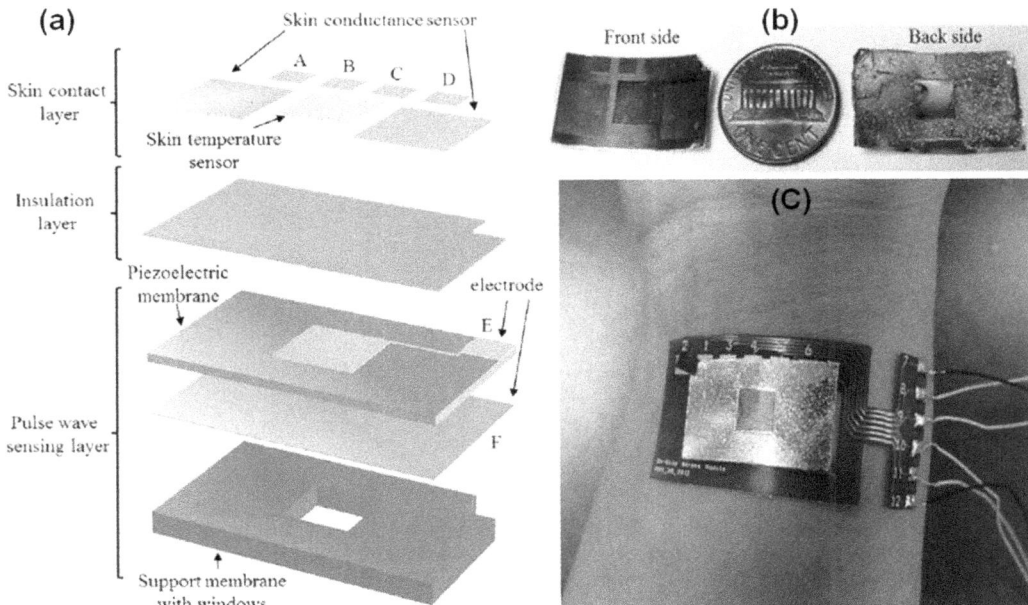

FIGURE 2.1
(a) Schematic representation of a multilayered flexible human stress monitoring patch. (b) Front and back sides of the developed stress monitoring patch with its comparison with a US penny. (c) Stress monitoring patch attached to the human wrist of the volunteer. Adapted with permission [17]. Copyright (2016) Copyright The Authors, some rights reserved; exclusive licensee Nature Springer. Distributed under a Creative Commons Attribution License 4.0 (CC BY) https://creativecommons.org/licenses/by/4.0/

The invention of a completely printed, wearable vital sensor based on a ferroelectric polymer for measuring the human pulse rate was presented by Sekini et al. [7]. Figure 2.2(a) schematically shows the fabrication process where the sensor was made on a PEN-layered substrate and attached to a glass carrier. As the planarization layer, a crosslinkable poly(4-vinyl-phenol) solution comprising PVP and melamine resin in 1-methoxy-2-propyl acetate solvent was spin-coated on the PEN layer (Figure 2.2(a)-(i)). The bottom electrode (500 nm thick), composed of PEDOT:PSS, was screen printed over the planarization layer followed by annealing (150°C; 30 minutes; Figure 2.2(a)(ii)). Screen printing was used to create a 2000-nm layer of P(VDF-TrFE) followed by annealing and dissolving in numerous polar solvents (Figure 2.2(a)(iii)). The sensor's top electrode was also developed using the screen

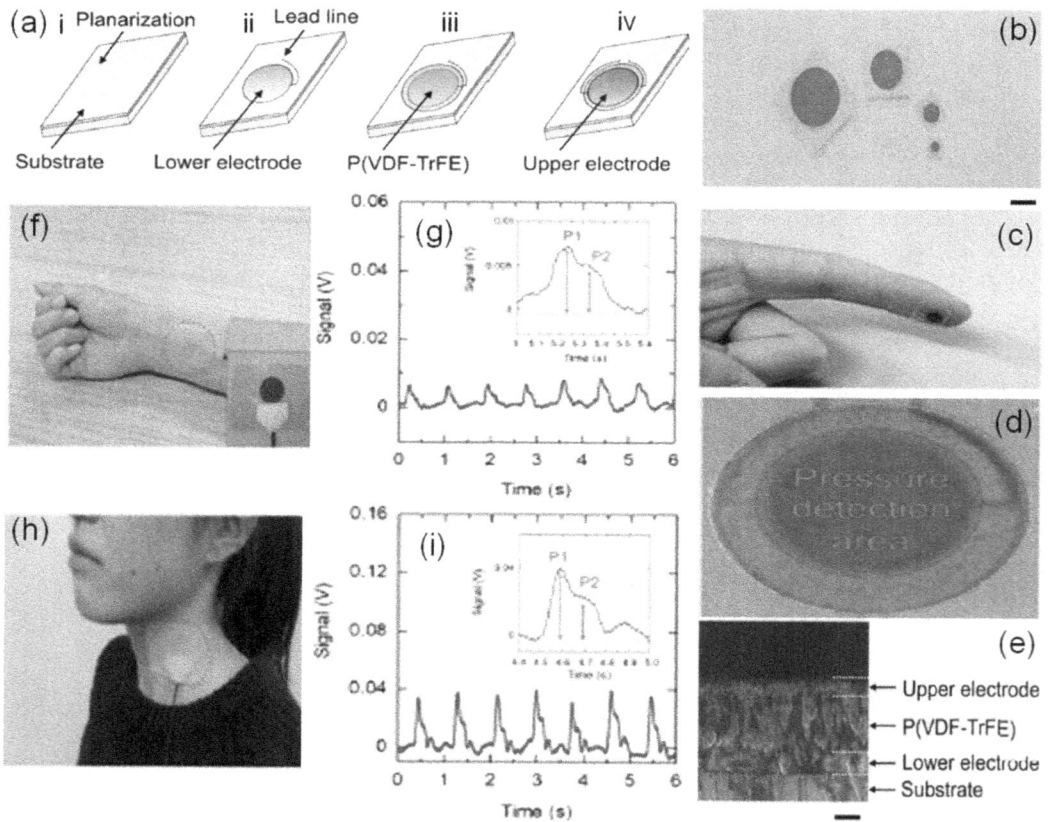

FIGURE 2.2
(a) Pictorial demonstration of a manufacturing procedure of fully printed vital sensors: (i) formation of the planarization layer on PEN substrate; (ii) formation of the lower electrode via screen printing of PEDOT:PSS; (iii) patterning of the P(VDF-TrFE) layer; (iv) formation of the upper electrode using screen printing of PEDOT:PSS. (b) Images of four different sized sensors (4, 20, 75, and 130 mm^2; (scale bar: 10 mm)). (c) Illustration of the flexible nature of the fabricated sensor by placing it on the forefinger. (d) Magnified view of the pressure sensor with an outer and inner circle indicating P(VDF-TrFE) and pressure detection areas, respectively (scale bar: 5 mm). (e) Cross-sectional SEM image of the printed sensor; scale bar: 1 μm. Photographs of the sensor attached to the wrist (f) and neck (h) of a volunteer employing a skin-friendly adhesive with real-time monitoring of the pulse-wave signal from the blood flow at the wrist (g) and neck (i). Adapted with permission [7]. Copyright (2018) Copyright The Authors, some rights reserved; exclusive licensee Nature Springer. Distributed under a Creative Commons Attribution License 4.0 (CC BY) https://creativecommons.org/licenses/by/4.0/

printing method using PEDOT:PSS. (Figure 2.2(a)(iv)). Figure 2.2(b) gives the photographs of four distinct constructed sensing devices of various sizes while its flexible nature is demonstrated in Figure 2.2(c) by placement on the forefinger. The corresponding magnified picture is depicted in Figure 2.2(d). Figure 2.2(e) provides the cross-sectional scanning electron microscope (SEM) image of the developed sensor. The performance of the developed sensor was analyzed by attaching the sensor to the wrist (Figure 2.2(f)) and neck (Figure 2.2(h)) of a volunteer via a skin-friendly adhesive. The instantaneous monitoring of the pulse-wave signal from the blood flow at the wrist (Figure 2.2(g)) and neck (Figure 2.2(i)) clearly shows the excellent behavior of the wearable sensor.

Apart from polymer-based materials for wearable and flexible sensors, carbon nanotubes (CNTs; single or multi-walled) [19], two-dimensional graphene-like materials [8] or h-BN [9], semiconductors [10], cotton fabric [11], or paper-based materials [20] are also employed for the fabrication of such sensors. The outstanding mechanical, electrical, and chemical characteristics of CNTs and graphene-based materials, high carrier mobility, and rich chemistry make them attractive electrode materials for such sensors. Because of their one- and two-dimensional nature, the bulk of carbon atoms in reported carbon nanomaterials are surface atoms, which are significantly influenced by environmental conditions. Various wearable and flexible sensors have benefitted from this property, as a greater surface area increases sensitivity.

Fully flexible silver nanoparticles (AgNPs)-all-carbon nanostructures [21] and single-walled CNTs on PTFE substrate [19] were reported to be used to create flexible and wearable NO_2 gas sensors. AgNPs-all-carbon hybrid nanostructure-based sprayed gas sensor array shows excellent sensitivity toward NO_2 response with high mechanical robustness, consisting of metallic single-walled CNTs and reduced graphene oxide decorated with AgNPs as conductive electrodes and sensing layers, respectively [21]. Amjadi et al. [22]. developed extremely flexible and skin-attachable strain sensors utilizing CNTs–Ecoflex nanocomposites. The fabricated sensor can measure the maximum induced strain of the finger (~42%), wrist (~45%), and elbow (~63%) under bending conditions. CNT/PDMS nanocomposites discussed by Li et al. [16]. resulted in a wearable motion sensor with increased sensitivity. Utilizing porous structured CNT/PDMS sponge has improved sensitivity in contrast to CNT/PDMS nanocomposite with no porous structure and demonstrates excellent monitoring ability of finger and elbow bending, speaking, drinking, and breathing [16]. A wide-area extremely capable flexible pressure sensor using a CNT active matrix for electronic skin has been described by Nela et al. [23]. Yu et al. [24] developed a wearable pressure sensor for electronic skins employing ultraviolet/ozone microstructured CNT/PDMS arrays. A wearable and sensitive heart-rate detector was built using a PbS quantum dot and a multi-walled CNT mix screen [25].

Moreover, gas sensors [8, 9], body motion sensors [26], and flexible wearable electronics [12] have also been reported to use two-dimensional layered materials, for example, graphene-based material or h-BN. Ayari et al. [9] created AlGaN/GaN-based gas sensors on nano-layered h-BN that were then transferred on metallic foil resulting in improved device performance with increased NO_2 gas sensitivity and response time. Boland et al. [26] reported on graphene–rubber composite-based sensitive human body motion sensors where the composites were prepared via infusing liquid-phase exfoliation-based graphene into natural rubber. Interestingly, these sensors can efficiently observe joint and muscle motion, along with breathing and a pulse, and can trail dynamic strain at vibration frequencies (160 Hz). Kabiri et al. [27] demonstrated wearable and flexible electronic tattoo-based sensors developed from graphene. A wearable electrochemical glucose sensor using a reduced graphene oxide nanocomposite electrode micropatterned on a flexible substrate and supported by a straightforward and cost-effective production approach has also been demonstrated [28]. Xu et al. [29] developed a multipurpose wearable sensor employing functionalized graphene sheets for concurrent observation of physiological parameters and volatile organic chemical biomarkers. Caccami et al. [30] developed a graphene oxide-based radiofrequency recognition wearable sensor for monitoring breathing.

A semiconducting material such as ZnO is also a potentially important material to fabricate such wearable sensors owing to its various fascinating properties. Breath-level acetone discrimination using a hierarchical ZnO gas sensor and temperature modulation has previously been described [10]. Zheng et al. [31] created a light-controlling, wearable, and highly transparent ethanol gas sensor using ZnO nanoparticles. It is worth mentioning here that ZnO is a well-known piezoelectric material. Employing this property, Park et al. [32] reported a simplistic and novel procedure for fabricating piezoelectric pressure sensors depending on the hybridization of ZnO nanotubes and the graphene layer. The sensor was developed using electron beam lithography and the chemical vapor deposition technique was used to grow graphene layers on copper foil which were then transferred to a silicon substrate using the wet chemical etching method. Metal-organic chemical vapor deposition methods were employed to develop vertically aligned and highly controllable ZnO nanotube arrays on graphene film. Here, the graphene film serves as flexible conduction channels linking ZnO nanotubes and metal electrodes, and finally, the piezoelectric devices were transferred onto a flexible substrate using the mechanical lift-off method. Figure 2.3(a) schematically illustrates the fabrication procedure of the ZnO

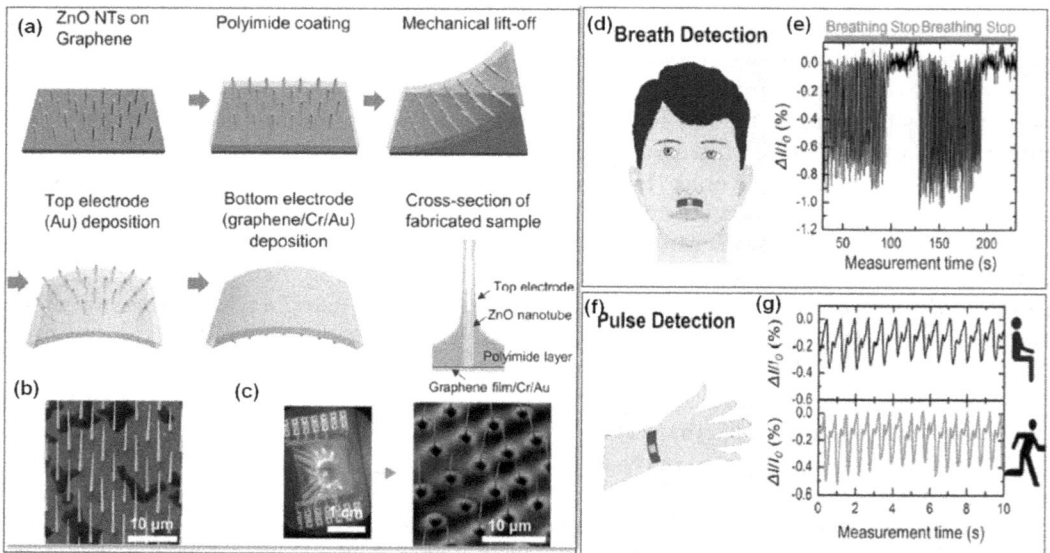

FIGURE 2.3

(a) Schematic for illustrating the fabrication process of ZnO nanotube/graphene-based pressure sensors. (b) FESEM image of ZnO nanotubes/graphene layers. (c) Image of fabricated pressure sensors on FPCB with the respective FESEM image. Schematic illustration of the fabricated sensor attached straightforwardly to the philtrum (d) and wrist (f) of a volunteer for breath and pulse detection, respectively. The real-time current response of the prepared sensor while breath monitoring (e) and pulse monitoring before and after running (g). Adapted with permission [32]. Copyright (2021) Copyright The Authors, some rights reserved; exclusive licensee Springer Nature. Distributed under a Creative Commons Attribution License 4.0 (CC BY) https://creativecommons.org /licenses/by/4.0/

nanotube-based piezoelectric pressure sensors where the last section represents the corresponding cross-sectional view. Figure 2.3(b) gives the field emission scanning electron microscope (FESEM) image of ZnO nanotube arrays/graphene and the fabricated flexible pressure sensors on a flexible printed circuit board (FPCB). The corresponding FESEM image is shown in Figure 2.3(c). The developed ZnO nanotube-based pressure sensors were then examined for biomedical applications and were straightforwardly fastened to the philtrum and wrist of a volunteer to monitor their breath and heart rate as schematically shown in Figures 2.3(d) and Figure 2.3(f), respectively. Figure 2.3(e) and Figure 2.3(g) show the real-time current behavior of the fabricated ZnO-based pressure sensor with breath and pulse monitoring before and after running, respectively.

Extremely responsive and wearable transistor biosensors utilizing In_2O_3 nanoribbon for glucose monitoring in body fluids have been developed by Liu et al. [33]. This biosensor can be laminated onto several surfaces including artificial arms and watches and facilitates glucose monitoring in sweat and saliva. Cotton cloth/fabric and paper-based materials are also important environmentally friendly materials utilized for constructing wearable sensors. Wearable lactate-sensing applications have also been developed using cotton cloth where carbon graphite (working electrode), Ag (reference electrode), and AgCl (counter electrode) are printed on the cotton fabric [11]. By covering conductive carbon thread (CT) with a PDMS polymer, Li et al. [34] constructed a wire-shaped flexible strain sensor. CT, the main strain-sensitive material, was made by pyrolyzing cotton thread in an N_2 environment. Figure 2.4(a) schematically depicts the fabrication of the wire-shaped strain sensor and the corresponding SEM image of the CT is shown in Figure 2.4(b). Figure 2.4(c) and Figure 2.4(d) show the applications of fabricated strain sensors in monitoring finger motion and blood pulse, respectively. Piezoresistive behavior of CT/PDMS composite wire is typical, with considerable strain sensitivity where the gauge factors (GF) determined under low and high strain conditions are 8.7 and 18.5, respectively. These values are much higher as compared with a typical metallic strain sensor (GF ~2). The developed wire-shaped CT/PDMS composite sensor demonstrates outstanding responsiveness to cyclic tensile load conditions.

Malon et al. [35] manufactured an electrochemical device using cotton fabric for measuring lactate in saliva. The wax patterning approach and template patterning methods were employed for preparing the reaction zone and electrodes, respectively. Zhan et al. [36] demonstrated the development of a paper/CNT-based wearable pressure sensor for soft robotic skin. Li et al. [20] created a flexible strain sensor for wearable electronics using tissue paper for monitoring breathing and controlling a robot. This sensor consists of carbon paper obtained from tissue paper via the high-temperature pyrolysis method and the PDMS elastomer and shows enhanced sensitivity under high applied strain.

2.3 Technologies for Flexible and Wearable Sensors

Nano/micro-fabrication is a critical stage and driving force in the development of flexible sensors and can be utilized to construct a circuit module on a flexible substrate, resulting in a downsized and highly integrated device. Lithography and printing are the two most commonly employed fabrication technologies. Because of precision manufacturing and the high resolution, lithography is generally utilized for the creation of microelectronic/nanoelectronic devices and can be either electron beam lithography or photolithography,

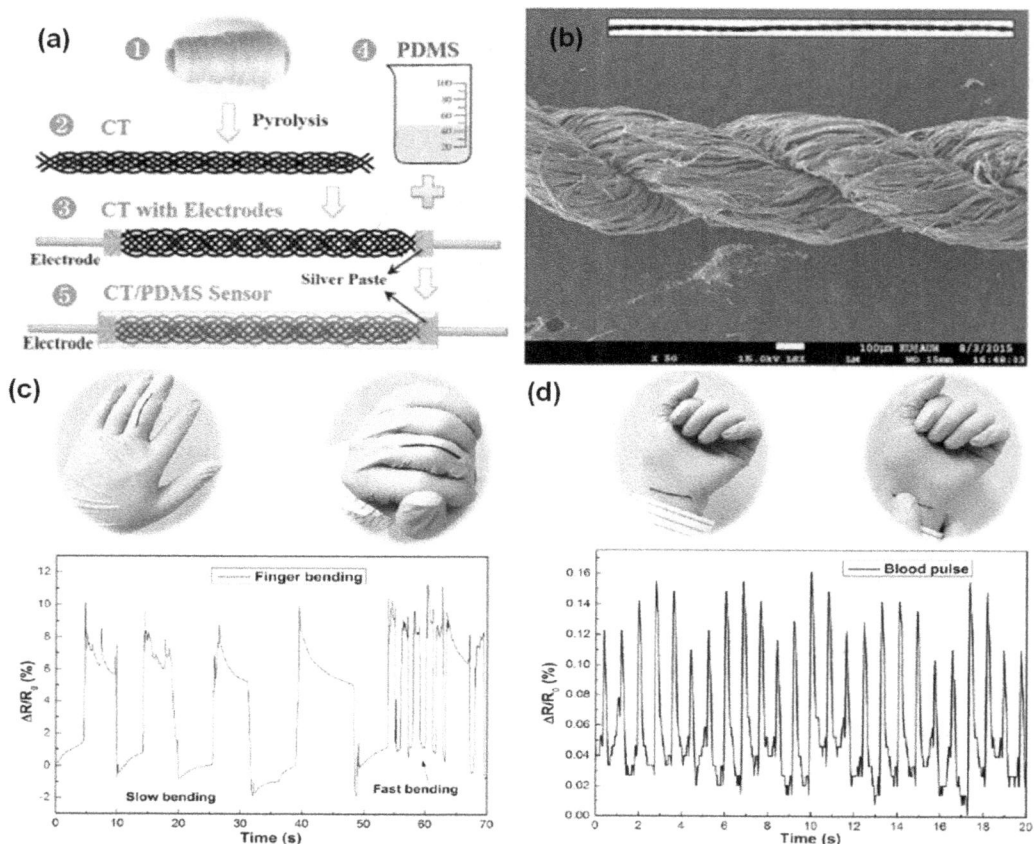

FIGURE 2.4
(a) Schematic illustration of the fabrication process of the wire-shaped strain sensor with (1), (2), (3), (4), and (5) being the cotton thread, CT, CT with electrodes, PDMS resin, and CT/PDMS sensor, respectively. (b) SEM images of CTs. The response of the fabricated sensor for monitoring finger motion (c) and blood pulse (d). Adapted with permission [34]. Copyright (2017) Copyright The Authors, some rights reserved; exclusive licensee Springer Nature. Distributed under a Creative Commons Attribution License 4.0 (CC BY) https://creativecommons.org /licenses/by/4.0/

depending on the exposure source. If an electron beam is exploited, then this is called electron beam lithography, while if an ultraviolet (UV) light source is used, then it is called photolithography, and this can further improve the resolution to micro- or nanoscale. Photolithography is much simpler in contrast to electron beam lithography, however, a mask is required in this process. The photolithography principle relies on the photochemical characteristics of the photoresist that are altered and display varying dissolubility with exposure to UV radiation. In this technique, the photoresist is first spin-coated onto a flexible substrate, followed by exposure to UV radiation. It is then developed to generate a patterned photoresist, as illustrated in Figure 2.5, and finally, the desired metal film is deposited via electron beam evaporation or thermal evaporation. Lift-off then takes place, which is achieved by submerging the sample in a polar solvent such as acetone. In this way, conducting electrode patterns can be obtained.

Accordingly, the production of flexible substrates warrants such substrate materials that are robust to high temperatures and polar solvents, and hence PI, PEN, and PET have

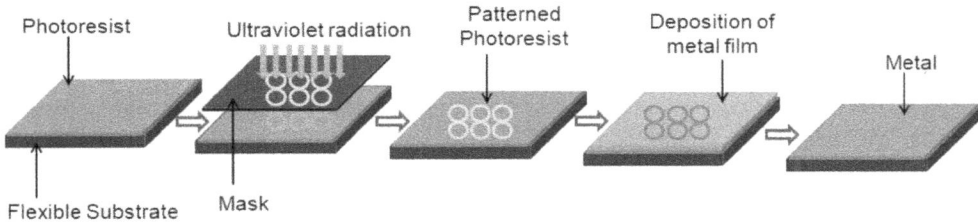

FIGURE 2.5
Schematic illustration of the photolithography process.

commonly employed substrate materials for fabricating flexible electrodes in wearable sensors. However, several issues remain with the lithography technique for fabricating wearable flexible sensors such as the thermal expansion of flexible substrates, achieving nanoscale alignment precision, and so on. Also, the polar solvent (e.g., acetone) utilized to remove the residual photoresist may damage the flexible material, reducing its qualities and performance in final applications.

Printing technology, contrary to lithography, is a cost-effective, large-scale approach for fabricating flexible wearable sensing devices. One of the significant advantages of this technology over lithography is the elimination of the requirement of polar solvents and a time-consuming lengthy fabrication process. By simply entering patterned pictures that could be easily updated, the technique could create a specific function circuit. For printed wearable sensing devices, various printing inks including conducting, insulating, and polymeric or semiconducting ink have been prepared.

It is worth mentioning here that wearable flexible sensors can be fabricated using various printing technologies, including transfer printing, inkjet printing, screen printing, double transfer printing, and aerosol-jet printing because they allow printing of any ink-type material on any target material along with their compatibility with printed electronics. Printed flexible electronics are extremely important for healthcare sensors because they are highly responsive and multifunctionality as a result of the latest development in conducting and sensing materials.

This printing technology has been employed by Shiwaku et al. [37] to fabricate a wearable amperometric electrochemical sensor via a printing organic circuit system. They fabricated the lactate sensor on a PEN substrate where a silver nanoparticle ink was inkjet printed on PEN substrates and then annealed (120°C for 30 minutes) in the air to form a 100-nm-thick silver electrode that was then coated with a carbon graphite ink containing Prussian blue, followed by air annealing. To specify the sensing area, a fluoropolymer was printed as a bank layer via a dispenser onto the substrate excluding the sensing area. Then a solution comprising of chitosan and lactate was dropped onto the area defined by the fluoropolymer bank, followed by air drying, and finally, phosphate-buffered saline was used to clean the sensor electrodes. Lorwongtragool et al. [38] also employed the same technology for developing a unique wearable electronic nose relying on a flexible printed chemical sensor array that is extensively used for healthcare purposes. This electronic nose is installed with a CNT/polymer sensor array using an inkjet printing method that can analyze armpit odor as well as more complex odors, while also classifying them.

Aerosol-jet printing has also been employed by various researchers to develop low-cost, flexible CNT-based pH sensors for live cell applications [39] and micro-hotplates of gold on the PI layer [15]. Jet printing is also a very popular technique and is used to develop temperature sensors on polyurethane substrate [40] and bifunctional CNTs for pH sensing [41].

Sensors are also created utilizing graphene/PEDOT: PSS ink (inkjet) is printed on top of a skin-relaxed polyurethane plaster and is characterized both in inert and ambient environments. The fabricated sensor can achieve greater sensitivity with negative temperature dependence [40]. Seifert et al. [13] presented a comparative study for printing silver inks using inkjet and aerosol-jet printing technologies. Khan et al. [42] reported flexible metal insulating-semiconductor field effect transistor devices fabricated using transfer-printed silicon microwires. Patterned graphene layers have been developed by employing transfer printing technology for flexible wearable electronics as well as optoelectronic systems by Choi et al. [43].

Screen printing is a widely accepted mask-necessary printing process where the functional ink is pushed into the aperture of the screen and moved to the substrate surface using a fill blade/squeegee. Upon removing the mask, functional ink is left over the substrate, forming a patterned layer. Such an approach is usually used to make working electrodes and sensing components in electromechanical sensors. To push the ink onto the substrate surface, various techniques such as vacuum filtering, spin/spray coating, or Mayer rod coating are employed. Tuteja et al. [14] reported the non-destructive analysis of cortisol and lactate via screen-printed graphene-embedded electrodes. Electrochemical sensors are also developed using layer structure Ag|AgCl|KCl reference electrodes prepared via screen printing technology [44]. Kim et al. [45] demonstrated a single wearable iontophoretic biosensor device to perform concurrent non-invasive collection and investigation of two separate biofluids, primarily sweat and interstitial fluid (ISF). Figure 2.6(a) depicts the concept of a wearable iontophoretic biosensor system for glucohol (glucose + alcohol) detection on a human body while consuming meals and alcoholic beverages, including synchronized transmission of ISF glucose and sweat alcohol behavior on a screen-printed tattoo platform (Papilio transfer base paper substrate). Ag acts as a reference electrode; AgCl serves as a counter electrode, while the functioning electrode was printed with conductive carbon ink. The image of the screen-printed tattoo electrode is shown in Figure 2.6(b). A pictorial representation of iontophoretic action is depicted in Figure 2.6(c), where sweat stimulation and ISF extraction occurred at the anode and cathode, respectively, which were screen printed on body-compatible tattoo materials. Conformal wireless readout circuits to permit instantaneous examination of biomarkers in epidermal biofluids were successfully realized.

Furtak et al. [46] developed screen-printed breathing rate sensors. They demonstrated a technique to screen-print a conductive CNT printing paste onto textile substrates to fabricate textronic strain sensors for measuring breathing rate which can also be integrated into garments after they have been crafted, demanding no significant changes to the manufacturing process. The garment is personalized to optimize sensor placement for enhanced system reliability. The breathing rate was determined by observing strain-induced modifications in the electrical resistance. Khan et al. [47] developed screen-printed P(VDF-TrFE) and P(VDF-TrFE)/multi-walled CNTs-based pressure sensors that are highly flexible. Apart from such advancements in technologies for developing flexible wearable sensors, there still exist certain challenges such as mass production, sensitivity, linearity, biodegradability, cost-effectiveness, durability, multifunctional sensing, and so on.

2.4 Conclusion

To conclude, the current chapter presents a detailed outline of several materials and technologies employed in the fabrication of flexible wearable sensors. Such materials include

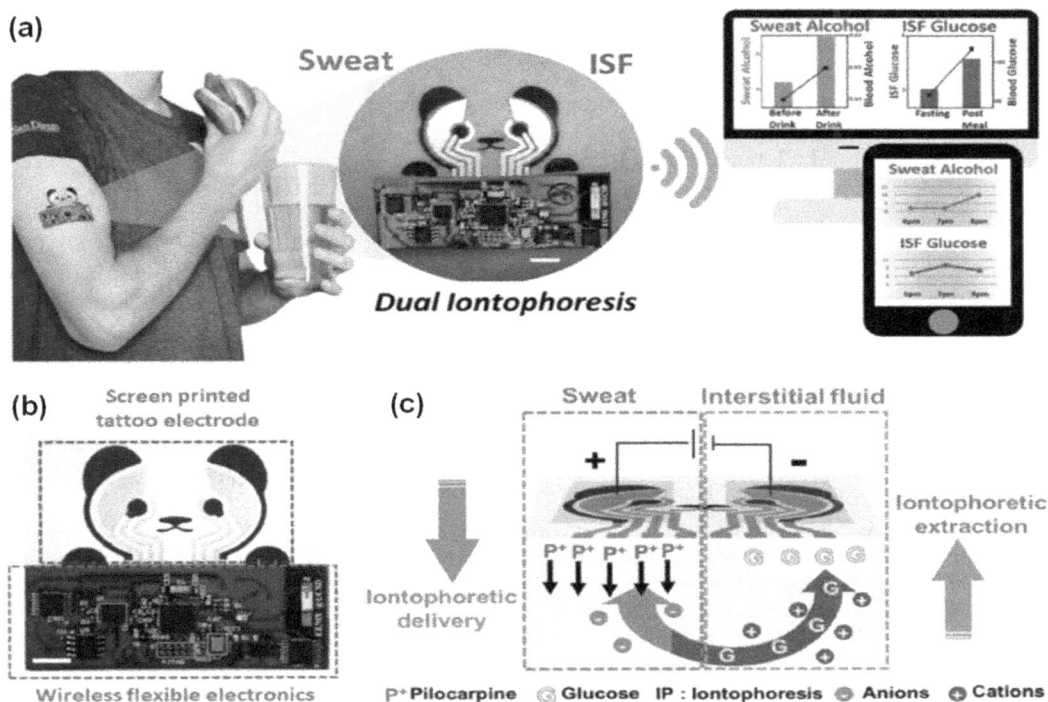

FIGURE 2.6
(a) Representation of wearable dual printed tattoo-based iontophoretic biosensor for simultaneous sensing of glucose and alcohol in the human body, along with wireless real-time monitoring of glucose and sweat alcohol behavior. (Scale bar: 7 mm). (b) Photograph of screen-printed electrodes for glucohol biosensor integrated with wireless FPCB. Scale bar: 7 mm. (c) A pictorial illustration of iontophoretic response. Adapted with permission [45]. Copyright (2018) Copyright The Authors, some rights reserved; exclusive licensee WILEY-VCH Verlag GmbH & Co. KGaA, Weinheim. Distributed under a Creative Commons Attribution License 4.0 (CC BY) https://creativecommons.org/licenses/by/4.0/

polymers, carbon nanotubes, graphene-based materials, ferroelectrics, semiconductors, and cotton fabric. The technologies are mainly lithography (electron beam lithography and photolithography) or printing (inkjet, screen printing, etc.) based processes. Despite the use of a variety of methods and materials to fabricate wearable and flexible sensors, mass manufacturing of these sensors remains problematic owing to their price, fabrication competence, and high-performance reliability issues. Building cost-effective, consistent wearable flexible sensors requires ongoing investment in the preparation of materials and the excellence of technologies. Novel materials and fabrication technologies can create a bright future for wearable and flexible sensors and they exhibit immense potential for commercialization.

References

1. A. Nyabadza, M. Vázquez, S. Coyle, B. Fitzpatrick, D. Brabazon, Review of materials and fabrication methods for flexible nano and micro-scale physical and chemical property sensors, *Appl. Sci.* 11 (2021) 8563.

2. Y. Ding, T. Xu, O. Onyilagha, H. Fong, Z. Zhu, Recent advances in flexible and wearable pressure sensors based on piezoresistive 3D monolithic conductive sponges, *ACS Appl. Mater. Interfaces* 11 (2019) 6685–6704.

3. A. Kristoffersson, M. Lindén, A systematic review of wearable sensors for monitoring physical activity, *Sensors (Basel)* 22 (2022) 573.

4. H. Zhao, R. Su, L. Teng, Q. Tian, F. Han, H. Li, Z. Cao, R. Xie, G. Li, X. Liu, Z. Liu, Recent advances in flexible and wearable sensors for monitoring chemical molecules, *Nanoscale* 14 (2022) 1653–1669.

5. Y. Aleeva, G. Maira, M. Scopelliti, V. Vinciguerra, G. Scandurra, G. Cannatà, G. Giusi, C. Ciofi, V. Figà, L.G. Occhipinti, B. Pignataro, Amperometric biosensor and front-end electronics for remote glucose monitoring by crosslinked PEDOT-glucose oxidase, *IEEE Sens. J.* 18 (2018) 4869–4878.

6. G.R. Manjunatha, K. Rajanna, D.R. Mahapatra, M. Nayak, U.M. Krishnaswamy, R. Srinivasa, Polyvinylidene fluoride film based nasal sensor to monitor human respiration pattern: An initial clinical study, *J. Clin. Monit. Comput.* 27 (2013) 647–657.

7. T. Sekine, R. Sugano, T. Tashiro, J. Sato, Y. Takeda, H. Matsui, D. Kumaki, F.D. Dos Santos, A. Miyabo, S. Tokito, Fully printed wearable vital sensor for human pulse rate monitoring using ferroelectric polymer, *Sci. Rep.* 8 (2018) 4442.

8. Y.J. Yun, W.G. Hong, N.-J. Choi, B.H. Kim, Y. Jun, H.-K. Lee, Ultrasensitive and highly selective graphene-based single yarn for use in wearable gas sensor, *Sci. Rep.* 5 (2015) 10904.

9. T. Ayari, C. Bishop, M.B. Jordan, S. Sundaram, X. Li, S. Alam, Y. ElGmili, G. Patriarche, P.L. Voss, J.P. Salvestrini, A. Ougazzaden, Gas sensors boosted by two-dimensional h-BN enabled transfer on thin substrate foils: Towards wearable and portable applications, *Sci. Rep.* 7 (2017) 15212.

10. J. Chen, X. Pan, F. Boussaid, A. McKinley, Z. Fan, A. Bermak, Breath level acetone discrimination through temperature modulation of a hierarchical ZnO gas sensor, *IEEE Sens. Lett.* 1 (2017) 1–4.

11. X. Luo, H. Yu, Y. Cui, A wearable amperometric biosensor on a cotton fabric for lactate, *IEEE Electron. Device Lett.* 39 (2018) 123–126.

12. H. Oh, J. Park, W. Choi, H. Kim, Y. Tchoe, A. Agrawal, G-C. Yi, Vertical ZnO nanotube transistor on graphene film for flexible inorganic electronics, *Small* 17 (2018) 1800240.

13. T. Seifert, E. Sowade, F. Roscher, M. Wiemer, T. Gessner, R.R. Baumann, Additive manufacturing technologies compared: Morphology of deposits of silver ink using inkjet and aerosol jet printing, *Ind. Eng. Chem. Res.* 54 (2015) 769–779.

14. S.K. Tuteja, C. Ormsby, S. Neethirajan, Noninvasive label-free detection of cortisol and lactate using graphene embedded screen-printed electrode, *Nano-Micro Lett.* 10 (2018) 41.

15. S. Khan, T. Nguyen, M. Lubej, L. Thiery, P. Vairac, D. Briand, Low-power printed micro-hot-plates through aerosol jetting of gold on thin polyimide membranes, *Microelectron. Eng.* 194 (2018) 71–78.

16. Q. Li, J. Li, D. Tran, C. Luo, Y. Gao, C. Yu, F. Xuan, Engineering of carbon nanotube/polydimethylsiloxane nanocomposites with enhanced sensitivity for wearable motion sensors, *J. Mater. Chem. C* 5 (2017) 11092–11099.

17. S. Yoon, J.K. Sim, Y.-H. Cho, A flexible and wearable human stress monitoring patch, *Sci. Rep.* 6 (2016) 23468.

18. K.-F. Lei, Y.-Z. Hsieh, Y.-Y. Chiu, M.-H. Wu, The structure design of piezoelectric poly (vinylidene fluoride)(PVDF) polymer-based sensor patch for the respiration monitoring under dynamic walking conditions, *Sensors* 15 (2015) 18801–18812.

19. P.B. Agarwal, B. Alam, D.S. Sharma, S. Mandal, A. Agarwal, Flexible NO_2 gas sensor based on single walled carbon nanotubes on PTFE substrate, *Flex. Print. Electron.* 3 (2018) 035001.

20. Y. Li, Y.A. Samad, T. Taha, G. Cai, S.-Y. Fu, K. Liao, Highly flexible strain sensor from tissue paper for wearable electronics, *ACS Sustain. Chem. Eng.* 4 (2016) 4288–4295.

21. W. Li, C. Teng, Y. Sun, L. Cai, J.L. Xu, M. Sun, X. Li, X. Yang, L. Xiang, D. Xie, T. Ren, Sprayed, scalable, wearable, and portable NO2 sensor array using fully flexible AgNPs-all-carbon nanostructures, *ACS Appl. Mater. Interfaces* 10 (2018) 34485–34493.

22. M. Amjadi, Y.J. Yoon, I. Park, Ultra-stretchable and skin-mountable strain sensors using carbon nanotubes–Ecoflex nanocomposites, *Nanotechnology* 26 (2015) 375501.
23. L. Nela, J. Tang, Q. Cao, G. Tulevski, S.J. Han, Large-area high-performance flexible pressure sensor with carbon nanotube active matrix for electronic skin, *Nano Lett.* 18 (2018) 2054–2059.
24. G. Yu, J. Hu, J. Tan, Y. Gao, Y. Lu, F. Xuan, A wearable pressure sensor based on ultra-violet/ozone microstructured carbon nanotube/polydimethylsiloxane arrays for electronic skins, *Nanotechnology* 29 (2018) 115502.
25. L. Gao, D. Dong, J. He, K. Qiao, F. Cao, M. Li, H. Liu, Y. Cheng, J. Tang, H. Song, Wearable and sensitive heart-rate detectors based on PbS quantum dot and multiwalled carbon nanotube blend film, *Appl. Phys. Lett.* 105 (2014) 153702.
26. C.S. Boland, U. Khan, C. Backes, A. O'Neill, J. McCauley, S. Duane, R. Shanker, Y. Liu, I. Jurewicz, A.B. Dalton, J.N. Coleman, Sensitive, high-strain, high-rate bodily motion sensors based on graphene–rubber composites, *ACS Nano* 8 (2014) 8819–8830.
27. S. Kabiri Ameri, R. Ho, H. Jang, L. Tao, Y. Wang, L. Wang, D.M. Schnyer, D. Akinwande, N. Lu, Graphene electronic tattoo sensors, *ACS Nano* 11 (2017) 7634–7641.
28. X. Xuan, H.S. Yoon, J.Y. Park, A wearable electrochemical glucose sensor based on simple and low-cost fabrication supported micro-patterned reduced graphene oxide nanocomposite electrode on flexible substrate, *Biosens. Bioelectron.* 109 (2018) 75–82.
29. H. Xu, J.X. Xiang, Y.F. Lu, M.K. Zhang, J.J. Li, B.B. Gao, Y.J. Zhao, Z.Z. Gu, Multifunctional wearable sensing devices based on functionalized graphene films for simultaneous monitoring of physiological signals and volatile organic compound biomarkers, *ACS Appl. Mater. Interfaces* 10 (2018) 11785–11793.
30. M.C. Caccami, M.Y. Mulla, C. di Natale, G. Marrocco, Graphene oxide-based radiofrequency identification wearable sensor for breath monitoring, *IET Microw. Antennas Propag.* 12 (2018) 467–471.
31. Z. Zheng, J. Yao, B. Wang, G. Yang, Light-controlling, flexible and transparent ethanol gas sensor based on ZnO nanoparticles for wearable devices, *Sci. Rep.* 5 (2015) 11070.
32. J.B. Park, M.S. Song, R. Ghosh, R.K. Saroj, Y. Hwang, Y. Tchoe, H. Oh, H. Baek, Y. Lim, B. Kim, S.-W. Kim, G.-C. Yi, Highly sensitive and flexible pressure sensors using position- and dimension-controlled ZnO nanotube arrays grown on graphene films, *NPG Asia Mater.* 13 (2021) 57.
33. Q. Liu, Y. Liu, F. Wu, X. Cao, Z. Li, M. Alharbi, A.N. Abbas, M.R. Amer, C. Zhou, Highly sensitive and wearable In_2O_3 nanoribbon transistor biosensors with integrated on-chip gate for glucose monitoring in body fluids, *ACS Nano* 12 (2018) 1170–1178.
34. Y.-Q Li, P. Huang, W.-B. Zhu, S.-Y. Fu, N. Hu, K. Liao, Flexible wire-shaped strain sensor from cotton thread for human health and motion detection, *Sci. Rep.* 7 (2017) 45013.
35. R.S. Malon, K. Chua, D.H. Wicaksono, E.P. Córcoles, Cotton fabric-based electrochemical device for lactate measurement in saliva, *Analyst* 139 (2014) 3009–3016.
36. Z. Zhan, R. Lin, V.-T. Tran, J. An, Y. Wei, H. Du, T. Tran, W. Lu, Paper/carbon nanotube-based wearable pressure sensor for physiological signal acquisition and soft robotic skin, *ACS Appl. Mater. Interfaces* 9 (2017) 37921–37928.
37. R. Shiwaku, H. Matsui, K. Nagamine, M. Uematsu, T. Mano, Y. Maruyama, A. Nomura, K. Tsuchiya, K. Hayasaka, Y. Takeda, T. Fukuda, D. Kumaki, S. Tokito, A printed organic circuit system for wearable amperometric electrochemical sensors, *Sci. Rep.* 8 (2018) 6368.
38. P. Lorwongtragool, E. Sowade, N. Watthanawisuth, R.R. Baumann, T. Kerdcharoen, A novel wearable electronic nose for healthcare based on flexible printed chemical sensor array, *Sensors* 14 (2014) 19700–19712.
39. G.L. Goh, S. Agarwala, Y.J. Tan, W.Y. Yeong, A low cost and flexible carbon nanotube pH sensor fabricated using aerosol jet technology for live cell applications, *Sens. Actuators B: Chem.* 260 (2018) 227–235.
40. T. Vuorinen, J. Niittynen, T. Kankkunen, T.M. Kraft, M. Mäntysalo, Inkjet-printed graphene/PEDOT: PSS temperature sensors on a skin-conformable polyurethane substrate, *Sci. Rep.* 6 (2016) 35289.

41. Y. Qin, H.-J. Kwon, A. Subrahmanyam, M.M. Howlader, P.R. Selvaganapathy, A. Adronov, M.J. Deen, Inkjet-printed bifunctional carbon nanotubes for pH sensing, *Mater. Lett.* 176 (2016) 68–70.
42. S. Khan, L. Lorenzelli, R. Dahiya, Flexible MISFET devices from transfer printed Si microwires and spray coating, *IEEE J. Electron Devices Soc.* 4 (2016) 189–196.
43. M.K. Choi, I.P. Dong, C. Kim, E. Joh, O.K. Park, J. Kim, M. Kim, C. Choi, J. Yang, K.W. Cho, J.-H. Hwang, J.-M. Nam, T. Hyeon, J.H. Kim, D.-H. Kim, Thermally controlled, patterned graphene transfer printing for transparent and wearable electronic/optoelectronic system, *Adv. Funct. Mater.* 25 (2015) 7109–7118.
44. L. Manjakkal, A. Vilouras, R. Dahiya, Screen printed thick film reference electrodes for electrochemical sensing, *IEEE Sens. J.* 18 (2018) 7779–7785.
45. J. Kim, J.R. Sempionatto, S. Imani, M.C. Hartel, A. Barfidokht, G. Tang, A.S. Campbell, P.P. Mercier, J. Wang, Simultaneous monitoring of sweat and interstitial fluid using a single wearable biosensor platform, *Adv Sci.* 5 (2018) 1800880.
46. N.T. Furtak, E. Skrzetuska, I. Kruci´nska, Development of screen-printed breathing rate sensors, *Fibres Text. Eastern Eur.* 21 (2013) 84–88.
47. S. Khan, W. Dang, L. Lorenzelli, R. Dahiya, Flexible pressure sensors based on screen-printed P (VDF-TrFE) and P(VDF–TrFE)/MWCNTs, *IEEE Trans. Semicond. Manuf.* 28 (2015) 486–493.

3

Design Principles of Flexible and Wearable Sensors

Samia Nawaz, Uswa Shafqat, and Komal Rizwan

CONTENTS

3.1 Introduction

Nanosensors are devices at the nanoscale that are used to detect specific molecules, pathogens, and organic and inorganic pollutants [1–4]. They are also used to monitor health problems. The development of telemonitoring portals and smart healthcare applications has opened up new avenues for people to purposefully engage in their healthcare and allows for the monitoring of clinically important parameters in non-clinical situations. Such tools can be included in normal treatment for both temporary and severe illnesses and give access to vital diagnostic data to patients and healthcare professionals According to studies, informed patients are more likely to engage in positive behavioral changes. To carry out routine medical surveillance, it is crucial to have extremely lightweight sensors that can be easily carried on the body. The use of multiple flexible and wearable sensors is becoming more and more popular in the world for exercise and training. These flexible and wearable sensors are simple to operate and improve accuracy and dependability for patient care. A sensor is a device that is used to transform energy from one type to another appropriate form for analysis, allowing someone to acquire information about their physical, chemical, and biological surroundings. Sensors are classified into flexible [5] and non-flexible sensors [6]. Flexible sensors are made of materials that may be bent to some extent without losing their qualities, whereas non-flexible sensors are hard and constructed of brittle materials. The material used to fabricate sensors depends on various factors, including the sensor's application, availability, total manufacturing price, and so on [7].

Different areas of the body can be used to wear wearable sensors including the wrist, waist, arms, ankle, legs, chest, and so on. Studs, necklaces, gloves, belts buckles, buttons, and rings are possible locations for wearable sensors. These sensors have the potential to measure a wide range of parameters including velocity, floors climbed, range, footsteps taken, and calorie intake and consumption. The introduction of sensors to the practical world has transformed human life. The ever-increasing modification of existing sensors has resulted from the dynamic use of sensors. Sensors have enabled us to measure some parameters, for example, perspiration rate, which were previously only achievable in laboratory-based systems. With the use of flexible and wearable sensors, it is feasible to analyze all physiological signs as well as the activities of a person. Sensors have been used in many applications including gas sensing, environmental observations [8, 9], and evaluating ingredients in different food products such as meat [10] and drinks [11, 12]. However, monitoring physiological signals is one of the most significant and essential uses of sensors. Wearable flexible sensors have completely changed the way people's actions are tracked [13]. They provide information about a person's behavior and actions in a precise and timely manner. Wearable sensors are employed in a variety of fields today including medicine, security, and communication [14]. During the activities of daily life, wearable flexible sensors have been used in a variety of detecting tasks. Their structure and qualities determine their role in different applications. Simple sensors on wearable devices, including biosensors for disease diagnostics, temperature sensors [15], strain sensors for body posture and body movement, and multipurpose sensors for facial expression identification [16], send these vital bodily characteristics to a central system in real-time. Wearable flexible sensors like smart textiles, power supplies, actuators, accelerometers, gyroscopes, wireless communication networks, and data capture technology for processing and decision support are all part of the sensing system. Flexible and wearable sensor devices do not affect the person's movements and daily activities, enabling monitoring in the patient's immediate environment as well as in the workplace [17].

3.2 Working Principle

Different types of sensors are utilized depending on the monitoring task. Sensor data is collected by the processor, and before being displayed on a screen, the data is processed. People use these types of affordable flexible and wearable sensor devices during walking, running, and other activities where it is simple to view the sensor's readings [13]. The tracking and transmitting of the sensor network recorded data is extremely important in real-time applications of physiological parameter monitoring. After being processed in the analog and digital divisions of the signal conditioning circuit, the data is transferred from the sensor node via a router to the monitoring unit for additional analysis. The setup cost, energy expenditure, variety of sensor modules, trans-reception range, and other factors all influence which communication network is chosen [18].

3.3 Nanomaterials and Techniques Used in the Synthesis of Wearable and Flexible Nanosensors

Wearable sensors are connected to the human body for highly precise monitoring, which is hindered by the low flexibility and stability of traditional old sensor systems based on rigid metal and semiconductor materials. Flexible and wearable sensors need specialized methods for high flexibility, sensitivity, stretchability, and a wide sensing range. As a result, scientists have been looking at new nanomaterials and nanocomposites for flexible and wearable sensors. The type of material and its qualities, processing technologies, and production on a large scale are all factors in the selection of nanomaterials for flexible and wearable sensors. Different nanomaterials like metallic nanowalls (NWs), carbon nanotubes (CNTs), and nanofibers are particularly popular because their large aspect ratio allows them to form a highly effective percolation conductive network with fewer components. One-dimensional nanomaterials are perfect for constructing sensors with both high transparency and stretchy conducting networks. Nowadays, flexible and wearable nanosensors are fabricated by using many of the latest techniques such as dip coating, spray coating, spin coating, drop coating, direct printing, layer-by-layer assembly, and writing techniques. Hsieh and colleagues synthesized a flexible and wearable pressure sensor by using a nanocomposite based on polydimethylsiloxane (PDMS) elastomer with zinc oxide (Zn-O) nanowire through a simple porogen-assisted process. This nanosensor has a high response rate and voice stimulation was from 0 to 50 kPa. A comparison of a flat PDMS device and the fabricated pressure nanosensor showed that the fabricated nanosensor had a high sensitivity of 0.717 kPa^{-1} (at 0–50 Pa), 0.360 kPa^{-1} (at 50–1000 Pa), and 0.200 kPa^{-1} (at 1000–3000 Pa), respectively. The nanocomposite dielectric layer has a low detection limit of about 1 Pa, as well as good stability and endurance after 4000 loading-unloading cycles allowing this PDMS-ZnO nanosensor to detect a variety of human actions such as throat vibrating, finger bending, airflow blowing, and calligraphy writing (Table 3.1) [19].

3.4 Types of Sensors

Several novel flexible and wearable sensors showed excellent performance and they have been synthesized by using different nanomaterials and nanocomposites. Different types of flexible and wearable sensors have been presented in Figure 3.1.

3.4.1 Wearable Temperature Sensors

A person's physiological status can be determined by their body temperature. Fever or infection is indicated by an elevated body temperature. On the other hand, decreased blood flow owing to circulatory shock indicates a decreased body temperature. Therefore, body temperature is considered one of the vital signals. Because body temperature changes based on the measurement site, the effect of the measurement site must be considered when measuring body temperature. Normal wrist temperature is approximately 32°C and body temperature is around 37°C. Temperature sensors are mostly wearable on the arms and chest, thus recorded temperatures are lower than body-core temperatures [20]. Zhang

TABLE 3.1

Synthesis, Characteristics, and Functions of Different Flexible and Wearable Nanosensors

Sr #	Material	Type	Wearable	Properties	Functions	Conditions
1	Zinc (ZnO)–porous sensor	Pressure sensor		Flexible polymer nanostructures have consciousness of temperature, moisture, light, touch, heat, electrical, optical, mechanical, and chemical properties	Detecting finger bending, calligraphy writing, audio detector, and health monitoring; hydro static underwater pressure	
2	MWCNT/Pd	Gas sensor		Influenced by the high temp. and humidity on the CH_4 sensor. Long stability and great self-life	For methane CH_4 detection at 25°C	CH_4 with its dilution varying from 0.5–100 ppm in air @ 25°C
3	Ag/AgCl ECG electrodes	ECG sensor	Wrist and right leg	Stable and uniform conductivity, very flexible, washable	To measure the ECG signals in humans	Conductive polymer has dispersed silver (Ag) nanoparticles (90–210nm) in a screen printable plastisol polymer
4	Na alginate/ polyacrylamide/ Laponite hydrogels			Great strength, stretchability, and Fatigue resistance	Analyze human motion, sensory to Skin, and healthcare detection	
5	ZnSe/CoSe$_2$// ECNT	Strain sensor	Fingers, wrists, and elbows	High-performance, flexible	Measuring joint movements of wrist, fingers, and elbow as a pressure sensor	
6	GNP based-PDMS nanocomposites	Motion sensor	Neck, wrist, and face		Human motion, neck movements, wrist and facial movements, strain, and also monitor body injuries	
7	PVA-prGO-PDA hydrogels (GO = graphene oxide)	Motion sensor		Self-healing, self-adhesive, conductive, stretchable	Detection of human motions such as bending and stretching of fingers, elbow, wrists, and knee and neck joints. Detection of small-scale movements and fractures such as swallowing and breathing	Tensile strength range of 146.5 KPa, adhesive strength of 23.04 KPa and elongation of 2580%
8	Graphene oxide/ polyacrylamide nanocomposite hydrogels	Dual strain-temperature sensor		Self-supporting, viscoelasticity, thixotropic	Potential material for human-machine interaction, artificial prosthetics, and electronic interconnects, robotics, personalized health monitoring	Maximum toughness up to 8.87 MJ/m^3 Maximum gauge factor (GF = 4.2), broad working range (0.02–2000%) Operate in pressure sensitivity
9	CNF/ZnO NR			Excellent mechanical, thermal, and piezoelectrical properties	Used in motion sensors, gas, and biosensors, also in photodetectors and photocatalyst	
10	Hybrid G/ZnO nanocomposites	Gas sensor		Excellent mechanical strength, good biocompatibility, high flexibility, and biodegradable. Unique nano structure, low-cost fabrication, highly flexible, and good sensitivity	To detect the concentration of carbon monoxide (CO) gas	Wet chemical deposition method and low cost and low temperatures
11	Ti$_3$C$_2$ NMC	Pressure and strain sensor	Wrist	Excellent conductivity, versatile surface chemistry	Monitor human health and body movements, also detect gas and humidity levels, voice detectors	
12	Ti$_3$C$_2$ /PVDF-TrFE	Pressure sensor		Mechanically stable	Analyze patients' physiological signals like muscle movements, pulse rate, eye motion, and respiration rate	GF = 3.6
13	Ti$_3$C$_2$ /PDMS	Strain sensor	Neck, skin	Flexibility, stretchability, capability of self-healing	Latest robots – electronic skin and human detection	Tensile strength = 0.51 MPa
14	BaTiO$_3$/ PVF	Motion sensor	Tape-like device attached to objects	Crystalline structures, high mechanical resistance, piezoelectric, pyroelectric, and good flexibility	Music sounds, ultrasound imaging, voice detection, speech therapy	Mechanical strength is up to 26.7 MPa, with an increase of 66% compared with PVDF

Sensitivity	Conductivity/ resistivity	Limit of detection (LOD)	Durability/ Stretchability	Response time/stability	Recovery time	Ref.
$S = 0.716$ kPa^{-1} in the regime of the 0–50 Pa		ultralow LOD of 1.0 Pa		Excellent operation stability around 4000 cycles		[19]
				Response of ~20-45.71% with a negligible change of ±2% towards 0.5–100 ppm for methane CH$_4$	(≈25 sec)	[36]
	2.12×10^{-6} Ω·m ±5.8% @ 70 wt-% of silver nanoparticles				Can use repeatedly over a duration of 10 days	[37]
			Tensile strength and stretchability of up to ~206 KPa and ~854% high levels of toughness (~134.78 kJ/m^3).			[38]
Highly sensitive and have sensing range up to 150% strain						[39]
	Conductivity ranges from 10^{-3} to 1 S/m for 5 to 11 wt.% samples			Gauge factor = approximately 15–25 and 25–50 for low and high levels of strain		[32]
	R = 740 Ω and C = 5 mS cm^{-1}				Recover for 10 min	[40]
2.0%/∘C			3370%	fast response time (~2.5 s) and long-time stability		[30]
			Highly sensitive			[41]
Sensing response by up to 40% (i.e., doubled sensitivity)	R = 0.26 MΩ to 3.1 KΩ			Response time = 435–370 s Shortest response 280 s	Recovery time = 45–115 s Shortest recovery time = 45 s	[42]
						[43]
High sensitivity of 0.51 kPa^{-1}		Low detection limit of 1.5 Pa				[44]
						[45]
$S = 18.0$ V/N with applied masses up to 1 g	Maximum output voltage of BT-10 was (50 V)					[46]

(Continued)

TABLE 3.1 (CONTINUED)

Synthesis, Characteristics, and Functions of Different Flexible and Wearable Nanosensors

Sr #	Material	Type	Wearable	Properties	Functions	Conditions
15	Ti_3C_2 and reduced graphene oxide	Pressure sensor	Hand	MXenes, MXene/polymer nanocomposites, and mixed-dimensional and 2D MXene-based structures	Human health, monitoring human body movements, recognizing voices, etc.	MXene in sensor applications depends on the concentration and functional groups (hydroxyl (OH) oxygen, fluorine, chlorine)
16	TiO_2-graphene quantum dots	Biosensor	Depends on the function	Physicochemical properties	Microbial infections, drug delivery, degradation of pollutants, biosensing, tissue engineering, and also in bioimaging	
17	Organohydrogels Polyacrylamide/montmorillonite/carbon nanotube organohydrogel (MMCOH)	Strain sensor	Dependent on function	Anti-freezing ability (−60°C), long-term environmental stability of up to (>30 days), and anti-drying ability (60°C), (GF = 8.5)	Flexible strain sensors, human motion detection, and power sources for sustainable electronics	Homogeneous and stable dispersion with N, N′-methylene bis(acrylamide) acting as the crosslinking agent. FTIR spectra were obtained – the band located at 3629 and 3404 cm⁻¹
18	Polyvinyl alcohol/cellulose nanofiber nanocomposite	Self-healing sensor	Arm, muscles of body, and foot		Self-healing ability and high toughness	Self-healing performance with a healing efficiency of nearly about 60%.
19	PVDF – nanocomposite of Praseodymium Oxide	Motion sensor			Force sensors for monitoring human physiological factors and the motion of muscles	
20	Graphene platelets (GnPs)	Temperature sensor	Tape-like device	Cost-effective nanomaterials-based strain sensors. Excellent performance over 35,000 cycles	Humidity, transportation, aerospace, and health monitoring response to temperature, rapid development of robotics, and damage evolution	Highly sensitive and excellent reliability
21	Kinesiology tape (K-Tape).	Motion sensor	Wear on legs, thighs, and arm	Repeatability, stability, and high sensitivity, Smart K-Tape sensors worn on different muscles	Motion primitive tool for assessment of sports coaching and athletic skills development	Tensile cyclic strains to a peak strain of 10% for 1000 cycles
22	Liquid metal-based nanocomposite	Flexible sensor		Highly conductivity smart footwear, being liquid state at room temperature	Polymer engineering, computational modeling, utilization in the fields of chemistry and nanotechnology	Pulse rate of 5 and 100 ns, scale bar: 200 nm
23	CNTs/PDMS	Pressure/temperature sensor	Back of neck, knee, elbow, foot, and head	High sensitivity, high stretchability Durability of wearable strain sensors under different atmospheric conditions	Detect human motion Soft robotics Facial expression recognition, heal healthcare monitoring	Detection of temperature and relative humidity
24	PDMS-MWCNT	Motion sensor	Skin attachable	Excellent curing time, electrical and mechanical properties High porosity, sensitivity, reliability, and biocompatibility	Human motion detection	Heart rate of 1 s⁻¹; measured viscosities of the 1.0 wt. %, 1.5 wt. %, 2.0 wt. %, and 2.5 wt. % nanohybrid pre-polymers ranges were 77.9 Pa, 271.9 Pa, 551.5 Pa, and 973.2 Pa, respectively
25	Polymer nanocomposite-based wearable strain sensors	Motion sensor.	Hand	Excellent mechanical, electrical, stretchability, flexible	Used in healthcare, medical diagnosis, robotic systems, professional support, and entertainment	
26	Nanocomposite hydrogel sensors 1. CNTs 2. GO 3. Ti_3C_2 4. polymer nanofillers based nanocomposite hydrogel	Strain and pressure sensor		Self-healing and self-adhesive	Monitor human motion, energy storage devices, physiological signals potential applications in wearable devices to detect electronic skin sensor	Nanocomposite hydrogel-based pressure and strain sensors that can change external stimulus to electrical signals
27	FD-BTO/PVDF composite film	Motion sensor			Monitor athlete motions; detect the pattern image of the external pressure input effectively	Based on the hydrophobic surface-FD-BTO/PVDF composite film

Sensitivity	Conductivity/ resistivity	Limit of detection (LOD)	Durability/ Stretchability	Response time/stability	Recovery time	Ref.
		A low detection limit of 6 Pa	Excellent durability over 6000 cycles	Slow recovery	Low response time of 230s	[47]
Sensitivity of 4.9 mA/mM (DA), with a range of 1–500 μM DA.		Low detection limit (0.015mM)		PEC response was (EC) response, 0.22 μM DA		[48]
Excellent reliability and highly sensitive		Have wide strain sensing range of 0–4196%.	Excellent durability of about 2000 cycles		Quick response time of (200 ms^{-1}) Real-time response at −60 °C for 6 h	[29]
Sensitivity range (~1.86) over large strain range of 0–100%				Fast response time (<240 ms^{-1}) and provide stable signal feedbacks to detect human motion		[49]
Higher sensitivity		Lower detection limits				[50]
Wide range of sensitivity from 1–1000	Electrical conductivity of 1460 S/cm			Fast		[22]
				Fast response directly fixed to the skin for monitoring		[51]
	High conductivity					[52]
High			High durability	maximum GF of 10, response time Of 204 ms		[21]
Sensitivity with GF = 7.9 and 10%	Highest piezoresistivity			quick linear response	Fast recovery for repetitive uses	[53]
High			High stretchability of up to 60%	Response time of 204 s		[54]
High sensitivity	Highly conductive					[35]
Sensitivity of 61.6 mV kPa^{-1}			Mechanical durability of 20,000 cyclic force tests			[55]

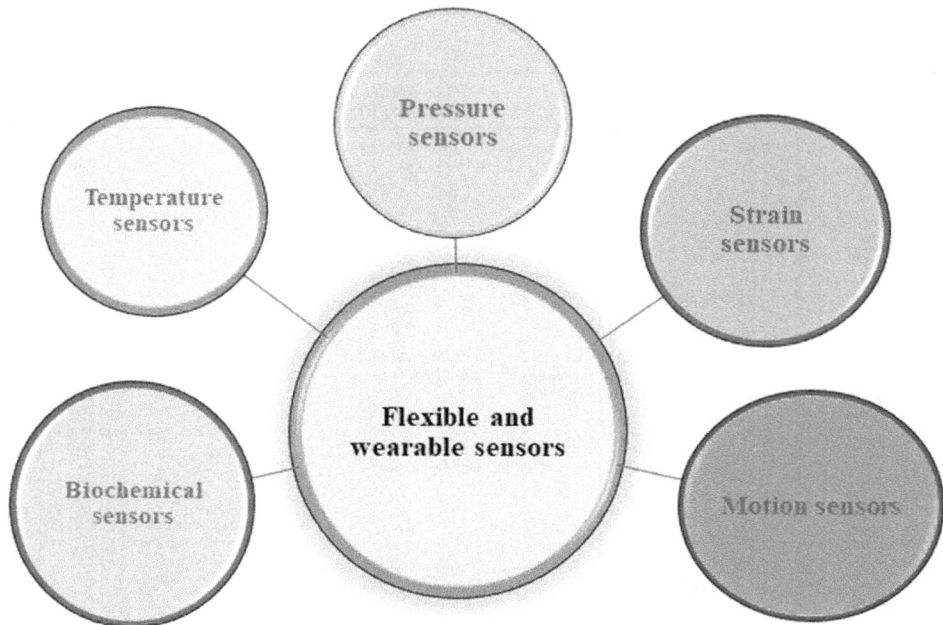

FIGURE 3.1
Various kinds of flexible and wearable sensors.

and his colleagues synthesized a flexible and wearable multipurpose sensor (loading composite/temperature/pressure) using the nanocomposite polydopamine-functionalized carbon nanotube/polydimethylsiloxane (PDA-CNT/PDMS) via a thiol-Michael addition click reaction followed by an inorganic salt sacrificial template approach. As strong covalent crosslinking develops between the matrix and the filler, this makes the resulting nanocomposite sensor highly stable when used to track both minute and large-scale scale movements of human body parts. Furthermore, the nanocomposite sensor performs the combined sensing for temperature (25–70°C), compressive strain (0.1–70%), and loading weight (55–150 g), which leads to high sensitivity and stability, indicating its practical use in temperature/pressure/loading sensing [21]. Han and colleagues synthesized wearable flexible nanocomposite strain sensors by using graphene platelets (GnPs) and epoxy (Epoxy/GnP). The high sensitivity and outstanding reliability in fabricated strain sensors were developed using graphene platelets (GnPs). Epoxy/GnP nanocomposites-based strain sensors have an electrical percolation threshold of 0.62 vol%. Young's modulus, fracture toughness, and energy release rate of fabricated nanosensor increased by 93%, 135%, and 215%, respectively, when 2.0 vol% GnPs was added. The nanocomposite-based films showed a positive temperature response from 30–100°C with a high-temperature coefficient of the resistance of 5800 ppm K. At 1–35°C, the film sensor demonstrated excellent thermal cycle stability and was able to identify variations in resistance brought by changes in humidity or fracture. In the strain range of 0–0.67%, the sensor showed high sensitivity, with a maximum range of 33 at 0.67% strain. Due to the homogeneous dispersion of 3 nm thick GnPs, the film sensor demonstrated great endurance up to 35,000 cycles, enhancing the load transmission over the composite surface and efficiently distributing the heat produced during fatigue testing [22].

3.4.2 Wearable Biochemical Sensors

The second-most used physiological parameter is heart rate. Heart rate is a tightly controlled feature that has a significant impact on human health and disease. Especially in people with persistent cardiac issues, wearable electrocardiogram (ECG) monitors are used to assess the real-time analysis of cardiovascular disorders. The ECG signal provides incredibly useful data on the frequency and regularity of heartbeats, which is employed in the treatment of cardiovascular disorders. A thoracic-impedance-variance and ECG monitoring system has been developed and incorporated into small plaster forms in flexible and wearable nanosensors [23]. An asynchronous analog to an information conversion system has been created to measure the RR intervals (time between two successive R waves) of ECG signals. The system includes a modified level-crossing analog-to-digital converter as well as a special algorithm for identifying R-peaks from level-crossing sampled data in a compressed volume of data [24]. Yida and colleagues synthesized a wearable heart monitoring sensor using a stretchable-hybrid ECG patch with microfluidic liquid-metal interconnected with carbon-black nanohybrid electrodes. Advanced power management and signal processing were possible because of a silicon-complementary metal-oxide-semiconductor device. This sensor's ECG signal response ranged from 100 mV$_{p-p}$ to 1 kHz. Good skin contact and low-noise-signal acquisition are possible by low profile, malleable soft carbon-black PDMS nanocomposite electrodes, which are superior to larger commercial wet electrodes [25]. A highly flexible and wearable electrochemical sensor was synthesized using Ni-Co metal-organic framework (MOF)-based nanosheet decorated with Ag/reduced graphene oxide/polyurethane (Ni-Co MOF/Ag/rGO/PU) fiber for glucose detection with a maximum sensitivity of 425.9 uAm/Mcm (Table 3.1, Figure 3.2) [26].

3.4.3 Wearable Strain Sensors

Strain sensors are one of the most often used flexible sensor applications. Various types of strain sensors have been fabricated to examine different physiological activities utilizing bandages, gloves, and other devices. These sensors differ in terms of gauge factor (GF) and the percentage of tensile and compressive strain they can withstand before breaking [27]. Traditional strain sensors are generally made of metal and semiconductor sheets and are delicate and inflexible. They have 5% stretchability, and hence cannot be connected to diverse textiles and soft skins [28]. Sun and colleagues fabricated an organohydrogel strain sensor with a wide sensing range up to 0 to 4196%, with great sensitivity and outstanding reliability and stability (almost 2000 cycles). Wearable devices could also take advantage of this anti-freezing sensor to identify and track human movements even in severe freezing environments. This adaptable, flexible, and wearable strain sensor demonstrated effective energy-harvesting capabilities with dependable stability throughout a broad temperature range (−60 to 60°C) or even under severe deformation circumstances as a self-charging power system (500% strain) [29].

A highly efficient dual (strain-temperature) sensor has been developed using alginate nanofibril/GO/polyacrylamide-based hydrogel nanocomposite. The hydrogel's excellent energy dissipation and ionic conductivity were attributed to an alginate nanofibril network while graphene oxide contributed great thermal sensitivity. The hydrogel has excellent stretchability up to 3370% and maximum toughness (8.87 MJ/m^3) due to various synergistic effects. The fabricated hydrogel strain sensor has a gauge factor of 4.2, a broad working range (0.02–2000%), a quick response time (2.5 s), and long-term stability [30].

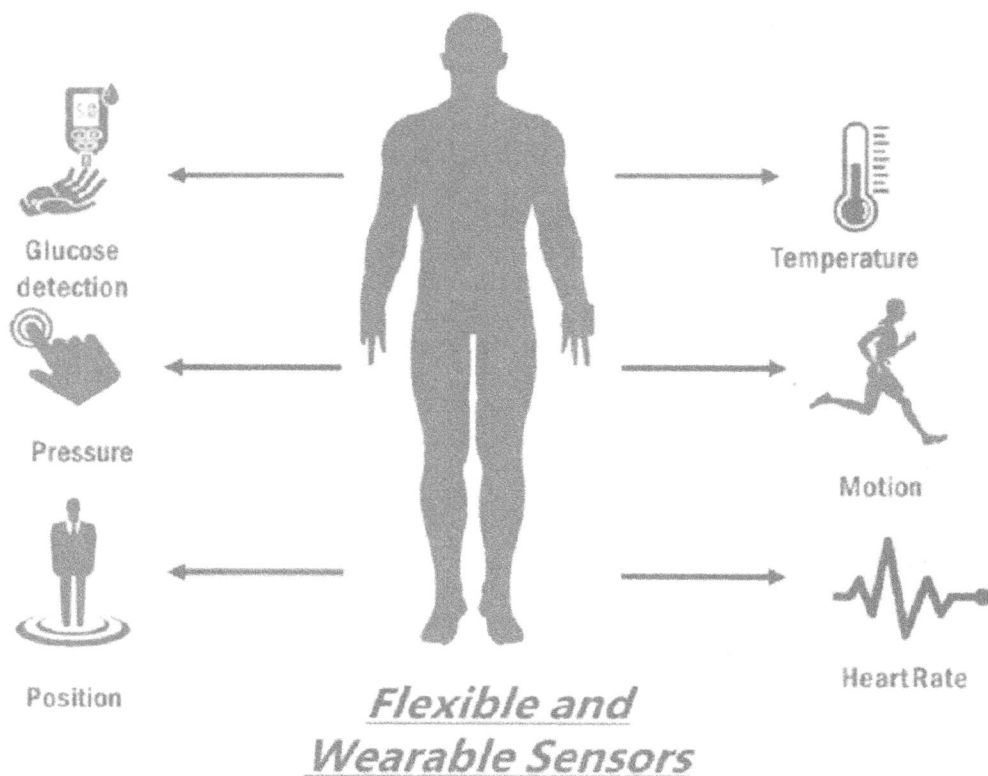

FIGURE 3.2
Different uses of flexible and wearable nanosensors in daily life.

3.4.4 Wearable Motion Sensors

Wearable sensors have been widely employed to detect human motions. Flexible and wearable motion sensors must possess great sensitivity and a broad sensing range to track both large- and small-scale movements such as delicate movements of the throat and chest during swallowing and breathing, as well as large-scale movements like the bending of hands, fingers, and knees. For instance, bandpass filters can be used to remove background noise brought on by speech and walking motion. The difference in timing between the chest and abdominal expansions can also be used to diagnose a partial airway obstruction using a flexible and wearable motion sensor. They also need to be highly stretchable for tracking human activities like stretching and bending, highly sensitive for detecting minute motions like pulse, heartbeat, swallowing, and facial micro-expressions, and be able to quickly respond for real-time analysis [31]. Del Bosque and colleagues fabricated a graphene nanoplatelet (GNP)-PDMS motion sensor. The suggested flexible sensors have an outstanding sensitivity to check propagation monitoring, with highly exponential behavior of electrical resistance because of the common breakdown of electrical channels due to crack propagation [32]. Microwaves were used to create piezoresistive sensors made of polydimethylsiloxane and MWCNT. The fabricated composites containing 2.5 wt.% MWCNTs had the maximum piezoresistive sensitivity, with a regular gauge factor of 7.9 at 10% applied strain, with the ability to compress a golf ball – the 2.5 wt%-nanocomposite sensors have been successfully employed as skin-attachable compression sensors for human movement sensing.

3.4.5 Wearable Pressure Sensors

Wearable pressure sensors with high sensitivity can mimic characteristics of the human skin by perceiving the applied pressure and producing electrical impulses. They have a large variety of possible applications in artificial smart systems and human physiological signal monitoring. Synthesizing nanoparticles and nanocomposites using existing nano-materials has resulted in a variety of new wearable sensors with good performance [33]. Resistive pressure sensors with remarkable performance have been constructed using a variety of nanomaterials. Hsieh and colleagues synthesized a flexible and wearable pressure sensor by using polydimethylsiloxane (PDMS) elastomer zinc oxide (Zn-O) nanowire using a simple porogen-assisted process. This nanosensor has a high response rate and voice stimulation (0–50 kPa). A comparison of a flat PDMS device and the fabricated pressure nanosensor showed that the fabricated nanosensor has a high sensitivity of 0.717 kPa^{-1} (at 0–50 Pa), 0.360 kPa^{-1} (at 50–1000 Pa), and 0.200 kPa^{-1} (at 1000–3000 Pa), respectively. The nanocomposite dielectric layer possesses a low detection limit of about 1.0 Pa, as well as good stability and endurance after 4000 loading-unloading cycles allowing this PDMS-ZnO nanosensor to detect a variety of human actions such as throat vibrating, finger bending, airflow blowing, and calligraphy writing [19].

3.5 Future Perspective and Challenges

The cost of manufacturing flexible sensing systems is expected to decrease because of nanoelectromechanical (NEMS) technology, opening up a broader range of applications. The cost of fabricating flexible and wearable sensing systems is also expected to decrease because nano-electrochemical technology opens up a wider range of applications. The use of residing and emerging manufacturing techniques will aid in the development of new sensing systems, enabling individuals to live better lives in the future [18]. It is still challenging to obtain greater stretchability, sensitivity, and strain adaptation in temperature sensors at the same time. Research efforts are still being made to enhance sensing efficiency and eliminate the impact of the sensor's elastic deformation on temperature detection. When compared with standard medical devices, the detection efficiency of flexible biochemical sensors is poor. Furthermore, the majority of physiological health data must be retrieved from internal body secretions. Improved biophilic implantable materials must be considered for the development of biochemical sensors to collect data from blood and muscles. Multifunctional sensors will require further research and development in novel materials, nanotechnology, and device structure design. Future wearable electronics will also require in-place data processing and real-time data transmission. It can be challenging to combine multiple functional moduli into a wearable system that can fully satisfy the demands of actual applications [34]. To summarize, the future of upcoming hydrogel-based pressure and strain nanosensors is largely dependent on the unique style and efficient use of conductive fillers, as well as a balance of conductivity, mechanical performance, functions, and stability [35].

3.6 Conclusion

A device that is used to transform energy from one type to another appropriate form for analysis, allowing someone to get information about their physical, chemical, and biological surroundings is called a sensor. The material used to fabricate sensors depends on various factors, including the sensor's application, availability, and total manufacturing price. Different types of sensors are utilized depending on the monitoring task. Sensor data is collected by the processor, and before being displayed on a screen, the data is processed. Various nanomaterials such as metallic NWs, CNTs, and nanofibers are particularly popular because their large aspect ratio allows them to form a highly sensitive and efficient percolation conductive network with fewer components. One-dimensional nanomaterials are perfect for constructing transparent sensors with both highly transparent and stretchy conducting networks. Nowadays, flexible and wearable nanosensors are created by using many of the latest techniques such as dip coating, spray coating, spin coating, drop coating, direct printing, layer-by-layer assembly, and writing techniques. This chapter has reported the selectivity, sensitivity, detection limit, reaction conditions, and sensing ranges of different flexible and wearable sensors.

References

1. Shakeel, A.; Rizwan, K.; Farooq, U.; Iqbal, S.; Altaf, A. A., Advanced polymeric/inorganic nanohybrids: An integrated platform for gas sensing applications. *Chemosphere* 2022, 294, 133772.
2. Rizwan, K.; Rahdar, A.; Bilal, M.; Iqbal, H. M. N., MXene-based electrochemical and biosensing platforms to detect toxic elements and pesticides pollutants from environmental matrices. *Chemosphere* 2021, 291, 132820.
3. Shakeel, A.; Rizwan, K.; Farooq, U.; Iqbal, S.; Iqbal, T.; Awwad, N. S.; Ibrahium, H. A., Polymer based nanocomposites: A strategic tool for detection of toxic pollutants in environmental matrices. *Chemosphere* 2022, 303, 134923.
4. Rizwan, K.; Bilal, M.; Iqbal, H. M., Role of nanomaterials in sensing air pollutants, in *Hybrid and Combined Processes for Air Pollution Control*, Elsevier: 2022; pp 1–17.
5. Segev-Bar, M.; Haick, H., Flexible sensors based on nanoparticles. *ACS Nano* 2013, 7, (10), 8366–8378.
6. Unno, Y.; Affolder, A.; Allport, P.; Bates, R.; Betancourt, C.; Bohm, J.; Brown, H.; Buttar, C.; Carter, J.; Casse, G., Development of n-on-p silicon sensors for very high radiation environments. *Nuclear Instruments and Methods in Physics Research Section A: Accelerators, Spectrometers, Detectors and Associated Equipment* 2011, 636, (1), S24–S30.
7. Liao, C.; Zhang, M.; Yao, M. Y.; Hua, T.; Li, L.; Yan, F., Flexible organic electronics in biology: Materials and devices. *Advanced Materials* 2015, 27, (46), 7493–7527.
8. Suryadevara, N.; Mukhopadhyay, S.; Rayudu, R.; Huang, Y., Sensor data fusion to determine wellness of an elderly in intelligent home monitoring environment, in *2012 IEEE International Instrumentation and Measurement Technology Conference Proceedings*, IEEE: 2012; pp 947–952.
9. Yunus, M. M.; Mukhopadhyay, S., Development of planar electromagnetic sensors for measurement and monitoring of environmental parameters. *Measurement Science and Technology* 2011, 22, (2), 025107.
10. Mukhopadhyay, S. C.; Gooneratne, C. P., A novel planar-type biosensor for noninvasive meat inspection. *IEEE Sensors Journal* 2007, 7, (9), 1340–1346.

11. Zia, A. I.; Mukhopadhyay, S. C.; Al-Bahadly, I. H.; Yu, P.-L.; Gooneratne, C. P.; Kosel, J., Introducing molecular selectivity in rapid impedimetric sensing of phthalates, in *2014 IEEE International Instrumentation and Measurement Technology Conference (I2MTC) Proceedings*, IEEE: 2014; pp 838–843.

12. Zia, A. I.; Rahman, M. S. A.; Mukhopadhyay, S. C.; Yu, P.-L.; Al-Bahadly, I. H.; Gooneratne, C. P.; Kosel, J.; Liao, T.-S., Technique for rapid detection of phthalates in water and beverages. *Journal of Food Engineering* 2013, 116, (2), 515–523.

13. Mukhopadhyay, S. C., Wearable sensors for human activity monitoring: A review. *IEEE Sensors Journal* 2014, 15, (3), 1321–1330.

14. Patel, S.; Park, H.; Bonato, P.; Chan, L.; Rodgers, M., A review of wearable sensors and systems with application in rehabilitation. *Journal of Neuroengineering and Rehabilitation* 2012, 9, (1), 1–17.

15. Hong, S. Y.; Lee, Y. H.; Park, H.; Jin, S. W.; Jeong, Y. R.; Yun, J.; You, I.; Zi, G.; Ha, J. S., Stretchable active matrix temperature sensor array of polyaniline nanofibers for electronic skin. *Advanced Materials* 2016, 28, (5), 930–935.

16. Su, M.; Li, F.; Chen, S.; Huang, Z.; Qin, M.; Li, W.; Zhang, X.; Song, Y., Nanoparticle based curve arrays for multirecognition flexible electronics. *Advanced Materials* 2016, 28, (7), 1369–1374.

17. Chan, M.; Estève, D.; Fourniols, J.-Y.; Escriba, C.; Campo, E., Smart wearable systems: Current status and future challenges. *Artificial Intelligence in Medicine* 2012, 56, (3), 137–156.

18. Nag, A.; Mukhopadhyay, S. C.; Kosel, J., Wearable flexible sensors: A review. *IEEE Sensors Journal* 2017, 17, (13), 3949–3960.

19. Hsieh, G.-W.; Shih, L.-C.; Chen, P.-Y., Porous polydimethylsiloxane elastomer hybrid with zinc oxide nanowire for wearable, wide-range, and low detection limit capacitive pressure sensor. *Nanomaterials* 2022, 12, (2), 256.

20. Olesen, B. W., *Thermal Comfort*, Bruel & Kjaer: 1982; Vol. 2.

21. Zhang, C.; Song, S.; Li, Q.; Wang, J.; Liu, Z.; Zhang, S.; Zhang, Y., One-pot facile fabrication of covalently cross-linked carbon nanotube/PDMS composite foam as a pressure/temperature sensor with high sensitivity and stability. *Journal of Materials Chemistry C* 2021, 9, (42), 15337–15345.

22. Han, S.; Zhang, X.; Wang, P.; Dai, J.; Guo, G.; Meng, Q.; Ma, J., Mechanically robust, highly sensitive and superior cycling performance nanocomposite strain sensors using 3-nm thick graphene platelets. *Polymer Testing* 2021, 98, 107178.

23. Yan, L.; Bae, J.; Lee, S.; Roh, T.; Song, K.; Yoo, H.-J., A 3.9 mW 25-electrode reconfigured sensor for wearable cardiac monitoring system. *IEEE Journal of Solid-State Circuits* 2010, 46, (1), 353–364.

24. Ravanshad, N.; Rezaee-Dehsorkh, H.; Lotfi, R.; Lian, Y., A level-crossing based QRS-detection algorithm for wearable ECG sensors. *IEEE Journal of Biomedical and Health Informatics* 2013, 18, (1), 183–192.

25. Li, Y.; Luo, Y.; Nayak, S.; Liu, Z.; Chichvarina, O.; Zamburg, E.; Zhang, X.; Liu, Y.; Heng, C. H.; Thean, A. V. Y., A stretchable-hybrid low-power monolithic ECG patch with microfluidic liquid-metal interconnects and stretchable carbon-black nanocomposite electrodes for wearable heart monitoring. *Advanced Electronic Materials* 2019, 5, (2), 1800463.

26. Shu, Y.; Su, T.; Lu, Q.; Shang, Z.; Xu, Q.; Hu, X., Highly stretchable wearable electrochemical sensor based on Ni-Co MOF nanosheet-decorated Ag/rGO/PU fiber for continuous sweat glucose detection. *Analytical Chemistry* 2021, 93, (48), 16222–16230.

27. Yamada, T.; Hayamizu, Y.; Yamamoto, Y.; Yomogida, Y.; Izadi-Najafabadi, A.; Futaba, D. N.; Hata, K., A stretchable carbon nanotube strain sensor for human-motion detection. *Nature Nanotechnology* 2011, 6, (5), 296–301.

28. Wagner, S.; Bauer, S., Materials for stretchable electronics. *Mrs Bulletin* 2012, 37, (3), 207–213.

29. Sun, H.; Zhao, Y.; Jiao, S.; Wang, C.; Jia, Y.; Dai, K.; Zheng, G.; Liu, C.; Wan, P.; Shen, C., Environment tolerant conductive nanocomposite organohydrogels as flexible strain sensors and power sources for sustainable electronics. *Advanced Functional Materials* 2021, 31, (24), 2101696.

30. Hou, W.; Luan, Z.; Xie, D.; Zhang, X.; Yu, T.; Sui, K., High performance dual strain-temperature sensor based on alginate nanofibril/graphene oxide/polyacrylamide nanocomposite hydrogel. *Composites Communications* 2021, 27, 100837.

31. Yan, C.; Wang, J.; Kang, W.; Cui, M.; Wang, X.; Foo, C. Y.; Chee, K. J.; Lee, P. S., Highly stretchable piezoresistive graphene–nanocellulose nanopaper for strain sensors. *Advanced Materials* 2014, 26, (13), 2022–2027.

32. del Bosque, A.; Sánchez-Romate, X. F.; Sánchez, M.; Ureña, A., Wearable sensors based on graphene nanoplatelets reinforced polydimethylsiloxane for human motion monitoring: Analysis of crack propagation and cycling load monitoring. *Chemosensors* 2022, 10, (2), 75.

33. Ai, Y.; Lou, Z.; Chen, S.; Chen, D.; Wang, Z. M.; Jiang, K.; Shen, G., All rGO-on-PVDF-nanofibers based self-powered electronic skins. *Nano Energy* 2017, 35, 121–127.

34. Gu, Y.; Zhang, T.; Chen, H.; Wang, F.; Pu, Y.; Gao, C.; Li, S., Mini review on flexible and wearable electronics for monitoring human health information. *Nanoscale Research Letters* 2019, 14, (1), 1–15.

35. Sun, X.; Yao, F.; Li, J., Nanocomposite hydrogel-based strain and pressure sensors: A review. *Journal of Materials Chemistry A* 2020, 8, (36), 18605–18623.

36. Shukla, P.; Saxena, P.; Madhwal, D.; Bhardwaj, N.; Jain, V., Electrostatically functionalized CVD grown multiwalled carbon nanotube/palladium nanocomposite (MWCNT/Pd) for methane detection at room temperature. 2022.

37. Chung, D.; Khosla, A.; Gray, B., Screen printable flexible conductive nanocomposite polymer with applications to wearable sensors, in *Nanosensors, Biosensors, and Info-Tech Sensors and Systems 2014*, International Society for Optics and Photonics: 2014, p 90600U.

38. Chen, Z.; Tang, J.; Zhang, N.; Chen, Y.; Chen, Y.; Li, H.; Liu, H., Dual-network sodium alginate/polyacrylamide/laponite nanocomposite hydrogels with high toughness and cyclic mechano-responsiveness. *Colloids and Surfaces A: Physicochemical and Engineering Aspects* 2022, 633, 127867.

39. Wang, Q.; Liu, J.; Ran, X.; Zhang, D.; Shen, G.; Miao, M., High-performance flexible self-powered strain sensor based on carbon nanotube/ZnSe/CoSe2 nanocomposite film electrodes. *Nano Research* 2022, 15, (1), 170–178.

40. Zhang, Y.; Liang, B.; Jiang, Q.; Li, Y.; Feng, Y.; Zhang, L.; Zhao, Y.; Xiong, X., Flexible and wearable sensor based on graphene nanocomposite hydrogels. *Smart Materials and Structures* 2020, 29, (7), 075027.

41. Park, T.; Kim, N.; Kim, D.; Kim, S.-W.; Oh, Y.; Yoo, J.-K.; You, J.; Um, M.-K., An organic/inorganic nanocomposite of cellulose nanofibers and ZnO nanorods for highly sensitive, reliable, wireless, and wearable multifunctional sensor applications. *ACS Applied Materials & Interfaces* 2019, 11, (51), 48239–48248.

42. Utari, L.; Septiani, N. L. W.; Nur, L. O.; Wasisto, H. S.; Yuliarto, B., Wearable carbon monoxide sensors based on hybrid graphene/ZnO nanocomposites. *IEEE Access* 2020, 8, 49169–49179.

43. Riazi, H.; Taghizadeh, G.; Soroush, M., MXene-based nanocomposite sensors. *ACS Omega* 2021, 6, (17), 11103–11112.

44. Sharma, S.; Chhetry, A.; Sharifuzzaman, M.; Yoon, H.; Park, J. Y., Wearable capacitive pressure sensor based on MXene composite nanofibrous scaffolds for reliable human physiological signal acquisition. *ACS Applied Materials & Interfaces* 2020, 12, (19), 22212–22224.

45. Zhang, K.; Sun, J.; Song, J.; Gao, C.; Wang, Z.; Song, C.; Wu, Y.; Liu, Y., Self-healing Ti3C2 MXene/PDMS supramolecular elastomers based on small biomolecules modification for wearable sensors. *ACS Applied Materials & Interfaces* 2020, 12, (40), 45306–45314.

46. Jiang, J.; Tu, S.; Fu, R.; Li, J.; Hu, F.; Yan, B.; Gu, Y.; Chen, S., Flexible piezoelectric pressure tactile sensor based on electrospun BaTiO3/poly (vinylidene fluoride) nanocomposite membrane. *ACS Applied Materials & Interfaces* 2020, 12, (30), 33989–33998.

47. Huang, W.; Hu, L.; Tang, Y.; Xie, Z.; Zhang, H., Recent advances in functional 2D Mxene-based nanostructures for next-generation devices. *Advanced Functional Materials* 2020, 30, (49), 2005223.

48. Bokare, A.; Chinnusamy, S.; Erogbogbo, F., TiO2–graphene quantum dots nanocomposites for photocatalysis in energy and biomedical applications. *Catalysts* 2021, 11, (3), 319.

49. Xu, K.; Wang, Y.; Zhang, B.; Zhang, C.; Liu, T., Stretchable and self-healing polyvinyl alcohol/cellulose nanofiber nanocomposite hydrogels for strain sensors with high sensitivity and linearity. *Composites Communications* 2021, 24, 100677.

50. Batra, A.; Sampson, J.; Arun, K.; Kassu, A., Polyvinylidene fluoride-praseodymium oxide nanocomposite force sensors for monitoring human physiological and muscle motion signs. *Sensors & Transducers* 2021, 250, (3), 26–31.
51. Lin, Y.-A.; Chiang, W.-H.; Loh, K. J., Wearable nanocomposite kinesiology tape for distributed muscle engagement monitoring. *MRS Advances* 2021, 6, (1), 6–13.
52. Ochirkhuyag, N.; Matsuda, R.; Song, Z.; Nakamura, F.; Endo, T.; Ota, H., Liquid metal-based nanocomposite materials: Fabrication technology and applications. *Nanoscale* 2021, 13, (4), 2113–2135.
53. Herren, B.; Charara, M.; Saha, M. C.; Altan, M. C.; Liu, Y., Rapid microwave polymerization of porous nanocomposites with piezoresistive sensing function. *Nanomaterials* 2020, 10, (2), 233.
54. Nankali, M.; Nouri, N. M.; Navidbakhsh, M.; Malek, N. G.; Amindehghan, M. A.; Shahtoori, A. M.; Karimi, M.; Amjadi, M., Highly stretchable and sensitive strain sensors based on carbon nanotube–elastomer nanocomposites: The effect of environmental factors on strain sensing performance. *Journal of Materials Chemistry C* 2020, 8, (18), 6185–6195.
55. Li, H.; Song, H.; Long, M.; Saeed, G.; Lim, S., Mortise–tenon joint structured hydrophobic surface-functionalized barium titanate/polyvinylidene fluoride nanocomposites for printed self-powered wearable sensors. *Nanoscale* 2021, 13, (4), 2542–2555.

4

Functionalized Materials for Improved Sensing

Selcan Karakuş

CONTENTS

4.1 Developed Sensor Systems

With rapid industrialization and the development of technology, the need for the detection of high-purity substances and pollutants has increased. As pollution in the ecosystem cycle increases with the increase of this pollution, it enters the food chain of living things and causes toxic effects on organisms. Various sources such as industrial activities, burning of fossil fuels, agricultural activities, mining activities of soil and rocks, and atmospheric emissions, with their negative effects on living things, cause the continuous release of trace metals into the environment. These reasons increase the importance of the development of functionalized sensing devices. Recently, there has been a significant trend in the fabrication of next-generation and functionalized sensing devices for target usage in monitoring water and food quality, safety, and biomedical applications [1–6].

The functionalized sensing material is an important component in the sensor system and plays a key role in multiple sensing mechanisms due to the different types of interactions (covalent bonding, electrostatic and ionic interactions) between analyte molecules and functionalized molecules on the surface of sensors. Because of their benefits in developed sensor systems, research into modified and functionalized sensing materials has grown in importance over the last decade. In smart sensor technologies, cost-effective manufacturing, wide concentration range, low detection limit, ultra-sensitivity, stability, selectivity, specificity, and rapid, ultra-trace level detection are the driving forces advocated for designing functional materials. With the advantages of smart sensors, a variety of sensing systems based on functionalized materials have been reported that focus on improved properties, for example, electrochemical, optical, enzymatic, photo-electrochemical, electrochemiluminescence, fluorescence, chemiresistive, calorimetric, colorimetric, and cataluminescence sensors. In sensor applications needing continuous monitoring of various target analytes, for example, heavy metals, pesticides, antibiotics, and antimicrobial residue detection in water and food quality, specific allergen (gluten and lactose) and biomarker detection in human health, blood glucose monitoring of diabetic patients, volatile organic compounds (VCOs), humidity, and gas detection in air quality have been investigated [7,8]. Specifically, previous studies have focused on the development of different functionalized material-based sensors such as inorganic functional nanomaterial-based sensors, polymer framework-based sensors, metal-organic framework-based sensors, fluorescent functional material-based sensors, functional nanomaterial-based sensors, two-dimensional (2D) nanomaterial-based sensors, and 2D metal carbide and nitride (MXenes)-based sensors. To make use of smart nanotechnology, the efficiency of these high-performance sensors in the sensing field and the effective sensing mechanism has been critically reviewed in this chapter.

4.2 Inorganic Functional Nanomaterial-Based Sensors

Recent developments in the field of metal/metal oxide nanostructures have led to a renewed interest in electrochemical sensors. The issue of inorganic functional nanomaterial-based sensors has received considerable attention due to their high electrical conductivity, high surface energy, unique surface, and good electronic, magnetic, optical, and chemical properties. Organic/inorganic hybrid materials, metal, and metal oxide nanomaterials, and carbon-based nanomaterials with various dimensions (zero-dimensional (0D), one-dimensional (1D), 2D, and three-dimensional (3D)) and morphologies (nanotubes, nanoparticles, and nanocomposites) have recently been used for the fabrication of multifunctional wearable and flexible devices. Based on the use of inorganic functional nanomaterials, these sensors showed high levels of detection performance for chemical and biological agents in previous studies due to their high electrical conductivity. So far, researchers have reported many inorganic functional nanomaterial-based sensor studies that have focused on flexible and wearable sensor applications to determine hazardous contaminants and physiological signals such as blood pressure, heart rate, pulse temperature, breathing rate, and biochemical analytes (creatinine, serum electrolyte, urea, albumin, and glucose).

Different metal-based nanoparticles such as gold (Au), copper (Cu), titanium (Ti), nickel (Ni), silver (Ag), and zinc (Zn) nanoparticle-based electrodes showed high selectivity, sensitivity, reactivity, signal-to-noise ratio (S/N), stability, and catalytic activity in inorganic functional nanomaterials-based sensor applications. Among them, the Au-based sensor has good biocompatibility, high chemical stability, high chemical inertness, and high conductivity. It exhibits excellent sensing performance with good selectivity and sensitivity, a short response time, room temperature operation, and good reproducibility to target analytes due to its excellent chemical and physical properties. In 2022, Zhao et al. developed a rapid Au nanoparticle-sensitized tungsten trioxide (WO_3) nanosheet-based trimethylamine gas sensor for the detection of seafood freshness. The trimethylamine gas sensing mechanism of Au nanoparticle-sensitized WO_3 nanosheets depends on the correlation of spillover effect and oxidation of trimethylamine with a low detection limit (0.5 ppm) and highly selective detection of trimethylamine in a rapid response time (8 s or 6 s). This study focused on the investigation of the redox mechanism in the trimethylamine sensing mechanism of Au nanoparticle-sensitized WO_3 nanosheet-based biosensors due to the oxidation and reduction electrochemical reactions in the trimethylamine sensing process and were attributed to the spillover effect [9]. Various Ag-based sensors have been reported for non-enzymatic or enzymatic detection of target analytes.

For instance, Lotfi et al. reported a novel ternary Ag nanoparticles/cadmium sulfide nanowires/reduced graphene oxide nanosystem-based electrochemical sensor for the detection of the antiviral medication acyclovir in human serum. The detection mechanism of acyclovir was demonstrated to be related to the oxidation signal of acyclovir and the role of Ag nanoparticles/cadmium sulfide nanowires/reduced graphene oxide nanostructure in increasing the electrochemical sensing performance of prepared modified electrodes using a cyclic voltammetric technique [10]. Considering the importance of the investigation of electrochemical nanosystems in advanced fields where ultrasensitive, flexible, and wearable sensors are required, nanomaterial-based sensors can provide suitable nanoplatforms for the targeted identification of biological analytes in biomedical applications. The most widely used sensors in the medical field are smart sensors.

The smart sensor with glucose oxidase is used for the diagnosis of diabetes, which has become widespread in the last half-century, and it is the most interesting non-enzymatic electrochemical sensor. For this reason, Zhai et al. prepared a non-enzymatic electrochemical Ag nanoparticle/N-doped-Co-metal-organic framework (MOF)@ polydopamine (core-shell) nanomaterial-based sensor for the detection of glucose. According to the electrochemical results, it was proved that the prepared sensor had good sensing performance with a low limit of detection (LOD) of 0.5 μM, a wide concentration range from 1 μM to 2 mM, good selectivity, reproducibility, and high stability for the electrocatalytic oxidation of glucose. The sensing mechanism of the core-shell nanomaterial-based sensor was based on its non-enzymatic electrochemical potential for selective glucose detection due to its excellent morphology, homogenous dispersion, high conductivity, and particle size dispersion [11]. In another study, Aparicio-Martinez et al. used a laser irradiation method to obtain a novel graphene/Ag nanoparticle-based flexible non-enzymatic electrochemical sensor, which demonstrated highly sensitive and selective detection of hydrogen peroxide. According to the experimental results, they observed that the sensor had a high catalytic response, a wide concentration range from 0.1 to 10 m, a low LOD of 7.9 μM, and a rapid response (3 s) due to the reduction of hydrogen peroxide [12].

Globally, water quality is one of the most significant issues for more than eight billion people as well as the aquatic ecosystem. For this purpose, Liang et al. developed a novel Cu nanowire-based electrochemical sensor for the detection of nitrate in water quality and safety applications. The prepared Cu-based sensors showed superior sensing performance with a wide concentration ranging from 8 to 5860 μM, high sensitivity of 1.375 μA/μM, and a low LOD of 1.35 μM for the detection of nitrate in real samples [13]. Hwa et al. investigated the sensitive electrochemical detection of phenolic hazards using an ultrasonic-synthesized CuO nanoflake-based sensor in water and pharmaceutical samples. The experimental results showed the prepared CuO nanoflake-based sensor had a wide linear range of 0.1 to 1400 μM and a low LOD of 10.4 nM with long-term stability. The sensing mechanism of the ultrasonic-synthesized CuO nanoflake-based sensor involves the ionization of hydroquinone to hydroquinone-, followed by an electron transfer between the analyte (hydroquinone−) and the sensor (Cu^{2+}) [14]. Based on the enhancing sensing performance of biosensors, Singh et al. evaluated cholesterol levels in human serum samples using a novel $NiFe_2O_4$/CuO/FeO-chitosan nanocomposite electrode. The electrochemical experimental findings indicated redox characteristics due to the electron transfer between the cholesterol oxidase and the modified $NiFe_2O_4$/CuO/FeO-chitosan nanocomposite/ITO electrode. Furthermore, the proposed sensor was found to have a low LOD of 313 mg/L, a good linearity range of 50 to 5000 mg/L, a rapid response time of 10 s, high sensitivity of

0.043 A/(mg/L.cm^2), and a long shelf-life of three months [15]. Consequently, the synergistic integration of electrochemical inorganic functional nanomaterial-based sensor sensing systems with devices in nanoplatforms provides selective and sensitive monitoring with flexible, wearable, disposable, and low-cost nanoplatforms for the detection of analytes in complex samples.

4.3 Polymer-Based Sensors

Polymer-based systems are sensors designed to monitor changes in molecules/biomolecules that encode information that helps us understand basic biological processes. When the application areas are examined, it is seen that different disciplines such as medicine, veterinary medicine, industry, pharmacy, defense, environmental protection, waste control, agriculture, and military fields attract attention. According to previous studies in the literature, polymer-based sensors are generally divided into two. The first of these is receptors, which are biological sensors, and the second is conductive polymer-based transducers, which are physical converters. Conductive polymer-based biosensors have been developed in the field of health and are used in *in vitro* and *in vivo* application studies where molecular or genetically encoded indicators are dominant. Biosensors have been investigated and have shown that sensing performance can be improved by varying the types, sizes, amounts, and chemical structures of nanostructures in the sensing layer.

In light of the conductive polymer-based nanosystems with physiochemical sensing performances, various nanoplatforms such as composite fibers, nanocomposites, and polymer/polymer composites have been developed as promising formulations in sensing applications. More in-depth, Gao prepared a novel wearable conductive poly(3,4-ethylenedioxythiophene): poly (styrene sulfonate) (PEDOT: PSS)/polyvinyl alcohol (PVA)/ethylene glycol (EG) composite fiber-based sensor to monitor the motion of the human body with good electrical conductivity [16]. In this study, wet spinning was used to create PEDOT: PSS/PVA conductive composite fibers, and solvent processing was used to add EG to the structure to create modified electrodes. PVA was used as a water-soluble synthetic polymer to improve the mechanical performance and mechanical properties of PEDOT: PSS fibers, taking advantage of the sensing performance of PEDOT: PSS/PVA conductive composite fibers. Obviously, with the addition of 20 wt% of PVA and 10 wt% of EG, a good recovery behavior, a hydrophilic and electroconductivity platform was achieved thanks to the functional groups on the composite fibers. In this pursuit, PEDOT: PSS fibers exhibited high stability in aqueous solution as a flexible intelligent textile for monitoring human body movements and a real-time monitoring system in the motion evaluation and muscle activity of the fingers, elbow, and neck. Another group, Almukhlifi et al., designed a conductive PEDOT-PSS nanocomposite film-based chemiresistive sensor for the determination of liquefied petroleum gas. The designed flexible PEDOT-PSS nanocomposite film-based sensor exhibited sensitive determination of liquefied petroleum gas with a concentration range from 10 ppm to 100 ppm, with a high sensitivity of 79% at room temperature and rapid response of 25s [17]. Lv et al. investigated polymer blend-based membrane sensors based on combining the polyaniline (PANI)/poly(styrene sulfonic acid) (PSS)/polyvinylidene-fluoride (PVDF) network. Briefly, the PSS-PANI/PVDF composite was examined with NH3 to enhance the detection performance at room temperature. According to the results, the amount of doped-PSS in the construction of the

PSS-PANI/PVDF composite-based structure inside the ternary polymer blend imparts the film platform with an excellent response of 9.4% and long-term stability after 30 days for detecting a sub-ppm level of NH_3. Furthermore, the exchange of protons between PANI and the adsorbed water molecule brings about the PSS-PANI/PVDF composite with a high NH_3 sensing performance in interfering gases [18]. To sum up, the proposed porous PSS-PANI/PVDF composite-based sensors demonstrated great potential in NH_3 sensing due to their electron transfers.

4.4 Metal-Organic Framework-Based Sensors

In recent years, various metal-organic frameworks (MOFs) such as luminescent MOFs (LMOFs), cataluminescence MOFs, and electrochemical MOF-based sensors have been reported. They exhibit unique properties such as high surface area, tunable pore size, crystallinity, porosity, luminescence, resistance, conductivity, catalysis, flexible functionality, magnetism, ferroelectricity, mechanical, and electrochemical properties, which could be used in various signal transduction strategies [19–21]. Given the many perfect sensing properties of MOFs, these MOF-based sensors have great potential for the design of optical, field-effect transistor, gas/vapor, and electrochemical sensors in favor of the ultrasensitive and selective binding of target molecules. In particular, LMOF-based sensors have unique systems for monitoring target analytes, water pollutants, drug molecules, hazardous materials, and biological species. These LMOFs have been reported from various strategies [22]: (i) the organic ligand-based MOF fluorescence emission due to the π-conjugation systems; (ii) the antenna effect of lanthanide (Ln)-based MOFs caused by the emission of Ln metal ions; (iii) charge transfers from metal to metal, a ligand to metal, metal to ligand, and ligand to ligand; and (iv) the emission color of the fluorescent guest molecules, including fluorescent guest molecules – centered or fluorescent guest molecules – sensitized emission.

Food pollutants such as pathogenic bacteria, heavy metal ions, mycotoxins, antibiotics, and pesticides pose serious risks to food safety and human health. The frequent occurrence of food safety issues as a result of food contamination has become a source of concern for both consumers and the food industry. LMOFs have been developed to control and prevent food contamination problems. Recent developments in the field of LMOFs have led to a renewed interest in the investigation of the mechanisms of fluorescent sensors. In 2022, Kajal et al. prepared a novel highly luminescent Zn-organic framework using a solvothermal method for the design of the LMOF nitroaromatic sensor. In this proposed luminescent Zn-organic framework sensor, it was seen that the mechanism of fluorescence quenching was dependent on the dipole-dipole and stacked interactions between the Zn-organic frameworks and nitroaromatics [23]. A considerable amount of literature has been published on pesticide sensors. These studies focused on the luminescent covalent organic framework for the detection of hazardous pesticides and heavy metal ions in aqueous solutions. Fang et al. prepared novel multifunctional luminescent covalent organic frameworks (LCOFs) with excellent chemical properties, thermal stability, and a large surface area ($750 \, m^2 \, g^{-1}$). The prepared multifunctional LCOFs-based sensors had a high chemo-sensing performance for p-nitroaniline due to the mechanism of electron transfer-induced fluorescence quenching with a fluorescent color change, high sensitivity, and a low LOD of 4.0×10^{-2} ppm [24].

The cataluminescence MOFs have been proposed from three strategies: (i) the recombination radiation in catalytically MOFs due to the holes and localized electrons; (ii) the radiation of the electronically excited state of the molecule; and (iii) the radiation from the energy transfer cataluminescence mechanism. Many researchers have reported cataluminescence MOF gas sensors with high catalytic activity, selectivity, sensitivity, and thermal stability. Motivated by these, Li et al. reported that the novel Ce(IV) MOF-based cataluminescence sensor had a potential for the detection of hydrogen sulfide. Due to the catalytic reaction, the Ce(IV)-MOFs-based cataluminescence sensor demonstrated superior hydrogen sulfide sensing performance in the concentration range of 1.04–21.0 g/mL and a low LOD of 0.035 g/mL with a good linear curve (R^2: 0.997). Consequently, the sensing mechanism of hydrogen sulfide was related to the catalytic oxidation process in the reaction between hydrogen sulfide and O_2, resulting in the generation of H_2O and SO_2 [25].

There are two basic strategies for electrochemical MOFs-based sensors that have been reported: (i) the redox-active conductive MOFs' design and (ii) the design of the conductive MOFs for monitoring target molecules. During the past 30 years, much more information has become available on the preparation, characterization, sensing behavior, and mechanism of electrochemical MOF-based sensors. Studies such as that conducted by Zhang et al. have shown that the metal-organic framework-based sensors could be an effective aptasensor for the detection of aflatoxin B1 in real samples. With this approach, the novel iron-porphyrinic metal-organic framework-based electrochemical aptasensor was developed and examined to detect aflatoxin B1 in peanut and milk samples. The prepared iron-porphyrinic metal-organic framework-based electrochemical aptasensor had high selectivity, stability, and reproducibility with a wide concentration range from 0.1 pM to 10 nM and a low LOD of 30 fM [26].

In another study, Chen et al. fabricated a novel cerium-based metal-organic framework on multi-walled carbon nanotubes by the *in situ* growth method. These nanoplatforms were used to prepare novel chemically modified electrodes for the detection of gallic acid, an antioxidant. According to the electrochemical results, the prepared cerium-organic framework-multiwalled carbon nanotubes-based sensor had good repeatability, a low LOD of 0.14 µM (S/N = 3) with a wide gallic acid range from 1.5 µM to 200 µM, stability, and anti-interference [27]. In 2012, Jiang et al. demonstrated the hydrogen peroxide sensing performance of a novel Cu_2O nanoparticle @CuHHTP heterojunction nanoarray-based sensor in urine and serum samples. As known, controlling the particle size and distribution of a sensing material could affect the electrocatalytic performance of hydrogen peroxide. In their work, the authors focused on the effect of surface functionalization on the catalytic performance of electrochemical MOF-based biosensors for the detection of hydrogen peroxide [28]. Pesticides are among the most toxic chemicals for any living creature, which can cause cancer, problems with the reproductive and central nervous systems, and eventually death if sustained exposure is involved.

Fernández et al. designed an electrochemical biosensor for the detection of imidacloprid (IMD), a pesticide with long-lasting effects in soil [29]. In this study, a direct competitive immunosensor for the electrochemical determination of IMD pesticide on the gold nanoparticle-modified electrode was reported for the first time. Self-obtained specific monoclonal antibodies were immobilized on the proposed gold nanoparticle-modified electrode by utilizing the biofunctionalization capabilities of gold nanoparticles. In the design of the biosensor, free IMD in the sample competed with horseradish peroxidase (IMD-HRP)-conjugated IMD for recognition by antibodies. Following that, HRP enzymatically oxidized and reduced 3,3′, 5,5′-tetramethylbenzidine (TMB) on the surface of screen-printed carbon electrodes. Despite these encouraging reports, MOF-based biosensors still have a great interest in further research into understanding the mechanism of nanosensors. In particular, the investigation of the fabrication of portable, flexible, ingestible, implantable, and wearable electrochemical MOF-based biosensors using large-scale production for novel sensing systems is a trending scientific issue [30]. The production and use of programmable portable sensing devices currently on the advanced sensor market are not sufficient. In addition, obtaining the combination of MOF-based detection systems with excellent detection performances and chemical properties for target molecules remains quite challenging. The improvement of green preparation methods for the sensitive and selective MOF-based detection systems that can eliminate these disadvantages is very important to increase the effective wearable detection ways of the target analyte for flexible sensing devices under optimal conditions such as pressure, temperature, humidity, and chemicals [31].

4.5 Functional Nanomaterial-Based Sensors

Large-scale creative functional nanomaterial-based sensor development opportunities have emerged in new multidisciplinary fields such as chemistry, physics, biology, computing, and engineering, with contributions from scientists from different fields [32]. Thanks to their good sensitivity, high selectivity, and sensing capabilities, these nanosensors are used in many applications such as clinical diagnostics, analysis of blood samples, diagnosis of infectious diseases, agricultural production, pharmacology, heavy metals, pesticide detection, water quality, organic pollutants, drug screening, food analysis, and environmental monitoring. It is successfully used in the detection of biological and chemical substances in the field. For monitoring, functional nanomaterial-based sensors are commonly used in two methods for developing advanced electrochemical sensors with the integration of the morphology, dimensional, shape, size, and structural properties of nanostructures:

(1) Enzymatic sensors by electron transfer and improving the sensing performance with an enzyme.
(2) Non-enzymatic sensors that undergo a catalytic reaction.

Considering these approaches, Jayaraman et al. developed dual thymine and carbohydrazide functionalized graphene oxide-based electrochemical sensors with high-performance multiplex sensing for Hg(II) and Cr(VI) [33]. In this work, thymine and carbohydrazide functionalized graphene oxide were prepared via a simultaneous reduction method. The proposed electrochemical sensor had good electrode stability and showed multiplex sensing performance for Hg(II) and Cr(VI) ions with low LODs of Hg(II) and Cr(VI) of 1 ppb and 20 ppb, respectively. The experimental results of the functionalized graphene oxide-based electrochemical sensor were high for the drinking-water quality standard recommended by the World Health Organization (WHO) authorities. Buledi et al. developed a sensitivity polyvinyl alcohol functionalized tungsten oxide and reduced graphene oxide nanocomposite-based electrochemical sensor for the detection of 4-aminophenol in tap water, mineral water, and river water [34].

Hui et al. investigated the sensing behavior of the novel polyaniline–antimony decorated laser-induced graphene-based multifunctional hybrid sensor for the detection of water quality against pesticides, heavy metals, and pH [35]. According to the experimental results, the authors assumed that the electrochemical behavior of the sensor involved the reduction of nitro compounds to the hydroxylamine group under electron transfers and the integration of the nitroso to the hydroxylamine group by two electrons and protons. In another work, Chen et al. pointed out that the multi-sensing mechanism of flexible ultrasensitive strain sensors should be investigated for the development of flexible and wearable electronics for monitoring human movements or healthcare [36]. The report highlighted the significance of creating a novel multilayer sensing structure with multiple response mechanisms as an advanced strategy to fabricate high-performance flexible strain sensors.

4.6 Two-Dimensional Nanomaterial-Based Sensors

In recent years, the medical methods used today have not been effective at the cell level. There are difficulties in the diagnosis and treatment of some diseases and injuries. For this reason, numerous studies have been conducted to improve the performance of advanced sensing materials by examining in detail the analyte-sensor interaction and subsequent signal transmission pathways. In particular, 2D nanomaterial-based sensors have been developed as a result of the use of nanotechnology for this purpose [37]; 2D nanoformulations have demonstrated a variety of superior properties relevant to analyte sensing, including optoelectronics, electronic, surface, structural, conductivity, and physical properties on charge transfer, adsorption, mass loading, surface modification, and the refractive index, among others. Generally, applications of nanomaterial-based electrical nano/biosensors are based on the mechanism of charge transfer, while analyte sensing with surface interaction and conduction of protons have been emphasized. Thanks to the use of 2D nanomaterial-based sensors in the field of medicine, the ability to prevent diseases or monitor the body's response to care are provided by keeping cell functions under control as a result of early diagnosis and treatment. Biosensors can also be used to detect bacteria, pathogens, and viral microorganisms. A novel non-enzymatic glucose sensor to monitor glucose based on 2D nitrogenated holey graphene for diabetes mellitus has been recently explored and reported by Panigrahi et al. [38]. In this research, the glucose-sensing mechanism of the 2D sensor is based on the interaction between glucose molecules with the nitrogenated holey graphene monolayer due to electronic transport. The resulting 2D nitrogenated holey graphene-based sensor allows the detection of glucose with high sensitivity (glucose, fructose, and xylose) for the non-invasive diagnosis of diabetes.

Hong et al. used novel 2D transition metal dichalcogenides (MoS_2)/graphene heterostructures as a gas sensing platform to detect nitrogen dioxide (NO_2) [39]. In their work, they demonstrated that the combined effect of transition metal dichalcogenides and graphene affected the detection performance of the proposed sensor with a wide concentration range of 0.2–100 ppm and a low LOD of 0.2 ppm at the working temperature range of 25–200°C. In another study, Khan et al. focused on the computational study of 2D borophene/boron nitride-based gas sensors for the different industrial affiliated gases such as CO, CO2, NO, NO_2, and NH_3 [40]. The gas sensing mechanism was related to the electronic structure of gases and chemisorption behavior of the analyte on the 2D borophene/boron nitride material associated with the adsorption energy. Congur et al. constructed a novel disposable 2D methyl germanane and activated a pencil graphite electrochemical sensor for the determination of phenol [41]. According to the experimental results, it was observed that the proposed electrochemical sensors had low LODs of 7.65 μM and 3.72 μM for chemically activated pencil graphite electrodes and 2D methyl germanane/activated pencil graphite-based electrochemical sensors, respectively. The sensing mechanisms of the phenol onto 2D methyl germanane and activated pencil graphite-based electrochemical sensors were evaluated and it was found that gas sensing was related to the 2D surface of the functionalized germanane/activated pencil graphite-based sensor. Huang et al. (2022) assessed the electrochemical biosensing performance of novel ultrathin conductive and 2D MOF nanosheets and gold nanoparticle-based sensors for detecting hydrogen peroxide as a cancer biomarker in cancer cells [42]. According to the electrochemical results, the proposed 2D sensor showed unique surface and sensing properties with a high sensitivity of 188.1 μA/cm^2.mM for hydrogen peroxide and a low LOD of 5.6 nM in different cancer cells as colon cells and normal colon epithelial cells. The sensing mechanism of the proposed 2D MOF nanosheets and the gold nanostructure-based sensor was based on the decomposition of hydrogen peroxide.

4.7 MXene-Based Sensors

MXene is a 2D transition metal carbide and nitride material with different chemical functional groups (-OH, -O, and -F) on its surface, which was discovered in 2011 and has attracted great attention in recent years, especially in advanced nanosensor technologies [43]. MXene and MXene-based composites attract attention with their excellent biological, electrical, and mechanical properties in energy storage materials, water treatment systems, photothermal materials, and advanced sensors in various application areas. Cancer is one of the diseases that pose a great problem for all countries in the world today. Cancer can be prevented if cells can be identified and intervened in the early stages of mutation [44]. Various sensors, such as electrochemical biosensors, nanosensors, biosensors, molecularly imprinted polymer-based chemiresistive sensors, exosomal microRNAs array sensors, luminescent biosensors, colorimetric sensors, quantum dot nanosensors, and potentiometric sensors, have been developed for early cancer detection. In particular, prostate cancer (PC) is one of the most common and dangerous types of cancer in middle-aged men, after lung cancer.

Soomro et al. developed novel NiWO$_4$ nanoparticles induced partial oxidation of MXene-based sensors using NiWO$_4$ nanoparticles and partially oxidized Ti$_3$C$_2$T$_x$ sheets as an example of MXene-based sensors in PC diagnosis [45]. The photo-electrochemical hybrid composite-based sensor, designed for prostate-specific antigen measurement, was applied to saliva samples containing prostate-specific antigen biomarkers, and signal changes were tested using a photo-electrochemical approach. The sensing mechanism of the MXene-based sensor was based on the synergic interaction between the strong Ni^{2+}/Ni^{3+} redox couple and the high surface area and high conductivity of the hybrid composite with electron transfer between the redox system and the modified electrode. According to a report from the World Health Organization, *Salmonella* is one of the main causes of diarrheal diseases and is the most common bacteria among foodborne pathogenic bacteria. As an example of a *Salmonella* sensor, Wang et al. described a bacteria-imprinted sensor by modifying the electrode with 2D-layered MXene-doped polypyrrole film [46]. The detection limit of the biosensor was found to be 23 CFU/mL of *Salmonella typhimurium* by artificially inoculating the *Salmonella typhimurium* pathogen into the drinking water. In this reported study, it was observed that the surface groups of *Salmonella* interacted with the functional groups of the MXene-based structure, increasing the selectivity of the MXene-doped polypyrrole film-based sensor against *Salmonella* bacteria. However, the selectivity, sensitivity, and response time of these proposed sensors should be improved, and novel MXene-based electrochemical nanosensors should be developed for food/water safety and pathogens monitoring. Regarding a solution to the COVID-19 pandemic problem, Mi et al. reported the sensing performance of the TDs/aptamer cardiac troponin I biosensors based on Au/Ti3C2-MXene for screening COVID-19 patient health with very low LODs of 0.04 fM–0.1 fM [47]. Consequently, the superiority in sensing performance of the MXene-based sensors in a wide spectrum has been proven.

4.8 Future Outlook of Advanced Sensing Materials

An example of a nanosensor developed for trace substance detection, a nano-enzyme system-based sensor, has been developed for the detection of nerve agents used in chemical weapon production [48]. In this respect, it was preferred because it resembles nerve agents. In the sensor system, a nano-enzyme system synthesized according to the photosensitive crosslinking method with ruthenium-based amino acid monomers was used as the sensing layer. The fluorescence spectrum of the obtained nano-enzyme was taken, and it was observed that it showed fluorescence. With this developed nano-enzyme system-based sensor, it is possible to analyze samples containing trace amounts of analytes. In another work, Madden et al. proposed a novel flexible and enzymatic electrochemical sensor using platinum, chitosan, and lactate oxidase-modified laser-scribed graphitic carbon for the detection of enzymatic lactate in real samples (artificial saliva and sterile human serum) [49]. The proposed chronoamperometric biosensor exhibited excellent sensing performance for the lactate biomarker with a high sensitivity of 35.8 $\mu A/mM/cm^2$ and a linear concentration range of 0.2–3 mM with a low LOD of 0.11 mM for the quantification of a non-invasive enzymatic lactate biomarker in biological fluids.

Despite efforts to enzymatically control the amount of analyte added to the test window, sensitive non-enzymatic nanosensors are rapidly gaining momentum with their economic and easy-to-use advantages. For example, Zhang et al. reported ultrasensitive and non-enzymatic electrochemical sensors using a self-assembled polyoxometalate/tris(2,2′-bipyridine) ruthenium (II)/chitosan (organic) and palladium (inorganic) film for the detection of ascorbic acid antioxidant [50]. Thus, they developed a novel non-enzymatic organic-inorganic sensor with high stability, ease of application/use, and low cost compared with enzymatic sensors.

In the literature, various advanced ultra-selective sensors with very low LOD values have been designed, and their sensor performances have been compared to detect the target molecule and illuminate the related problem. Accordingly, the studies conducted in the last ten years on flexible sensors with different morphological and chemical properties, which are more economical, easier to apply, and more sensitive, are evaluated in this chapter. Furthermore, the fabrication methods of large-scale nanostructures-based sensors are gradual and difficult. However, it has been demonstrated that several biological agents can be detected rapidly at the nanomolar level, and research into specific detection in complex environments is gaining momentum. It has been observed that successful experimental data has been obtained in the field of multiplex detection of 2D-layered nano/biosensors for biomedical and environmental applications. Consequently, not enough research has been conducted on flexible-layer nanostructure-based sensors. For this reason, 2D and 3D flexible layered nanostructure-based sensors are being investigated in high-performance sensing systems, which will guide future studies.

References

1. S. Vallejos, I. Gràcia, J. Bravo, E. Figueras, J. Hubálek, C. Cané, Detection of volatile organic compounds using flexible gas sensing devices based on tungsten oxide nanostructures functionalized with Au and Pt nanoparticles, *Talanta*. 139 (2015) 27–34.
2. Z. Shen, F. Liu, S. Huang, H. Wang, C. Yang, T. Hang, J. Tao, W. Xia, X. Xie, Progress of flexible strain sensors for physiological signal monitoring, *Biosensors and Bioelectronics*. 211 (2022) 114298.
3. C. Mu, X. Guo, T. Zhu, S. Lou, W. Tian, Z. Liu, W. Jiao, B. Wu, Y. Liu, L. Yin, X. Jian, Y. Song, Flexible strain/pressure sensor with good sensitivity and broad detection range by coupling PDMS and carbon nanocapsules, *Journal of Alloys and Compounds*. 918 (2022) 165696.
4. Y. Yang, H. Wang, Y. Hou, S. Nan, Y. Di, Y. Dai, F. Li, J. Zhang, MWCNTs/PDMS composite enabled printed flexible omnidirectional strain sensors for wearable electronics, *Composites Science and Technology*. 226 (2022) 109518.
5. R. Qin, X. Li, M. Hu, G. Shan, R. Seeram, M. Yin, Preparation of high-performance MXene/ PVA-based flexible pressure sensors with adjustable sensitivity and sensing range, *Sensors and Actuators A: Physical*. 338 (2022) 113458.
6. G. Gilanizadehdizaj, K.C. Aw, J. Stringer, D. Bhattacharyya, Facile fabrication of flexible piezoresistive pressure sensor array using reduced graphene oxide foam and silicone elastomer, *Sensors and Actuators A: Physical*. 340 (2022) 113549.
7. J. Prakash, P.T. Rao, S. Ghorui, J. Bahadur, V. Jain, K. Dasgupta, Tailoring surface properties with O/N doping in CNT aerogel film to obtain sensitive and selective sensor for volatile organic compounds detection, *Sensors and Actuators B: Chemical*. 359 (2022) 131606.
8. N. Taşaltın, C. Taşaltın, S. Karakuş, A. Kilislioğlu, Cu core shell nanosphere based electrochemical non-enzymatic sensing of glucose, *Inorganic Chemistry Communications*. 118 (2020) 107991.
9. C. Zhao, J. Shen, S. Xu, J. Wei, H. Liu, S. Xie, Y. Pan, Y. Zhao, Y. Zhu, Ultra-efficient trimethylamine gas sensor based on Au nanoparticles sensitized WO3 nanosheets for rapid assessment of seafood freshness, *Food Chemistry*. 392 (2022) 133318.
10. Z. Lotfi, M.B. Gholivand, M. Shamsipur, M. Mirzaei, An electrochemical sensor based on Ag nanoparticles decorated on cadmium sulfide nanowires/reduced graphene oxide for the determination of acyclovir, *Journal of Alloys and Compounds*. 903 (2022) 163912.
11. X. Zhai, Y. Cao, W. Sun, S. Cao, Y. Wang, L. He, N. Yao, D. Zhao, Core-shell composite N-doped-Co-MOF@polydopamine decorated with Ag nanoparticles for nonenzymatic glucose sensors, *Journal of Electroanalytical Chemistry*. 918 (2022) 116491.
12. E. Aparicio-Martínez, A. Ibarra, I.A. Estrada-Moreno, V. Osuna, R.B. Dominguez, Flexible electrochemical sensor based on laser scribed graphene/Ag nanoparticles for non-enzymatic hydrogen peroxide detection, *Sensors and Actuators B: Chemical*. 301 (2019) 127101.
13. J. Liang, Y. Zheng, Z. Liu, Nanowire-based Cu electrode as electrochemical sensor for detection of nitrate in water, *Sensors and Actuators B: Chemical*. 232 (2016) 336–344.
14. P. Karuppaiah, N.S.K. Gowthaman, V. Balakumar, S. Shankar, H.N. Lim, Ultrasonic synthesis of CuO nanoflakes: A robust electrochemical scaffold for the sensitive detection of phenolic hazard in water and pharmaceutical samples, *Ultrasonics Sonochemistry*. 58 (2019) 104649.
15. J. Singh, M. Srivastava, P. Kalita, B.D. Malhotra, A novel ternary NiFe2O4/CuO/FeO-chitosan nanocomposite as a cholesterol biosensor, *Process Biochemistry*. 47 (2012) 2189–2198.

16. Q. Gao, M. Wang, X. Kang, C. Zhu, M. Ge, Continuous wet-spinning of flexible and water-stable conductive PEDOT: PSS/PVA composite fibers for wearable sensors, *Composites Communications*. 17 (2020) 134–140.

17. H.A. Almukhlifi, S. Khasim, A. Pasha, Fabrication and testing of low-cost and flexible smart sensors based on conductive PEDOT-PSS nanocomposite films for the detection of liquefied petroleum gas (LPG) at room temperature, *Materials Chemistry and Physics*. 263 (2021) 124414.

18. D. Lv, W. Shen, W. Chen, R. Tan, L. Xu, W. Song, PSS-PANI/PVDF composite based flexible NH3 sensors with sub-ppm detection at room temperature, *Sensors and Actuators B: Chemical*. 328 (2021) 129085.

19. S. Das, S. Mojumder, D. Saha, M. Pal, Influence of major parameters on the sensing mechanism of semiconductor metal oxide based chemiresistive gas sensors: A review focused on personalized healthcare, *Sensors and Actuators B: Chemical*. 352 (2022) 131066.

20. D. Han, X. Li, F. Zhang, F. Gu, Z. Wang, Ultrahigh sensitivity and surface mechanism of gas sensing process in composite material of combining In2O3 with metal-organic frameworks derived Co3O4, *Sensors and Actuators B: Chemical*. 340 (2021) 129990.

21. J. Xu, J. Ma, Y. Peng, S. Cao, S. Zhang, H. Pang, Applications of metal nanoparticles/metal-organic frameworks composites in sensing field, *Chinese Chemical Letters*. (2022). https://doi.org/10.1016/j.cclet.2022.05.041

22. X. Huang, Z. Gong, Y. Lv, Advances in metal-organic frameworks-based gas sensors for hazardous substances, *TrAC Trends in Analytical Chemistry*. 153 (2022) 116644.

23. N. Kajal, S. Gautam, Efficient nitroaromatic sensor via highly luminescent Zn-based metal-organic frameworks, *Chemical Engineering Journal Advances*. 11 (2022) 100348.

24. X. Fang, Y. Liu, W.-K. Han, X. Yan, Y.-X. Shi, L.-H. Chen, Y. Jiang, J. Zhang, Z.-G. Gu, A luminescent covalent organic framework with recognition traps for nitro-pesticides detection, pH sensing and metal ions identification, *Dyes and Pigments*. 205 (2022) 110507.

25. Q. Li, M. Sun, L. Zhang, H. Song, Y. Lv, A novel Ce(IV)-MOF-based cataluminescence sensor for detection of hydrogen sulfide, *Sensors and Actuators B: Chemical*. 362 (2022) 131746.

26. J. Zhang, L. Gao, B. Chai, J. Zhao, Z. Yang, K. Yang, Electrochemical aptasensor for aflatoxin B1 detection using cerium dioxide nanoparticle supported on iron-porphyrinic metal–organic framework as signal probes, *Microchemical Journal*. 181 (2022) 107716.

27. J. Chen, Y. Chen, S. Li, J. Yang, J. Dong, In-situ growth of cerium-based metal organic framework on multi-walled carbon nanotubes for electrochemical detection of gallic acid, *Colloids and Surfaces A: Physicochemical and Engineering Aspects*. 650 (2022) 129318.

28. L. Jiang, H. Wang, Z. Rao, J. Zhu, G. Li, Q. Huang, Z. Wang, H. Liu, In situ electrochemical reductive construction of metal oxide/metal-organic framework heterojunction nanoarrays for hydrogen peroxide sensing, *Journal of Colloid and Interface Science*. 622 (2022) 871–879.

29. B. Pérez-Fernández, J. v. Mercader, A. Abad-Fuentes, B.I. Checa-Orrego, A. Costa-García, A. de la Escosura-Muñiz, Direct competitive immunosensor for imidacloprid pesticide detection on gold nanoparticle-modified electrodes, *Talanta*. 209 (2020) 120465.

30. S.E.A. Elashery, N.F. Attia, H. Oh, Design and fabrication of novel flexible sensor based on 2D Ni-MOF nanosheets as a preliminary step toward wearable sensor for onsite Ni (II) ions detection in biological and environmental samples, *Analytica Chimica Acta*. 1197 (2022) 339518.

31. X. Huang, Z. Gong, Y. Lv, Advances in metal-organic frameworks-based gas sensors for hazardous substances, *TrAC Trends in Analytical Chemistry*. 153 (2022) 116644.

32. Disha, M.K. Nayak, P. Kumari, M.K. Patel, P. Kumar, Functional nanomaterials based opto-electrochemical sensors for the detection of gonadal steroid hormones, *TrAC Trends in Analytical Chemistry*. 150 (2022) 116571.

33. N. Jayaraman, Y. Palani, R.R. Jonnalagadda, E. Shanmugam, Covalently dual functionalized graphene oxide-based multiplex electrochemical sensor for Hg(II) and Cr(VI) detection, *Sensors and Actuators B: Chemical*. 367 (2022) 132165.

34. J.J.A. Buledi, A.R. Solangi, A. Hyder, M. Batool, N. Mahar, A. Mallah, H. Karimi-Maleh, O. Karaman, C. Karaman, M. Ghalkhani, Fabrication of sensor based on polyvinyl alcohol functionalized tungsten oxide/reduced graphene oxide nanocomposite for electrochemical monitoring of 4-aminophenol, *Environmental Research*. 212 (2022) 113372.

35. X. Hui, M. Sharifuzzaman, S. Sharma, C.I. Park, S. Yoon, D.H. Kim, J.Y. Park, A nanocomposite-decorated laser-induced graphene-based multi-functional hybrid sensor for simultaneous detection of water contaminants, *Analytica Chimica Acta*. 1209 (2022) 339872.
36. J. Chen, G. Zhu, F. Wang, Y. Xu, C. Wang, Y. Zhu, W. Jiang, Design of flexible strain sensor with both ultralow detection limit and wide sensing range via the multiple sensing mechanisms, *Composites Science and Technology*. 213 (2021) 108932.
37. S. Su, S. Chen, C. Fan, Recent advances in two-dimensional nanomaterials-based electrochemical sensors for environmental analysis, *Green Energy & Environment*. 3 (2018) 97–106.
38. P. Panigrahi, M. Sajjad, D. Singh, T. Hussain, J. Andreas Larsson, R. Ahuja, N. Singh, Two-dimensional nitrogenated holey graphene (C2N) monolayer based glucose sensor for diabetes mellitus, *Applied Surface Science*. 573 (2022) 151579.
39. H.S. Hong, N.H. Phuong, N.T. Huong, N.H. Nam, N.T. Hue, Highly sensitive and low detection limit of resistive NO2 gas sensor based on a MoS2/graphene two-dimensional heterostructures, *Applied Surface Science*. 492 (2019) 449–454.
40. M.I. Khan, S.H. Aziz, A. Majid, M. Rizwan, Computational study of borophene/boron nitride (B/BN) interface as a promising gas sensor for industrial affiliated gasses, *Physica E: Low-Dimensional Systems and Nanostructures*. 130 (2021) 114692.
41. G. Congur, Development of a novel methyl germanane modified disposable sensor and its application for voltammetric phenol detection, *Surfaces and Interfaces*. 25 (2021) 101268.
42. W. Huang, Y. Xu, Z. Wang, K. Liao, Y. Zhang, Y. Sun, Dual nanozyme based on ultrathin 2D conductive MOF nanosheets intergraded with gold nanoparticles for electrochemical biosensing of H2O2 in cancer cells, *Talanta*. 249 (2022) 123612.
43. S. Alwarappan, N. Nesakumar, D. Sun, T.Y. Hu, C.Z. Li, 2D metal carbides and nitrides (MXenes) for sensors and biosensors, *Biosensors and Bioelectronics*. 205 (2022) 113943.
44. B. Mohan, S. Kumar, H. Xi, S. Ma, Z. Tao, T. Xing, H. You, Y. Zhang, P. Ren, Fabricated metal-organic frameworks (MOFs) as luminescent and electrochemical biosensors for cancer biomarkers detection, *Biosensors and Bioelectronics*. 197 (2022) 113738.
45. R.A. Soomro, S. Jawaid, P. Zhang, X. Han, K.R. Hallam, S. Karakuş, A. Kilislioğlu, B. Xu, M. Willander, NiWO4-induced partial oxidation of MXene for photo-electrochemical detection of prostate-specific antigen, *Sensors and Actuators B: Chemical*. 328 (2021) 129074.
46. O. Wang, X. Jia, J. Liu, M. Sun, J. Wu, Rapid and simple preparation of an MXene/polypyrrole-based bacteria imprinted sensor for ultrasensitive Salmonella detection, *Journal of Electroanalytical Chemistry*. 918 (2022) 116513.
47. X. Mi, H. Li, R. Tan, B. Feng, Y. Tu, The TDs/aptamer cTnI biosensors based on HCR and Au/Ti3C2-MXene amplification for screening serious patient in COVID-19 pandemic, *Biosensors and Bioelectronics*. 192 (2021) 113482.
48. G.Ü. Fen, B. Dergisi, O. Yağmuroğlu, S.E. Diltemiz, Development of nano ache enzyme based sensor for determination of nerve agents used in chemical weapons, *Gazi University Journal of Science, Part C*. 8 (2020) 205–223.
49. J. Madden, E. Vaughan, M. Thompson, A. O' Riordan, P. Galvin, D. Iacopino, S. Rodrigues Teixeira, Electrochemical sensor for enzymatic lactate detection based on laser-scribed graphitic carbon modified with platinum, chitosan and lactate oxidase, *Talanta*. 246 (2022) 123492.
50. L. Zhang, S. Li, K.P. O'Halloran, Z. Zhang, H. Ma, X. Wang, L. Tan, H. Pang, A highly sensitive non-enzymatic ascorbic acid electrochemical sensor based on polyoxometalate/tris(2,2′-bipy ridine)ruthenium (II)/chitosan-palladium inorganic-organic self-assembled film, *Colloids and Surfaces A: Physicochemical and Engineering Aspects*. 614 (2021) 126184.

5

Recent Advances in Flexible and Wearable Sensors

Özgür Yasin Keskin, Erdem Tevfik Özdemir, Mustafa Erol, and Aylin Altınışık Tağaç

CONTENTS

5.1 Introduction

Flexible and wearable sensors, which combine materials science, physics, mechanics, and other disciplines, have undergone significant advances recently. Their ability to fit many surfaces and mechanical deformations without notable changes in their properties makes them attractive in many applications. Different types of sensors have been used depending on the type of application. They have great potential in many application areas such as biomedicine, soft robotics, sports monitoring, virtual reality, and healthcare. While they have been used for different purposes, demand for more sensitive, flexible, and reliable sensors has increased. The sensing performance and flexibility of sensors are generally dominated by sensing material and design. Within this framework, researchers are trying to improve their properties by using different types of matrix and sensing materials and designs. Recently, the usage of inorganic materials is a new trend to improve the flexibility of sensors. Also, developing composite structures with inorganic materials such as metal and semiconductors is another approach for increased flexibility. Additionally, carbon-based materials have attracted more attention due to their excellent electrical and mechanical properties and suitability with flexible structures [1].

In this chapter, recent advances in flexible and wearable sensors are reviewed. Within this context, recently used materials, design approaches, and applications of wearable and flexible sensors are summarized using their design and development. Initially, the types of sensors will be classified according to the materials used in the applications. Second, it will be shown how the materials used are improved and modified depending on material type. Finally, the applications of flexible and wearable sensors are summarized.

DOI: 10.1201/9781003299455-5

5.2 Biomedical Sensors

Conventional healthcare systems for monitoring the physiological parameters of patients require complex, invasive, and costly devices that restrict the daily life of patients due to the requirements of continuous monitoring. To overcome the drawbacks of these systems and the need for more suitable and easy implantable devices for monitoring physiological signals, researchers have made efforts to miniaturize and develop more suitable devices. Flexible and wearable sensors have great potential for biomedical applications with their ability to fit all surfaces without any characteristic change in their structure. Different types of biomedical sensors for different analytes have been developed [2,3]. Gao et al. developed a sensing system for simultaneously monitoring glucose, lactic acid, sodium, potassium, and skin temperature. Figure 5.1a shows the wearable flexible integrated sensing array. An integrated sensor was designed on polyethylene terephthalate (PET) to give sensor flexibility. Sensor arrays patterned by a photolithographic process and a microcontroller were used for wireless communication [4].

Koo et al. formed a wearable/implantable sensor with a display for mobile health monitoring. The sensor structure consists of carbon nanotube electronics and color-tunable organic light-emitting diodes (Figure 5.1b). Researchers aimed to fabricate wearable electrocardiograms for heart rhythm monitoring. They form a layered structure with a charge-blocking layer between emitting layers. Color changes were obtained by using increasing voltages. To simulate heart rate in the device, a carbon nanotube (CNT)-based amplifier is used to amplify the heart rate to desired voltages to obtain color changes in the flexible display. This device can detect and display electrocardiography (ECG) signals simultaneously with color changes according to variable heart rates [5]. Wearable and flexible sensors for cardiovascular diseases were constructed by Pang et al. [6]. They form a wearable multilayered capacitive pressure sensor with a pyramidal-shaped polydimethylsiloxane (PDMS) structure (Figure 5.1c). They demonstrated a potential wearable sensor that is capable of detecting weak signals from blood vessels with an increased aspect ratio of micro hair structure amplification of signals. This device can be enabled to detect early cardiovascular disease according to the waveform obtained from patients [6]. Luo et al. designed a flexible e-skin sensor with multimodal sensing properties. The designed sensor consists of polyaniline hollow nanospheres composite films (PANI-HNSCF) (Figure 5.1d). The authors used multi-walled carbon nanotubes to construct conductive pathways in composite structures. The fabricated sensor showed superior flexibility and mechanical stability. The flexible sensor is suitable for human skin and it can be used for real-time monitoring of heart rate, temperature, and arterial blood pressure [7].

5.3 Biosensors

One of the most effective methods of treating illnesses is the timely diagnosis of disease. Thus, flexible and wearable biosensors have received appreciable attention because of their real-time continuous monitoring capability, quick response, non-invasiveness, and compatibility with flexible substrates. Unlike invasive conventional silicon-based biosensors, flexible and wearable biosensors are preferred in epidermal and implantable medical

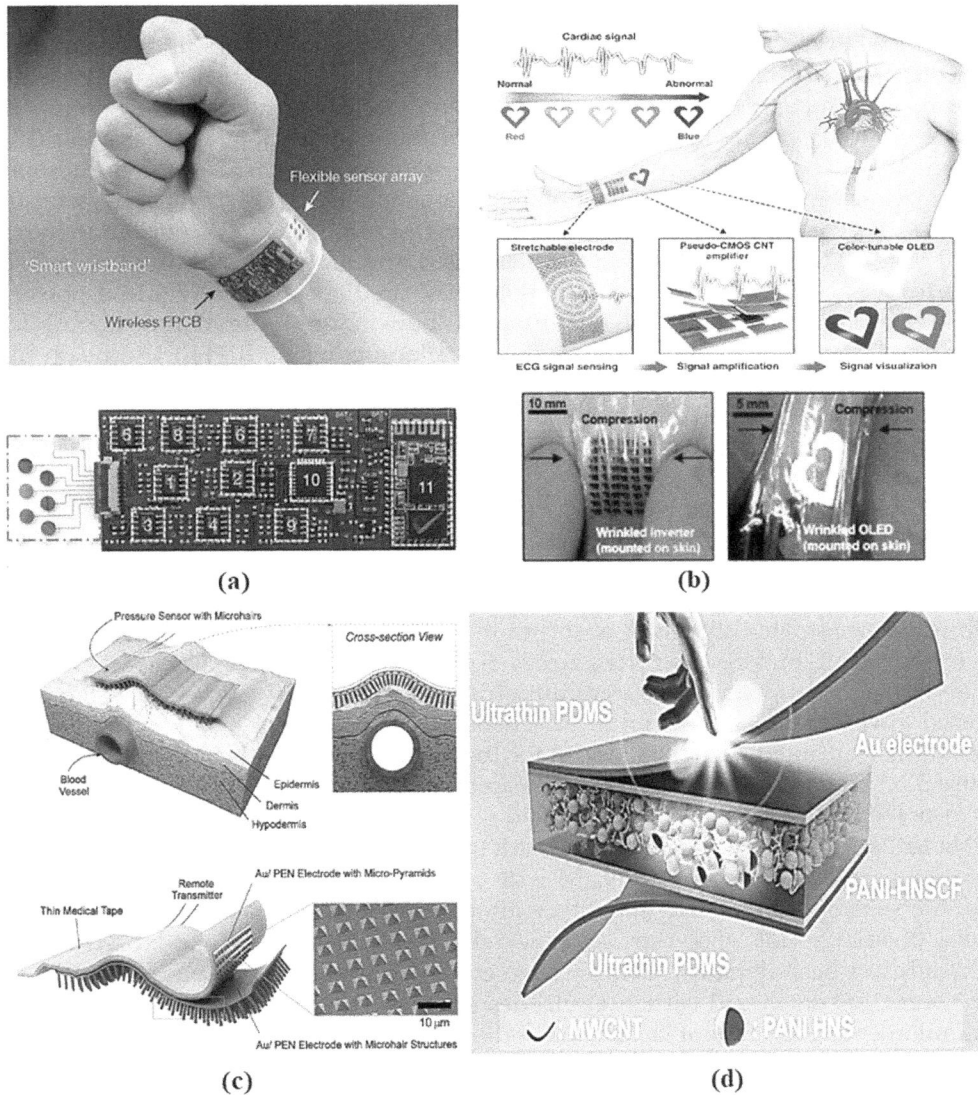

FIGURE 5.1
a) Structure of flexible printed circuit board. Adapted with permission [4]. Copyright (2016), Springer Nature. b) Presentation of real-time monitoring of the cardiac system. Adapted with permission [5]. Copyright (2017), American Chemical Society. c) Schematic illustration and cross-section of the designed sensor and sensor structure. Adapted with permission [6]. Copyright (2014), Wiley. d) Designed wearable and flexible e-skin sensor structure. Adapted with permission [7]. Copyright (2017), Elsevier.

applications because of their ability to make *in situ*, accurate, and early detection of biomarkers including electrolytes (e.g., Ca^{2+}, Na^+, and pH) and molecules (e.g., lactate, glucose, and alcohol) in body fluids. In recent years, the suitability and sensitivity of these biosensors have been improved by using various nanomaterials such as reduced graphene oxide (rGO), metal nanostructures, and different designs. As of late, various types of flexible and wearable body sensors have been developed to improve the quality of life in humans [8,9].

Mishra et al. developed a flexible and skin-worn potentiometric biosensor that has a tattoo form for G-type nerve agent detection (Figure 5.2a). The authors first screen-printed the electrodes that were coated with polyaniline (PANI) as a pH-sensitive layer on a tattoo paper and subsequently compacted them with an electronic interface for wireless data transmission. To provide surface distribution of the analyte, poly(vinyl alcohol) (PVA)-acrylamide hydrogel was covered over the transducer surface. Diisopropyl fluorophosphate (DFP) was used as a fluorine-containing organophosphates (OP) nerve agent. This epidermal biosensor exhibits good sensitivity and fast response in the concentration range of DFP from 10 to 120 mM [10]. Su et al. fabricated a wearable self-powered acetone biosensor for a painless method for the diagnosis of prediabetes. The required power for this biosensor is generated by the breathing gas flow-induced vibration of the triboelectric charged membranes that consist of polytetrafluoroethylene (PTFE) and nylon films. They developed chitosan and reduced graphene oxide (rGO) composite film-based sensors showing an excellent sensing response of 27.89% under 97.3% relative humidity at room temperature, which is approximately five times higher than the sensitivity of pure chitosan film-based devices [11].

Kim et al. studied a wearable biosensor consisting of nanofibrous hydrogel for determining the glucose level in blood. The glucose biosensor was produced through electrospinning of PVA/1,2,3,4-butanetetracarboxylic acid (BTCA)/β-cyclodextrin (β-CD)/glucose oxidase (Gox)/gold nanoparticle (AuNP) nanofibrous hydrogels, which showed flexibility, biocompatibility, good absorptivity, mechanical strength, and high enzyme activity (Figure 5.2b). Moreover, the glucose biosensor exhibited a wide linear range, high sensitivity, high rapid time, and a minimum sensing limit of 0.01 mM [12]. Zhang et al. reported a wearable electrochemical biosensor based on silver nanowires (AgNWs) and molecularly imprinted polymers (MIPs) for the determination of lactate in human sweat (Figure 5.2c). The authors fabricated MIP based on 3-aminophenylboronic acid (3-APBA) electro-polymerization over AgNWs coated on the screen-printed carbon electrode. The flexible MIPs-AgNWs exhibited high sensitivity and specificity for detecting lactate concentrations ranging from 10^{-6} to 0.1 M with a detection limit of 0.22 µM. On the other hand, the sensor showed high stability with a 99.8% ± 1.7% of sensitivity during seven months of storage [13]. A non-invasive wearable electrochemical sweat biosensor based on threads has been demonstrated by Zhao et al. (Figure 5.2d). The authors synthesized ZnO nanowires on thread carbon electrodes by low-temperature hydrothermal growth for simultaneous detection of lactate and sodium ions in human sweat in linear ranges of 0–25 mM and 0.1–100 mM, respectively. The limits of detection of the biosensors are 3.61 mM and 0.16 mM, and the sensitivities of the biosensors are 0.94 mV/decade and 42.9 mV/decade, respectively [14].

5.4 Strain Sensors

Demand for strain sensors has increased in the last decades because of their applicability in many areas such as health monitoring, smart garnets, and robotics. Particularly with the improvement in flexibility and sensitivity of strain sensors, they have started to be used in human bodies for monitoring activities by attaching strain sensors onto clothing or integrating them into electronic devices. To increase the sensitivity of sensors, more recently, different types of materials and approaches have been developed. Carbon nanomaterials

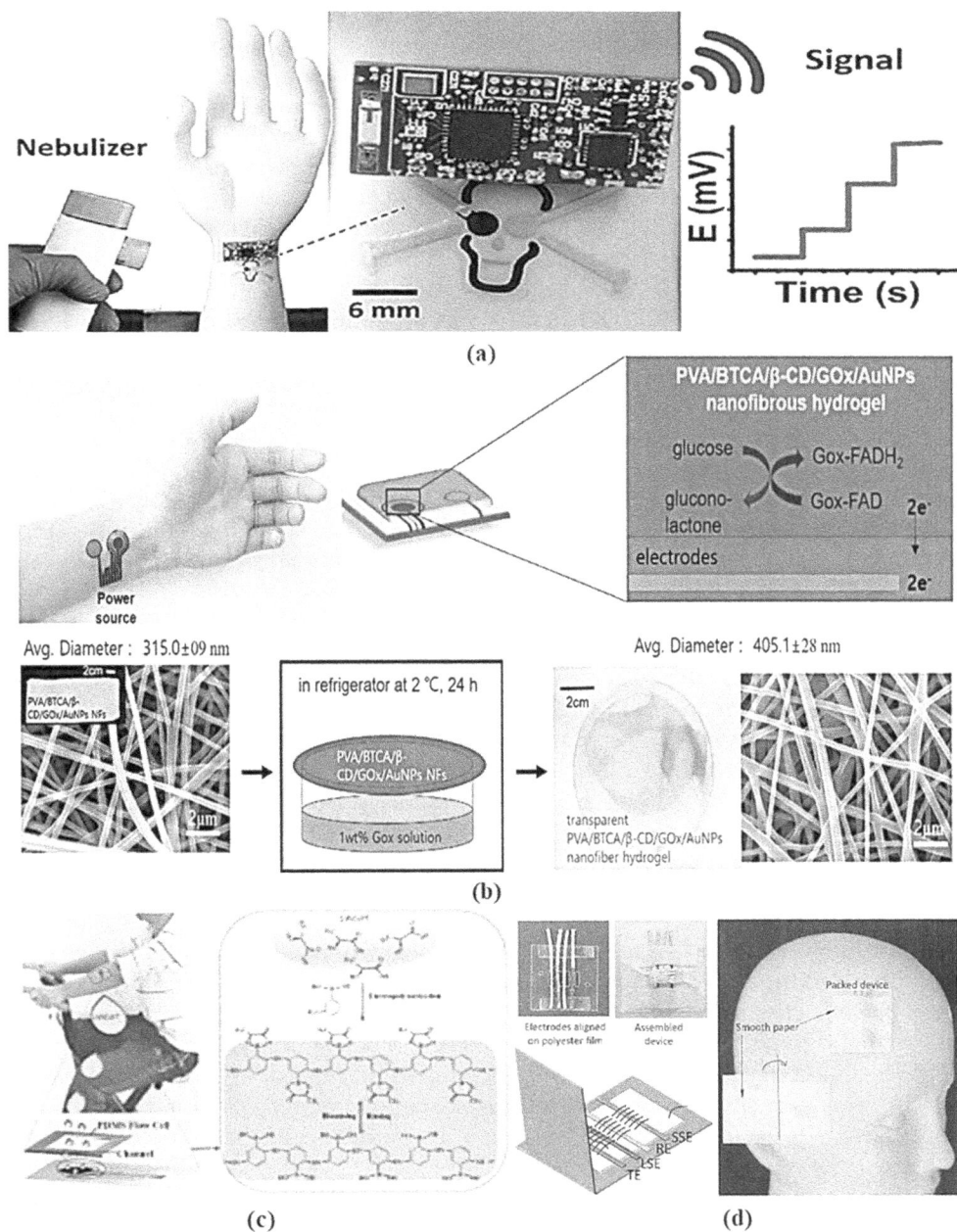

FIGURE 5.2
a) Skin-adapted potentiometric biosensor system. Adapted with permission [10]. Copyright (2018), Elsevier. b) Illustration of real-time monitoring of the glucose sensors – SEM images of transparent nanofibrous hydrogels. Adapted with permission [12]. Copyright The Authors, some rights reserved; exclusive licensee Nature. Distributed under a Creative Commons Attribution License 4.0 (CC BY). c) Silver nanowires as electrochemical biosensors with MIPs. Adapted with permission [13]. Copyright (2020), Elsevier. d) Illustration of the biosensor patch. Adapted with permission [14]. Copyright (2021), Elsevier.

are considered next-generation sensing materials, given excellent conductivity, chemical, thermal stability, and high sensitivity. Carbon nanomaterial-based strain sensors have overcome the limitations of traditional strain sensors such as low sensitivity, linearity, and durability. On the other hand, metal oxides give functionalities to sensors such as temperature monitoring and pH sensing [15,16]. Xu et al. generated a strain sensor using Ecoflex rubber filled with rGO (reduce graphene oxide)/DI (deionized water) mixture. The authors used Ecoflex to template a wire in the system and located a cylindric cavity filled with a conductive rGO/DI mixture. The produced sensor is sensitive to both stretching and compressing with a high sensitivity range from 0.1% to 400% strain. The sensor also showed a 31.6 gauge factor (GF) with very high stability (>10,000 cycles for stretching and >15,000 cycles for compression) (Figure 5.3a). On the other hand, the sensor is capable of distinguishing compression and tensile motions with decreasing and increasing resistivity change, respectively. Also, templating a wire to form a cavity enables a change of conductive material when the sensor loses its sensing properties [17].

Eom et al. reported multifunctional strain sensors using poly(3,4-ethylenedioxythiophene) (PEDOT)-coated fibers (Figure 5.3b). The authors used the PEDOT layer as a conducting layer on polyester fiber, and they applied a fabricated strain sensor to the glove to demonstrate the designed sensor can be used in textiles as a wearable human-computer interface device. Fabricated devices are also designed to express sign languages by adjusting predefined hand gestures. The changing resistance with different fingers, which is predefined by the user interface (UI) system, changes the letters on the screen to make a word [18]. Xia et al. used carbonized Chinese art paper to make a low-cost strain sensor. The authors carbonized the paper in a furnace to obtain conductive films and cut them into strips and placed them onto PDMS substrates. Encapsulations were also carried out with liquid PDMS (Figure 5.3c). The fabricated strain sensor is capable of giving a response on a large scale from 0.01% to 120%. Also, sensors show good sensitivity with a GF of 68 at 100% strain and 248 at 100–120% strain. The designed sensor is also able to distinguish facial expressions and phonations [19]. Wang et al. fabricated a strain sensor using gold nanowires. The authors used an elastomeric substrate to grow Au nanowires and the obtained structure is shown in Figure 5.3d. The designed sensor showed non-linear properties that can be stretched up to 900% and can be nearly 93% stable and conductive after 2000 stretching and release cycles. The authors also designed a mobile device interface to detect childhood autism disorder and demonstrated the capability of the sensor for real-time facial expressions monitoring. The authors showed that the fabricated sensor is capable of distinguishing different facial expressions such as anger, fear, and surprise [20]. Xu et al. designed a highly sensitive and biocompatible capacitive-type strain sensor for monitoring physiological signals. Within this scope, the authors used a mixture of ionic hydrogel and metallic nanofibers. The usage of Ag nanofibers in the structure increased the sensitivity of the sensor by three times. The sensor exhibited very high stretchability up to 1000% with a GF of 165 (Figure 5.3e). Also, the authors applied the sensors in the body for different actions such as breathing and speaking [21].

5.5 Environment Sensors

Environmental pollution and the reckless use of resources, which have increased worldwide in recent years, seriously affect human health and the environment. To control and

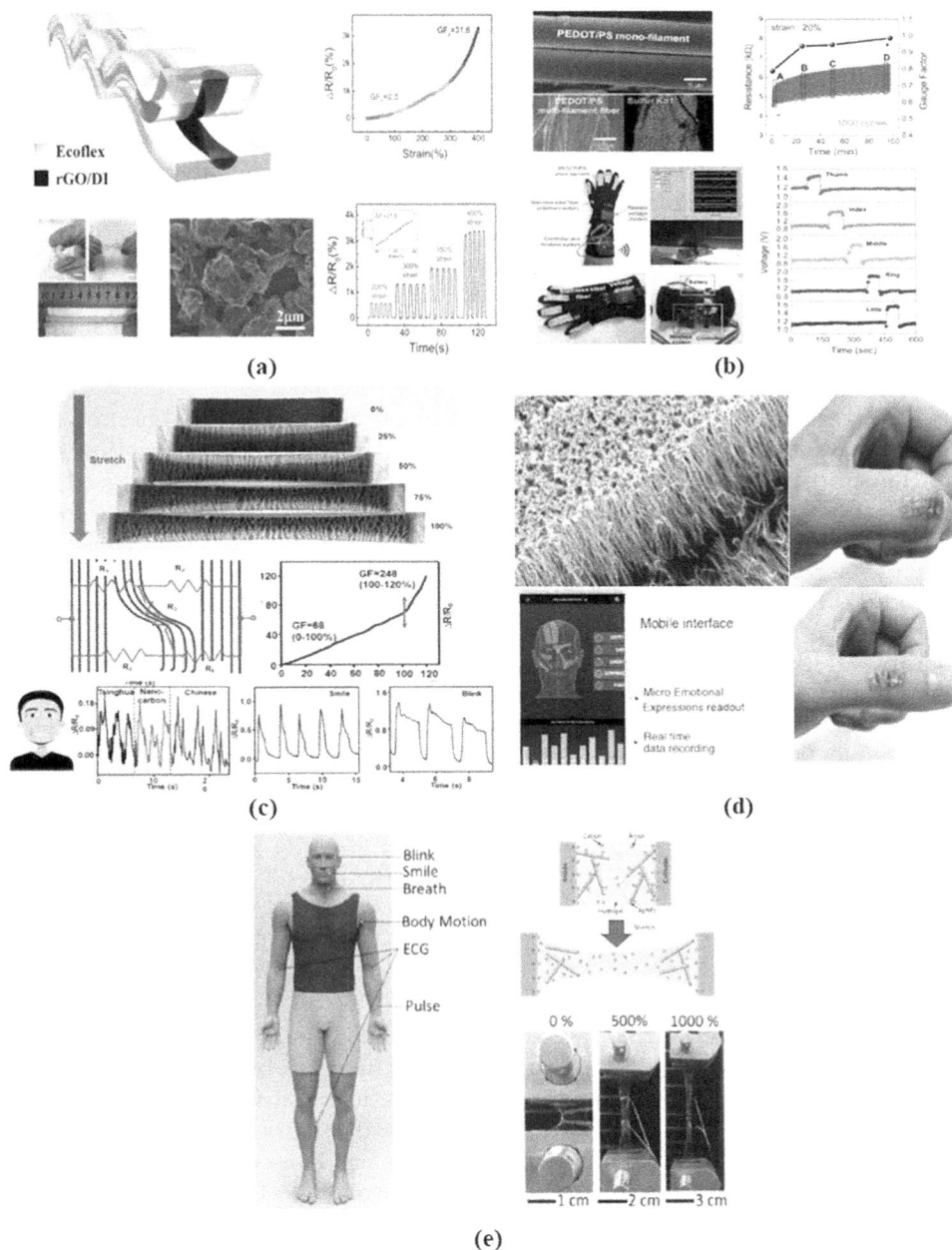

(a)

(b)

(c)

(d)

(e)

FIGURE 5.3
a) Structure of sensor. Adapted with permission [17]. Copyright (2017), Royal Society of Chemistry. b) Images of the textile-based sensor. Adapted with permission [18]. Copyright (2017), American Chemical Society. c) Carbonized strain sensor working mechanism. Adapted with permission [19]. Copyright The Authors, some rights reserved; exclusive licensee American Chemical Society. Distributed under a Creative Commons Attribution License 4.0 (CC BY). d) SEM image of enokitake-like nanowire structure. Adapted with permission [20]. Copyright The Authors, some rights reserved; exclusive licensee American Chemical Society. Distributed under a Creative Commons Attribution License 4.0 (CC BY). e) Working mechanism and stretchability of sensor under different strain rates. Adapted with permission [21]. Copyright (2019), Royal Society of Chemistry.

measure the quality of the surrounding environment in which people live, environmental sensors have been developed. In particular, flexible and wearable environmental sensors are used for the easy and timely monitoring of harmful chemicals to prevent potential hazards and threats to environmental protection, food processing, medical diagnosis, and so on. To increase the sensitivity of these sensors, various new designs have been developed with flexible/stretchable 2D materials, polymers, and biomaterials instead of traditional metal or semiconductor materials [22,23]. Ko et al. developed a biodegradable, flexible silicon-based NOx gas detection system. The nanomembrane consisted of a magnesium electrode and a semi-permeable membrane (silicon dioxide and PDMS) (Figure 5.4a). Under 5 ppm NO_2 and RT conditions, the environmental sensor exhibited an excellent sensitivity of 136 Rs. While the sensor has lower response (30 s) and recovery (60 s) times, the response signal was also proportional to the decrease in concentration from 5 to 0.1 ppm of NO_2 gas, obtaining the lowest limit of detection of ~20 ppb [24].

Chen et al. constructed a wearable gas sensor inspired by an artificial neuron network. CuS quantum dots and Bi_2S_3 nanosheets were used as the main adsorption sites and charge transport pathways, respectively (Figure 5.4b). Thanks to the improved sensitivity, the sensor showed a response time of 18 s, a recovery rate of 338 s, and a theoretical limit of detection of 78 ppb [25]. A wearable non-enzymatic electrochemical sensor was reported by Paschoalin et al. The screen-printing technique was chosen for the fabrication of electrodes on solution-blown spun poly (lactic acid) (PLA) mats for on-site non-enzymatic detection of pesticides on the flat, curved, and irregular surfaces of agricultural and food samples (Figure 5.4c). Detection of carbendazim and diquat pesticides in apple and cabbage samples was performed with detection limits of 43 nM and 57 nM, respectively [26]. Zhang et al. presented an ionic polyimine network (IPIN)-based electrochemical sensor for the detection of exogenous volatile iodine. The sensor, which included malleable, self-reparable, and recyclable conductive polymers, can be combined with flexible devices. The synthesis and designed structure of IPIN films are illustrated in Figure 5.4d [27].

5.6 Photosensors

Photosensors (PD) are devices that can convert detected lights into electrical signals. They have been used in many daily life applications such as imaging, biomedical, solar cells, and telecommunications. Introducing photosensors to flexible and wearable forms is a great challenge for scientists. Conventional PDs are generally fabricated based on inorganic semiconductors, which have some drawbacks such as brittleness, rigidity, and expansiveness. Non-invasive and flexible integration of photosensors to the human body requires new materials and designs for sustainable performance under severe mechanical deformations. To overcome the drawbacks of conventional materials, organic semiconductors are promising candidates to make photosensors suitable for wearable and flexible applications. Recently, many different materials and strategies have been demonstrated [28,29].

Jang et al. designed a flexible photodetector using CNT conductive layers. In conventional photodetectors, indium-tin-oxide (ITO) is generally used as an electrode, but the brittle nature of metal oxides restricts its practicality in wearable applications. To exceed this disadvantage of metal oxide, the authors replaced the ITO with a CNT transparent electrode. To show the effectiveness of the CNT electrode, the authors compared the CNT electrode-based flexible photosensor with ITO and PEDOT: PSS-based photosensors. According to

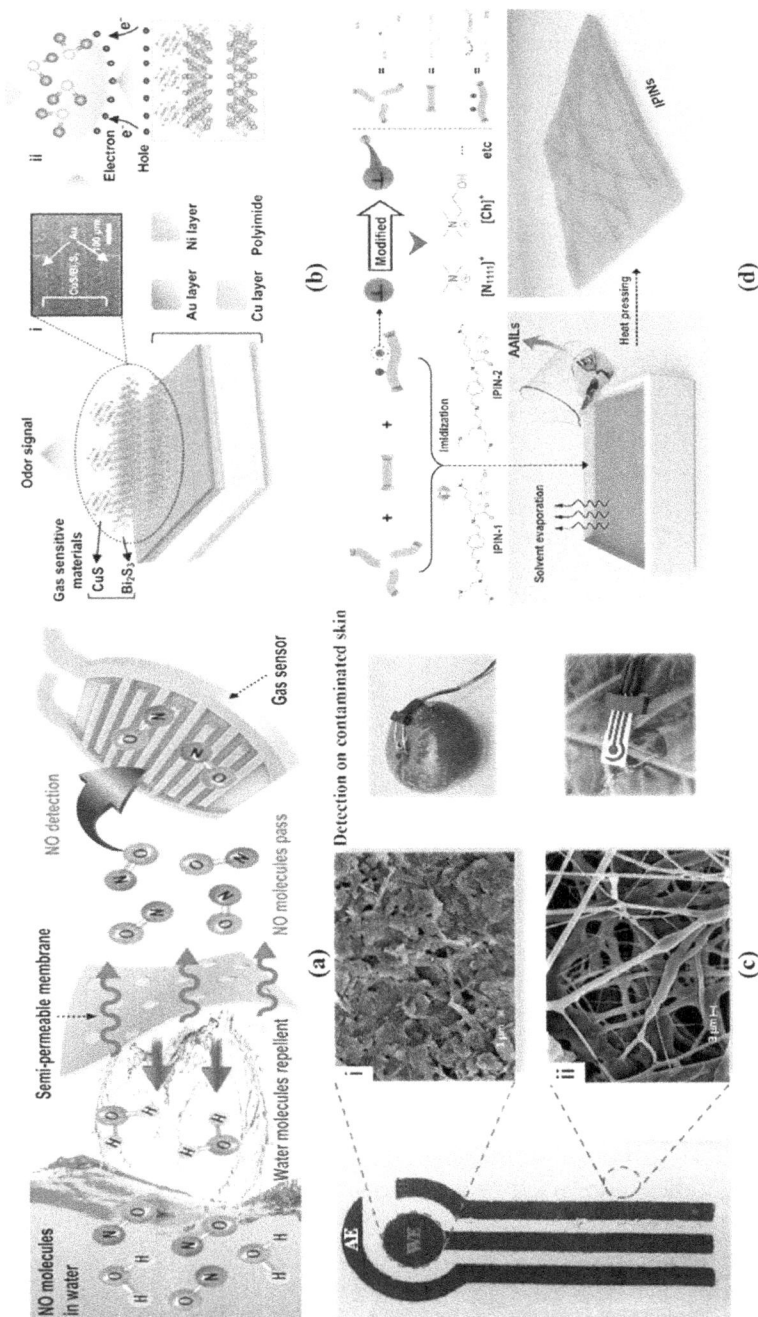

FIGURE 5.4

a) PDMS membranes as gas sensors. Adapted with permission [24]. Copyright The Authors, some rights reserved; exclusive licensee American Association for the Advancement of Science. Distributed under a Creative Commons Attribution License 4.0 (CC BY). b) Schematics of wearable gas sensors based on artificial networks. Adapted with permission [25]. Copyright The Authors, some rights reserved; exclusive licensee American Chemical Society. Distributed under a Creative Commons Attribution License 4.0 (CC BY). c) SEM images of the under various magnifications, and pesticide detection on apple and cabbage skin. Adapted with permission [26]. Copyright (2022), Elsevier d) Flowchart for the synthesis of IPIN films. Adapted with permission [27]. Copyright (2022), American Chemical Society.

the results, electrodes showed 90% transparency, and very high detectivity was achieved in 2.07×10^{14} jones, which is 100 times higher than ITO-based photosensors (Figure 5.5a). Also, the CNT electrode base photosensor exhibited 100,000 times higher dark current density than ITO-based sensors [30]. An et al. fabricated a flexible reduced graphene oxide (rGO)–zinc oxide-based photodetector. With different production approaches, the authors achieved hierarchical hybrid nanocomposites. Figure 5.5b represents the single-step production of photosensors. First, GO nanosheet aqueous suspension was mixed with zinc oxide precursor (zinc acetate dihydrate) and stirred to obtain a homogeneous mixture. The uniform mixture was coated onto a flexible substrate and femtosecond laser direct writing was performed to obtain rGO and zinc oxide simultaneously. The fabricated structure is presented in Figure 5.5b. The authors aimed to use fabricated devices for monitoring UV light exposure to human skin. As illustrated in Figure 5.5b, increasing UV light intensities increased photocurrent densities showing the facile usage of the designed photosensor in the human body for monitoring UV exposure. In addition, the authors demonstrated efficient and multiscale manufacturing methods for flexible and wearable photosensors [31].

Zhang et al. designed a graphene-based flexible and transparent photosensor. The authors created a graphene/PbI_2/graphene structure, where graphene films were used as transparent electrodes and PbI_2 was used as a photoactive layer and fabricated sensors were encapsulated with polyethylene terephthalate (PET) (Figure 5.5c). Transparent photosensors exhibit high responsivity (45 A/W), high-resolution image capability (1 μm), and fast response (35 μm rise and 20 μm decay time). This study demonstrated that carbon derivatives are proper candidates as the electrode in flexible sensors instead of ITO electrode layers [32]. Li et al. fabricated a layered textile-based photosensor using carbon fiber fabric (CFF). The authors proposed a self-powered mechano-optical communication system based on a Te@TeSe nanowire structure, which is capable of converting near-infrared radiation (NIR) signals to electrical signals (Figure 5.5d). The authors designed a photosensor that can encode NIR light and modulate robotic actions. These functions are facile methods for manipulating robots remotely and quickly. Also, the authors aimed to use the fabricated sensor in textiles for optical communication [33].

5.7 Humidity Sensors

In general, humidity is defined as vapor concentration in the environment and an important parameter for biological life. Measurement, detection, and control of humidity are important for environmental, agricultural, and healthcare applications. An expectation for the constant measurement of humidity under different environments and conditions increases the demand for more sensitive, flexible, and non-invasive humidity sensors. Recently, wearable and flexible humidity sensors gained attention due to different applications such as healthcare and human-machine interfaces. To achieve durable, sensitive, and flexible humidity sensors, several types of materials, including metal oxides, carbon nanomaterials, ceramics, and polymers, have been used extensively [34,35]. Gong et al. demonstrated a flexible and wearable humidity sensor for human respiratory detection and skin dryness monitoring, which is self-powered by a motion-driven alternator (Figure 5.6a). The sensor was fabricated through screen-printing of a cerium oxide/graphitic carbon nitride (CeO_2/g-C_3N_4) solution with a film thickness of 100 μm. In the relative humidity range of 0–97%, the response value of the sensor reached 6573. Compared with pristine

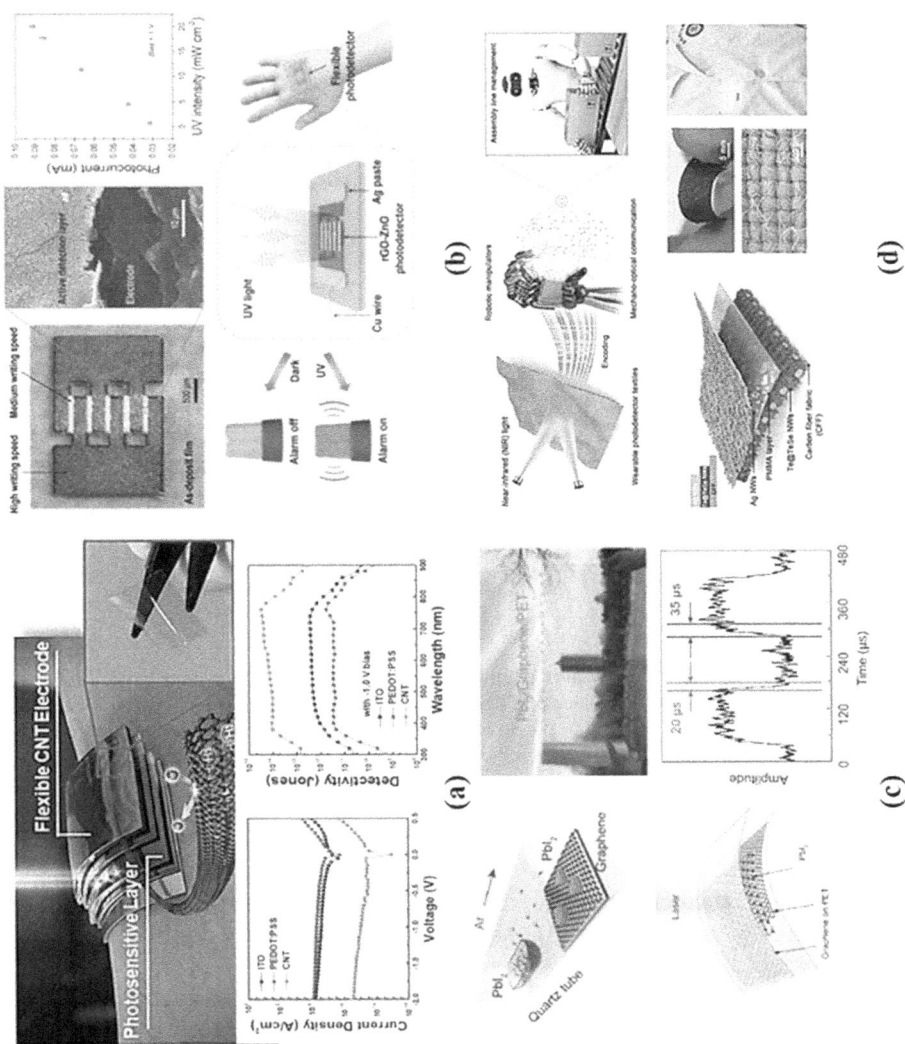

FIGURE 5.5

a) Layered sensor structure. Adapted with permission [30]. Copyright (2021), Elsevier. b) Fabricated sensor and SEM image of an electrode and active detection layer. Adapted with permission [31]. Copyright The Authors, some rights reserved; exclusive license Wiley. Distributed under a Creative Commons Attribution License 4.0 (CC BY). c) Illustration of the growing process of base material and an image of the sensor. Adapted with permission [32]. Copyright (2018), Wiley. d) Designed sensor and its layered structure. Adapted with permission [33]. Copyright (2021), Wiley.

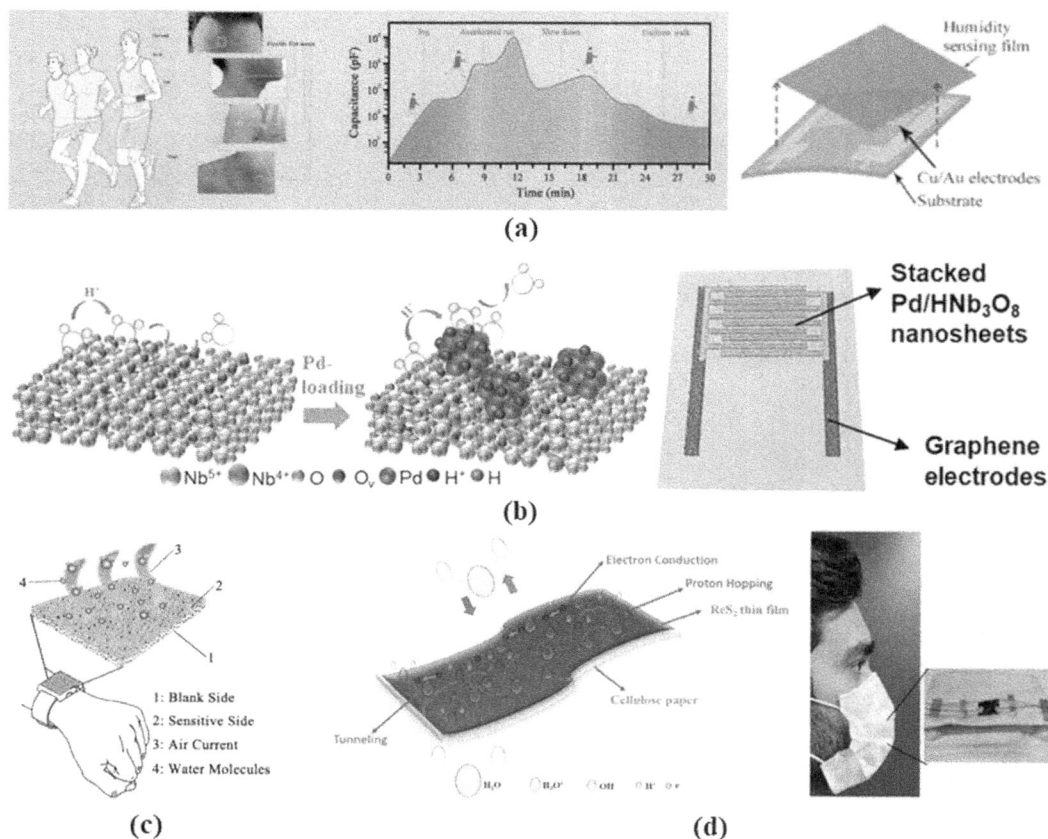

FIGURE 5.6
a) CeO$_2$/g-C$_3$N$_4$ hybrid sensor application on the human body and capacitance changes over time. Adapted with permission [36]. Copyright (2022), Royal Society of Chemistry. B) Presentation of the chemical structure of HNb$_3$O$_8$ and Pd-modified HNb$_3$O$_8$, and the illustration of the Pd/HNb$_3$O$_8$ humidity sensor. Adapted with permission [37]. Copyright (2021), Royal Society of Chemistry. C) Schematic demonstration of the working principle of the single-sided integrated flexible humidity sensor. Adapted with permission [38]. Copyright (2022), Elsevier. d) ReS$_2$-based wearable and flexible sensor structure. Adapted with permission [39]. Copyright (2021), Royal Society of Chemistry.

humidity sensors, which used CeO$_2$ or g-C$_3$N$_4$ materials, this sensor has excellent properties such as high sensitivity (959.5 pF/% RH), fast responsiveness, high stability, and low recovery time [36].

Lu et al. developed a highly stable, flexible, and wearable humidity sensor for dehydration diagnosis. The authors investigated the advantages of Pd-modified HNb$_3$O$_8$ nanosheets on the sensor, which showed superior stability and reproducibility for over 100 h of cycle tests at a high humidity range (>70% RH) (Figure 5.6b). Moreover, a fast response time (0.2 s) and recovery time (3 s) were obtained via this sensor, which is comparable to the traditional humidity sensors [37]. Zhao et al. reported a harmless and breathable flexible single-sided integrated flexible humidity sensor (Figure 5.6c). They unilaterally deposited polyaniline (PANI) on hydrophobic poly (vinylidene fluoride) (PVDF) microporous membrane, which has the advantages of non-toxicity, being lightweight, good chemical stability, flexibility, and air permeability. Additionally, the sensor remained stable in humidity responsiveness under bending deformation [38]. Adepu et al. demonstrated a flexible, cost-effective, and environmentally friendly humidity sensor for real-time applications (respiration monitoring, skin hydration, and baby diaper wetting). The researchers effectively deposited a flexible ReS2 thin film on a decomposable cellulose paper and connected the film with a Cu wire (Figure 5.6d). The sensor displayed a response time of 142.94 s and a sensitivity of 18.26 [39].

5.8 Conclusion

Some prominent works about recent advances in flexible and wearable sensors were evaluated in this chapter. Different types of sensors with a variety of advanced materials have been used to make wearable sensors more durable, sensitive, and adaptive to the human body and textiles. Recent studies show that the need for flexible properties of sensors forced researchers to design different structures and materials. Also, studies have shown that they have been used in many application areas from virtual reality to healthcare, and will practicality be used in different applications such as controlling robots and converting human motions to words. For healthcare applications in particular, wearable sensors can change our perspective on the early diagnosis and tracking of diseases. From this, it can be expressed that in the near future, flexible and wearable sensors with higher capabilities of detection and a wide range of application areas will be a part of our daily life once the challenges in the transfer of know-how to scale production are accomplished.

References

1. R.K.Y. Tong, *Wearable technology in medicine and health care*, 2018. https://doi.org/10.1016/C2016-0-01277-9.
2. D. Vilela, A. Romeo, S. Sánchez, Flexible sensors for biomedical technology, *Lab on a Chip*. 16 (2016) 402–408. https://doi.org/10.1039/c5lc90136g.
3. N. Zhou, T. Liu, B. Wen, C. Gong, G. Wei, Z. Su, Recent advances in the construction of flexible sensors for biomedical applications, *Biotechnology Journal*. 15 (2020). https://doi.org/10.1002/biot.202000094.
4. W. Gao, S. Emaminejad, H.Y.Y. Nyein, S. Challa, K. Chen, A. Peck, H.M. Fahad, H. Ota, H. Shiraki, D. Kiriya, D.H. Lien, G.A. Brooks, R.W. Davis, A. Javey, Fully integrated wearable sensor arrays for multiplexed in situ perspiration analysis, *Nature*. 529 (2016) 509–514. https://doi.org/10.1038/nature16521.
5. J.H. Koo, S. Jeong, H.J. Shim, D. Son, J. Kim, D.C. Kim, S. Choi, J.I. Hong, D.H. Kim, Wearable electrocardiogram monitor using carbon nanotube electronics and color-tunable organic light-emitting diodes, *ACS Nano*. 11 (2017) 10032–10041. https://doi.org/10.1021/acsnano.7b04292.
6. C. Pang, J.H. Koo, A. Nguyen, J.M. Caves, M.G. Kim, A. Chortos, K. Kim, P.J. Wang, J.B.H. Tok, Z. Bao, Highly skin-conformal microhairy sensor for pulse signal amplification, *Advanced Materials*. 27 (2015) 634–640. https://doi.org/10.1002/adma.201403807.
7. Z. Lou, S. Chen, L. Wang, R. Shi, L. Li, K. Jiang, D. Chen, G. Shen, Ultrasensitive and ultraflexible e-skins with dual functionalities for wearable electronics, *Nano Energy*. 38 (2017) 28–35. https://doi.org/10.1016/j.nanoen.2017.05.024.

8. A. Salim, S. Lim, Recent advances in noninvasive flexible and wearable wireless biosensors, *Biosensors and Bioelectronics*. 141 (2019) 111422. https://doi.org/10.1016/J.BIOS.2019.111422.

9. G. Ge, W. Huang, J. Shao, al-, H. Hussin, N. Soin, S. Fatmadiana Wan Muhamad Hatta, Y. Wang, B. Yang, Z. Hua, J. Zhang, P. Guo, D. Hao, Y. Gao, J. Huang, Recent advancements in flexible and wearable sensors for biomedical and healthcare applications, *Journal of Physics D: Applied Physics*. 55 (2021) 134001. https://doi.org/10.1088/1361-6463/AC3C73.

10. R.K. Mishra, A. Barfidokht, A. Karajic, J.R. Sempionatto, J. Wang, J. Wang, Wearable potentiometric tattoo biosensor for on-body detection of G-type nerve agents simulants, *Sensors and Actuators, B: Chemical*. 273 (2018) 966–972. https://doi.org/10.1016/j.snb.2018.07.001.

11. Y. Su, T. Yang, X. Zhao, Z. Cai, G. Chen, M. Yao, K. Chen, M. Bick, J. Wang, S. Li, G. Xie, H. Tai, X. Du, Y. Jiang, J. Chen, A wireless energy transmission enabled wearable active acetone biosensor for non-invasive prediabetes diagnosis, *Nano Energy*. 74 (2020). https://doi.org/10.1016/j.nanoen.2020.104941.

12. G.J. Kim, K.O. Kim, Novel glucose-responsive of the transparent nanofiber hydrogel patches as a wearable biosensor via electrospinning, *Scientific Reports*. 10 (2020). https://doi.org/10.1038/s41598-020-75906-9.

13. Q. Zhang, D. Jiang, C. Xu, Y. Ge, X. Liu, Q. Wei, L. Huang, X. Ren, C. Wang, Y. Wang, Wearable electrochemical biosensor based on molecularly imprinted Ag nanowires for noninvasive monitoring lactate in human sweat, *Sensors and Actuators, B: Chemical*. 320 (2020). https://doi.org/10.1016/j.snb.2020.128325.

14. C. Zhao, X. Li, Q. Wu, X. Liu, A thread-based wearable sweat nanobiosensor, *Biosensors and Bioelectronics*. 188 (2021). https://doi.org/10.1016/j.bios.2021.113270.

15. T. Yan, Z. Wang, Z.J. Pan, Flexible strain sensors fabricated using carbon-based nanomaterials: A review, *Current Opinion in Solid State and Materials Science*. 22 (2018) 213–228. https://doi.org/10.1016/j.cossms.2018.11.001.

16. H. Souri, H. Banerjee, A. Jusufi, N. Radacsi, A.A. Stokes, I. Park, M. Sitti, M. Amjadi, Wearable and stretchable strain sensors: Materials, sensing mechanisms, and applications, *Advanced Intelligent Systems*. 2 (2020) 2000039. https://doi.org/10.1002/aisy.202000039.

17. M. Xu, J. Qi, F. Li, Y. Zhang, Highly stretchable strain sensors with reduced graphene oxide sensing liquids for wearable electronics, *Nanoscale*. 10 (2018) 5264–5271. https://doi.org/10.1039/c7nr09022f.

18. J. Eom, R. Jaisutti, H. Lee, W. Lee, J.S. Heo, J.Y. Lee, S.K. Park, Y.H. Kim, Highly sensitive textile strain sensors and wireless user-interface devices using all-polymeric conducting fibers, *ACS Applied Materials and Interfaces*. 9 (2017) 10190–10197. https://doi.org/10.1021/acsami.7b01771.

19. K. Xia, X. Chen, X. Shen, S. Li, Z. Yin, M. Zhang, X. Liang, Y. Zhang, Carbonized Chinese art paper-based high-performance wearable strain sensor for human activity monitoring, *ACS Applied Electronic Materials*. 1 (2019) 2415–2421. https://doi.org/10.1021/acsaelm.9b00564.

20. Y. Wang, S. Gong, S.J. Wang, X. Yang, Y. Ling, L.W. Yap, D. Dong, G.P. Simon, W. Cheng, Standing enokitake-like nanowire films for highly stretchable elastronics, *ACS Nano*. 12 (2018) 9742–9749. https://doi.org/10.1021/acsnano.8b05019.

21. H. Xu, Y. Lv, D. Qiu, Y. Zhou, H. Zeng, Y. Chu, An ultra-stretchable, highly sensitive and biocompatible capacitive strain sensor from an ionic nanocomposite for on-skin monitoring, *Nanoscale*. 11 (2019) 1570–1578. https://doi.org/10.1039/c8nr08589g.

22. F. Wang, S. Liu, L. Shu, X.M. Tao, Low-dimensional carbon based sensors and sensing network for wearable health and environmental monitoring, *Carbon N Y*. 121 (2017) 353–367. https://doi.org/10.1016/j.carbon.2017.06.006.

23. F. Wen, T. He, H. Liu, H.Y. Chen, T. Zhang, C. Lee, Advances in chemical sensing technology for enabling the next-generation self-sustainable integrated wearable system in the IoT era, *Nano Energy*. 78 (2020). https://doi.org/10.1016/j.nanoen.2020.105155.

24. G.J. Ko, S.D. Han, J.K. Kim, J. Zhu, W.B. Han, J. Chung, S.M. Yang, H. Cheng, D.H. Kim, C.Y. Kang, S.W. Hwang, Biodegradable, flexible silicon nanomembrane-based NOx gas sensor system with record-high performance for transient environmental monitors and medical implants, *NPG Asia Materials*. 12 (2020). https://doi.org/10.1038/s41427-020-00253-0.

25. X. Chen, T. Wang, J. Shi, W. Lv, Y. Han, M. Zeng, J. Yang, N. Hu, Y. Su, H. Wei, Z. Zhou, Z. Yang, Y. Zhang, A novel artificial neuron-like gas sensor constructed from CuS quantum dots/Bi2S3 nanosheets, *Nano-Micro Letters*. 14 (2022). https://doi.org/10.1007/s40820-021-00740-1.

26. R.T. Paschoalin, N.O. Gomes, G.F. Almeida, S. Bilatto, C.S. Farinas, S.A.S. Machado, L.H.C. Mattoso, O.N. Oliveira, P.A. Raymundo-Pereira, Wearable sensors made with solution-blow spinning poly(lactic acid) for non-enzymatic pesticide detection in agriculture and food safety, *Biosensors and Bioelectronics*. 199 (2022). https://doi.org/10.1016/j.bios.2021.113875.

27. G.H. Zhang, L. Zhang, Q.H. Zhu, H. Chen, W.L. Yuan, J. Fu, S.L. Wang, L. He, G.H. Tao, Self-healable, malleable, and flexible ionic polyimine as an environmental sensor for portable exogenous pollutant detection, *ACS Materials Letters*. 4 (2022) 136–144. https://doi.org/10.1021/acsmaterialslett.1c00687.

28. T. Dong, J. Simões, Z. Yang, Flexible photodetector based on 2D materials: Processing, architectures, and applications, *Advanced Materials Interfaces*. 7 (2020). https://doi.org/10.1002/admi.201901657.

29. S. Lim, M. Ha, Y. Lee, H. Ko, Large-area, solution-processed, hierarchical MAPbI3 nanoribbon arrays for self-powered flexible photodetectors, *Advanced Optical Materials*. 6 (2018). https://doi.org/10.1002/adom.201800615.

30. W. Jang, B.G. Kim, S. Seo, A. Shawky, M.S. Kim, K. Kim, B. Mikladal, E.I. Kauppinen, S. Maruyama, I. Jeon, D.H. Wang, Strong dark current suppression in flexible organic photodetectors by carbon nanotube transparent electrodes, *Nano Today*. 37 (2021). https://doi.org/10.1016/j.nantod.2021.101081.

31. J. An, T.S.D. Le, C.H.J. Lim, V.T. Tran, Z. Zhan, Y. Gao, L. Zheng, G. Sun, Y.J. Kim, Single-step selective laser writing of flexible photodetectors for wearable optoelectronics, *Advanced Science*. 5 (2018). https://doi.org/10.1002/advs.201800496.

32. J. Zhang, Y. Huang, Z. Tan, T. Li, Y. Zhang, K. Jia, L. Lin, L. Sun, X. Chen, Z. Li, C. Tan, J. Zhang, L. Zheng, Y. Wu, B. Deng, Z. Chen, Z. Liu, H. Peng, Low-temperature heteroepitaxy of 2D PbI2/graphene for large-area flexible photodetectors, *Advanced Materials*. 30 (2018) 1803194. https://doi.org/10.1002/adma.201803194.

33. L. Li, D. Wang, D. Zhang, W. Ran, Y. Yan, Z. Li, L. Wang, G. Shen, Near-infrared light triggered self-powered mechano-optical communication system using wearable photodetector textile, *Advanced Functional Materials*. 31 (2021). https://doi.org/10.1002/adfm.202104782.

34. M. Shojaee, S. Nasresfahani, M.K. Dordane, M.H. Sheikhi, Fully integrated wearable humidity sensor based on hydrothermally synthesized partially reduced graphene oxide, *Sensors and Actuators A: Physical*. 279 (2018) 448–456. https://doi.org/10.1016/J.SNA.2018.06.052.

35. T. Delipinar, A. Shafique, M.S. Gohar, M.K. Yapici, Fabrication and materials integration of flexible humidity sensors for emerging applications, *ACS Omega*. 6 (2021) 8744–8753. https://doi.org/10.1021/ACSOMEGA.0C06106/ASSET/IMAGES/LARGE/AO0C06106_0004.JPEG.

36. L. Gong, X. Wang, D. Zhang, X. Ma, S. Yu, Flexible wearable humidity sensor based on cerium oxide/graphitic carbon nitride nanocomposite self-powered by motion-driven alternator and its application for human physiological detection, *Journal of Materials Chemistry A*. 9 (2021) 5619–5629. https://doi.org/10.1039/d0ta11578a.

37. Y. Lu, K. Xu, M.Q. Yang, S.Y. Tang, T.Y. Yang, Y. Fujita, S. Honda, T. Arie, S. Akita, Y.L. Chueh, K. Takei, Highly stable Pd/HNb3O8-based flexible humidity sensor for perdurable wireless wearable applications, *Nanoscale Horizons*. 6 (2021) 260–270. https://doi.org/10.1039/d0nh00594k.

38. H. Zhao, Z. Wang, Y. Li, M. Yang, Single-sided and integrated polyaniline/poly(vinylidene fluoride) flexible membrane with micro/nanostructures as breathable, nontoxic and fast response wearable humidity sensor, *Journal of Colloid and Interface Science*. 607 (2022) 367–377. https://doi.org/10.1016/j.jcis.2021.08.214.

39. V. Adepu, N. Bokka, V. Mattela, P. Sahatiya, A highly electropositive ReS2based ultra-sensitive flexible humidity sensor for multifunctional applications, *New Journal of Chemistry*. 45 (2021) 5855–5862. https://doi.org/10.1039/d1nj00064k.

6

Recent Advances in Flexible, Stretchable, and Self-Healing Sensors

Hai Wang, Meijie Qu, Rui Zhang, and Yuezhen Bin

CONTENTS

6.1 Introduction

Sensors have been widely used in a variety of fields, including human sports equipment, biomedical devices, environmental alarm, and so on. They play extremely important roles in human life and industrial production. In particular, sensors with high flexibility and stretchability expand their applications to an even wider extent, especially when these advantages are strengthened by self-healing performance. To be flexible, stretchable, and healable, polymer composites are the most commonly used materials. They are light-weight, easily processable, diverse in properties, and compatible with the human body. Thus, these sensors can be widely used in mechanical engineering, the environment, and human health monitoring, as well as in a multimodal sensing platform that integrates several sensing functions. Well-designed structures of polymers, conductive fillers, ion carriers, and composites are frequently used in the preparation of sensors by taking advantage of different self-healing mechanisms, including hydrogen bonding and metal-ligand interaction. Conductive fillers and ionic carriers are mainly used to construct electron and ion transport networks for rapid and sensitive detection of deformation, compression, temperature, humidity, pH value, ion concentration, and so on. These sensors with specially designed chemical and physical structures provide a bright future for physiological and environmental monitoring. In this chapter, typical healing mechanisms, design concepts, and applications of flexible, stretchable, and healable sensors are summarized to give a brief outline of the recent developments in this field.

6.2 Self-Healing Materials and Healing Mechanism for Flexible and Stretchable Sensors

Sensors with self-healing characteristics are mainly based on substrates that have both intrinsic healing and conductive properties, or a combination of self-healable substrates and conductive fillers. Self-healing materials take advantage of the physical interaction between molecules in most cases, which includes hydrogen bonding [1–5], meta-ligand [6–8], ionic dipole [9–11], and host-guest complexation [12–14]. In some cases, quasi-covalent [15] or dynamic imide [16,17], boronate ester [12,18], and disulfide [2,19] covalent bonds are designed and applied to fulfill the healing purpose. At the level of macromolecules, polymer chain diffusion at the temperature above the glass transition temperatures and melting points also enables self-healing in the condition of additional heating [20,21]. Thus, polymers and polymer composites in the form of hydrogel, organogel, ionogel, and elastomer are the most frequently reported matrix for self-healing sensors. Liquid metal (LM) as a novel conductive filler also exhibits healing performance, but in a manner of reconfiguration in composite [22,23]. Meanwhile, MXene, as another newly discovered 2D filler, could not only form hydrogen bonding with polymer chains but also interact with ions, which is attributed to its self-healing property [24]. Figure 6.1 illustrates a summary of the self-healing mechanisms commonly used as strategies in sensors. Polymers, including polyvinyl alcohol (PVA), polyacrylamide (PAM), polyacrylic acid (PAA), polyurea (PUrea),

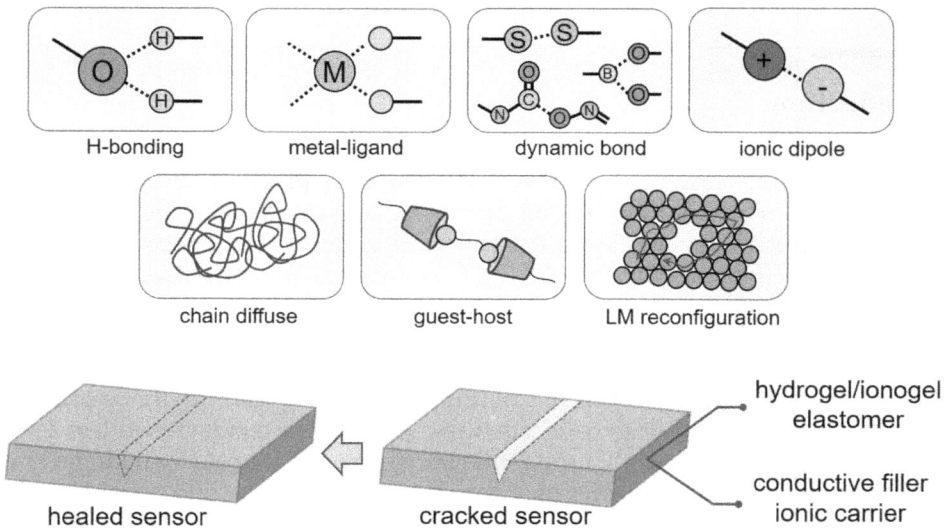

FIGURE 6.1
Self-healing mechanisms and materials used as healing strategies in flexible, stretchable, and self-healable sensors.

polyurethane (PU), polydimethylsiloxane (PDMS), chitosan, cellulose, sodium alginate (SA), and κ-carrageenan are commonly used as matrices. Conductive species like polyaniline (PANI), poly(3,4-ethylene dioxythiophene): polystyrene sulfonate (PEDOT:PSS), carbon nanotube (CNT), Ag nanoparticles (NP), LM, and MXene, as well as ionic providers like ionic liquid (IL), metal ion, and organic salt, are used as fillers or coatings to construct conductive pathways.

6.3 Design and Preparation of Flexible, Stretchable, and Self-Healing Sensors

Preparing a sensor with flexibility and stretchability requires elastomer and hydrogel as substrates. Meanwhile, physical and chemical interaction should be designed to realize healing functions in response to the fracture that occurred in sensors during application. Thus, the preparation of sensors that have both robust mechanical properties and excellent sensing performance is relatively complicated and needs different physical and chemical processes, or a combination of the two when necessary.

6.3.1 Chemical Structure Design

Preparation of traditional homopolymer or copolymer involves a polymerization process, but it is difficult to obtain a highly stretchable and healable substrate without a special chemical structure design. A typical example is the preparation of PUrea and PU elastomer with self-healability. Soft chain segments encapsulated with OH- or NH_2-groups can react with diisocyanate species, forming a highly rubbery PU or PUrea main chain with dimethylglyoxime–urethane dynamic interaction or intermolecular hydrogen bond for healing [16,19]. By incorporating disulfide in the main chain, these PU and PUrea with intramolecular -S-S- dynamic covalent bonds can also exhibit healing performance [2,19]. In some cases, a PDMS segment was also introduced to strengthen the flexibility of the main chain [25,26]. A similar design concept was used to form a crosslinked network for highly robust elastomers that have a large number of intermolecular hydrogen bond interactions for self-healing, such as carboxyl [27,28], amino [8,26], and aldehyde [17]. A gel-like substrate in sensors usually takes advantage of the numerous hydrogen bonds formed between polymer chains, providing excellent stretchability and healability.

Chemical structure design also involves the introduction of ionic carriers and metal-ligand to the main chain structure and intermolecular interaction. Using thiol-terminated PDMS oligomer and silver trifluoroacetate, an Ag-S-metal-ligand interaction between Ag NPs and PDMS chains was formed during one-step radical polymerization, forming a self-healing Ag-metal-ligand and highly flexible main chain [29]. Another typical work used Ga liquid metal as both metal-ligand and catalyst in PAA polymerization and GO reduction. The resultant hydrogel not only formed Ga^{3+}/PAA interaction to enhance the healing performance but also has Ga^{3+} and LM/rGO ionic transport pathways [30].

6.3.2 Physical Structure Design

The physical structure discussed here mainly refers to conductive filler and ionic carrier interaction in flexible and healable sensors for conductive pathways and self-healing besides intermolecular interaction. A continuous conductive pathway is a network for electron transport, which can be built using inorganic fillers, such as CNT [31], AgNWs [32], MXenes [5], LM [23], and polymers like PANI [4] and PEDOT:PSS [28]. At the same time, IL [33] and metal ions [34] as ionic carriers were applied, forming a network for cation and anion transport, especially in a matrix of hydrogels and ionogels. Another important function of ion species in these sensors is the metal-ligand interaction for healing properties as mentioned earlier, which is also related to physical structure design since simple ion-polymer interaction does not involve a covalent bond. Metal-ligands could form between organic functional groups and ions, such as Ca^{2+}/-COO$^-$ [6], Fe^{3+}/-COO$^-$ [34], and $Fe^{2+}(Zn^{2+})$/bipyridine [8]. In some of the cases, host-guest structure as a physical interaction was introduced between main chains to fulfill the healing purpose, such as β-cyclodextrin/adamantine interaction [14].

Although an elastomer or gel-like sensor in an integral form is most widely used, a multi-layer or complex structure is also designed for sensors to obtain a better sensitivity. For example, a fiber-like sweat sensor was designed, which was composed of a carbon fiber core, PEDOT:PSS flexible conductive coating, an ion-selective membrane, and a self-healing polymer skin, which were assembled layer by layer [28]. Another popular way was encapsulating conductive paste or film in a self-healing substrate, forming a sandwich structure for better sensing and healing properties [27,35]. Healable substrates in the form of fiber or film dip-coated with conductive filler like CNT would also provide the sensor with flexible and stretchable properties without sacrificing the very high conductivity that is ascribed to the densely entangled CNT [21,36]. Meanwhile, integrating several single-mode sensors on one healable substrate is another option that will not result in interference between different sensing units [32]. The several commonly used chemical and physical strategies for sensor design are illustrated in Figure 6.2.

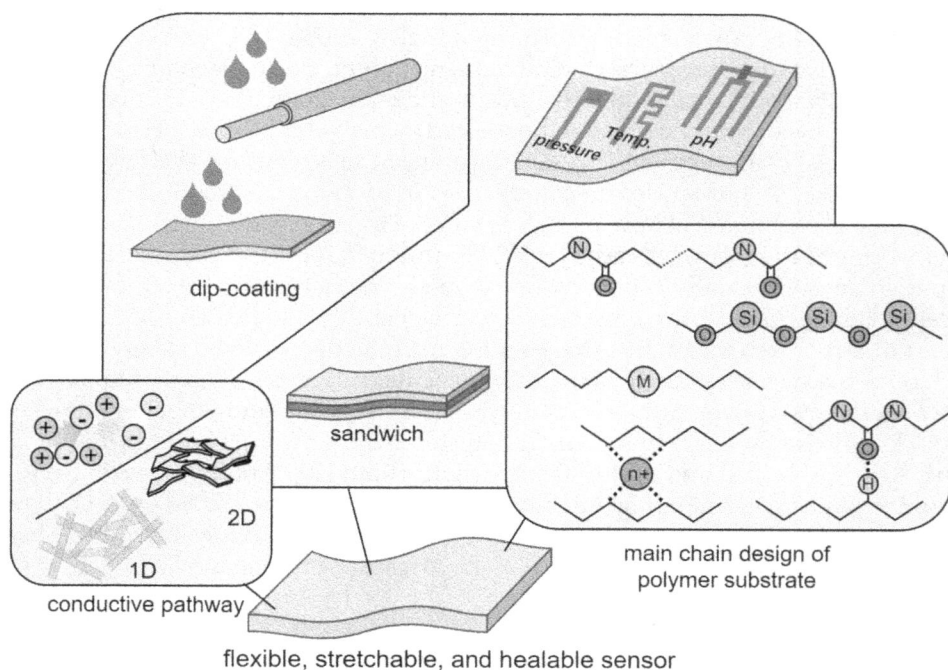

FIGURE 6.2
Typical flexible, stretchable, and self-healable sensor designs.

6.4 Flexible, Stretchable, and Self-Healing Sensors in Applications

6.4.1 Flexible, Stretchable, and Self-Healing Sensors in Mechanical Engineering

Strain and compression (pressure) monitoring are the most widespread applications of flexible and stretchable sensors, the lifetime of which can be elongated by self-healable properties. In brief, the external force that induces the change of the conductive pathway is transformed into an electric signal in the aspects of resistance, capacitance, current, or dipole, which reflects the strength and frequency change of strain or compression.

Elastomers, such as PUrea, PU, and PDMS, are frequently used as the flexible substrate for strain sensors, which are incorporated with conductive fillers to fulfill sensing performance. In general, PUrea forms a hydrogen bond between urea groups (-NH-CO-NH-), which are used as a healable component and coated by or blended with AgNW [37], Au nanoparticles (NP) [19], or nanosheet (NS) [38], and CNT [26] to form compression and strain sensing devices. A dynamic disulfide bond was also designed in polymer main chains, which would recover after breaking to enhance the self-healing property [19,37]. In other cases of compression, bending, and strain sensors, PDMS capped with NH_2-groups were covalently bonded with PUrea chains by terephthalaldehyde (TPA) as a chain extender that forms a dynamic imine bond (-CH=N-) to strengthen the self-healing property [25]. These elastomer-based sensors usually have a tensile strength of 0.1~10 MPa, and an elongation at break of more than 1000% depending on conductive filler content. These sensors were applied in compression and bending sensing and could respond to a strain of less than 50%, which is a relatively small strain extent, considering long-term repeatability and stability. LM as a conductive and self-healable species could extend the application of elastomer-based sensors to monitor a very large strain. Styrene-isoprene-styrene (SIS) block copolymer mixed with EGaIn exhibited excellent strain sensing performance even at a strain of 800% and after several hundred cyclic stretches [22]. A puncture on an LM/SIS sensor did not affect much sensing property due to the reconfiguration manner of LM in the elastomer [23].

Hydrogels and ionogels are another widely investigated category of soft matrix used in strain sensing devices. Hydrogel is a water-soluble or dispersible polymer species with strong intermolecular interaction used to form a physical crosslinking network, which includes PVA, PAA, PAM, and so on. PANI [4,34,39] and CNT [12,13,38] are the most popular and easily accessible conductive fillers, which could be blended with or dip-coated on hydrogels. Ionogels use similar polymer matrix with hydrogels, but a large number of metal ions or IL were applied for conductive network construction, which includes Fe^{3+} [6,7,40], Ca^{2+} [6], [BMIM][BF4] [40], [BMIM][TFSI] [9,40], and so forth. To enhance the conductivity of the strain sensor, a double conductive network was built by LM/rGO and Ga^{3+} ion pathways [30]. MXene is a kind of 2D transition metal carbide, nitride, or carbonitride, which has very high electrical conductivity and could provide a highly efficient electron transport network [5,24]. Similarly, a double conductive network could also be built by a combination of IL and MXene [41]. Hydrogels and ionogels usually exhibit large extension under stretching, the upper limit of which could be from 400% to 6000% [30], but a relatively low tensile strength of less than 1000 kPa. A cycling stretch test of these gel-based sensors could be conducted stably at a strain of as high as 500% [4]. These hydrogel- and ionogel-based sensors exhibited almost no loss of mechanical and electrical properties after they were fully recovered, including from repeatedly cutting [11,42]. However, healing time played an important role during the recovery process, which could be in seconds, minutes, or hours for different healing systems [4,6,30].

Physical parameters monitored in hydrogel, organogel, and ionogel strain sensors are in the form of capacitance and resistance in most cases, which is very similar to elastomer sensors. Human knot bending, face motion, and even throat vibration evoked by speaking can be sensed. In particular, pressure sensing is also reported for several ionogel-based sensors as an important function for small strain sensing. One kind of pressure-sensing mechanism relies on the capacitance change of the sensor due to the change of conductive area and distance between electrodes under external pressure [6,43]. This will result in a pressure sensitivity ranging from 0.01 to 10 kPa^{-1} under a pressure of 0.1~5 kPa [30,44], which would become lower in a condition of high pressure. By coating PEDOT:PSS on the IL-based ionogel with a pyramid micropatterned surface, a very high sensitivity of 164 kPa^{-1} can be obtained under a pressure of 20~80 kPa, which was ascribed to the profoundly increased contact area between sensing and electrode layers [11]. This design also enhanced the response time to 19 ms, which is obviously shorter than reported results in other literature. In general, the response time is a shortcoming of the flexible strain sensor, which is usually several tens or even hundreds of milliseconds.

6.4.2 Flexible, Stretchable, and Self-Healing Sensors in Environment Monitoring

Temperature variation is a highly concerned environmental parameter, which is important for daily life. Elastomers and gel-like substrates can both be used for the flexible and stretchable substrates of sensors with a temperature response function. There are two types of thermosensitive species that are used in the preparation of temperature sensors, which include conductive fillers and ionic carriers. AgNWs [32], CNT [9,45], PANI [28], and LM [30] are the most widely used conductive fillers that will change their carrier mobility and numbers upon temperature change. But ionic conductors have a different mechanism of thermal sensing, which relies on the variation of ionic transportation during heating or cooling. Thus, IL [33] and cations [10,44,46] are widely employed. Sometimes, a combination of conductive filler and ion carrier is employed to enhance the carrier hopping between conductive pathways [30,34]. In general, ion transportation results in a higher thermosensitivity, ranging from 1%/°C to even 19%/°C [33,46]. But for conductive fillers, thermosensitivity is usually less than 1%/°C.

Several cases of self-healable humidity sensors have been reported, especially when flexibility and stretchability are required at the same time. As for a conductive elastomer composite based on poly(2-hydroxyethyl methacrylate) (PHEMA)/CNT, the hydrophilicity of polymer chains with hydroxyl groups allows the absorption of water molecules at different humidity [14], which changes the resistance of the conductive composite. There are different mechanisms of humidity sensing depending on different polymer matrices and charge carriers, which are closely related to ion transport. For the PVA/CNF composite, a gradient distribution of the oxygen-containing group created an H_3O^+ diffusion driving force at a different humidity [3]. However, for ionogel, ions are the main charge carriers that transport in a gel matrix, forming a continuous current. Therefore, low humidity will increase polymer concentration, which hinders the movement of cations and weakens the solubility of cations, thus the resistance of ionogel increases. On the contrary, a higher humidity results in the swelling of ionogel, which strengthens the cation mobility and concentration in the gel matrix [1]. If IL was used as a charge carrier, hygroscopicity plays an important role in absorbing and releasing moisture, which changes ionogel resistance [40].

Gas sensing is another very important aspect of environmental monitoring, especially for the detection of toxic gas NH_3 and NO_2 from wastewater and chemicals. The mechanism of gas monitoring is different for conductive polymers and ionogels. For p-type semiconductor PANI, NH_3 as an electron donor could increase the conductance of PANI-based materials after molecular interaction. Using modified PDMS as a flexible and healable substrate and PANI/CNT as a conductive filler, an NH_3 gas sensor senses NH_3 at a concentration of 5–50 ppm [17], which could keep the sensing stability after cutting and healing. Ionogel with a certain amount of water in the gel matrix could absorb NO_2 and NH_3 which have good solubility in water. The absorbed gas molecules would hinder electron transport to some extent, thus changing the conductivity of ionogels. In a PAM/κ-carrageenan ionogel with self-healing property, NO_2 with a concentration of 0.8 to 4 ppm could be detected at a humidity of 10% to 98% and a temperature of –10°C to 45°C, with a sensitivity of 7%/ppm to 120%/ppm [47]. The sensitivity of this ionogel did not change upon tensile deformation that increased from 0% to 100%, which was stable at around 7.5%/ppm at a humidity of 98%. The PAM/κ-carrageenan ionogel system could also be used in the detection of NH_3, which also exhibited high sensitivity even in a deformed state and after cut-healing treatment [48].

Organic species sense is also one type of chemical sensing like gas sensing, but a rare case has been reported for the flexible and healable sensor. Here is one example of a healable PU substrate embedded with AuNP that could detect different kinds of organic solvents, such as benzaldehyde, hexanol, octanol, trimethylbenzene, and so on [19]. The interaction between organic molecules with AuNP is the main reason to change the conductance of the elastomer composite.

As an environmental factor, UV light is harmful if the illumination is too high. There is also a report of a healable UV photosensor, which is made of a PEDOT:PSS/ZnO bilayer. ZnO is the main functional layer for UV detection due to the so-called acoustoelectric effect. A five-cutting-healing cycle would not profoundly affect the UV-induced current change of this sensor due to the healing performance of the PVA/agarose layer [49].

In general, the self-healing property endows the above-mentioned sensors with excellent conductivity and sensitivity even after several cutting-healing cycles. But the healing time required to recover the original sensing performance depends on the polymer matrix with various healable properties, which could be in a time scale of seconds or even days.

6.4.3 Flexible, Stretchable, and Self-Healing Sensors in Human Health

Temperature sensors with flexible, stretchable, and healable properties were reviewed in Section 6.4.2, which could be used for body temperature monitoring. It was also shown that strain sensors have an essential function in human health monitoring. A strain sensor in stretching mode can monitor large deformations, such as knot and limb movements, and small deformations, including face motion and throat vibration. A more vital mode is pressure or compression sensing, which could be used to monitor the pulse and heartbeat status of the human body, giving feedback on human health status. Flexibility and healability endow these sensors with excellent adherence to skin and reusability even after cutting and scratching.

pH value and ion concentration in sweat are important criteria for human health, which can be monitored when pH and sweat sensors are attached to the epidermis. Conductive fillers are used as a pH-sensitive component for healable pH sensors, including commonly applied CNT and PANI. When CNTs are employed, they are coated on or sandwiched between healable PU films that are kept in a water solution with various pH values during measurement [32]. The surface charge of CNT changes with the increasing pH from 4 to 9, increasing current, even after the sensor was cut and healed. The pH-sensing mechanism is a little different for PANI, which forms a state of redox equilibrium after contacting

H_3O^+ [50]. By incorporating flexible and healable polyester elastomer, this pH sensor can give a stable electrical response after four cutting-healing cycles, and sense the pH of fruit and human fluid, which shows great potential in human health conditions and sports monitoring. Using the same healable polyester elastomer, pH sensors can also be transformed into K^+ and Na^+ ion sensors for sweat monitoring using dip-coating of ion-selective membrane on the conductive layer [28]. K^+ and Na^+ with a concentration of 0.1 mM to 100 mM are detected and can still revoke almost the same electrical signal after the sweat sensor undergoes four cutting-healing cycles. Bending and temperature increase do not result in obvious signal fluctuation for this sweat sensor.

Glucose metabolism plays an irreplaceable role in human life. Thus, it is necessary to monitor glucose levels in a straightforward way. A flexible and healable glucose sensor has been designed to attain this goal. One way is to make a combination of glucose oxidase with hydrogel and use an electrochemical medium like CeO_2/MnO_2 to sense the product of the reaction between glucose oxidase and O_2 [15]. This type of glucose sensor can detect glucose with a concentration of 0.2 μM to 20 mM. A hydrogel bond is the driving force of healing in this hydrogel-based glucose sensor, but a slight decrease in the current signal was observed after the hydrogel-based glucose sensor was healed for 2 h from multi-position scratch damage. However, hydrogel made of chitosan derivatives was not highly stretchable and had to be encapsulated in a flexible substrate. Another way of glucose detection by the healable sensor is by using the dynamic boronate ester bond [18]. The hydroxyl group from the dopamine part formed a dynamic bond with a polymer chain containing boronic acid. This dynamic chemical bond breaks after glucose replaces the dopamine–boronate ester bond, forming a new molecular interaction between glucose and boron. This healable sensor can respond to glucose with a concentration of 0~16 mM, and also to fructose, mannose, sucrose, and galactose.

6.4.4 Flexible, Stretchable, and Self-Healing Sensors in Multimodal Sensing

The review of the above-mentioned flexible and healable sensors is categorized by physical and chemical parameters that are sensed in the monitoring process, including deformation, temperature, humidity, pH, gas, ions, and so forth. While the detection of a single environmental or biological signal is not enough so far, the integration of multiple sensing functions on one sensor with a compact structure is the future trend for environmental and human health monitoring. Figure 6.3 is a summary of the applications of multi-model sensing platform with flexible, stretchable and healable properties.

Temperature and strain sensing are the most widely investigated dual-mode sensing platform in the literature. As for strain sensing, there are two different mechanisms: conductive pathway changes of the conductive filler and ion transport and mobility changes of ionic species as discussed previously. In a PAM/CNF organogel, Na^+ and Cl^- can interact with COO^- of CNF and NH_{2-} of PAM, forming an ion transport tunnel for electric conductivity [44], which is sensitive to both strain and temperature. Bending and finger touching with different rates (30~180 mm/min) and extents (1%~400%) as well as compression strength (3 kPa~12 kPa) can be monitored accurately, and the sensitivity is 1.32. Temperatures ranging from 25 to 68°C can be detected with a temperature coefficient of resistance (TCR) of –1.49%/°C. Another dual-mode sensor of temperature and strain used PAA and PANI as polymer networks and Fe^{3+} as ion carriers in a hydrogel. Fe^{3+} can hop between PANI and transport in the PAA network, forming two kinds of conductive pathway [34]. This PAA/PANI/Fe^{3+} sensor had a temperature sensitivity of –1.64%/°C in a range of 50 to 90°C, which also showed an abrupt increase of current when the temperature was

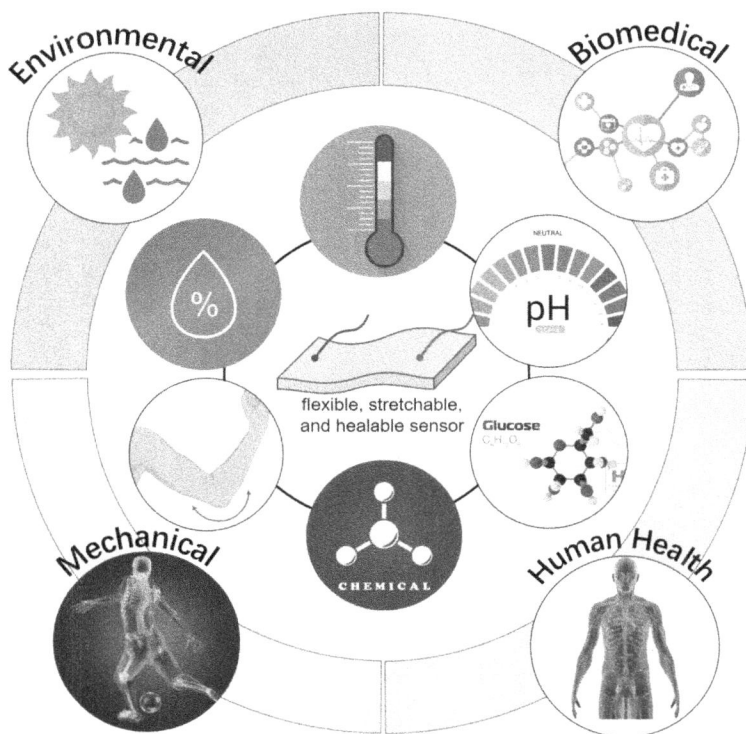

FIGURE 6.3
A summary of the application of flexible, stretchable, and healable sensors.

increased from the body temperature of 36.4°C to a fever temperature 39.1°C. Compression, tensile strength, and throat vibration could be sensed using this sensor, but the long-term use might be a concerning aspect for further improvement. By taking advantage of the expansion and contraction of PNIPAM during temperature fluctuation between 20 and 45°C, the PVA/SA/Na+ hydrogel sensor could sense temperature change due to the condensed ion concentration at a higher temperature, which ascribed to the shrinkage of PNIPAM [42]. Bending and finger touching were also sensed due to the ion transport pathway change under external force. A PU elastomer encapsulated with CNT was also used for temperature sensing and strain dual-mode sensing, but based on a different mechanism as discussed in Section 6.4.2 [37].

Besides temperature and strain sensing, a multi-mode sensor can integrate other sensing functions. Several triple-mode sensors with flexibility, stretchability, and healability were reported. One strategy is integrating different sensors on one substrate, and another way is using the same sensor for different sensing functions. For the former, the strain sensor made of graphene-PVA/CNF ink, the temperature sensor made of CNT-PVA/CNF ink, and the humidity sensor made of carbon black/graphene-PVA/CNF ink were all pasted on a CNF/PVA healable substrate, forming an integrated circuit [3]. A similar structure was made, using AgNWs as a conductive paste for temperature-sensitive ink, carbon black for pressure-sensing ink, and SWCNT as pH-sensing ink, which was all integrated on a healable PU substrate [32]. The advantage of this type of sensor is that signals would not interfere with each other. A temperature range of 12 to 50°C, pressure of 200 Pa to 12.8 kPa,

TABLE 6.1

A Summary of Flexible, Stretchable, and Healable Sensors in Aspects of a Polymer Matrix, Fillers, Healing Mechanisms, and Applications

Polymer matrix	Conductive/ionic	Healing mechanism	Application	Reference
PUrea elastomer	AgNWs, CNT	H-bonding dynamic disulfide bond	Strain	[2]
PUrea-PDMS elastomer	CNT	H-bonding dynamic imine bond	Strain	[25]
PUrea/PDMS elastomer	MWCNT	H-bonding dynamic disulfide bond	Strain	[26]
PUrea elastomer	AgNW, SWCNT	H-bonding	Pressure, T, pH	[32]
PU elastomer	AuNP	H-bonding metal-ligand	Strain, T, chemical	[19]
PU	Gr	H-bonding dimethylglyoxime–urethane dynamic bond	Strain	[16]
PU-Urea/PS–b-PI–b-PS	CNT	H-bonding	Strain, T	[37]
epoxy nature rubber/CNC@histidine elastomer	ZnCl$_2$, CNT,	Zn-histidine bond	Strain	[31]
SIS rubber	LM	LM reconfiguration	Strain	[22]
Silicone-thiol/Ag NPs	MWCNT	Meta-ligand	Strain	[29]
Silicone elastomer	LM	LM reconfiguration	Strain, pressure	[23]
PDMS-bipyridine elastomer	Fe^{2+}, Zn^{2+}	Meta-ligand	Strain	[8]
Polyamide composite	CF, AgNWs	H-bonding	Strain	[27]
PVA/CNF composite	CB, Gr, CNT	H-bonding	Strain, T, humidity	[3]
PCL/TPU composite	CNT	Chain diffusion	Strain	[21]
PAM/alginate ionogel	Fe^{3+}, Ca^{2+}	Metal-ligand	Strain, pressure	[6]
PAM/CNF	Na$^+$	H-bonding	Strain, pressure, T	[44]
PAM/picolinium-containing Adamantanes/β-cyclodextrin hydrogel	H$_3$PO$_4$	Host-guest complex	Strain	[13]
PAM/PVA	MXene	H bond borate ester bond	Strain	[5]
Acylate copolymer ionogel	IL	Ion dipole	Strain	[9]
PVA/borax+poly(β-cyclodextrin)/polyferrocene hydrogel	CNT	Host-guest complex borate ester bond	Strain	[12]
PAA/phytic acid hydrogel	PANI	H bond	Strain	[4]
PAA hydrogel	LM/rGO, Ga^{3+}	H bond metal-ligand	Strain, pressure, T, humidity, solvent	[30]
PAA/CNC@Tannic acid	IL, MXene	H bond borate ester bond	Strain, pressure, T, photo	[41]

(Continued)

TABLE 6.1 (CONTINUED)

A Summary of Flexible, Stretchable, and Healable Sensors in Aspects of a Polymer Matrix, Fillers, Healing Mechanisms, and Applications

Polymer matrix	Conductive/ionic	Healing mechanism	Application	Reference
PAA	MXene, Ca^{2+}	H-bonding metal-ligand	Strain, pressure	[24]
PAA/PANI/glycerol organogel	Fe^{3+}, PANI	H-bonding metal-ligand	Strain, T	[34]
PAA ionogel	IL, Fe^{3+}	H-bonding metal-ligand	Strain, pressure, humidity	[40]
PAA/SA hydrogel	CNT, Ca^{2+}	H bond metal-ligand	Pressure	[43]
PAA/PVA organogel	Gr, Fe3+	H bond metal-ligand	Strain	[35]
PAA/PVA hydrogel	CNT, Fe3+	H bond metal-ligand	Strain	[36]
Poly(2-acrylamido-2-methyl-1-propanesulfonic acid) ionogel	PEDOT:PSS, IL	Ionic dipole, H bond	Strain, pressure	[11]
PVA/PDA@Ag NPs hydrogel	PANI	H bond	Strain	[39]
PVA/PEG ionogel	Fe^{3+}	Metal-ligand	Pressure	[7]
PVA/SA/borax/PNIPAM ionogel	Na^+	Borate ester bond	Pressure, T	[42]
PVA/CNC/AuNS composite	PANI, MWCNT, AuNS	H bond (H_2O spray)	Pressure, T	[38]
Acrylate copolymer-sulfonate/glycerol ionogel	K^+	Ion dipole	T	[10]
Poly(diacids/triacids/diethylenetriamine) elastomer	SWCNT	H bond	T	[45]
PMA/polyundecylenyl alcohol modified CNTs composite	modified CNT	Chain diffusion	T	[20]
PAM organogel	κ-carrageenan	H-bonding	Humidity	[1]
CNT-β-cyclodextrin/PMA-adamantane elastomer	SWCNT	Host-guest complex	Humidity	[14]
Diamino-PDMS/benzene-tricarbaldehyde elastomer	PANI, CNT	Dynamic imide bond	Gas (NH_3)	[17]
PAM ionogel	κ-carrageenan, Ca^{2+}	H bond ionic dipole	Gas (NO2)	[47]
PAM/ethyl glycol organogel	κ-carrageenan,	H bond	Gas (NO2, NH3)	[48]
PVA/Agarose hydrogel	AgNW, ZnO, PEDOT:PSS	H bond	UV	[49]
Poly(citric acid/succinic acid/1,4-cyclohexanedimethanol) ester	PEDOT:PSS	H bond	Sweat (ion)	[28]
Poly(citric acid/succinic acid/1,4-cyclohexanedimethanol) ester	CF, PANI	H bond	pH	[50]
Quaternized chitosan/oxidized dextran hydrogel	CeO_2/MnO_2	Quasi-covalent Schiff base bond	Glucose	[15]
PMA-dopamine/phenylboronic acid hydrogel	Phenylboronic acid	Borate ester bonds	Saccharides	[18]

and a pH of 4 to 9 were detected, respectively, which is not affected by the cutting-healing cycle.

Integrating temperature, strain, and solvent sensing functions on one flexible and healable sensor made of composite hydrogel was also reported in a rare case. This hydrogel sensor was made of PAA/LM-rGO/Ga^{3+}, in which LM/rGO formed a conductive pathway and PAA/Ga^{3+} formed an ionic carrier [30]. Simultaneously, this sensor could sense small and large strains ranging from 5% to 400%, compression up to 5 kPa, and bending from 15° to 90°. Meanwhile, a temperature fluctuation ranging from 20 to 30°C was detected with a sensitivity of 2%/°C, and an ethyl alcohol/water solvent exchange was also sensed with obvious resistance change.

Another interesting E-skin-like sensor combined the sensing of stain, temperature, and optical change using PAA, tannic-coated CNC, borax, and MXene [41]. This self-healable ionic gel-based sensor can respond to a strain of up to 270%, a pressure of 0~12 kPa, and a stretch frequency of 0.024~0.192 Hz. Human motion, including knee bending, finger bending, lip movement, and compression from pen writing can be monitored sensitively. The temperature sensing performance of this sensor can respond to a temperature change from 11 to 90°C, which is a very good candidate for a fever sensor. It can also be sensitive to a light power change from 0.2 W/cm^2 to 1.0 W/cm^2 as a photo intensity monitoring device.

6.5 Challenges and Future Perspectives

Flexible, stretchable, and self-healing sensors have obtained considerable progress due to the rapid development of polymer synthesis and composite preparation techniques. However, polymer-based sensors usually have a longer response time than metal-based ones, which limits their application when a fast response is required. The major advantage of flexible sensors is that they are compatible with the human body due to their soft texture, and the healable property makes these sensors have tissue-like repairing ability. This could be a great opportunity for human-machine interaction to have more intelligent monitoring of human health and environmental parameters. But the long-term use stability should be a concern, especially for gel-based sensors. Multi-mode sensing is another big challenge that requires highly integrated devices with no signal interference. Since current flexible sensors are compact, portable, and lightweight, a persistent power supply is required that satisfies the long-term use of these sensors. Table 6.1 is a summary of flexible, stretchable, and healable sensors discussed and reviewed in this paper.

References

1. J. Wu, Z. Wu, H. Xu, Q. Wu, C. Liu, B.R. Yang, X. Gui, X. Xie, K. Tao, Y. Shen, J. Miao, L.K. Norford, An intrinsically stretchable humidity sensor based on anti-drying, self-healing and transparent organohydrogels, *Mater. Horizons*. 6 (2019) 595–603.
2. M. Khatib, O. Zohar, W. Saliba, S. Srebnik, H. Haick, Highly efficient and water-insensitive self-healing elastomer for wet and underwater electronics, *Adv. Funct. Mater*. 30 (2020) 1910196.

3. X. Lin, F. Li, Y. Bing, T. Fei, S. Liu, H. Zhao, T. Zhang, Biocompatible multifunctional E-skins with excellent self-healing ability enabled by clean and scalable fabrication, *Nano-Micro Lett.* 13 (2021) 200.

4. G. Su, S. Yin, Y. Guo, F. Zhao, Q. Guo, X. Zhang, T. Zhou, G. Yu, Balancing the mechanical, electronic, and self-healing properties in conductive self-healing hydrogel for wearable sensor applications, *Mater. Horizons.* 8 (2021) 1795–1804.

5. H. Liao, X. Guo, P. Wan, G. Yu, Conductive MXene nanocomposite organohydrogel for flexible, healable, low-temperature tolerant strain sensors, *Adv. Funct. Mater.* 29 (2019) 1904507.

6. Y. Wang, M. Tebyetekerwa, Y. Liu, M. Wang, J. Zhu, J. Xu, C. Zhang, T. Liu, Extremely stretchable and healable ionic conductive hydrogels fabricated by surface competitive coordination for human-motion detection, *Chem. Eng. J.* 420 (2021) 127637.

7. S. Park, B.G. Shin, S. Jang, K. Chung, Three-dimensional self-healable touch sensing artificial skin device, *ACS Appl. Mater. Interfaces.* 12 (2020) 3953–3960.

8. Y.L. Rao, A. Chortos, R. Pfattner, F. Lissel, Y.C. Chiu, V. Feig, J. Xu, T. Kurosawa, X. Gu, C. Wang, M. He, J.W. Chung, Z. Bao, Stretchable self-healing polymeric dielectrics cross-linked through metal-ligand coordination, *J. Am. Chem. Soc.* 138 (2016) 6020–6027.

9. P. Shi, Y. Wang, W.W. Tjiu, C. Zhang, T. Liu, Highly stretchable, fast self-healing, and waterproof fluorinated copolymer ionogels with selectively enriched ionic liquids for human-motion detection, *ACS Appl. Mater. Interfaces.* 13 (2021) 49358–49368.

10. J. Lee, M.W.M. Tan, K. Parida, G. Thangavel, S.A. Park, T. Park, P.S. Lee, Water-processable, stretchable, self-healable, thermally stable, and transparent ionic conductors for actuators and sensors, *Adv. Mater.* 32 (2020) 1906679.

11. X. Su, X. Wu, S. Chen, A.M. Nedumaran, M. Stephen, K. Hou, B. Czarny, W.L. Leong, A highly conducting polymer for self-healable, printable, and stretchable organic electrochemical transistor arrays and near hysteresis-free soft tactile sensors, *Adv. Mater.* 34 (2022) 2200682.

12. X. Liu, Z. Ren, F. Liu, L. Zhao, Q. Ling, H. Gu, Multifunctional self-healing dual network hydrogels constructed via host-guest interaction and dynamic covalent bond as wearable strain sensors for monitoring human and organ motions, *ACS Appl. Mater. Interfaces.* 13 (2021) 14612–14622.

13. X. Li, H. Zhang, P. Zhang, Y. Yu, A Sunlight-degradable autonomous self-Healing supramolecular elastomer for flexible electronic devices, *Chem. Mater.* 30 (2018) 3752–3758.

14. K. Guo, D.L. Zhang, X.M. Zhang, J. Zhang, L.S. Ding, B.J. Li, S. Zhang, Conductive elastomers with autonomic self-healing properties, *Angew. Chemie - Int. Ed.* 54 (2015) 12127–12133.

15. Z. Liang, J. Zhang, C. Wu, X. Hu, Y. Lu, G. Wang, F. Yu, X. Zhang, Y. Wang, Flexible and self-healing electrochemical hydrogel sensor with high efficiency toward glucose monitoring, *Biosens. Bioelectron.* 155 (2020) 112105.

16. H. Gao, J. Xu, S. Liu, Z. Song, M. Zhou, S. Liu, F. Li, F. Li, X. Wang, Z. Wang, Q. Zhang, Stretchable, self-healable integrated conductor based on mechanical reinforced graphene/ polyurethane composites, *J. Colloid Interface Sci.* 597 (2021) 393–400.

17. B. Zhang, P. Zhang, H. Zhang, C. Yan, Z. Zheng, B. Wu, Y. Yu, A transparent, highly stretchable, autonomous self-healing poly(dimethyl siloxane) elastomer, *Macromol. Rapid Commun.* 38 (2017) 1–9.

18. Q. Chen, Z. Wei, S. Wang, J. Zhou, Z. Wu, A self-healing smart photonic crystal hydrogel sensor for glucose and related saccharides, *Microchim. Acta.* 188 (2021) 210.

19. T.P. Huynh, H. Haick, Self-healing, fully functional, and multiparametric flexible sensing platform, *Adv. Mater.* 28 (2016) 138–143.

20. Q. Zhang, L. Liu, C. Pan, D. Li, G. Gai, Thermally sensitive, adhesive, injectable, multiwalled carbon nanotube covalently reinforced polymer conductors with self-healing capabilities, *J. Mater. Chem. C.* 6 (2018) 1746–1752.

21. M. Qu, H. Wang, Q. Chen, L. Wu, P. Tang, M. Fan, Y. Guo, H. Fan, Y. Bin, A thermally-electrically double-responsive polycaprolactone – thermoplastic polyurethane/multi-walled carbon nanotube fiber assisted with highly effective shape memory and strain sensing performance, *Chem. Eng. J.* 427 (2022) 131648.

22. R. Tutika, A.B.M.T. Haque, M.D. Bartlett, Self-healing liquid metal composite for reconfigurable and recyclable soft electronics, *Commun. Mater.* 2 (2021) 64.
23. E.J. Markvicka, M.D. Bartlett, X. Huang, C. Majidi, An autonomously electrically self-healing liquid metal-elastomer composite for robust soft-matter robotics and electronics, *Nat. Mater.* 17 (2018) 618–624.
24. X. Li, L. He, Y. Li, M. Chao, M. Li, P. Wan, L. Zhang, Healable, degradable, and conductive MXene nanocomposite hydrogel for multifunctional epidermal sensors, *ACS Nano.* 15 (2021) 7765–7773.
25. H. Yan, S. Dai, Y. Chen, J. Ding, N. Yuan, A high stretchable and self–healing silicone rubber with double reversible bonds, *ChemistrySelect.* 4 (2019) 10719–10725.
26. X. Dai, L.B. Huang, Y. Du, J. Han, Q. Zheng, J. Kong, J. Hao, Self-Healing, Flexible, and tailorable triboelectric nanogenerators for self-powered sensors based on thermal effect of infrared radiation, *Adv. Funct. Mater.* 30 (2020) 1–10.
27. D. Jiang, Y. Wang, B. Li, C. Sun, Z. Wu, H. Yan, L. Xing, S. Qi, Y. Li, H. Liu, W. Xie, X. Wang, T. Ding, Z. Guo, Flexible sandwich structural strain sensor based on silver nanowires decorated with self-healing substrate, *Macromol. Mater. Eng.* 304 (2019) 1900074.
28. J.H. Yoon, S.M. Kim, Y. Eom, J.M. Koo, H.W. Cho, T.J. Lee, K.G. Lee, H.J. Park, Y.K. Kim, H.J. Yoo, S.Y. Hwang, J. Park, B.G. Choi, Extremely fast self-healable bio-based supramolecular polymer for wearable real-time sweat-monitoring sensor, *ACS Appl. Mater. Interfaces.* 11 (2019) 46165–46175.
29. T. Searle, V. Sencadas, J. Greaves, G. Alici, Room-temperature self-healing piezoresistive sensors, *Compos. Sci. Technol.* 211 (2021) 108856.
30. Z. Zhang, L. Tang, C. Chen, H. Yu, H. Bai, L. Wang, M. Qin, Y. Feng, W. Feng, Liquid metal-created macroporous composite hydrogels with self-healing ability and multiple sensations as artificial flexible sensors, *J. Mater. Chem. A.* 9 (2021) 875–883.
31. X. Liu, G. Su, Q. Guo, C. Lu, T. Zhou, C. Zhou, X. Zhang, Hierarchically structured self-healing sensors with tunable positive/negative piezoresistivity, *Adv. Funct. Mater.* 28 (2018) 1–10.
32. M. Khatib, O. Zohar, W. Saliba, H. Haick, A multifunctional electronic skin empowered with damage mapping and autonomic acceleration of self-healing in designated locations, *Adv. Mater.* 32 (2020) 1–7.
33. Y. He, S. Liao, H. Jia, Y. Cao, Z. Wang, Y. Wang, A self-healing electronic sensor based on thermal-sensitive fluids, *Adv. Mater.* 27 (2015) 4622–4627.
34. G. Ge, Y. Lu, X. Qu, W. Zhao, Y. Ren, W. Wang, Q. Wang, W. Huang, X. Dong, Muscle-inspired self-healing hydrogels for strain and temperature sensor, *ACS Nano.* 14 (2020) 218–228.
35. L. Wu, M. Fan, M. Qu, S. Yang, J. Nie, P. Tang, L. Pan, H. Wang, Y. Bin, Self-healing and anti-freezing graphene-hydrogel-graphene sandwich strain sensor with ultrahigh sensitivity, *J. Mater. Chem. B.* 9 (2021) 3088–3096.
36. L. Wu, L. Li, M. Fan, P. Tang, S. Yang, L. Pan, H. Wang, Y. Bin, Strong and tough PVA/PAA hydrogel fiber with highly strain sensitivity enabled by coating MWCNTs, *Compos. Part A Appl. Sci. Manuf.* 138 (2020) 106050.
37. M. Khatib, T.P. Huynh, Y. Deng, Y.D. Horev, W. Saliba, W. Wu, H. Haick, A freestanding stretchable and multifunctional transistor with Intrinsic self-healing properties of all device components, *Small.* 15 (2019) 1803939.
38. J.W. Kim, H. Park, G. Lee, Y.R. Jeong, S.Y. Hong, K. Keum, J. Yoon, M.S. Kim, J.S. Ha, Paper-like, thin, foldable, and self-healable electronics based on PVA/CNC nanocomposite film, *Adv. Funct. Mater.* 29 (2019) 1905968.
39. Y. Zhao, Z. Li, S. Song, K. Yang, H. Liu, Z. Yang, J. Wang, B. Yang, Q. Lin, Skin-inspired antibacterial conductive hydrogels for epidermal sensors and diabetic foot wound dressings, *Adv. Funct. Mater.* 29 (2019) 1901474.
40. Y. Wang, Y. Liu, N. Hu, P. Shi, C. Zhang, T. Liu, Highly stretchable and self-healable ionogels with multiple sensitivity towards compression, strain and moisture for skin-inspired ionic sensors, *Sci. China Mater.* (2022). https://doi.org/10.1007/s40843-021-1977-5.

41. Y. Lu, X. Qu, S. Wang, Y. Zhao, Y. Ren, W. Zhao, Q. Wang, Ultradurable, freeze-resistant, and healable MXene-based ionic gels for multi-functional electronic skin, *Nano Res.* 15 (2022) 4421–4430.

42. O.Y. Kweon, S.K. Samanta, Y. Won, J.H. Yoo, J.H. Oh, Stretchable and self-healable conductive hydrogels for wearable multimodal touch sensors with thermoresponsive behavior, *ACS Appl. Mater. Interfaces.* 11 (2019) 26134–26143.

43. J. Wei, J. Xie, P. Zhang, Z. Zou, H. Ping, W. Wang, H. Xie, J.Z. Shen, L. Lei, Z. Fu, Bioinspired 3D printable, self-healable, and stretchable hydrogels with multiple conductivities for skin-like wearable strain sensors, *ACS Appl. Mater. Interfaces.* 13 (2021) 2952–2960.

44. Y. Wei, L. Xiang, P. Zhu, Y. Qian, B. Zhao, G. Chen, Multifunctional organohydrogel-based ionic skin for capacitance and temperature sensing toward intelligent skin-like devices, *Chem. Mater.* 33 (2021) 8623–8634.

45. H. Yang, D. Qi, Z. Liu, B.K. Chandran, T. Wang, J. Yu, X. Chen, Soft thermal sensor with mechanical adaptability, *Adv. Mater.* 28 (2016) 9175–9181.

46. J. Wu, Z. Wu, Y. Wei, H. Ding, W. Huang, X. Gui, W. Shi, Y. Shen, K. Tao, X. Xie, Ultrasensitive and stretchable temperature sensors based on thermally stable and self-healing organohydrogels, *ACS Appl. Mater. Interfaces.* 12 (2020) 19069–19079.

47. Z. Wu, L. Rong, J. Yang, Y. Wei, K. Tao, Y. Zhou, B.R. Yang, X. Xie, J. Wu, Ion-conductive hydrogel-based stretchable, self-healing, and transparent NO2 sensor with high sensitivity and selectivity at room temperature, *Small.* 17 (2021) 1–12.

48. J. Wu, Z. Wu, W. Huang, X. Yang, Y. Liang, K. Tao, B.R. Yang, W. Shi, X. Xie, Stretchable, stable, and room-temperature gas sensors based on self-healing and transparent organohydrogels, *ACS Appl. Mater. Interfaces.* 12 (2020) 52070–52081.

49. M.S. Tsai, T.L. Shen, H.M. Wu, Y.M. Liao, Y.K. Liao, W.Y. Lee, H.C. Kuo, Y.C. Lai, Y.F. Chen, Self-powered, self-healed, and shape-adaptive ultraviolet photodetectors, *ACS Appl. Mater. Interfaces.* 12 (2020) 9755–9765.

50. J.H. Yoon, S.M. Kim, H.J. Park, Y.K. Kim, D.X. Oh, H.W. Cho, K.G. Lee, S.Y. Hwang, J. Park, B.G. Choi, Highly self-healable and flexible cable-type pH sensors for real-time monitoring of human fluids, *Biosens. Bioelectron.* 150 (2020) 111946.

7

Properties and Applications of Conjugated Polymers for Flexible Electronics: Current Trends and Perspectives

Yub N. Thapa, Bhim P. Kafle, and Rameshwar Adhikari

CONTENTS

7.1 Introduction and Overview

Plastics have invaded every aspect of modern society. It is not possible to take a single step in a day without encountering some kind of plastic. Such plastics developed from organic polymers and are substantially insulators. However, in the 1960s, a new class of organic polymers was demonstrated that showed semiconducting potential. In the late 1970s, such conducting polymers were found to be as conductive as metals [1,2]. MacDiarmid [3] were one of the pioneers in developing polyacetylene film which could, later on, be oxidized by iodine converting it into a conductive one. This astounding finding won them the Nobel Prize in Chemistry in 2000 "for the discovery and development of conductive polymers". A conjugated polymer (CP) alternately contains single and double bonds along the polymeric chain. This conjugation enables π electrons to be delocalized throughout the system – responsible for its optical and conductive properties. Such polymers with conducting capacity are intrinsic conducting polymers. Nevertheless, only a conjugation is not sufficient to give conductive properties to the polymer. Conjugated polymers can be

DOI: 10.1201/9781003299455-7

transformed into conducting polymers by doping either in the gas phase or in the solution state that provides charge carriers, namely extra electrons or holes, into the polymer, making them conductive [3]. The function of a dopant is to add or remove electrons from the polymer. Doping can be of two types – p-type and n-type that can be carried out by two methods: chemical or electrochemical. P-type doping or oxidation leads to the removal of an electron from the highest occupied molecular orbital (HOMO) of the polymer forming holes. In contrast, n-type doping or reduction leads to the gain of an electron at the lowest unoccupied molecular orbital (LUMO) of the polymer thereby increasing electron density. The doping mechanism of p-type doping can be observed as shown in Figure 7.1 [4]. A polaron (radical cation) with a positively charged hole site is formed in p-type doping as stated earlier. Polaron movement and the electrical conductivity mechanism along the PA are shown in Figure 7.2.

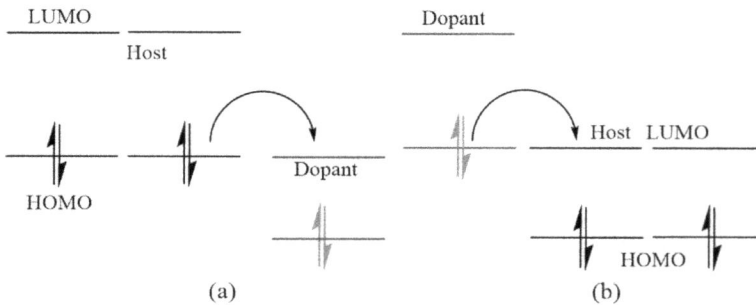

FIGURE 7.1
(a) Mechanism of p-type doping (schematic only). The molecular dopant acts as an acceptor in p-type dopants. (b) Mechanism of n-type doping (schematic). The molecular dopant acts as a donor in n-type doping. Adapted with permission [4], Copyright © The Authors, some rights reserved, exclusive license [Hindawi]. Distributed under a Creative Common Attribution License 4.0 (CCBY) https://creativecommons.org/licenses/by/4.0/.

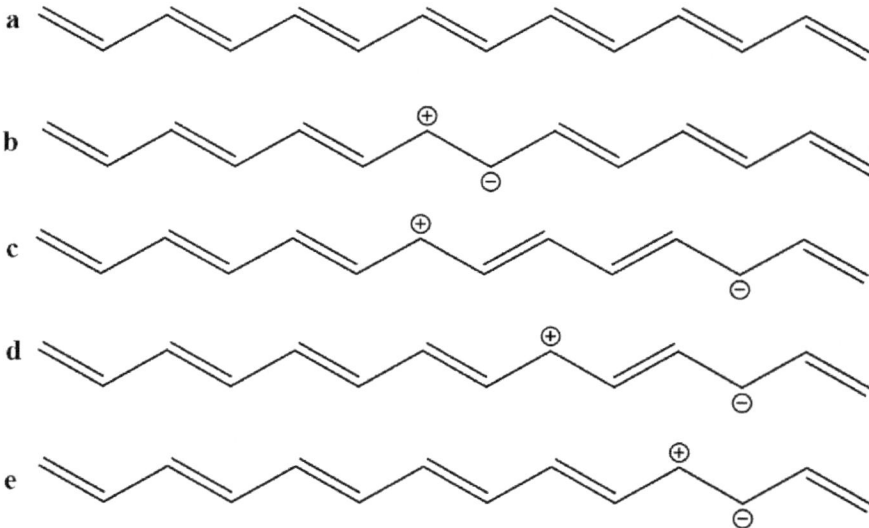

FIGURE 7.2
Conductivity mechanism. Formation of polarons in polyacetylene by removal of electrons (a→b). Movement of polarons along the chain (c→e).

Polyacetylene (PA), poly(3,4-ethylene dioxythiophene) (PEDOT), polypyrrole (PPY), polyaniline (PANI), poly(p-phenylene-vinylene) (PPV), polyfuran (PF), and polycarbazole (PCz) are some of the conjugated polymers that have been extensively studied. Owing to the wide range of unique properties (discussed in the next section), CPs have captivated the world. Replacing microsystem technologies with nanosystems is now demanded globally. CPs could be the game-changer in the search for substances that are lightweight, conductive, flexible, and durable. They have opened a new horizon in material science. This brief review addresses the current trends in the application of conjugated polymers and their future perspectives.

7.2 Properties of Conjugated Polymers

7.2.1 Electrical Properties

Solids are classified into insulators, conductors, or semiconductors based on energy band structure, arrangement of electrons in these bands, and the gap of forbidden energy. Free movement of valence electrons on the conduction band makes substances conductive. However, this band theory does not distinctly explain the conductive nature of conjugated polymers. Conjugated double bonds present in the backbone attribute a conductive property to CPs; σ bond electrons, due to their low mobility, have no contribution to electrical conductivity. However, delocalized π electrons are shifted between the adjacent σ bonded carbon atoms, which compel immediate vicinal π electrons to shift ahead. This continues throughout the chain making CP conductive. However, this does not make CPs highly conductive. Electrical conductivity greatly relies on conjugation length, the concentration of charge carriers, temperature, and the level of doping [5, 6]. Undoped CPs are electrical insulators. The conductivity of undoped polymers generally ranges from 10^{-6} to 10^{-10} S/cm, which lies between semiconductors and insulators. Doping to CPs increases electrical property drastically, even to ten or more orders of magnitude. Electrons are deliberately introduced or knocked off from the backbone of the delocalized π bond polymer, that is, CPs can be doped by n-type and p-type methods. However, the p-type is more common, whereas the n-type is not very stable and is in its infancy [4, 7]. These dopants undergo redox processes that produce charge carriers in the polymer by adding or removing electrons from the backbone of CP. Elevated electrical conductivity by doping can be explained by the increased charge carriers in the polymer. Electrons are withdrawn from the HOMO of the valence band of the polymer creating a hole in P-type doping (oxidation) or transmitted to the LUMO of the conduction band that increases electron density (reduction) [7]. These redox processes produce charge carriers – polarons or bipolarons in the polymer as shown in Figure 7.3.

The effect of doping on enhanced conductivity can be illustrated by considering polypyrrole (PPy). Undoped PPy with a high bandgap of 3.16 eV is a semiconductor but when adulterated with p-type substances, it oxidizes the polymer and the π electron is lost. This loss of the electron results in the deformation of the structure of the polymer from benzenoid to quinoid that yields the polaron as the backbone of PPy [8, 9]. As a result, the localized electronic level within the bandgap is created. Successive oxidation generates bipolarons, and this benzenoid-quinoid transformation is faster in bipolarons. Subsequent oxidation results in the overlapping of bipolarons, resulting in diminished bipolaronic bands and the bandgap is reduced to 1.4 eV from 3.16 eV as shown in Figure 7.4.

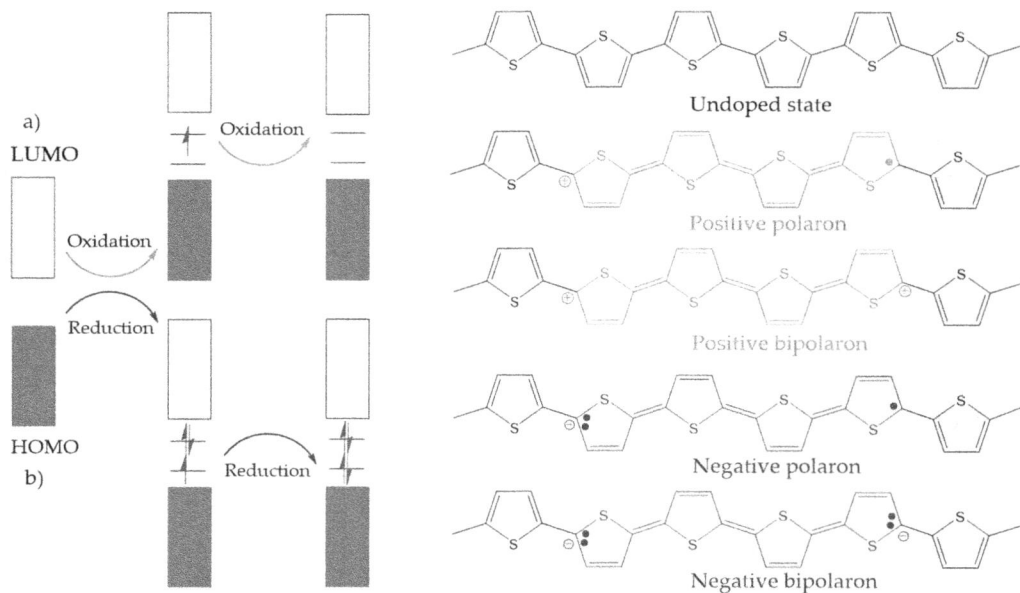

FIGURE 7.3
The electronic band and chemical structures of polythiophene (PTh) with (a) p-type doping and (b) n-type doping. Adapted with permission [7], Copyright © The Authors, some rights reserved, exclusive license [MDPI]. Distributed under a Creative Common Attribution License 4.0 (CCBY) https://creativecommons.org/licenses/by/4.0/.

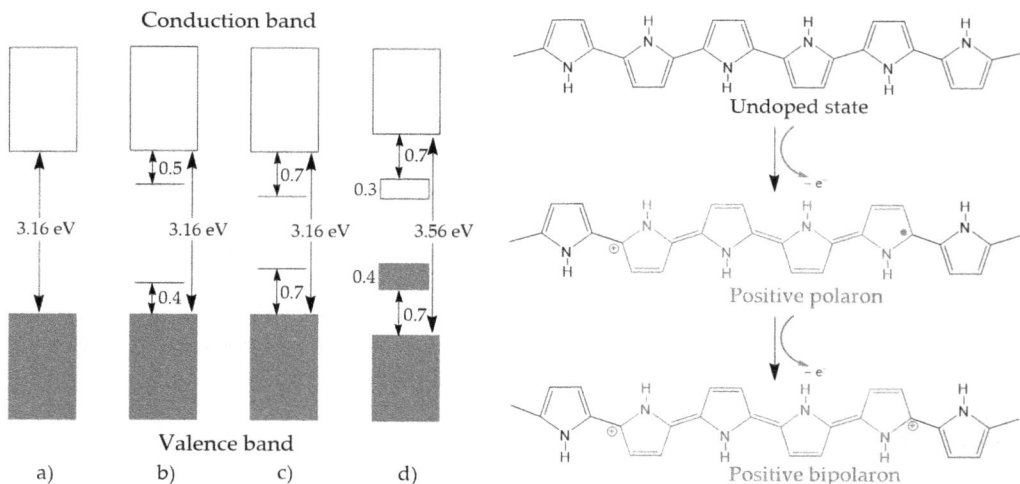

FIGURE 7.4
Electronic bands and chemical structures illustrate (a) undoped; (b) polarons; (c) bipolarons, and (d) fully doped states of PPy. Adapted with permission [7], Copyright © The Authors, some rights reserved, exclusive license [MDPI]. Distributed under a Creative Common Attribution License 4.0 (CCBY) https://creativecommons.org/licenses/by/4.0/.

The electrical properties of various conjugated polymers are being studied and PANI is considered to have the most potential than any other CPs [10]. Composites are formed by blending nanoparticles into the polymer to enhance electrical conductivity [11]. Such nanocomposites of CPs could meet the demand for flexible, highly conductive lightweight electronic devices.

7.2.2 Optical Properties

The optical properties of CPs are attributed to the delocalized π electron. When the semiconducting polymer absorbs a photon of sufficient energy, these π electrons are promoted from the HOMO to the LUMO, which forms exciton, electron-hole pairs. These excited state species relax by the emission of light and show the phenomenon of luminescence as shown in Figure 7.5 [4, 12]. The optical properties of conjugated polymers are affected by topochemical reaction, degree of anisotropy, and the length of conjugation. Conducting polymers in optical devices seems to be propitious due to the high visual figure of merit, excellent threshold of optical damage, flexibility, energy gap, and high optical response [2]. Nanocomposites with enhanced optical properties can be prepared that have high-end potential.

7.2.3 Mechanical Properties

One of the most crucial but less studied properties of CPs is the mechanical property. Mechanical deformability supports many other properties as the durability of products is a matter of concern. Low modulus, outstanding adhesion, and flexibility provide robust mechanical deformability to CPs. Crystalline polymer is considered to have better mechanical properties than an amorphous semi-crystalline one. Polymers with higher

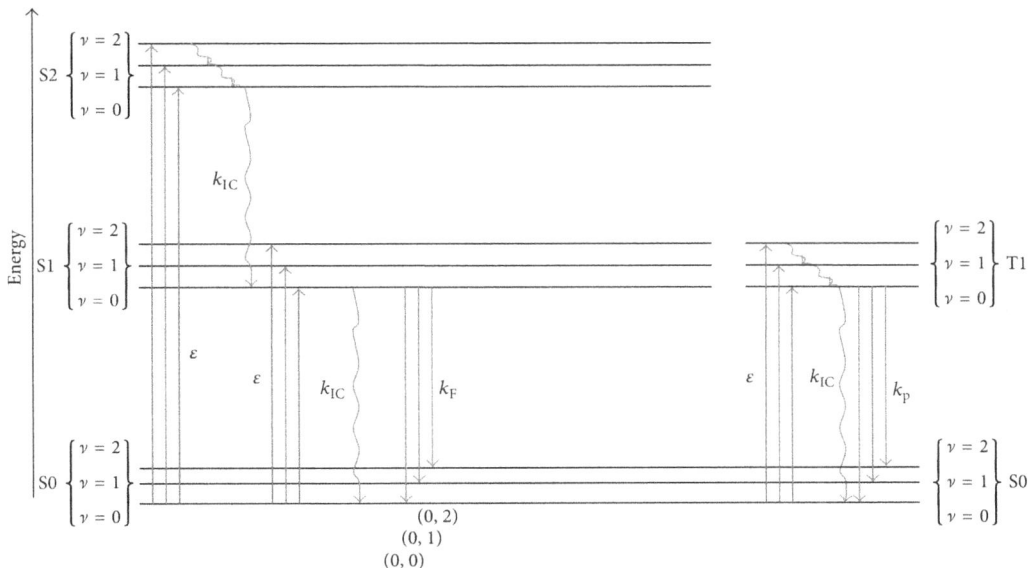

FIGURE 7.5
Energy state diagram exhibiting the phenomenon of luminescence. Adapted with permission [4], Copyright © The Authors, some rights reserved, exclusive license [Hindawi]. Distributed under a Creative Common Attribution License 4.0 (CCBY) https://creativecommons.org/licenses/by/4.0/.

molecular weight have enhanced toughness and strength parameters [13]. A molecular structure like the length of side chains, the existence of non-conjugated monomer units, the extent of polymerization, regioregularity, conditions like self-aggregation, and spin coating determine the mechanical properties of CP [14–16].

Mechanical flexibility required for flexible electronics can generally be achieved in two ways; adding secondary components such as elastomer matrixes, crosslinkers, and small molecular additives and the next more successful route of molecular structural modification of the CPs. Flexibility in CPs can be enhanced without a significant drop in electrical properties by adding flexible conjugation breakers, incorporating longer and softer side chains, or by using conjugated backbones that have high planarity but low packing tendency. D. Pei et al. [17] showed that hole mobility and elastic mobility are dependent on molecular weight and observed that a molecular weight of 88 kg/mol exhibited the best hole mobility and elastic mobility in a diketopyrrolopyrrole (DPP) based conjugated polymer. This infers that the polymer at the critical molecular weight is the key to achieving flexible and conductive polymers of interest for practical applications.

Various non-conjugated spacers like dodecyl segments incorporated in the backbone of DPP increased ductility without affecting charge mobility. It was observed that the polymer maintained the electrical mobility of >0.36 cm^2/V.s at 100% strain and after 100 cycles at 50% strain [18]. Moreover, hydrogen bonding in conjugation breakers incorporated in DPP increases packing, chain aggregation, and crystallinity. Increased strength of hydrogen bonding increases the modulus and crack on-set strain. The interaction of hydrogen bonding increased the degree of crystallinity and stretchability as well, which was predicted due to greater strain energy dissipation [19].

An increment in the length of the alkyl side chain decreased the tensile modulus of poly(3-alkylthiophenes) (P3HT) and [6,6]-phenyl C$_{61}$ butyric acid methyl ester (PCBM) [20]. Nano-structured composites of CPs have drawn much attention lately. Kalakonda et al. [21] fabricated nanoporous composites of single-walled carbon nanotubes (SWCNT) and a polyaniline polymer and found an increase in tensile modulus by 200% compared with the undoped PANI polymer at 25% weight SWCNT loading. Table 7.1 gives a summary of some important CPs with their unique properties.

7.3 Applications of Conjugated Polymers in Flexible Electronics

Technology has a great impact on society that transforms our lifestyles. A new revolution in the field of electronics is expected soon that will transform the fundamental interactions between humans and technology. Wearable and flexible electronics have already taken momentum in the field of entertainment, communication, and medical science. Flexibility is the virtue of a material to withstand bending deformation and cylindrical or conical form factors. Silicon is widely used as a semiconducting layer owing to its electrical properties, however, its rigidity and low tolerance to mechanical stress limit it from use in flexible electronics. This limitation of silicon can be outweighed by CPs. One of the promising advantages of CPs is their tunable properties. Various chemical structure engineering can be done to modify glass transition temperature and crystallinity without significant degradation of electrical properties. These modifications make them deformable and highly electrically conductive, therefore, suitable for flexible electronics. Flexible organic electronics have the advantages of being low cost, eco-friendly, ultra-thin, and ultra-lightweight,

TABLE 7.1

Different CPs with Their Unique Properties

SN	Polymer	Unique properties	Drawbacks
1	Polyacetylene (PA)	Flexible, stretchable up to three times its original length Electrical, optical, magnetic anisotropy	Processing difficulty, instability in the air, insoluble in solvents
2	Polyaniline (PANI)	pH changes properties, ease synthesis, low cost, environmentally stable, highly conductive, ease doped by protonic acids	Low processing capacity, inflexibility, lack of biodegradability, poor solubility
3	Polypyrrole (PPy)	Chemical and electrochemical stability, insoluble in solvents but swellable, ease synthesis and formation of composites, excellent stimulus-responsive	Mechanically weak
4	Polythiophenes (PTh)	High environmental stability, high optical properties, ease of solution processability	Poor solubility
5	Poly(3,4-ethylene dioxythiophene) (PEDOT)	Thermoelectric polymer, high stability, optically transparent (conducting state)	Poor solubility
6	Poly(p-phenylene-vinylene) (PPV)	Tunable high optical property, good reactivity, high electrical conductivity, mechanically strong	Ease of aggregation, fluorescence quenching
7	Polycarbazole (PCz)	Excellent morphological stability, higher redox potential, good electro and photoactive properties	Difficult process, less efficient OPV devices
8	Polyfuran (PF)	Thermally stable, smooth morphology, rigidity of backbone	Low reduction potential, difficult to synthesize, lower stability than other heterocyclic polymers

which opens a new dimension for developing novel devices in the future. Figure 7.6 displays the various applications of CPs in the field of flexible electronics.

7.3.1 Supercapacitor

Supercapacitors are energy storage devices that accumulate charge through agile reversible redox reactions that are best suited for high-power applications and restoration energy harvest. Supercapacitors are the propitious next-generation device for storing energy in portable electronic devices and electronic motors. They can store 1000 times more energy than dielectric capacitors and possess a simple principle, a rapid charge-discharge cycle, good life cycle, high charge density, and high dynamic charge propagation [22, 23]. Conjugated polymers having high specific capacitance can be suitable for supercapacitors. Conjugated polymers like polyaniline and polypyrrole with high electrical conductivity, good redox reversibility, mechanical strength, and eco-friendly properties have made them applicable in preparing supercapacitors. J. Zhang et al. [24] prepared flexible 2D PEDOT and WO_3 PEDOT sheets and used them as an electrode in Swagelok and pouch-type supercapacitors, which exhibited excellent energy storage characteristics, high-rate capacity, and thickness-independent energy storage. They claimed areal capacitance of 701 mF/cm^2 with a 1 mm thick film and energy density of 0.083 mWh/cm^2 at a power density of 10 mW/cm^2. Z. Huang et al. [25] synthesized a flexible and stretchable PANI supercapacitor that exhibited high-rate capability. Figure 7.7(a) shows the galvanostatic charge-discharge (GCD) curves at different stretching states ranging from 5% to 20%. Initially, it exhibited

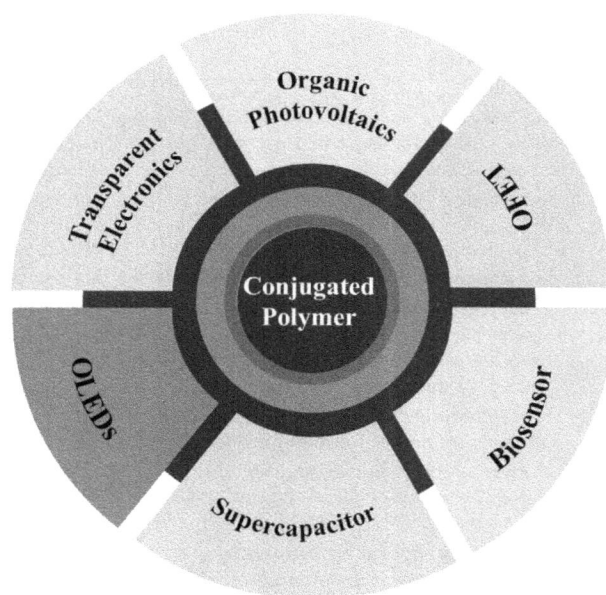

FIGURE 7.6
Applications of CPs in stretchable electronics.

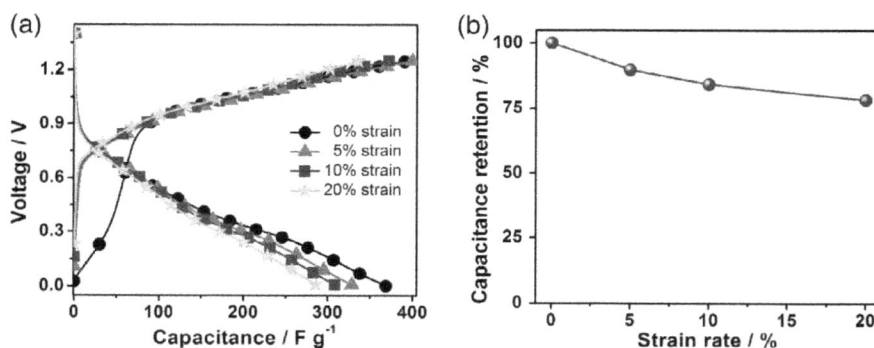

FIGURE 7.7
Electrochemical performance of the PANI supercapacitor. (a) GCD curve at different strains (b) Capacitance retention at various strain rates. Adapted with permission from Reference [25], Copyright © (2021), John Wiley and Sons.

a capacitance of 369 F/g without strain, which reduced marginally to 331 F/g at 5% strain. Even under a higher strain of 20%, it exhibited a satisfactory result of 290 F/g. Figure 7.7(b) shows the capacitance retention rate at different strain levels that exhibit a slight reduction in the capacitance value but at a satisfactory level.

7.3.2 Light-Emitting Diodes

A light-emitting diode (LED) is a device that emits light in response to an electric current. Essentially, in semiconductors, electrons recombine with holes that release energy in the form of photons that gives the specific color of light. OLED is constructed using organic carbon-based film made up of conjugated polymers sandwiched between two electrodes. Various colored OLEDS like blue, red, and multicolored have been synthesized using different CPs, carbazole derivatives for blue, red, and thiophene derivatives for multicolored [26]. Low driving voltage, low consumption of power, and flexible fabrication make OLEDs solid thin film luminescent devices that are propitious for panel displays and wearable devices. The emergence of polymers as polymer light-emitting diodes (PLEDS) is a revolution in electronics. Their advantages over traditional incandescent light-emitting devices are remarkable. Low power consumption, eco-friendly, and durability are their striking aspects [27–29]. The basic principle involves injecting an electron from a metal cathode to the LUMO into the emitting polymer and a hole from the anode to the HOMO in the conducting polymer. The convergence of holes and electrons in the semiconducting polymer produces light. Polymers like PANI, poly (9-vinylcarbacazole), poly (p-phenylene vinylene), and so on are commonly used. In the quest for efficient and economic PLEDs, research has focused on developing new materials like layered silicate nanocomposites and metallic nanocomposites that can provide a better and more durable alternative to traditional materials [30, 31]. Zhong et al. [32] synthesized a perfluorinated ionomer (PFI) modified poly(3,4-ethylenedioxythiophene) polystyrene sulfonate (PEDOT: PSS) layer, which exhibited higher work function than the regular PEDOT: PSS. D.H. Jiang [33] demonstrated that a polyfluorene-based light-emitting diode device made by conjugated polymer poly(9,9-di-n-octyl-2,7-fluorene) (PFO) as the emissive material emitted light under 20% stretch and bending, which revealed the potential of CPs in the field of flexible organic LEDs.

7.3.3 Transparent Electronics

Various conventional electrical appliances using available silicon technologies are opaque and rigid. A versatile design of transparent and flexible smart next-generation electronics have emerged as a formidable challenge. Truly flexible and perfectly transparent high-mobility semiconducting material has been demonstrated based on a polymeric semiconductor system by K. Yu et al. [34]. They developed a blended diketopyrrolopyrrole (DPP)-based semiconducting polymer (about 15%) into an inert polystyrene matrix to obtain a polymer blend system that exhibited field-effect transistor mobility and optical transparency of 100%.

7.3.4 Sensors

New trends in the analysis of environmental, biomedical, food, and beverages have emerged that involve conjugated polymers that have high sensitivity and a short response time. Conjugated polymers have applications in chemical sensors, optical sensors, and biosensors. Poly (vinylidene fluoride-co-hexafluoropropene) (PVDF-HFP)/PEDOT nanofibers were fabricated by O. Young Kweon and his team [35], which was used to construct a pressure sensor that exhibited an impressive pressure sensitivity of 13.5 kPa^{-1}. A prototype wireless blood pressure sensor from these fibers was devised in the form of a wristband that sensed the blood pressure and displayed diagnostic signals on a smartphone as shown in Figure 7.8.

7.3.5 Chemical Sensors

Various gases like H_2, NH_3, oxides of nitrogen, and toxic gases released from different industries have created environmental and human health issues. Analyzing their

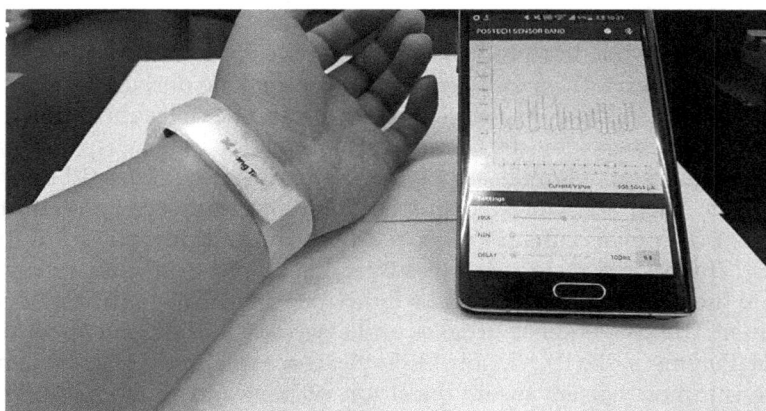

FIGURE 7.8
Photograph of PVDF-HFP/PEDOT-based wireless wrist blood pressure sensor displaying diagnostic signals on a smartphone. Adapted with permission [35], Copyright © The Authors, some rights reserved, exclusive license [Springer Nature]. Distributed under a Creative Common Attribution License 4.0 (CCBY) https://creativecommons.org/licenses/by/4.0/.

concentration requires sensitive sensors that can be made from CPs like PANI, PPY, PTh, and so on [6]. The basic working of such sensors relies on the reaction with the targeted gas that produces an electrical signal that is directly proportional to the concentration of the gas. Among different conjugated polymers, PANI has been used effectively as a sensor due to its ease of synthesis, doping and de-doping behavior, reactivity toward gases, and deposition on substrates [36]. Different composite nanomaterials of CPs are more effective than conventional materials [37]. Polypyrrole/Fe_2O_3 nanocomposites were synthesized by Wang et al. [38] by the one-step hydrothermal route, which exhibited high selectivity and response to NO_2 in the range of 0.1–10 ppm at 50°C. The composite displayed an excellent sensitivity of 220.7% compared with existing records. He et al. [39] developed a PANI/WO_3 cotton thread-based flexible sensor able to sense NH_3 at room temperature. The result revealed that the optimal sensor (10 mol% WO_3) responded to a 6.0–100 ppm concentration of NH_3 that was higher than the sensors from undoped PANI and other composites.

7.3.6 Biosensors

Conjugated polymers are flexible, biocompatible, and can mold into a desired shape and structure. They can be integrated into micro or nanoscale for in vivo sensing and tracking bio analytes [40]. They deliver excellent matrixes for the immobilization of various biorecognition elements, with molecules keeping their redox sites in good communication with transducing electrodes, thereby enhancing the efficiency of biosensors [41]. Different tools and methodologies are in trend for making biosensors extend from conducting polymers and their nanowires, with polymeric films lodged with metal nanoparticles and conducting polymer blends in ionic fluids and polymeric ionic fluids [42]. PPy-based biocomposite materials synthesized by enzymatic and microbial synthesis are suited to make biosensors [43]. Gokoglan et al. [44] synthesized a composite of vertically aligned carbon nanotubes and CP as the immobilization matrix to construct a flexible glucose sensor with high sensitivity of 65.816 µA/mMcm². PANI-NiCo$_2$O$_4$ nanocomposite was synthesized by Yu et al. [45] to determine glucose concentration, which showed higher electrochemical stability, good reproducibility, and excellent selectivity. Belgis [46] developed a cholesterol analysis method using an enzyme immobilization technique on chitosan-polyaniline film. The study revealed it as an eco-friendly material for cholesterol biosensors.

7.3.7 Organic Photovoltaic Solar Cells

Rising energy prices, dwindling fossil fuels, elevated environmental pollution, and the increasing need for energy alternatives drive the demand for green energy, including solar energy. Photovoltaics (PV), changing sunlight into electrical power, technology is anticipated to play a leading role in the renewable energy supply industry in the future. Silicon-based solar cells are more efficient (η=25%) than polymer solar cells. However, they are more expensive, and the global concern is to construct cost-effective solar cells. CPs are a promising material for PV due to their flexibility, low cost, and lightweight nature [47–49]. Polymer solar cells comprise a bilayer structure with p-type and n-type layers for conducting holes and electrons, respectively sandwiched between two conductive electrodes. A bulk heterojunction layer, a blend of p- and n-type materials, is used to make efficient polymer solar cells [12, 50]. Recent progress has over 13% efficiency of organic solar cells. Fluorination of conjugated polymers has been observed to increase efficiency [51]. Flexible photovoltaics that convert light into electricity have now emerged as wearable, sustainable, and eco-friendly energy sources. Among various CPs, PEDOT: PSS has abilities such as solution processability, high optical transparency, cost-effective fabrication, flexibility, and excellent thermal stability. These organic flexible photovoltaics can have applications in the construction of solar trees, self-powered skin wearable sensors, and washable electric textiles for self-powered smart clothing [52]. A new type of polymer based on indaceno-dithiophene in conjugation with dicyanomethylene indanone motifs has been developed that exceeded the 10% power conversion efficiency benchmark [53]. Table 7.2 shows various CPs that have potential in flexible polymer solar cells with electrical conductivity and mechanical properties.

TABLE 7.2

Different Types of CPs and Their Physical Properties

Material	Type	Conductivity	Mechanical properties	Contribution
PA	Conducting/ semiconducting	1.7×10^{-11}–3.8×10^{-2} S/m	Bendable (in nanocomposite film) [55]	Active material and electrode material
PPy	Conducting	0.02-1 S/m	No difference upon application of 13% bending strain (in planar supercapacitor) [56]	Active material and electrode material
PEDOT	Conducting	550–1000 S/m	Slightly decreased after bending 1000 cycles (in fiber-shaped PSC) [57]	Active material and electrode material
PPV	Conducting	10^{-5}–1 S/m	Bendable (in organic planar solar cell) [58]	Active material and electrode material
PANI	Conducting/ semiconducting	6.3×10^{-9}–4.6×10^{-5} S/m	Well maintained after being stretched for 100 cycles to a strain of 100% (in fiber-shaped supercapacitor) [59]	Active material and electrode material

Adapted with permission [54], Copyright © (2018), John Wiley and Sons.

7.3.8 Organic Field-Effect Transistors

Organic field-effect transistors (OFETs) are organic materials-based electronic devices that utilize electric fields to curb the conductance of the material. CP-based OFETs possess unique qualities like mechanical flexibility, solution processability, and tunable optoelectronic properties compared with traditional silicon electronics. Contrary to traditional silicon-based field-effect transistors, OFETs use organic semiconductors like conjugated polymers as the active material, which helps to process the devices at a lower temperature and cost. The performance of OFETs mainly depends on charge transfer mobility. Liu et al. developed a novice method that improved the charge carrier mobility by the synergetic effect of pyridine and selenophene in the backbone of a DPP-based copolymer [60]. The research revealed that the pyridine DPP and selenophene-based copolymer showed low LUMO and HOMO energy levels as a consequence of the packing of molecules that elevated the hole transfer barrier and decreased the electron transfer barrier. Mobility of more than $10 \text{ cm}^2 \text{ V}^{-1} \text{ S}^{-1}$ has been obtained in OFETs, opening doors for their applications in flexible displays and wearable devices [61]. Zhang et al. synthesized stretchable transistor arrays based on P3HT-PDMS IPN blend films and devices that exhibited transparency and stretchability as shown in Figure 7.9 [62]. The device exhibited average mobility of $0.12 \text{ cm}^2 \text{ V}^{-1} \text{ S}^{-1}$ with only a minute fall in mobility on 100% strain. The mobility was recovered to its initial value on releasing strain over 300 stretching cycles.

7.4 Challenges

One of the major challenges in obtaining a flexible, stretchable, and conducting polymer is its solid-state morphology. Higher electrical conductivity is expected in the polymer with a higher degree of crystallinity as the charge transport occurs by the charge hopping mechanism that demands closer conjugated backbones. However, this rigid and increased order of morphology due to strong π stacking in between the polymeric chain decreases the flexibility of the polymer (Figure 7.10) [63]. In contrast, the amorphous morphology of the polymer enables free movement of the polymer chain under strain making them more stretchable and flexible. However, this structure with better mechanical compliance is not favorable for charge transport.

FIGURE 7.9
Photograph showing high transparency of a transistor without (a) and with (b) strain. Adapted with permission from Reference [61], Copyright © (2017), American Chemical Society.

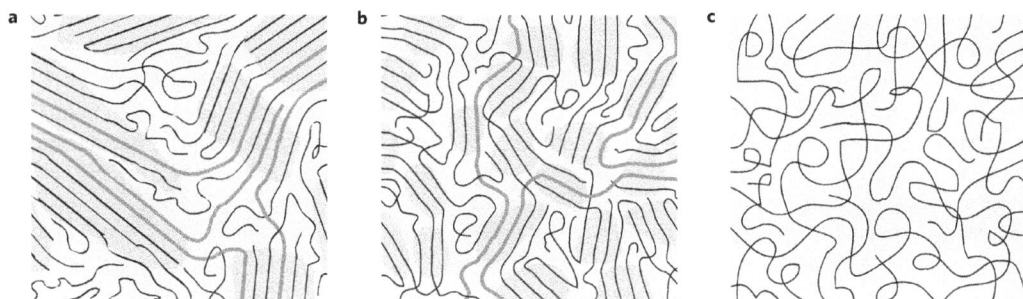

FIGURE 7.10
Schematic microstructure of thin film of CPs. a) Semi-crystalline structure suitable for charge transport; b) semi-crystalline disordered structure supposed to be ideal for mechanical and electrical properties; c) amorphous morphology favorable for good mechanical compliance. Adapted with permission from Reference [62], Copyright © (2013), Springer Nature.

The mediocre performance of CPs, the cost benefits of envisioned technology, and the suitable infrastructure for their production in the field of supercapacitors, lighting, and transistors make them struggle compared with the current commercial reign of silicon-based electronics.

Deeper knowledge of the time-dependent mechanical behavior of these organic semiconductors is essential. How the microstructure of an organic semiconductor emerges over time when strained must be known for the viability of flexible electronics.

Longevity, stability, and durability – surviving many cycles of bending, rolling, or flexing without critical failures and degradation of the CP-based material – in the production of flexible electronics is also a challenge for their commercial application.

Finally, it may be concluded that the stiffness of the conjugated backbone, high molecular weight, regioregularity, conjugation breakers, and the side chain are now known to play a pivotal role in the flexibility and electrical properties of conjugated polymers. In a special and intriguing charge transport regime seen in CPs, structure and charge dynamics are tightly connected. Compact packing arrangements along with the degree of crystallinity are disagreeable for flexibility but are critical for charge transport, however, it is a challenge to find an avenue perfectly balancing them. Although some progress has been achieved in the development of components in flexible electronics, the design of flexible commercial CPs is still a big challenge [64].

7.5 Conclusion and Future Perspectives

Conjugated polymers have drawn global attention since their advent and have made significant progress in synthesis techniques, structural and morphological characterization, physical and chemical properties, and commercial applications. Recent trends in the enhancement of properties of CPs include the fabrication of composites of polymer nanocomposites with other materials like graphene, carbon nanotubes, or inorganic compounds. The future of CPs is expected to expand via interlinked activities of enhanced molecular design, consummate understanding of the material, and exploration of the commercialization sector.

Owing to high electrical conductivity, ease of fabrication, flexibility, versatile bandgap properties, and sizeable specific capacitance, CPs have enormous potential in supercapacitor applications. However, they have a low cycle life compared with their rivals, and carbonaceous-based supercapacitors drag them behind. To overcome this limitation, a hybrid nanostructure of conjugated polymers with carbonaceous materials like CNT, graphene, and other metal oxides Co_3O_4, RuO_2 can be prepared [47, 65]. Viable commercial applications in supercapacitors still have a long journey ahead. CP-based supercapacitors' cycle life is expected to be increased by compositing CPs with carbonaceous materials and metal oxides.

Conjugated polymers have brought a paradigm shift in the field of optoelectronics. Flexible polymer light-emitting diodes have applications in foldable electronic gadgets and curved and ultra-thin high-definition electronic displays. However, cost, low light efficiency, shorter lifetime, and durability make them challenging to existing technologies. Modification of polymers, copolymerization, doping by nanoparticles, phosphorescent emitters, and other techniques will do away with these issues. Organic photovoltaic (OPV) solar cells have been shown as an alternative source of energy to fossil fuels due to their flexibility and low weight and cost. Transparent solar windows, a cutting-edge technology possible by CPs, are expected to transform large cities from energy consumers to energy suppliers. The efficiencies in the conversion of power of such OPV solar cells can be increased by lowering the bandgap, which can be achieved by chemical modification of the polymer structure. The challenge to efficient organic solar cells lies in the fabrication of a polymer material with high solar spectrum absorbance and ease of processing methods. Incongruity in the absorption spectra of the polymer and solar irradiance spectrum has forced a lower efficiency in solar cells [49]. Cumulative efforts to increase the efficiency of economic OPV solar cells will surpass the existing expensive silicon-based solar cells.

Conjugated polymer nanotubes and nanofibers are magnificent candidates for biochemical sensors. Agile responsivity and recovery time, superb sensitivity, stability, and selectivity are the characteristics of an exemplary sensor. Constant efforts by different researchers with developed nanotechnology have grown the potential of CPs as sensors. However, there is more room for selectivity in the future. Composites of polymer with suitable receptors may be helpful to overcome this limitation. Flexible, even wearable, highly efficient biosensors will shortly be available [42]. Polyaniline and polypyrrole have proven to be efficient in electrochemical sensors due to their surface area [36]. Other potential polymers must be studied, and they could prove to be superior to the current ones.

Conjugated polymers are novice materials in the biomedical engineering field. However, they are promising in tissue engineering, actuators, drug delivery, diagnosis, molecular imaging, and cancer therapy in their early stages. The central issue of their implementation in the field of tissue engineering is biodegradability. Attempts have been made to overcome this issue by blending CPs with biodegradable polymers like polylactide, polycaprolactone, polyurethane, and so on, and integrating modified CP monomers with erodible monomer units. Future works in biodegradable CPs will ensure safety and quality assurance for their implementations in the clinical phase.

On the whole, conjugated polymers with good mechanical and electrical properties and ease of processability will find their best application in a commercial setting. The deeper and entirely novel knowledge of transport phenomena can also be explored to understand the synergy between mechanical and electrical properties. Flexible and mechanically robust electronics will undoubtedly open a new horizon in the future. Although commercial applications of CPs in various fields are struggling, collaborative works will undoubtedly lead to a breakthrough.

References

1. S. C. Rasmussen, "Conjugated and conducting organic polymers: The first 150 years," *Chempluschem*, vol. 85, no. 7, pp. 1412–1429, 2020.
2. M. S. AlSalhi, J. Alam, L. A. Dass and M. Raja, "Recent advances in conjugated polymers for light emitting devices," *Int. J. Mol. Sci.*, vol. 12, no. 3, pp. 2036–2054, 2011.
3. A. G. MacDiarmid, "Synthetic metals: A novel role for organic polymers (Nobel Lecture)," *Angew. Chem. Int. Ed.*, vol. 40, pp. 2581–2590, 2001.
4. M. Bajpai, R. Srivastava, R. Dhar and R. S. Tiwari, "Review on optical and electrical properties of conducting polymers," *Indian J. Mater. Sci.*, vol. 2016, pp. 1–8, 2016.
5. A. K. Bakhshi, "Electrically conducting polymers: From fundamental to applied research," *Bull. Mater. Sci.*, vol. 18, no. 5, pp. 469–495, 1995.
6. R. Kumar, S. Singh and B. C. Yadav, "Conducting polymers: Synthesis, properties and applications," *Int. Adv. Res. J. Sci. Eng. Technol.*, vol. 2, no. 11, pp. 110–124, 2015.
7. T. H. Le, Y. Kim and H. Yoon, "Electrical and electrochemical properties of conducting polymers," *Polymers (Basel).*, vol. 9, no. 4, 2017.
8. A. O. Patil, A. J. Heeger and F. Wudl, "Optical properties of conducting polymers," *Chem. Rev.*, vol. 88, no. 1, pp. 183–200, 1988.
9. J. L. Bredas and G. B. Street, "Polarons, bipolarons, and solitons in conducting polymers," *Acc. Chem. Res.*, vol. 18, no. 10, pp. 309–315, 1985.
10. A. Kausar, "Review on structure, properties and appliance of essential conjugated polymers," *Am. J. Polym. Sci. Eng.*, vol. 4, no. 1, pp. 91–102, 2016. Accessed: Jun. 23, 2021.
11. S. K. Pillalamarri, F. D. Blum, A. T. Tokuhiro and M. F. Bertino, "One-pot synthesis of polyaniline - Metal nanocomposites," *Chem. Mater.*, vol. 17, no. 24, pp. 5941–5944, 2005.
12. S. Hossain, "Optical properties of polymers and their applications," Master's Thesis; New Jersey Institute of Technology University, 2019.
13. X. S. Wang, H. P. Tang, X. D. Li and X. Hua, "Investigations on the mechanical properties of conducting polymer coating-substrate structures and their influencing factors," *Int. J. Mol. Sci.*, vol. 10, no. 12, pp. 5257–5284, 2009.
14. A. X. Chen, A. T. Kleinschmidt, K. Choudhary and D. J. Lipomi, "Beyond stretchability: Strength, toughness, and elastic range in semiconducting polymers," *Chem. Mater.*, vol. 32, no. 18, pp. 7582–7601, 2020.
15. R. Xie, R. H. Colby and E. D. Gomez, "Connecting the mechanical and conductive properties of conjugated polymers," *Adv. Electron. Mater.*, vol. 4, no. 10, 2017.
16. S. E. Root, S. Savagatrup, A. D. Printz, D. Rodriquez and D. J. Lipomi, "Mechanical properties of organic semiconductors for stretchable, highly flexible, and mechanically robust electronics," *Chem. Rev.*, vol. 117, no. 9, pp. 6467–6499, 2017.
17. D. Pei, Z. Wang, Z. Peng, J. Zhang, Y. Deng, Y. Han, L. Ye and Y. Geng, "Impact of molecular weight on the mechanical and electrical properties of a high-mobility diketopyrrolopyrrole-based conjugated polymer," *Macromolecules*, vol. 53, no. 11, pp. 4490–4500, 2020.
18. J. Mun, G. J. N. Wang, J. Y. Oh, T. Katsumata, F. L. Lee, J. Kang, H. C. Wu, F. Lissel, S. Rondeau-Gagné, J. B. H. Tok and Z. Bao, "Effect of nonconjugated spacers on mechanical properties of semiconducting polymers for stretchable transistors," *Adv. Funct. Mater.*, vol. 28, no. 43, pp. 1–10, 2018.
19. Y. Zheng, M. Ashizawa, S. Zhang, J. Kang, S. Nikzad, Z. Yu, Y. Ochiai, H. C. Wu, H. Tran, J. Mun, Y. Q. Zheng, J. B. H. Tok, X. Gu and Z. Bao, "Tuning the mechanical properties of a polymer semiconductor by modulating hydrogen bonding interactions," *Chem. Mater.*, vol. 32, no. 13, pp. 5700–5714, 2020.
20. S. Savagatrup, A. S. Makaram, D. J. Burke and D. J. Lipomi, "Mechanical properties of conjugated polymers and polymer-fullerene composites as a function of molecular structure," *Adv. Funct. Mater.*, vol. 24, no. 8, pp. 1169–1181, 2014.

21. P. Kalakonda, P. B. Kalakonda and S. Banne, "Studies of electrical, thermal, and mechanical properties of single-walled carbon nanotube and polyaniline of nanoporous nanocomposites," *Nanomater. Nanotechnol.*, vol. 11, pp. 1–8, 2021.
22. C. Meng, O. Z. Gall and P. P. Irazoqui, "A flexible super-capacitive solid-state power supply for miniature implantable medical devices," *Biomed. Microdevices*, vol. 15, no. 6, pp. 973–983, 2013.
23. J. R. Miller and P. Simon, "Materials science: Electrochemical capacitors for energy management," *Science (80-.).*, vol. 321, no. 5889, pp. 651–652, 2008.
24. J. Zhang, X. Fan, X. Meng, J. Zhou, M. Wang, S. Chen, Y. Cao, Y. Chen, C. W. Bielawski and J. Geng, "Ice-templated large-scale preparation of two-dimensional sheets of conjugated polymers: Thickness-independent flexible supercapacitance," *ACS Nano*, vol. 15, no. 5, pp. 8870–8882, 2021.
25. Z. Huang, Z. Ji, Y. Feng, P. Wang and Y. Huang, "Flexible and stretchable polyaniline supercapacitor with a high rate capability," *Polym. Int.*, vol. 70, no. 4, pp. 437–422, 2020.
26. S. Jadoun and U. Riaz, "Conjugated polymer light-emitting diodes," in *Polymers for Light-Emitting Devices and Displays*, John Wiley & Sons, 2020, pp. 77–98.
27. J. Y. Low, Z. M. Aljunid Merican and M. F. Hamza, "Polymer light emitting diodes (PLEDs): An update review on current innovation and performance of material properties," *Mater. Today Proc.*, vol. 16, pp. 1909–1918, 2019.
28. X. Guo and A. Facchetti, "The journey of conducting polymers from discovery to application," *Nat. Mater.*, vol. 19, no. 9, pp. 922–928, 2020.
29. F. Bekkar, F. Bettahar, I. Moreno, R. Meghabar, M. Hamadouche, E. Hernaez, J. L. Vilas-Vilela and L. Ruiz-Rubio, "Polycarbazole and its derivatives: Synthesis and applications. A review of the last 10 years," *Polymers (Basel).*, vol. 27, no. 10, 2020.
30. O. Y. Posudievsky, D. A. Lypenko, O. A. Khazieieva, O. L. Gribkova, A. A. Nekrasov, A. V. Vannikov, V. M. Sorokin, V. G. Koshechko and V. D. Pokhodenko, "Nanocomposite of polyaniline with partially oxidized graphene as the transport layer of light-emitting polymer diodes," *Theor. Exp. Chem.*, vol. 50, no. 2, pp. 96–102, 2014.
31. L. Qian, Y. Zheng, K. R. Choudhury, D. Bera, F. So, J. Xue and P. H. Holloway, "Electroluminescence from light-emitting polymer/ZnO nanoparticle heterojunctions at sub-bandgap voltages," *Nano Today*, vol. 5, no. 5, pp. 384–389, 2010.
32. Z. Zhong, Y. Ma, H. Liu, F. Peng, L. Ying, S. Wang, X. Li, J. Peng and Y. Cao, "Improving the performance of blue polymer light-emitting diodes using a hole injection layer with a high work function and nanotexture," *ACS Appl. Mater. Interfaces*, vol. 12, no. 18, pp. 20750–20756, 2020.
33. D.-H. Jiang, "Flexible light-emitting diode application of polyfluorene-based conjugated polymers," Ph.D. dissertation, Hokkaido University, Sapporo, 2021.
34. K. Yu, B. Park, G. Kim, C. H. Kim, S. Park, J. Kim, S. Jung, S. Jeong, S. Kwon, H. Kang, J. Kim, M. H. Yoon and K. Lee, "Optically transparent semiconducting polymer nanonetwork for flexible and transparent electronics," *Proc. Natl. Acad. Sci. U. S. A.*, vol. 113, no. 50, pp. 14261–14266, 2016.
35. O. Y. Kweon, S. J. Lee and J. H. Oh, "Wearable high-performance pressure sensors based on three-dimensional electrospun conductive nanofibers," *NPG Asia Mater.*, vol. 10, no. 6, pp. 540–551, 2018.
36. K. Namsheer and C. S. Rout, "Conducting polymers: A comprehensive review on recent advances in synthesis, properties and applications," *RSC Adv.*, vol. 11, no. 10, pp. 5659–5697, 2021.
37. T. A. Rajesh and D. Kumar, "Recent progress in the development of nano-structured conducting polymers/nanocomposites for sensor applications," *Sensors Actuators, B Chem.*, vol. 136, no. 1, pp. 275–286, 2009.
38. C. Wang, M. Yang, L. Liu, Y. Xu, X. Zhang, X. Cheng, S. Gao, Y. Gao and L. Huo, "One-step synthesis of polypyrrole/Fe2O3 nanocomposite and the enhanced response of NO2 at low temperature," *J. Colloid Interface Sci.*, vol. 560, no. 2, pp. 312–320, 2020.
39. M. He, L. Xie, G. Luo, Z. Li, J. Wright and Z. Zhu, "Flexible fabric gas sensors based on PANI/WO3 p–n heterojunction for high performance NH3 detection at room temperature," *Sci. China Mater.*, vol. 63, no. 10, pp. 2028–2039, 2020.

40. S. Nambiar and J. T. W. Yeow, "Conductive polymer-based sensors for biomedical applications," *Biosens. Bioelectron.*, vol. 26, no. 5, pp. 1825–1832, 2011.

41. W. A. El-Said, M. Abdelshakour, J. H. Choi and J. W. Choi, "Application of conducting polymer nanostructures to electrochemical biosensors," *Molecules*, vol. 25, no. 2, pp. 1–11, 2020.

42. F. Ghorbani Zamani, H. Moulahoum, M. Ak, D. Odaci Demirkol and S. Timur, "Current trends in the development of conducting polymers-based biosensors," *TrAC - Trends Anal. Chem.*, vol. 118, pp. 264–276, 2019.

43. S. Ramanavicius and A. Ramanavicius, "Conducting polymers in the design of biosensors and biofuel cells," *Polymers (Basel).*, vol. 13, no. 1, pp. 1–19, 2021.

44. T. C. Gokoglan, S. Soylemez, M. Kesik, I. B. Dogru, O. Turel, R. Yuksel, H. E. Unalan and L. Toppare, "A novel approach for the fabrication of a flexible glucose biosensor: The combination of vertically aligned CNTs and a conjugated polymer," *Food Chem.*, vol. 220, pp. 299–305, 2017.

45. Z. Yu, H. Li, X. Zhang, N. Liu, W. Tan, X. Zhang and L. Zhang, "Facile synthesis of NiCo2O4@ Polyaniline core-shell nanocomposite for sensitive determination of glucose," *Biosens. Bioelectron.*, vol. 75, pp. 161–165, 2016.

46. Belgis, "Cholesterol biosensor based on polyaniline conducting chitosan film," *Malaysian J. Med. Heal. Sci.*, vol. 15, no. 17, pp. 1–6, 2019.

47. M. Ates, T. Karazehir and A. S. Sarac, "Conducting polymers and their applications," *Curr. Phys. Chem.*, vol. 2, no. 3, pp. 224–240, 2012.

48. T. Tadesse, "Application of conjugated organic polymers for photovoltaic's: Review," *J. Phys. Chem. Biophys.*, vol. 8, no. 1, 2018.

49. A. R. Murad, A. Iraqi, S. B. Aziz, S. N. Abdullah and M. A. Brza, "Conducting polymers for optoelectronic devices and organic solar cells: A review," *Polymers (Basel).*, vol. 12, no. 11, pp. 1–47, 2020.

50. Y. J. Cheng, S. H. Yang and C. S. Hsu, "Synthesis of conjugated polymers for organic solar cell applications," *Chem. Rev.*, vol. 109, no. 11, pp. 5868–5923, 2009.

51. Q. Zhang, M. A. Kelly, N. Bauer and W. You, "The curious case of fluorination of conjugated polymers for solar cells," *Acc. Chem. Res.*, vol. 50, no. 9, pp. 2401–2409, 2017.

52. C. Liu, C. Xiao, C. Xie and W. Li, "Flexible organic solar cells: Materials, large-area fabrication techniques and potential applications," *Nano Energy*, vol. 89, p. 106399, 2021.

53. M. Sommer, "Development of conjugated polymers for organic flexible electronics," in *Organic Flexible Electronics*, P. Cosseddu and M. Caironi, Eds. Woodhead Publishing, 2021, pp. 27–70.

54. Z. Zhang, M. Liao, H. Lou, Y. Hu, X. Sun and H. Peng, "Conjugated polymers for flexible energy harvesting and storage," *Adv. Mater.*, vol. 30, no. 13, pp. 1–19, 2018.

55. W. Cai, M. Li, S. Wang, Y. Gu, Q. Li and Z. Zhang, "Strong, flexible and thermal-resistant CNT/ polyarylacetylene nanocomposite films," *RSC Adv.*, vol. 6, no. 5, pp. 4077–4084, 2016.

56. T. G. Yun, B. Hwang, D. Kim, S. Hyun and S. M. Han, "Polypyrrole-MnO2-coated textile-based flexible-stretchable supercapacitor with high electrochemical and mechanical reliability," *ACS Appl. Mater. Interfaces*, vol. 7, no. 17, pp. 9228–9234, 2015.

57. Z. Zhang, X. Chen, P. Chen, G. Guan, L. Qiu, H. Lin, Z. Yang, W. Bai, Y. Luo and H. Peng, "Integrated polymer solar cell and electrochemical supercapacitor in a flexible and stable fiber format," *Adv. Mater.*, vol. 26, no. 3, pp. 466–470, 2014.

58. G. Dennler, C. Lungenschmied, H. Neugebauer, N. S. Sariciftci, M. Latrèche, G. Czeremuszkin and M. R. Wertheimer, "A new encapsulation solution for flexible organic solar cells," *Thin Solid Films*, vol. 511–512, pp. 349–353, 2006.

59. X. Chen, H. Lin, J. Deng, Y. Zhang, X. Sun, P. Chen, X. Fang, Z. Zhang, G. Guan and H. Peng, "Electrochromic fiber-shaped supercapacitors," *Adv. Mater.*, vol. 26, no. 48, pp. 8126–8132, 2014.

60. Q. Liu, S. Kumagai, S. Manzhos, Y. Chen, I. Angunawela, M. M. Nahid, K. Feron, S. E. Bottle, J. Bell, H. Ade, J. Takeya and P. Sonar, "Synergistic use of pyridine and selenophene in a diketo-pyrrolopyrrole-based conjugated polymer enhances the electron mobility in organic transistors," *Adv. Funct. Mater.*, vol. 30, no. 34, pp. 1–10, 2020.

61. J. Yang, Z. Zhao, S. Wang, Y. Guo and Y. Liu, "Insight into high-performance conjugated polymers for organic field-effect transistors," *Chem*, vol. 4, no. 12, pp. 2748–2785, 2018.

62. G. Zhang, M. McBride, N. Persson, S. Lee, T. J. Dunn, M. F. Toney, Z. Yuan, Y. H. Kwon, P. H. Chu, B. Risteen and E. Reichmanis, "Versatile interpenetrating polymer network approach to robust stretchable electronic devices," *Chem. Mater.*, vol. 29, no. 18, pp. 7645–7652, 2017.

63. R. Noriega, J. Rivnay, K. Vandewal, F. P. V. Koch, N. Stingelin, P. Smith, M. F. Toney and A. Salleo, "A general relationship between disorder, aggregation and charge transport in conjugated polymers," *Nat. Mater.*, vol. 12, no. 11, pp. 1038–1044, 2013.

64. M. U. Ocheje, B. P. Charron, A. Nyayachavadi and S. Rondeau-Gagne, "Stretchable electronics: Recent progress in the preparation of stretchable and self-healing semiconducting conjugated polymers," *Flex. Print. Electron.*, vol. 2, no. 4, pp. 0–31, 2017.

65. S. Sopcic, M. Kraljic Rokovic and Z. Mandic, "Preparation and characterization of RuO2/polyaniline/polymer binder composite electrodes for supercapacitor application," *J. Electrochem. Sci. Eng.*, vol. 2, no. 1, pp. 41–52, 2012.

8

Flexible and Wearable Strain Sensors

Ömer F. Ünsal and Ayşe Çelik Bedeloğlu

CONTENTS

8.1 Introduction

Flexible and wearable sensor applications are important and a growth area for advanced materials. Physical sensors (pressure, strain, etc.) are also an important subgroup of sensors [1]. Previously, sensor devices were based on brittle semiconducting materials like silicon or different types of inorganic crystals [2], which are not convenient for intensification elongation or compression operations. Further, flexible structures in physical sensor applications have increased rapidly not only for operating conditions but also for wearing comfortability [3,4]. Material selection in sensor manufacturing trends changed from metal-crystal-based materials to polymers thanks to the lightweight, ductile, flexible, and easy-to-process properties of polymers [5]. As sensor applications started to be polymer-based, flexible, and wearable human motion sensor applications, healthcare sensor applications (blood pressure, respiration monitoring, heartbeat, etc.), soft robotic applications, and human-machine interface applications began to appear more frequently [3,4,6–9].

DOI: 10.1201/9781003299455-8

Strain sensors are the largest area among sensors due to the great interest in wearable electronics [1]. A strain sensor also has the capability to measure the loaded stress onto the sensing component, related to the design of the device and type of sensing mechanism. For instance, the design (porous, helical, wrinkled, etc.) and material scale (macro-micro-nano) of the active material of the strain sensor should be suitable for the operating conditions. Furthermore, the stimulation threshold and error rate of strain sensors with good flexibility are quite low, therefore, flexible strain sensor devices are preferable compared with conventional strain sensor devices [3].

Strain sensors convert the degree of deformation into electrical signals [10]. They also consist of sensing material, at least two electrodes, which are mounted onto the sensing material, and contact wires. The active material of strain sensors can be varied according to the sensing principle of the device. Piezoresistive, piezoelectric, and capacitive mechanisms are the current sensing mechanisms for strain detection [5]. Furthermore, material selection and fabrication methods should be chosen according to the working mechanism and operating conditions of the strain sensor. Thermoplastic polymers, rubbers, carbon-based nano-micro particles, metal nano-micro particles, metal oxide nano-micro particles, conducting polymers, and so on, and their combinations can be formed from the active material of a strain sensor [7]. Furthermore, as the fabrication processes of the active layer can be solution based, they can melt polymers or resin-based polymers [8]. In this chapter, sensing mechanisms for strain sensors, materials for strain sensing layers, and fabrication methods of sensing layers are summarized, in addition, a challenge-opportunity analysis is conducted.

8.2 Sensing Mechanisms

8.2.1 Piezoresistive Principle

A piezoresistive mechanism is the electrical resistance change under mechanical stress for a homogeneous material. In other words, geometric deformation of the sensing layer ensures the resistance change, which is a proportional rate of deformation [3]. In piezoresistive sensors, resistance change can be detected with a basic circuit. The simplicity of the measurement paves the way for piezoresistive sensor usage for various applications [5]. Further, according to an expression of resistance (Equation 1), resistance is a function of a volume of material.

$$R = \frac{\rho L}{A} \tag{1}$$

where the ρ is the resistivity of the material, L is the length of the material, and A is the cross-sectional area of the material. Thus, geometric changes, and volumetric changes, in other words, cause resistance changes [10]. The gauge factor is the factor that defines the sensitivity of the sensing layer and can be expressed as:

$$GF = \left| \frac{\Delta R}{R_0 \varepsilon} \right| \tag{2}$$

where the ΔR is resistance change under ε mechanical strain, and R_0 is the initial resistance of the sensing layer [5].

As can be understood above, there are two main requirements for a piezoresistive strain sensor: stretchability and electrical conductivity. Therefore, piezoresistive strain sensors are usually polymer-conductive particle composites or nanocomposites. In an active layer of the piezoresistive strain sensor system, the matrix material is a polymer, which provides stretchability, and the filler material is the conductive particles or nanoparticles that bring electrical conductivity. Here, percolation theory is the actual mechanism that makes the piezoresistive principle work. Conducting particles form conducting pathways by contacting each other. In the case of geometrical deformation, contacting particles move away from each other, and conductive pathways are broken (Figure 8.1). Thus, resistance change is observed as proportional to applied stress.

8.2.2 Piezoelectric Principle

Similar to the piezoresistive mechanism, the piezoelectric principle consists of a voltage difference occurrence under mechanical stress. Piezoelectric materials are non-centrosymmetric materials. The lack of a central atom indicates the piezoelectric effect. The deformation of non-centrosymmetric crystal generates electrical dipoles (Figure 8.2) [11]. The stimulation threshold is important for strain sensors to detect even small magnitudes of deformation. Thanks to the low stimuli threshold, piezoelectric materials are promising materials for sensor applications [12]. As piezoelectric materials can be used as sensing materials, they can also be used as energy harvesting materials. Further, thanks to the energy-generation ability of piezoelectric materials, they can be used as self-powered sensors [13].

Additionally, the sensitivity of piezoelectric strain sensors is defined as the following equation (3), where V is the output voltage of the piezoelectric component and Δd is the change of displacement.

$$\text{Sensitivity} = \frac{V}{\Delta d} \tag{3}$$

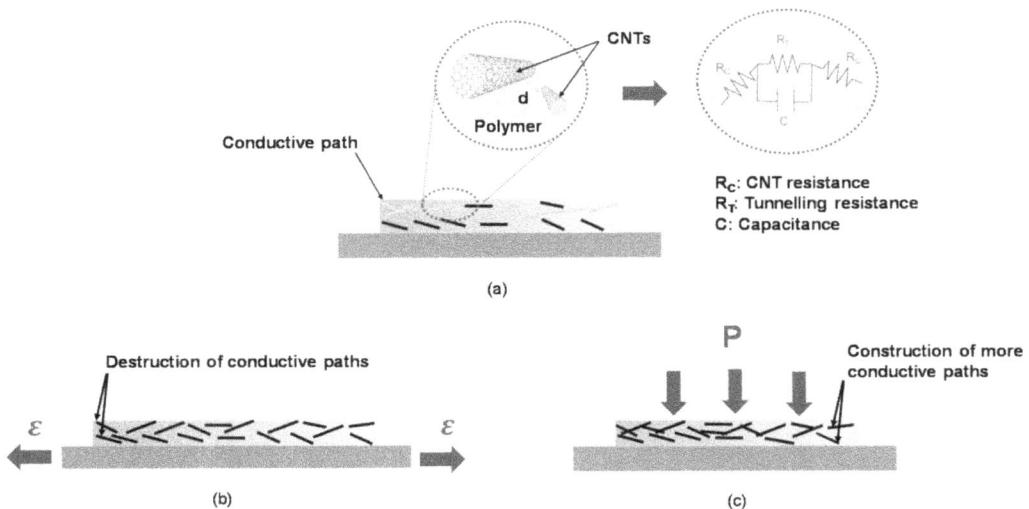

FIGURE 8.1
Schematic illustration of a) conductive pathway formed polymer composite, b) breakage of the conductive pathway under elongation, and c) the formation of more intense pathways under pressure. Adapted with permission [8]. Copyright The Authors, some rights reserved; exclusive licensee MDPI. Distributed under a Creative Commons Attribution License 4.0 (CC BY).

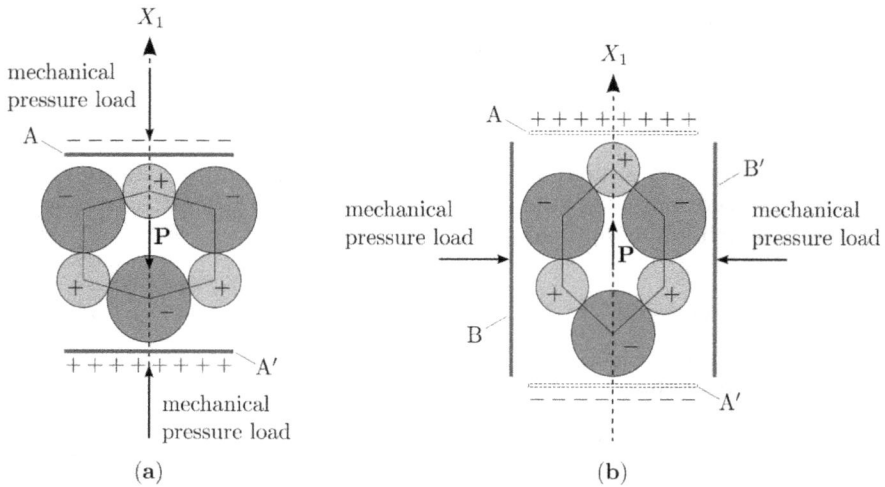

FIGURE 8.2
Schematic illustration of deformation of non-centrosymmetric quartz crystal. Adapted with permission [14]. Copyright (2021) IOP Publishing.

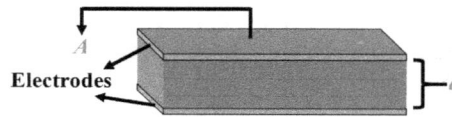

FIGURE 8.3
Schematic description of *A* and *d* factors in Equation 4.

8.2.3 Piezocapacitive Principle

Capacitance is defined as a measure of electrical energy storage ability. The piezocapacitive effect, on the other hand, can be defined as the capacitance change of a dielectric material when the material is geometrically deformed [10]. In other words, piezocapacitive strain sensors are flexible capacitors [11]. According to the formula of capacitance (Equation 4), shape and/or size dependence can be clearly understood.

$$C = \frac{\varepsilon_0 \varepsilon_r S}{d} \tag{4}$$

where the ε_0 and ε_r are dielectric permittivities of vacuum and the energy-storing material, S is the electrode integrated area of energy storing material, and d is the thickness of energy storing material, respectively (Figure 8.3). The S/d ratio is the relation between capacitance and when a material is exposed to geometrical deformation – the ratio is changed easily.

Piezocapacitive components have some advantages, however, the linearity of output signals is the most important one for sensing accuracy. Another great advantage of piezocapacitive strain sensors is low hysteresis, which can be defined as the effect of previous conditions on material [15]. Additionally, piezocapacitive sensors can be divided into two classes, according to electrode configuration: sandwich structure and parallel-plate structure [15]. Since geometric deformation will cause the fracture of electrodes, a parallel-plate configuration is more suitable for strain sensors.

8.3 Materials

8.3.1 Conducting Particles

Nanocomposite materials are the biggest class of strain sensors. Here, conductive particles are a common reinforcement phase for nanocomposite-based sensors. Conductive material addition is a necessity for piezoresistive systems due to providing mechanically manipulative resistance. As mentioned in Section 8.2.1, conductive particles form pathways, which can easily be broken with mechanical force. In piezoelectric sensors, although it is not necessary, conductive addition to the piezoelectric component improves the sensitivity and signal magnitude. Piezoelectric materials are included in the semiconductor materials class due to their relatively high conductivity. Although they have noticeable conductivity, they have a high electrical resistance nature compared with conductors. This phenomenon causes a loss of piezoelectrically generated electricity between the geometric center of the material and the electrodes. The particles form pathways and generated electricity can reach electrodes through the pathways without energy loss. On the other hand, the performance improvement mechanism in piezocapacitive sensors is different from piezoelectric and piezoresistive mechanisms. As mentioned in Section 8.2.3, piezocapacitive sensors are flexible capacitors that consist of a dielectric layer and two electrodes. In conductive-filled capacitors, conducting materials are not only employed for charge transportation but are also employed to enhance the energy storage ability of the sensor. In other words, the conductive reinforcement phase forms micro-capacitors in the dielectric matrix [16]. Thus, the capacitance and the sensitivity of the nanocomposite increase.

Conductive nanoparticles for sensor applications have a wide range of materials from metals to carbonaceous particles and even polymers. Nano/sub-micron scale conductive material usage is one of the key points of sensitivity of strain sensors. The required uniformity of particles in the matrix phase provides uniform electrical properties to the resulting composite. Reduced particle size also provides a high specific surface area, which is quite important, especially for piezocapacitive devices because of the S/d ratio. Another important property of nanomaterials is in advanced applications in geometry. It has been shown that the high aspect ratio of reinforcement particles helps to form pathways easier (Figure 8.4) [15].

Metal nanoparticles offer some advantages like high conductivity and easy controllable geometry and size of particles. Therefore, metal particles of various shapes (wire,

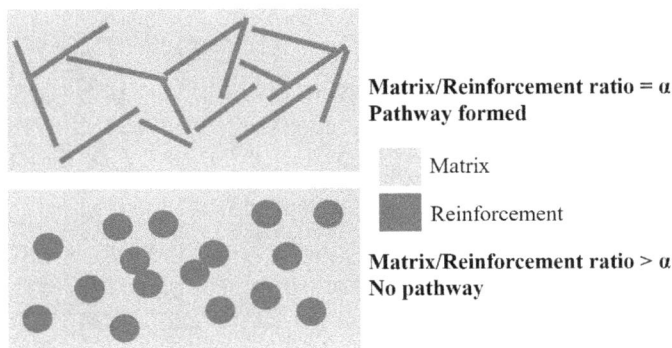

Matrix/Reinforcement ratio = α
Pathway formed

Matrix

Reinforcement

Matrix/Reinforcement ratio > α
No pathway

FIGURE 8.4
Schematic description of the effect of the aspect ratio of conductive fillers.

rod, sphere, cube, etc.) and sizes are available for sensor applications [6]. Besides metal particles, nanoscale carbonaceous materials have had increased attention in the last few decades for advanced material applications. Although micro-scale carbon materials like graphite and carbon black can be used for sensor applications, smaller-scale graphene, fullerene, and carbon nanotubes show better performance thanks to their unique electrical properties [7]. Conjugated conductive polymers are also a huge class of materials for sensing applications. Poly(3,4-ethylenedioxythiophene) (PEDOT), polyaniline (PANI), and polypyrrole (PPy) are the most known conducting polymers. Even the general "polymer" concept includes ductile and thermally processable, conducting polymers, which are usually brittle, and particle-formed materials [17].

8.3.2 Functional Particles

Particles that are not used for conductivity in composite-based strain sensors can be called functional particles. This type of filler is usually used for piezoelectric systems. As mentioned in Section 8.2.2, piezoelectric materials are non-centrosymmetric molecules. In this regard, some inorganic and organic materials can provide for this condition (Figure 8.5). Organic-based piezoelectric materials are synthetic or natural polymers that have low piezoelectric properties. Conversely, inorganic piezoelectric materials have better piezoelectric properties [18]. Furthermore, inorganic piezoelectric materials, which consist of wurtzite and perovskite crystals, can be produced in desired geometry even at the nanoscale. ZnO, ZnS, GaN, CdS, and so on are the most known wurtzite piezoelectric materials. On the other hand, PZT, $BaTiO_3$, and $LiNbO_3$ are the most common perovskite-based piezoelectric materials for energy harvesting and sensing applications. Conducting fillers, particle size, and aspect ratio are important for piezoelectric filler materials. Piezoelectric fillers (usually in nanowire form), can be used with or without conducting fillers in a strain-sensing composite.

8.3.3 Flexible Matrix Materials

The matrix phase is continuous and encapsulated in composites. Matrix phase materials are used for their desired durability and to protect the formed shape of the material. In addition, material selection should be convenient for the purpose of composite usage. Because of that, matrix materials for strain sensors should be flexible and stretchable. Additionally, the elastomeric behavior of matrix material provides cyclic mechanical

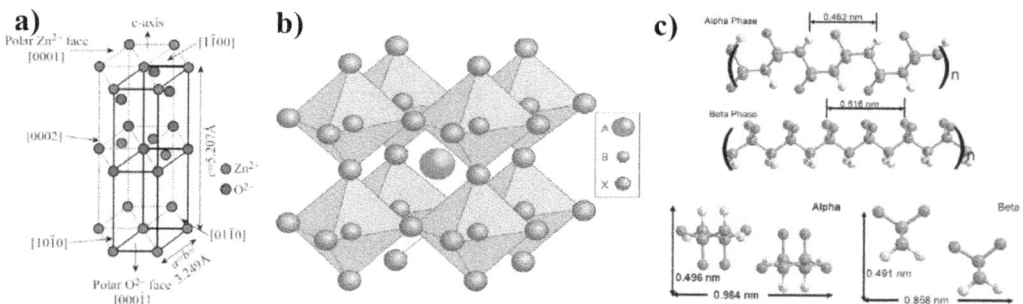

FIGURE 8.5
Schematics of a) wurtzite crystal structure, b) perovskite crystal structure, and c) the molecular structure of poly(vinylidene fluoride). Adapted with permission [19]. Copyright (2020) Elsevier.

detection and multiple uses for a sensor device. In other words, composite-based strain sensors should be used in the elastic region of matrix materials.

Flexible matrix materials for strain sensors can be thermoplastic or thermoset materials. Common examples of these materials are rubbers, silicone-based thermoset resins (siloxane-based resins), and thermoplastic polyurethane (TPU) [3]. Poly(dimethyl siloxane) (PDMS) has been the most favorable elastomer for the last decade in electronic applications due to its flexibility with a low Young's Modulus, transparency, and unique electrical properties [7]. Particularly in energy storage (capacitor) applications, excellent properties of PDMS have been reported [16]. TPU is also a candidate as a matrix material for sensing applications, as it is a thermoplastic polymer with similar mechanical properties to PDMS. PDMS or derivative silicone rubbers are resin-based materials with at least two components. Therefore, thermoset molding fabrication techniques can be used for strain-sensing composites. On the other hand, because TPU is a thermoplastic, TPU-based sensor composites can be fabricated by melt or solution-based processes [6].

PVDF is different from the composite mechanism, which includes a functional material as a reinforcement phase and a non-functional matrix material, as it can be used as functional matrix material thanks to its piezoelectric and flexible properties [12]. PVDF is also thermoplastic as TPU. Although the melting process is possible for PVDF, solution processing for sensor composites is more popular. Further, PVDF nanofibers are the most common form of PVDF in sensor and energy harvesting applications [3].

8.3.4 Flexible Electrodes

Electrodes are not only used for sensors but for various types of devices (energy generators, batteries, capacitors, etc.) they are a requirement. There is no challenge for electrode integration onto a rigid device (like a conventional battery). However, electrode stability for a flexible device is very important for device efficiency. Flexible electrodes are usually durable for bending or compressing forces. In this regard, various techniques were adapted to flexible electrode preparation. Conducting polymers can be coated on the desired surface with complex geometry, although they are rigid materials, as mentioned in Section 8.3.1. Thus, conductive polymers can be used as flexible electrodes by the in-situ polymerization method [20]. In another approach, graphene, graphene oxide, or carbon nanotube can be coated onto sensing layer surfaces with spraying methods (airbrush, electrospray, etc.) [12]. Furthermore, flexible composites of carbon-based materials can also be used as flexible electrodes [21]. Metal thin films are also easy to apply and are quite durable structures, which can be obtained by sputtering methods [22].

8.4 Fabrication Techniques

Flexible and stretchable sensor manufacturing is based on polymer materials, as mentioned in Section 8.3. Therefore, conventional or novel polymer processing methods are used for flexible strain sensor fabrication. Furthermore, the preference for manufacturing techniques directly depends on the processing abilities of the material. Additionally, the design of the sensor device determines the fabrication technique [6]. In this section, flexible strain sensor fabrication techniques based on melt-based techniques, solution-based techniques, and in-situ polymerization are discussed.

8.4.1 Melt Fabrication Techniques

Melting is described as the liquidation of a thermoplastic polymer with increasing temperature by the removal of intermolecular forces between crystallized polymer chains. In other words, molten polymer is a polymer solution without any solvent. Polymer composite fabrication by melt processing is a common method for commercial thermoplastic products. Various types of thermoplastic composite materials are manufactured by extrusion or injection methods in various forms such as fiber, film, and granule. In this method, the reinforcement phase (including conductive or functional nanoparticles) is blended into the molten polymer with screws, and the sensing layer is formed in the desired geometry [8]. Furthermore, composite granules, which can also be obtained with melt processes, can be used for sensing layer fabrication in melt or solution processing. For a complex example, conductive or functional nanoparticle-loaded composite granules can be employed as raw materials to obtain a 3D printing filament. Thus, a 3D-printed sensing layer can be fabricated (Figure 8.6).

8.4.2 Solution-Based Fabrication Techniques

Solution-based techniques are usually used for thermoplastic-based sensing layers, such as melt processing, however, wet processing has more alternative techniques than the melt method. In wet processing techniques, the key point is the uniform dispersion of nanomaterials. Although stirring and mixing are easy and cheap methods, ultrasonication is more convenient to disperse a bulk nanomaterial [24]. In the ultrasonication technique, the bulk material is exposed to a soundwave with a high magnitude. The soundwave causes shear stress on the bulky particles. Thus, the bulk nanomaterial is disintegrated into nanoparticles, and nanoparticles are easily dispersed in convenient solvents. Fabrication of a solution-processed nanocomposite consists of a dispersion of nanomaterial in a suitable solvent and polymer addition into the nanomaterial dispersion to obtain a composite solution. Solution processing provides advantages such as application convenience at low viscosity, less polymer degradation or chain cracking at low temperature, and better dispersion in the reinforcement phase [6].

Here, we can count various techniques from spin coating to the doctor blade method, or electrospinning (Figure 8.7). However, the casting method is the most basic method to

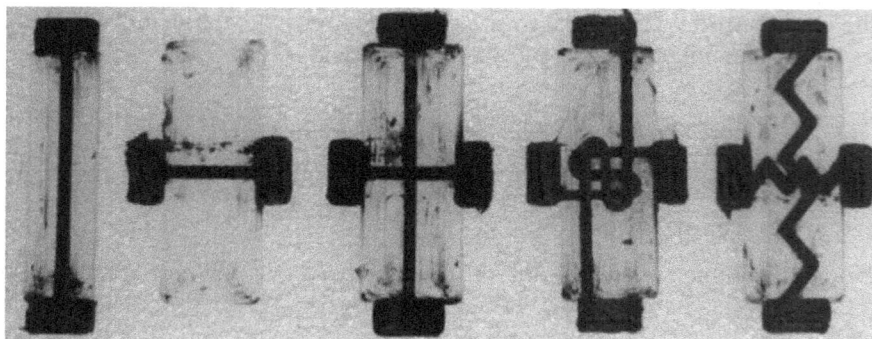

FIGURE 8.6
Image of 3D printed CNT/TPU composite sensor device. Adapted with permission [23]. Copyright The Authors, some rights reserved; exclusive licensee MDPI. Distributed under a Creative Commons Attribution License 4.0 (CC BY)

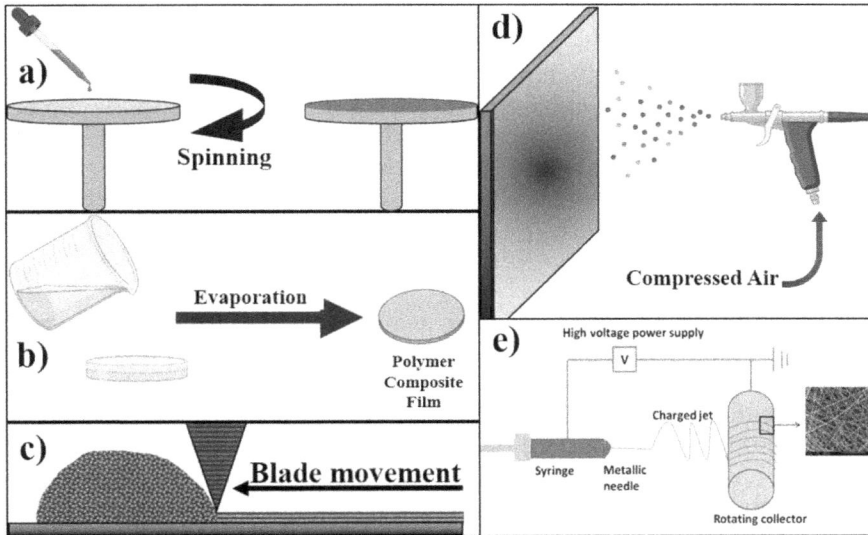

FIGURE 8.7
Schematic illustrations of a) spin coating, b) solution casting, c) doctor blade method, d) airbrush, and e) electrospinning. Adapted with permission [27]. Copyright The Authors, some rights reserved; exclusive licensee MDPI. Distributed under a Creative Commons Attribution License 4.0 (CC BY).

fabricate a nanocomposite structure, which means pouring the solution into the mold and obtaining the composite after evaporation of the solvent [25]. Similarly, dip coating and spin-coating methods are based on similar evaporation principles, with only the coating technique different. In spin coating, enough composite solution is dropped onto a planar substrate and the substrate is spun at a very high rotational speed (a few tens to thousands of RPMs). Thus, an excess amount of solution is removed from the substrate and the solution with uniform thickness is coated onto the surface. Similarly, in the dip coating method, a substrate with desired geometry is dipped into the composite solution. After both coating processes, coated substrates are left to lose all of the solvent [6]. Spinning time, spinning velocity, dipping time, and the speed of the substrate are the most important process parameters for these methods. The doctor blade method is basic and popular. In this method, the solution is poured onto a flat substrate and the blade, which has a definite and constant height from the surface, is moved to the substrate. In other words, the blade removes the excess composite solution [11]. All of these methods, such as the removal of the solvent, are possible at room temperature or at higher temperature evaporation, as well as with the non-solvent method (or coagulation). However, some methods only use the coagulation method. For example, the wet spinning method directly works with the coagulation of a polymer jet into the non-solvent to form composite fibers [26].

Additionally, a novel technique is also very popular in smart electronic materials: electrospinning. In this technique, the polymer solution is fed onto a needle or nozzle with a constant feeding rate and it is exposed to a high-magnitude electrical field. The composite solution is oriented from the nozzle to other poles of the electrical field with continuous polymer jets. Most of the solvent of the solution is evaporated until the jets reach the collecting surface and nano or sub-micro-scale polymer fibers are obtained [7]. Solvents should be more volatile in electrospinning techniques and solidify easier. Printing techniques are also a huge area for strain-sensing composite layer fabrication. Ink-jet printing, gravure printing, or screed printing are the most common printing techniques for sensor applications [11]. The printing method is preferred for conductive component integration onto a flexible substrate compared with composite fabrication. Similarly, nanomaterial deposition processes are included in solution-based processes and the same methods can be used to obtain deposited nanomaterial surfaces, especially for electrode applications of sensors [25]. For instance, if a spraying method is preferred in the dispersion of a nanomaterial, a nanostructured conducting surface is obtained with the spray coating method.

8.4.3 In-Situ Polymerization

Polymerization of matrix materials is also an available method for polymer nanocomposites, which can be employed as the sensing layer of a sensor. In-situ polymerized polymer composite fabrication is based on the polymerization of monomers in the presence of the reinforcement conducting phase. The composites produced with this method provide a strong matrix-reinforcement interface owing to higher thermodynamic compatibility of functional groups of these phases before polymerization compared with polymer chains [8]. Furthermore, thermoset resin (e.g., PDMS) based composite films, which are suitable for strain-sensing ability, can be mixed before curing, then the curing process is progressed with in-situ polymerization and a composite is obtained [25]. Moreover, in-situ polymerization can be used for the surface modification of reinforcement particles for various purposes such as improving electrical properties, enhancing the interphase between matrix and reinforcement phases, and even preventing the agglomeration of nanomaterials [28].

8.5 Designing Procedure of Flexible Strain Sensors

Flexible sensor devices have challenges such as sensitivity, hysteresis, linearity, and durability [7]. Here, sensitivity is a difference of a variable in the unit stimulus. Hysteresis is related to the relaxation of a flexible and/or stretchable structure, which is caused by the wrong signals due to unsynchronized relaxation or stimuli of sensing layers. On the other hand, linearity is stable regarding sensitivity independent of the intensity of the stimulus. Durability has both electrical and mechanical stability of the sensing layer during its life [26].

All of these challenges can be handled by material selection, fabrication techniques, and fabrication quality to some extent. Design is another opportunity to overcome the usual challenges of flexible sensors. Design parameters determine the stretchability and/or bending ability of flexible sensors, not only in the macro design of the device but the micro-nano structure preference directly affects the mechanical abilities of the device [6]. As an example, polyethylene terephthalate is a common material with high toughness and low flexibility. However, knitted polyethylene terephthalate fabric is a flexible material. Wrinkled, buckled, helical, and porous designs are common for strain sensors.

Porous structures are useful materials for strain sensing due to their wide surface area, good deformation-relaxation stability, and low density (Figure 8.8a) [29]. Porous sensing layers can be obtained from composite sensing precursors, which have been discussed previously, and they can also be obtained by a conductive or functional material coating of a porous conventional material. Conductive hydrogel structures have been used as sensing layers in strain sensors [22]. These materials are made from conductive-functional particle/polymer composites or naturally conductive materials. On the other hand, as a

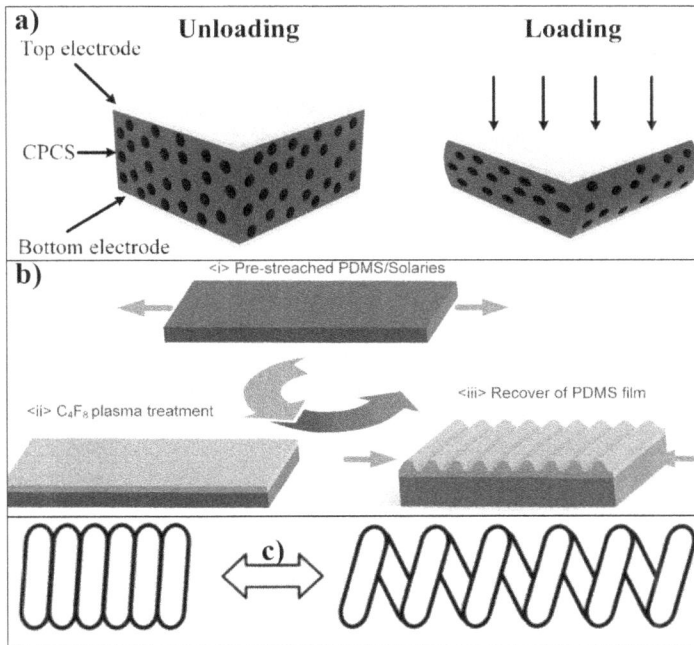

FIGURE 8.8
Schematic illustrations of a) a porous structure. Adapted with permission [30]. Copyright The Authors, some rights reserved; exclusive licensee MDPI. Distributed under a Creative Commons Attribution License 4.0 (CC BY). b) Wrinkled structure. Adapted with permission [31]. Copyright The Authors, some rights reserved; exclusive licensee Nature. Distributed under a Creative Commons Attribution License 4.0 (CC BY). c) Spring structure. Adapted with permission [32]. Copyright The Authors, some rights reserved; exclusive licensee MDPI. Distributed under a Creative Commons Attribution License 4.0.

common and cheap porous structure, a PU sponge can be coated with convenient materials to gain sensing ability. Wrinkled structures are also popular candidates for strain sensing (Figure 8.8b). In a wrinkled design, each wrinkle contributes to the "healthy deformation" of the sensing layer [6]. Each wrinkle provides a deformation without a mechanical fracture or tears in the sensing layer [29]. Furthermore, serpentine, coil, and helical designs are commonly used for strain sensing due to the stress-absorbing ability of these designs, which works with the spring mechanism (Figure 8.8c) [6]. Improved designs and structures are also available for stretchable and flexible sensors for improved gauge factor and sensitivity. Various shapes of cavities on the surface are common to sense quite low stimuli. The cavity can be a channel, V-shape, hemisphere, and so on. Another mechanism is overlapped planar structures. In this mechanism, contact areas of overlapped layers or sheets with electrodes are increased, which changes the signal intensity [29].

8.6 Wearability

Demand for flexible electronics caused a rapid growth of wearable sensors from chemical sensors to physical sensors. The mechanical sensor is a subclass of the physical sensor, which includes flexible and wearable sensors. Sensors consist of a sensing layer and at least two electrodes, as described in the Introduction. Furthermore, the material and fabrication techniques of a flexible sensor device can be preferred to a wearable version of the sensor. In Sections 8.3 and 8.4, composite-based strain sensor devices were discussed in detail. However, some differences in fabrication techniques or resulting structures for textile applications are possible. Fibers, yarns, and fabrics are the main textile substrates and at least one of these substrates should be employed in a wearable sensor [33]. We will mention two approaches for wearable sensor fabrication. First, sensing materials and electrodes can be coated onto non-sensing and mechanically convenient textile substrates by suitable methods. Here, the substrate material works as a support and flexible part of the sensor device [34,35]. Second, sensing materials can be used to produce textile substrates. In other words, the fabrication of intrinsically sensing yarns, fibers, or fabrics is possible for wearable sensors.

The first step to manufacturing a flexible and wearable strain sensor is the bottom electrode. A single fiber, yarn, or fabric can be intrinsically electrically conductive. However conductive materials, such as carbon or metal-based nanoparticles and conductive polymers, are brittle and low-strength materials. Therefore, conductive material coating on conventional polymers (such as polyester, polyamide, wool, etc.) is more favorable to bottom electrode integration on textile substrates [36]. This type of conductive fiber also can be a direct-sensing layer. Furthermore, various advanced coating techniques can be used, but dip coating is the preferred method thanks to its simplicity. Except for coating applications, in multi-fibrous textile substrates (fabrics and yarns), integration of a conductive single fiber or yarn during the twisting or weaving process is another alternative to transform the substrate [33]. Furthermore, instead of composite films, composite fibers and fabrics (e.g., elastomer/conductive) can be used for strain-sensing textile materials.

The same methods are also available for non-woven, woven, or knitted fabrics. In a fabric-based strain sensor, the filament and coating materials should at least be flexible as a primary requirement. Conductivity is another requirement. Dip coating, spraying, printing techniques, and so on are the most common methods to coat the fabric with conductive materials. As the sensing layer can be the material of the fabric, a sensing composite film can be coated onto the fabric substrate. Nanofibers are the most frequently used non-woven sensor material, which allows both fiber material compositing and coating processes [36]. The low cost and fast production process of conventional non-woven fabrics are also important points for sensing applications. On the other hand, woven fabrics have been used for strain sensors as wearable devices with many advantages. The greatest advantage of woven fabrics for sensing applications is mechanical stability, compared with non-wovens. Moreover, material selection and sensing layer fabrication techniques are not the only parameters to adjust the degree of sensing ability. Weaving parameters and woven structure design easily affect the performance of textile-based sensors [33]. As another fabric-based substrate, knitted fabrics are preferable materials for sensing applications with their excessively stretchable and multidirectional elongation abilities [36]. However, cyclic stability, which is one of the decisive factors in the performance of strain sensors, is low for knitted fabrics, alongside high stretchability [36].

8.7 Applications

Textile applications of strain-sensing devices include both conventional textile processes and advanced material manufacturing techniques. However, particularly for laboratory scale works, combining these two technologies may be more difficult. In this regard, studies on textile implantable advanced materials are preferable for researchers.

Nanofibers and nanofiber yarns are mostly used for filtration and biomedical studies. However, advanced electronic applications are also a big consumer of nanofibers or nanofiber yarns thanks to their relatively high surface area and porosity. Further, polyacrylonitrile (PAN) is one of the preferred polymers for nanofiber-based electronic applications [37]. PAN is also the precursor material for carbon fiber, which is another motivation for electronic applications. According to this approach, in a study to take advantage of the nanofibrous structure of PAN and compare the nanofibrous structure with conventional fibers [38], various core-sheath-wrapped yarns were fabricated. Core-sheath yarns consisted of "silk core/PAN nanofiber yarn sheath" (S/N; Sensor 1), "cotton core/PAN nanofiber yarn sheath" (C/N; Sensor 2), and "silk core/cotton sheath" (S/C; Sensor 3) structures. All samples were carbonized and attached to the textile substrate for strain sensitivity measurements. The yarns capsulated in PDMS and cyclic piezoresistive tests were performed at a rate of 7.5 mm/min. According to the piezoresistive test results, the sensors can even sense 0.05% strain (Figure 8.9).

Graphene-doped carbon nanofiber yarns with an elastomeric support layer are also good candidates for strain sensors. In a similar study, graphene/PAN nanofiber yarns were carbonized [39]. The carbon-graphene composite nanofiber yarns were lined up horizontally at a constant distance from each other and encapsulated with thermoplastic polyurethane (Figure 8.10). Cyclic piezoresistive tests were performed under a 2% strain rate and the maximum average gauge factor was shown to be 1724.

Besides the use of advanced materials for textile implantable sensors, more conventional materials can be used to fabricate textile implantable sensing components. In a study,

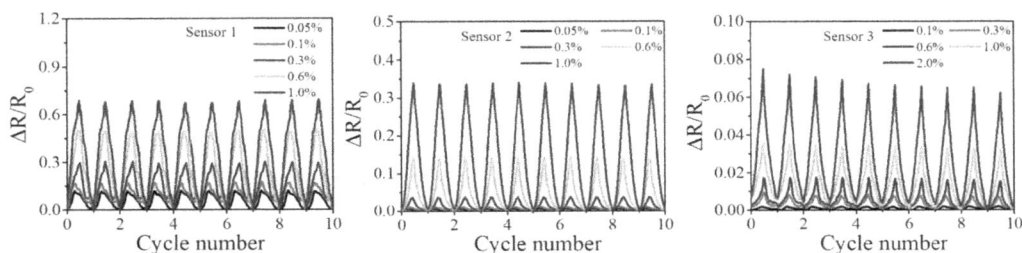

FIGURE 8.9
Cyclic piezoresistive test results of sensors 1, 2, and 3. Adapted with permission [38]. Copyright (2021) Elsevier.

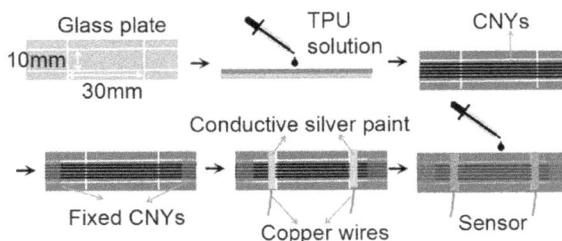

FIGURE 8.10
Schematic illustration of nanofibrous yarn-based strain sensor fabrication steps. Adapted with permission [39]. Copyright (2018) Elsevier.

using this approach, thermoplastic elastomers and carbon black were compounded and composite fibers were obtained by the extrusion method [40]. The obtained single fibers were used as sensing components with contact with conductive yarns as electrodes. The sensor thread is integrated into a textile product with a silicone coating (Figure 8.11). Thus, sensor-integrated textile has gained a washable and highly cyclic stable character. Further, sensor performance was not changed in up to eight washing cycles.

As single fibers can be the sensing layer, so can multifilament structures. A novel compositing of a thermoplastic elastomer and silver has been reported [41]. In this study, multi-filament elastomeric yarns absorbed Ag^+ ions in a specified solvent by immersing method. Then the silver ions were reduced by using hydrazine hydrate and transformed into silver particles. Here, the surface of the filaments were Ag-rich regions due to the better absorption abilities of surface polymer chains (Figure 8.12a). The multifilament sensor was mounted onto a glove by sewing, and output resistance values by human motion were visualized (Figure 8.12b)

On the other hand, the coating of fibers, yarns, and fabrics is a popular method in the fabrication of sensing components in textile-based sensors. Various types of coating material previously. For instance, a graphene-coated textile substrate-based strain sensor was reported in 2017 [42]. In this example, 60% stretchable polyester-based threads were dip-coated with graphene. After graphene coating, the PDMS resin coating process was performed on the thread. The study proved that the PDMS capsulation of sensing textile substrate decreased crack formation and provided durability in the device.

Conducting polymers are also frequently used materials for textile coating-based sensor applications. PPy and PANI are the most common conducting polymers for flexible electronics. As described in Section 8.3, the in-situ polymerization technique provides well-coated textile substrates, although conducting polymers are brittle materials. Furthermore, as a more suitable method for porous structures, vapor phase polymerization is another common method. According to a reported study with this approach [43], polypyrrole coating by vapor phase polymerization was quite uniform. In this study, PDMS nanofibers were produced as textile substrates, then the nanofiber mat was dipped in $FeCl_3$ solution and dried to obtain smaller crystals on the nanofibers. The nanofiber was subjected to cold pyrrole vapor for 24 hours and SEM images showed the uniform coating of PPy. The electromechanical tests were performed as cyclic strain tests with high strain rates.

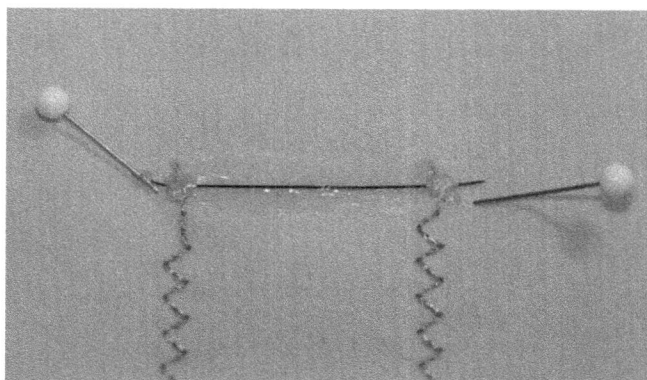

FIGURE 8.11
Textile integrated single fiber-based strain sensor. Adapted with permission [40]. Copyright The Authors, some rights reserved; exclusive licensee MDPI. Distributed under a Creative Commons Attribution License 3.0 (CC BY).

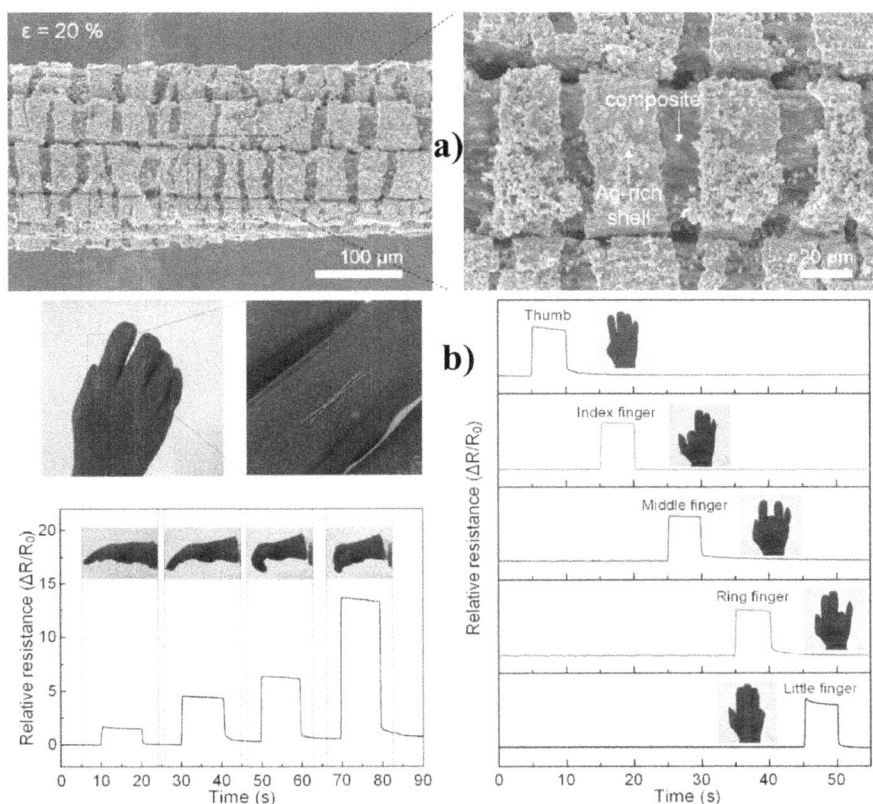

FIGURE 8.12
a) SEM images of Ag-rich regions of elastomer multifilament and b) human motion tests of multifilament sensor mounted glove. Adapted with permission [41]. Copyright (2018), American Chemical Society.

In a similar study, the PPy coating process of fabrics was performed by chemical vapor deposition (CVD) and solution-based methods separately [44]. Piezoresistive tests were performed under 50% strain cycles. The coating uniformity of fibers was not only proven by the SEM images but the resistance change signal intensity results were supported by this deduction (Figure 8.13).

8.8 Future Prospects

In this chapter, we have described sensors, strain-sensing mechanisms, and the importance of the wearability of electronic devices. Polymer composites, which commonly consist of the mixing of conductive or functional particles with elastomeric matrix materials, are the most important materials for sensing applications. Furthermore, the development of fabrication techniques and novel nanomaterials (such as MXenes) becoming cheaper than conventional nanomaterials render polymer composite-based wearable electronics more reachable for consumer electronics. Besides composite-based wearable sensors, the coating of textile substrates has a significant role in making wearable electronics accessible. Coated textile-based

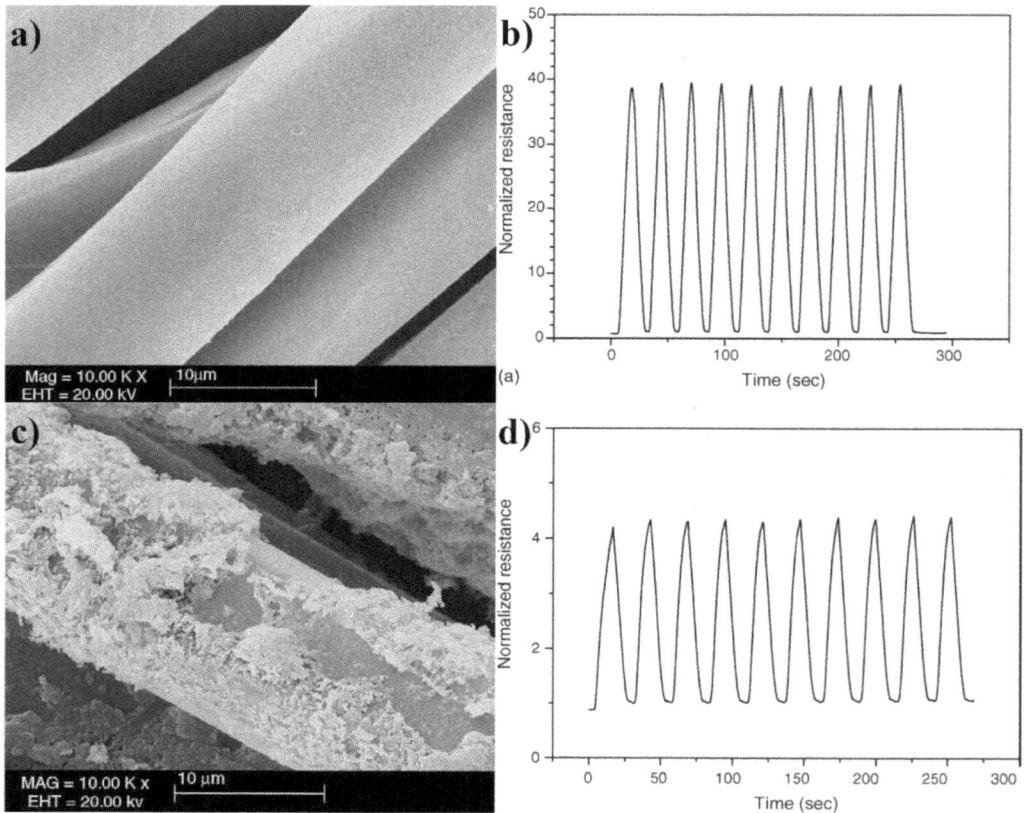

FIGURE 8.13
a) SEM image of CVD coated fabric, b) electromechanical test results under 50% strain of CVD coated fabric, c) SEM image of the in-situ polymerized fabric, and d) electromechanical test results of in-situ polymerized fabrics. Adapted with permission [44]. Copyright (2005), Elsevier.

wearable sensors are preferable because of their simple application processes, low cost, and repeatability advantages. Additionally, textile coating processes are established methods due to the deep-rooted history of textile science, which is another strong side of coating-based wearable sensors. Furthermore, structural engineering-based solutions have become more visible. Structural engineering provides more complex raw material processing and less use of functional or advanced materials. Thanks to developments in composites, the simplicity of coated textile-based wearable sensors, and structural-engineered solutions, a technological revolution is inevitable in healthcare, human motion detecting, human-machine interface studies, robotics, and performance data in sports.

Acknowledgement

This study was supported by Turkish Scientific and Technical Research Council, TUBITAK, project no: 219M103. This chapter is based upon work from COST Action

"High-performance Carbon-based composites with Smart properties for Advanced Sensing Applications" (EsSENce Cost Action CA19118, https://www.context-cost.eu) supported by COST (European Cooperation in Science and Technology, https://www.cost.eu)

References

1. D. Barmpakos, G. Kaltsas, A review on humidity, temperature and strain printed sensors—Current trends and future perspectives, *Sensors (Switzerland)*. 21 (2021) 1–24.
2. A. Nag, M.E.E. Alahi, S.C. Mukhopadhyay, Z. Liu, Multi-walled carbon nanotubes-based sensors for strain sensing applications, *Sensors (Switzerland)*. 21 (2021) 1–22.
3. X. Liu, Y. Wei, Y. Qiu, Advanced flexible skin-like pressure and strain sensors for human health monitoring, *Micromachines*. 12 (2021) 695.
4. L. Tang, S. Wu, J. Qu, L. Gong, J. Tang, A review of conductive hydrogel used in flexible strain sensor, *Materials*. 13 (2020) 1–17.
5. Z. Luo, X. Hu, X. Tian, C. Luo, H. Xu, Q. Li, Q. Li, J. Zhang, F. Qiao, X. Wu, V.E. Borisenko, J. Chu, Structure-property relationships in graphene-based strain and pressure sensors for potential artificial intelligence applications, *Sensors (Switzerland)*. 19 (2019) 1–27.
6. K. Singh, S. Sharma, S. Shriwastava, P. Singla, M. Gupta, C.C. Tripathi, Significance of nanomaterials, designs consideration and fabrication techniques on performances of strain sensors - A review, *Materials Science in Semiconductor Processing*. 123 (2021) 105581.
7. F. Han, M. Li, H. Ye, G. Zhang, Materials, electrical performance, mechanisms, applications, and manufacturing approaches for flexible strain sensors, *Nanomaterials*. 11 (2021) 1–31.
8. O. Kanoun, A. Bouhamed, R. Ramalingame, J.R. Bautista-Quijano, Review on conductive polymer/CNTs nanocomposites based flexible and stretchable strain and pressure sensors, *Sensors (Switzerland)*. 21 (2021) 341.
9. X. Chang, L. Chen, J. Chen, Y. Zhu, Z. Guo, Advances in transparent and stretchable strain sensors, *Advanced Composites and Hybrid Materials*. 4 (2021) 435–450.
10. S. Li, X. Xiao, J. Hu, M. Dong, Y. Zhang, R. Xu, X. Wang, J. Islam, Recent advances of carbon-based flexible strain sensors in physiological signal monitoring, *ACS Applied Electronic Materials*. 2 (2020) 2282–2300.
11. Y. Li, Y. Liu, S.R.A. Bhuiyan, Y. Zhu, S. Yao, Printed strain sensors for on-skin electronics, *Small Structures*. 3 (2022) 2100131.
12. Ö.F. Ünsal, Y. Altın, A. Çelik Bedeloğlu, Poly(vinylidene fluoride) nanofiber-based piezoelectric nanogenerators using reduced graphene oxide/polyaniline, *Journal of Applied Polymer Science*. 137 (2020) 1–14.
13. S. Azimi, A. Golabchi, A. Nekookar, S. Rabbani, M.H. Amiri, K. Asadi, M.M. Abolhasani, Self-powered cardiac pacemaker by piezoelectric polymer nanogenerator implant, *Nano Energy*. 83 (2021) 105781.
14. R.G. Ballas, The piezoelectric effect - An indispensable solid state effect for contemporary actuator and sensor technologies, *IOP Conference Series: Earth and Environmental Science*. 1775 (2021) 012012.
15. G. Li, C. Li, G. Li, D. Yu, Z. Song, H. Wang, X. Liu, H. Liu, W. Liu, Development of conductive hydrogels for fabricating flexible strain sensors, *Small*. 18 (2022) 1–37.
16. M. Li, X. Huang, C. Wu, H. Xu, P. Jiang, T. Tanaka, Fabrication of two-dimensional hybrid sheets by decorating insulating PANI on reduced graphene oxide for polymer nanocomposites with low dielectric loss and high dielectric constant, *Journal of Materials Chemistry*. 22 (2012) 23477–23484.
17. Ö.F. Ünsal, A.Ç. Bedeloğlu, Recent trends in flexible nanogenerators: A review, *Material Science Research India*. 15 (2018) 114–130.

18. T.E. Hooper, J.I. Roscow, A. Mathieson, H. Khanbareh, A.J. Goetzee-Barral, A.J. Bell, High voltage coefficient piezoelectric materials and their applications, *Journal of the European Ceramic Society*. 41 (2021) 6115–6129.
19. Ö.F. Ünsal, A.S. Hiçyilmaz, A.N.Y. Yilmaz, Y. Altin, İ. Borazan, A.Ç. Bedeloğlu, Energy-generating textiles. Woodhead Publishing, Sawstoni, 2020, 415–455.
20. B. Yue, C. Wang, X. Ding, G.G. Wallace, Polypyrrole coated nylon lycra fabric as stretchable electrode for supercapacitor applications, *Electrochimica Acta*. 68 (2012) 18–24.
21. W.S. Kim, S.Y. Moon, H.J. Kim, S. Park, J. Koyanagi, H. Huh, Large-scale graphene-based composite films for flexible transparent electrodes fabricated by electrospray deposition, *Materials Research Express*. 1 (2015) 046404.
22. L. Wang, T. Xu, X. Zhang, Multifunctional conductive hydrogel-based flexible wearable sensors, *TrAC - Trends in Analytical Chemistry*. 134 (2021) 116130.
23. J.F. Christ, N. Aliheidari, P. Pötschke, A. Ameli, Bidirectional and stretchable piezoresistive sensors enabled by multimaterial 3D printing of carbon nanotube/thermoplastic polyurethane nanocomposites, *Polymers*. 11 (2019) 11.
24. M.A. Kassim, T.K. Meng, R. Kamaludin, A.H. Hussain, N.A. Bukhari, Bioprocessing of sustainable renewable biomass for bioethanol production, *Value-Chain of Biofuels*. Elsevier, Amsterdam, Netherlands, (2022) 195–234.
25. O. Kanoun, C. Müller, A. Benchirouf, A. Sanli, T.N. Dinh, A. Al-Hamry, L. Bu, C. Gerlach, A. Bouhamed, Flexible carbon nanotube films for high performance strain sensors, *Sensors (Switzerland)*. 14 (2014) 10042–10071.
26. Y. Yi, B. Wang, X. Liu, C. Li, Flexible piezoresistive strain sensor based on CNTs–polymer composites: A brief review, *Carbon Letters*. 32 (2022) 713–726.
27. H. Kadavil, M. Zagho, A. Elzatahry, T. Altahtamouni, Sputtering of electrospun polymer-based nanofibers for biomedical applications: A perspective, *Nanomaterials*. 9 (2019) 77.
28. Y. Altin, O.F. Unsal, A.C. Bedeloglu, Fabrication and characterization of polyaniline functionalized graphene nanosheets (GNSs)/polydimethylsiloxane (PDMS) nanocomposite films. *Polymers and Polymer Composites*. (2021). https://doi.org/10.1177/09673911211023941.
29. X. Huang, Z. Yin, H. Wu, Structural engineering for high-performance flexible and stretchable strain sensors, *Advanced Intelligent Systems*. 3 (2021) 2000194.
30. Y. Jung, K.K. Jung, D.H. Kim, D.H. Kwak, J.S. Ko, Linearly sensitive and flexible pressure sensor based on porous carbon nanotube/polydimethylsiloxane composite structure, *Polymers*. 12 (2020) 1–12.
31. X. Cheng, L. Miao, Z. Su, H. Chen, Y. Song, X. Chen, H. Zhang, Controlled fabrication of nanoscale wrinkle structure by fluorocarbon plasma for highly transparent triboelectric nanogenerator, *Microsystems and Nanoengineering*. 3 (2017) 1–9.
32. K. Choi, S.J. Park, M. Won, C.H. Park, Soft inductive coil spring strain sensor integrated with sma spring bundle actuator, *Sensors*. 21 (2021) 1–11.
33. J. Wang, C. Lu, K. Zhang, Textile-based strain sensor for human motion detection, *Energy and Environmental Materials*. 3 (2020) 80–100.
34. N. Afsarimanesh, A. Nag, S. Sarkar, G.S. Sabet, T. Han, S.C. Mukhopadhyay, A review on fabrication, characterization and implementation of wearable strain sensors, *Sensors and Actuators, A: Physical*. 315 (2020) 112355.
35. A. Bedeloğlu, G. Findik, M. Çetinoğlu, Ö.F. Ünsal, Stretchable piezoresistive sensors with graphene and polyaniline coated woven polyester fabrics, *Eskişehir Technical University Journal of Science and Technology A - Applied Sciences and Engineering*. 22 (2021) 28–38.
36. R. Yu, C. Zhu, J. Wan, Y. Li, X. Hong, Review of graphene-based textile strain sensors, with emphasis on structure activity relationship, *Polymers*. 13 (2021) 1–22.
37. Y. Altin, A. Celik Bedeloglu, Polyacrylonitrile/polyvinyl alcohol-based porous carbon nanofiber electrodes for supercapacitor applications, *International Journal of Energy Research*. 45 (2021) 16497–16510.
38. T. Yan, Y. Wu, J. Tang, Z. Pan, Highly sensitive strain sensor with wide strain range fabricated using carbonized natural wrapping yarns, *Materials Research Bulletin*. 143 (2021) 111452.

39. T. Yan, Z. Wang, Y.Q. Wang, Z.J. Pan, Carbon/graphene composite nanofiber yarns for highly sensitive strain sensors, *Materials and Design*. 143 (2018) 214–223.
40. C. Mattmann, F. Clemens, G. Tröster, Sensor for measuring strain in textile, *Sensors*. 8 (2008) 3719–3732.
41. J. Lee, S. Shin, S. Lee, J. Song, S. Kang, H. Han, S. Kim, S. Kim, J. Seo, D. Kim, T. Lee, Highly sensitive multifilament fiber strain sensors with ultrabroad sensing range for textile electronics, *ACS Nano*. 12 (2018) 4259–4268.
42. H.R. Nejad, M.P. Punjiya, S. Sonkusale, Washable thread based strain sensor for smart textile, in *TRANSDUCERS 2017–19th International Conference on Solid-State Sensors, Actuators and Microsystems*. (2017) 1183–1186.
43. H. Niu, H. Zhou, H. Wang, T. Lin, Polypyrrole-coated PDMS fibrous membrane: Flexible strain sensor with distinctive resistance responses at different strain ranges, *Macromolecular Materials and Engineering*. 301 (2016) 707–713.
44. Y. Li, X.Y. Cheng, M.Y. Leung, J. Tsang, X.M. Tao, M.C.W. Yuen, A flexible strain sensor from polypyrrole-coated fabrics, *Synthetic Metals*. 155 (2005) 89–94.

9

Metal Oxide-Based Flexible and Wearable Sensors

Bakhytzhan Baptayev, Yerbolat Tashenov, and Mannix P. Balanay

CONTENTS

9.1 Introduction

In recent years, flexible and wearable sensors have been extensively studied due to their reduced weight, flexibility, and durability compared with rigid analogs. Many different sensing materials and composites have been developed that include conducting polymers, carbonaceous materials, biomolecules, metals, and their oxides. Among these materials, metal oxides are promising candidates as sensing materials as a result of their ideal electronic structure, inherent semiconducting property, abundance, and low cost. The preparation method of metal oxide nanostructures includes sol-gel, hydro- and solvothermal, wet chemistry, spin coating, electrodeposition, and sputtering. For flexibility purposes, the focus is made on the substrates. Polymer substrates such as polyimide, polyethylene terephthalate (PET), polyethylene naphthalene (PEN), and polydimethylsiloxane (PDMS), among others, are the materials of choice for flexible sensor preparation. Paper and clothing are also used as substrates for wearable sensing applications. The main parameters when choosing a substrate are its bending, transparency, dielectric, thermal stability, and gas permeation properties. Metal oxides can be directly grown on the substrates or post-transferred onto the substrates. The direct growing method requires a low temperature to prevent polymer substrate thermal degradation or glass transition. However, low-temperature processes restrict the number of metal oxides because most metal oxides require high-temperature treatment. For this reason, many metal oxides are post-transferred onto a polymer substrate after their synthesis. Different methods of the transfer of metal oxides to mechanically flexible substrates have been developed. In a typical transfer process, a sacrificial layer is used for the initial growth step. Drop casting is another method of transferring the dispersion of metal oxide materials onto flexible substrates.

DOI: 10.1201/9781003299455-9

9.2 Gas Sensors

Gas sensors are devices capable of measuring the concentrations of specific gases, volatile compounds, and vapors. Gas sensors have wide applications, these include the detection of toxic and explosive gases, the prevention of specific gas leakages, and the measurement of gases from the human organism and the environment. Metal oxides as gas sensors have gained considerable attraction due to their low cost, abundance, ease of manufacturing, and adjustable band gap. The electronic structure of metals, especially transition metals with unfilled d-shells, can offer excellent electrical characteristics, high dielectric constants, and wide band gaps. Therefore, metal oxides may have tunable electronic and optical properties.

The task of the gas sensor is to detect the gas analyte and convert it into a measurable signal. The main principle of gas sensing by metal oxides relies on a change of resistance. When the metal oxide sensor is exposed to a gas analyte, the latter interacts with the adsorbed oxygen on the surface of the sensor that is causing a change in the resistance of the metal oxide. The oxygen from the air is adsorbed on the surface of the metal oxide sensor causing the extraction of electrons as shown in Equations 1 to 3.

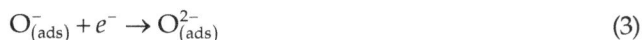

$$O_{2(ads)} + e^- \rightarrow O_{2(ads)}^- \tag{1}$$

$$O_{2(ads)}^- + e^- \rightarrow 2O_{(ads)}^- \tag{2}$$

$$O_{(ads)}^- + e^- \rightarrow O_{(ads)}^{2-} \tag{3}$$

This interaction causes a change in conductivity. If the semiconductor is an n-type, then its conductivity decreases because the electrons are captured by oxygen. However, if the semiconductor is a p-type, then its conductivity increases due to the generation of extra hole carriers. When a target gas comes into contact with the metal oxide surface, the interaction between adsorbed oxygen and target gas causes a change in the number of charge carriers in the metal oxide. Thus, the conductivity of the sensor also changes, and this allows for the detection of a target gas. When the metal oxide sensor is exposed to reducing target gas (H_2, H_2S, NH_3, CO, CH_3CHO, C_2H_5OH, etc.), the adsorbed oxygen ions react with the target gas, and reinjection of electrons into metal oxide conduction band occurs. On the contrary, exposure to oxidizing gas (NO, NO_2, etc.) leads to the extraction of more electrons from the metal oxide by adsorbed oxygen. The reinjection of electrons into the metal oxide conduction band decreases the number of hole carriers (p-type) or increases the number of electron carriers (n-type). Likewise, the extraction of electrons from the metal oxide conduction band increases the number of hole carriers (p-type) or decreases the number of electron carriers (n-type). Consequently, the change in charge carrier concentrations causes a change in conductivity, which is read as a signal.

Various metal oxides have been studied as gas sensors. The most common examples are binary oxides such as CuO, Cu_2O, In_2O_3, ZnO, TiO_2, WO_3, V_2O_3, and SnO_2. Among the gas-sensing binary metal oxides, ZnO is the most studied and widely utilized semiconductor due to its suitable physicochemical properties. ZnO is an n-type semiconductor with a wide bandgap of 3.37 eV at the bulk state. Moreover, it is cheap, abundant, eco-friendly, and chemically stable. Furthermore, its exciton binding energy is large at about 60 meV and it has a remarkable electron mobility of 400 cm^2/Vs. Due to the crystalline nature of ZnO, it can grow into different 1D, 2D, and 3D nanostructures, which is beneficial to further tune the band gap. In 2011, a group of researchers presented vertically aligned ZnO nanorods and graphene composite-based flexible high-sensitive flexible gas sensors [1]. The hybrid flexible sensor was fabricated on a metal foil. Structurally, the film was composed of a bottom ZnO conductive layer, vertically grown ZnO nanorods, and a conductive top layer based on graphene. Repeated tests of up to 100 cycles showed that the hybrid structure withstood the flexural deformation (with a bending radius of 0.8 cm) with no mechanical and electrical failure. Additionally, the hybrid sensor demonstrated excellent transparency (>70%) on a glass substrate. This 1D ZnO/graphene-based sensor was used to detect ethanol gas vapor with a concentration range of 10 to 40 ppm. Lee and colleagues prepared porous 2D ZnO nanosheets with a thickness of 80 nm using the solvothermal method. The porosity of the obtained nanosheet was found to be around 16% with an average pore size of 60 nm. This porous ZnO-based sensor demonstrated highly sensitive detection of NO_2 gas with 0.5 to 3 ppm and 10 to 75 ppm concentrations at 200°C. The authors concluded that the remarkable NO_2 detection by the porous ZnO nanosheets was due to the synergistic effect of high surface area, ZnO/ZnO homojunction, and structural defects [2]. Tonezzer et al. compared the performance of 1D and 2D ZnO nanostructures in gas sensing [3]. ZnO nanowires with an average diameter of 80–250 nm and nanosheets with a thickness of around 70–360 nm were used to fabricate conductometric gas sensors for liquid petroleum gas detection. They found that the depletion layer affected two dimensions of 1D ZnO and only one dimension of 2D ZnO, which improves the sensor response in the 1D sensor. On the other hand, the greater cross-section of 2D ZnO increases the base current and lowers the limit of detection. Another group of researchers presented a 3D ZnO-based flexible gas sensor [4]. The sensor is micropatternable double-faced ZnO nanoflowers on a polyimide substrate with single-walled carbon nanotubes as an electrode. The 3D structure provides multiple gas diffusion channels, whereas the micropatterned island structure disperses the strain and provides stability to the structure. The sensor demonstrated high selectivity to NO_2 gas with fast rising and decay times of 25.0 and 14.1 s, respectively. The percent recovery was also high at 98%. After 10,000 bending cycles with a curvature radius of 5 mm, the sensor still exhibited excellent sensing abilities.

TiO_2 is another semiconductor oxide used for gas sensing applications. In a work presented by Perillo and Rodriguez, TiO_2 membrane nanotubes were used as a flexible trimethylamine sensor [5]. The detection of trimethylamine is very important in determining fish freshness since this compound is responsible for its fishy odor. Kapton, a flexible material, with 150 μm thickness was used as a substrate due to its high thermal stability up to 300°C, wide solvent resistance, low cost, and good mechanical properties. The TiO_2 membrane nanotube array was grown via a two-step electrochemical anodization method. The schematic representation of the sensor is presented in Figure 9.1. Based on their results, the sensor detected trimethylamine with a concentration range of 40 to 400 ppm at low temperatures. Sensors that work at low and ambient temperatures are in high demand for wearable applications.

Some researchers opted to develop a composite to properly enhance detecting ability such as the NiO@CuO sensor for ammonia detection at room temperature. The sensor was coupled with a flexible poly(para-phenylene terephthalamide) fiber substrate that exhibited remarkable thermal stability up to 450°C with almost no weight change. It demonstrated a low limit of detection of 46.5 ppb for NH_3 at room temperature [6].

For flexible and wearable applications, metal oxide gas sensors should operate at ambient temperatures. Therefore, new materials and composites that can provide room temperature gas sensing should be studied. Moreover, many practical issues, which include

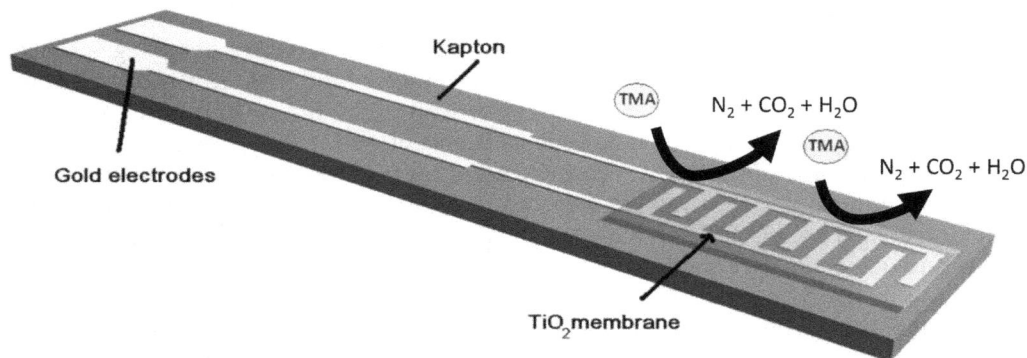

FIGURE 9.1
TiO_2 nanotube membrane on flexible Kapton substrate for a trimethylamine gas sensor. Adapted with permission from Reference [5], Copyright (2016), Elsevier.

the high reproducibility of flexible sensors, the cost-effectiveness of fabrication methods, long-term stability, large-scale production, and so on, must be thoroughly considered.

9.3 pH Sensors

The demand for wearable sensors that can be used in medicine for health monitoring and non-invasive analysis is growing. Fast and accurate real-time detection of biological and chemical markers of diseases allows early and efficient treatment of them. Traditional methods of human health analysis and diagnosis of diseases are characterized by a long detection time of up to several days. Time is an important parameter for medical treatment, especially when it comes to fast-moving diseases. Rapid diagnostic methods can guarantee efficient treatment. In this regard, the study and development of wearable human health monitoring devices have gained interest. These devices can be attached to the human body and/or clothing and in real time can perform fast biochemical analysis. The determination of pH level for human health monitoring is important as it gives information about biochemical reactions; therefore, the development of wearable pH sensors is of utmost importance.

The measurement of the pH level of human biofluids can provide extensive health-related information. The pH level of the skin, for example, can be related to the antimicrobial effect that can prevent the growth of microorganisms. The proper function of many cellular processes also depends on pH level. It was found that when the pH level of body fluid decreases from 6.9 to 6.5, there will most probably be damage to the living cells and there is an increased potential for tumor formation [7]. Other studies have shown that acidic pH leads to the inflammation of blood cells and may reduce oxygen levels. These may promote the failure of the liver, kidney, and sweat glands [8]. To conclude, inadequate pH levels in the human body may cause serious health problems. Wearable pH sensors can provide rapid and real-time monitoring of pH and early detection of related diseases.

Recently, metal oxides have gained great interest as pH sensors due to their excellent electrochemical and highly sensing characteristics. These properties enable metal oxide-based sensors to be used in human health monitoring, food quality control, and environmental applications. Metal oxides have a high surface-to-volume ratio that enhances their selectivity, sensitivity, and response time. Moreover, metal oxides are abundant, cost-effective, and possess high thermal stability and mechanical strength. This ensures fast responses in different environments, prolonged exploitation time, and easy miniaturization. Therefore, metal oxides are convenient materials for flexible and wearable pH sensors.

The sensing mechanism of metal oxide pH sensors is complex. Fog and Buck proposed five possible working mechanisms of metal oxide pH sensors [9]: (1) ion exchange via a surface layer containing OH groups; (2) redox equilibrium between two different solid phases; (3) solid solution or intercalation reaction, wherein there is a redox equilibrium involving only one solid phase, whose hydrogen content can be varied continuously by passing a current through the electrode; (4) single phase oxygen intercalation of the electrode; and (5) steady-state corrosion of the electrode. Other researchers also explained the working principle of the metal oxide pH sensors by the first mechanism as suggested by Fog and Buck where ion exchange occurs in surface layers containing OH groups [10,11]. When the metal oxide sensor comes into contact with the solution, it is covered by water molecules due to dissociative adsorption and the surface is covered with hydroxyl groups. When the proton is lost, the oxide layer is formed with new valence layers which creates a potential difference. The magnitude of the potential difference is proportional to the pH of the solution. McMurray et al., on the other hand, claimed that the fourth mechanism was the most adequate explanation of metal oxide pH response [12]. Most of the reported metal oxide pH sensors are based on the potentiometric method. In this configuration, the sensor is composed of a metal oxide-sensitive electrode and a reference electrode. The performance of the sensor is compared with that of a glass-based pH electrode. Various films of metal oxides including thin and thick films can be utilized as sensing electrodes.

Iridium oxide (IrO_x) is the most extensively studied wearable pH sensor. This is because of its excellent sensitivity, quick response, and biocompatibility. Zamora et al. developed a potentiometric textile-based pH sensor using electrodeposited IrO_2 films [13]. In comparison with the commercial pH test strip, the developed sensor produced a relative error of 4%. Recently, Cao and colleagues presented a flexible potentiometric pH sensor with IrO_x as the sensing electrode and a cured Ag/AgCl paste as the pseudo-reference electrode [14]. The sensor was fabricated by electrodeposition with inductively coupled wireless transmission. The sensitivity of the device was in the range of 65–75 mV/pH with a response time of 2 s. The difference from commercial pH electrode when measuring "artificial sweat" pH was 0.4 to 0.8 pH. The performance of the device did not show any significant difference after different bending conditions (0°, 30° at 55 mm radius, 45° at 37 mm, and 90° at 20 mm).

Zinc oxide is another potential sensing material for pH sensor application, and due to its amphoteric nature, it can interact both with H^+ and OH^- ions. These characteristics make zinc oxide a suitable candidate for pH-sensing applications. On the other hand, the presence of defect sites such as zinc vacancies, oxygen vacancies, zinc/oxygen interstitials, and free OH groups may trap charge carriers and hinder their transportation to active sites. Therefore, lowering the number of defects is important for ZnO. For example, Hilal and Yang reported doped ZnO with Co to make a defect-free ZnO nanostructure and decorated it with CoO clusters. This heterostructure was grown on a flexible PET substrate and used as a dual-functional flexible sensor for pH and glucose monitoring of fruit juices and human fluids [15]. The presence of CoO prevents corrosion, increases active sites, and enhances ultimate tensile strength. The heterostructural pH sensor demonstrated high sensitivity of 52 mV/pH and good chemical stability. It has a response time of 19 s and the sensor retained 52% of its initial sensitivity after mechanical deformation for 500 cycles. Another method of improving the defect sites of ZnO is adding a second metal to form ternary oxide. Liu et al. reported flexible indium-zinc-oxide (IZO) based neuromorphic transistors on a plastic substrate [16]. The sensor demonstrated a remarkable 105 mV/ pH for quasi-static dual-gate synergic sensing mode. The authors demonstrated a short response time of 5 ms with a single-spike dynamic sensing mode that also showed good reproducibility.

The fabrication of flexible pH sensors is usually done by microfabrication methods such as thin film deposition, photolithography, and wet or dry etching on a flexible substrate. These methods are more expensive compared with electrodeposition, screen printing, and sol-gel coating. Fu et al. reported flexible inkjet-printed WO_3-based pH sensors with Ag/AgCl as a reference electrode [17]. The inkjet-printing method is cost-effective and simple which makes it suitable for in-hospital application. The sensor is composed of WO_3 nanoparticles printed on Ag lines at a low temperature of 120°C. The sensing material and the reference electrode were supported by a flexible polyimide substrate. The sensitivity of the pH sensor was around 24.4 mV/pH. Another group of researchers used electrodeposition to fabricate a WO_3-based pH sensor [18]. Hydrothermally synthesized WO_3 nanoparticles were electrodeposited on a flexible polyimide substrate of 50 μm thickness over Ti/ Au metal electrodes with a sensing area of 1 mm^2. The electrodeposition method allowed the fabrication of a thin WO_3 layer with improved surface area, and good sensitivity of 56.7 mV/pH in the pH range of 9 and 5.

Other metal oxides used in pH sensing are RuO_2, Ta_2O_5, TiO_2, SnO_2, PbO_2, CeO_2, and others. However, only a few of them were studied as flexible and wearable pH sensors. For the further development of flexible and wearable metal oxide pH sensors, novel nanomaterials must be synthesized and new methods of fabrication should be explored. Oxide materials like indium-gallium-oxide and gallium-tin-zinc-oxide should be carefully investigated.

9.4 Humidity Sensors

Humidity sensing has become an important monitoring method in medicine, environmental analysis, the food industry, and agriculture, among others. In addition, the control of the indoor microclimate of houses, museums, and research laboratories also relies on the proper detection of humidity levels. All of these led to the development of humidity-sensing materials and devices that have become a key component in developing smart monitoring systems in recent years, especially with the advent of the Internet of Things technologies. Therefore, many different functional nanomaterials have been explored and developed for the fabrication of flexible and wearable humidity sensors in recent years.

The working mechanism of humidity sensors is based on the detection of changes in the electrical current or temperature. Humidity sensors can be capacitive, resistive, and thermal. A capacitive humidity sensor measures the change of electrical capacity relative to the atmosphere's humidity level. The working principle of resistive humidity sensors is like that of gas sensors. When humidity changes, the resistance of the electrodes also changes, which is detected by the sensor. In thermal humidity sensors, two electrode sensors conduct electricity where one of them is under an inert nitrogen atmosphere, whereas the other measures the ambient air. The difference between these two electrodes is detected by the humidity sensor. Recently, other types of humidity sensors have also been developed. They use field-effect transistors [19], optical fibers [20], surface acoustic waves [21], and so on. Among these, resistive humidity sensors are more attractive and extensively studied due to their high response, better stability, simple manufacturing, and cost-effectiveness.

A variety of metal oxides have been studied as humidity-sensing materials. Some of them are NiO, SrO, WO_3, V_2O_3, MoO_3, La_2O_3, CeO_2, In_2O_3, Ta_2O_5, and Co_3O_4 [22]. Molybdenum oxide is one of the metal oxide semiconductors successfully used as a humidity sensor. Yang et al. presented a transparent and flexible MoO_3 nanosheets-based humidity sensor fabricated with simple spin coating and then annealing at 60°C [23]. The reported humidity sensor exhibited ultra-high sensitivity with a rapid response time (<0.3 s) and short recovery time (<0.5 s). The selectivity of the sensor exceeded that of the state-of-the-art sensors. Due to its remarkable response to fingertip humidity, the MoO_3 humidity sensor was also applied in non-contact mode, wherein if the finger is placed 5 mm away from the sensor, only one LED light is turned on, while when the distance was reduced to 3 mm, two LED lights turned on. This demonstrates the potential of the sensor to be used in human-machine interaction systems.

In recent years, TiO_2 nanostructures have found applications in many promising areas including photovoltaics [24], photocatalysis [25], and sensors [5]. Dubourg and his associates presented a miniaturized resistive-type TiO_2-based humidity sensor printed on a flexible PET substrate in a 3×3 matrix [26]. The fabrication process involves screen printing of TiO_2 nanoparticle paste on PET electrodes designed by laser ablation. The sensor demonstrated excellent electrical and humidity response stability after mechanical testing. The good repeatability, relatively fast, and linear response in the range between 5% and 70 % relative humidity showed the great potential of the TiO_2 sensor for environmental monitoring and humidity sensing applications.

Flexible and wearable humidity sensors have attracted wide attention and developed very fast due to their potential application in industry, environmental monitoring, medicine, and the Internet of Things. Although great advances were made in humidity sensors, there are still several unsolved issues. Future research efforts should be focused on the development of new humidity-sensing materials, easy and cost-effective fabrication methods, and eco-friendly manufacturing processes.

9.5 Biomaterial Sensors

The operational principle of the majority of biosensors is based on the detection of primary and secondary metabolites of biological bodies through the use of specific bioreceptors (enzyme, DNA probe, antibody) as active sensitive sites chemically linked to signal detection transducers. Metal oxides are widely used due to their distinctive advantages such as high conductivity and stability, low cost, ease of preparation, and biocompatibility. In most cases, target analytes for biosensing are glucose, uric acid, and lactate since information about their concentration in the human body is a marker of some important diseases [27].

Two mechanisms can be distinguished for monitoring the glucose content in bioassays: (1) enzymatic and (2) non-enzymatic electrochemical detection. The enzymatic method involves the use of glucose oxidase (GOx) as a biocatalyst for the oxidation of the given carbohydrate. In this case, the concentration of the by-products of the glucose conversion reaction [H_2O_2] and [H^+] can be exploited as analytes and correlated with the amount of the biomolecule according to the following equations:

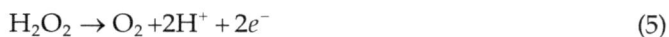

$$D\text{-glucose} + H_2O + O_2 + GOx \rightarrow \text{Gluconic acid} + H_2O_2 \qquad (4)$$

$$H_2O_2 \rightarrow O_2 + 2H^+ + 2e^- \qquad (5)$$

Rim et al. [28] demonstrated a facile solvothermal method for fabricating an ultrathin and highly sensitive In_2O_3 semiconductor film on polyimide/glass substrate by spin-coating technique and utilized it as a flexible biosensor to detect D-glucose in biofluids (Figure 9.2a). Functionalization of the In_2O_3 surface with glucose oxidase was achieved by the introduction of an amine-terminated (3-aminopropyl) triethoxysilane and glutaraldehyde linkers

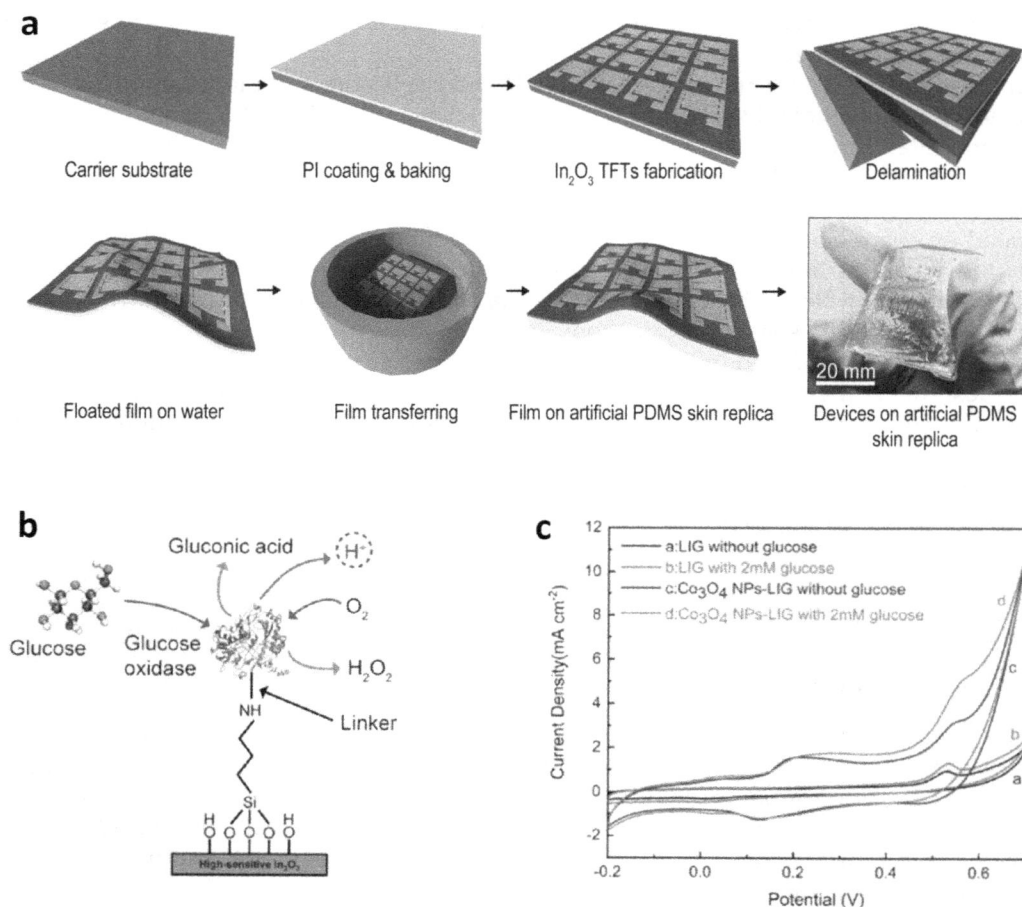

FIGURE 9.2

a) Schematic illustration of the flexible biosensor fabrication procedure. b) Enzymatic oxidation of D-glucose by glucose oxidase to produce gluconic acid and hydrogen peroxide. Protons are generated during this oxidation process and protonation of the In_2O_3 surfaces is manifested. Adapted with permission from Reference [28], Copyright (2015), American Chemical Society. c) CV curves of pristine LIG and Co_3O_4 NPs-LIG electrodes with and without glucose. Adapted with permission from Reference [35], Copyright (2021), Elsevier.

(Figure 9.2b). The wearability of the designed biosensor based on In_2O_3 was illustrated by a conformal connection with polydimethylsiloxane (PDMS) artificial human skin using van der Waals forces.

Multifunctional biosensors based on a piezo-biosensing unit matrix of enzyme/ZnO nanoarrays on Kapton substrate were also tested for enzymatic detection of lactate, glucose, uric acid, and urea in sweat. Modification of the ZnO surface with four different enzymes (lactate oxidase (LOx), GOx, uricase, and urease) was carried out by direct absorption of biocatalysts via impregnating the oxide in their solutions. They manufactured a wearable biosensing device with an autonomous power supply generated by the piezoelectric impulse of the piezo-biosensing units [29]. In a similar work, Prasad and his research group developed an electrochemical sensor containing a thin film of Au/ZnO on a flexible PI substrate for detecting glucose in human perspiration with a concentration range from 0.01 to 200 mg/dL [30]. In 2020, Nien et al. also reported an enzymatic urea biosensor with a sensor film made from an Au-NPs/graphene oxide (GO)/NiO system to absorb urease, and device fabrication was performed on a flexible polyethylene terephthalate substrate using screen printing technology [31]. The urea detection mechanism is based on the enzymatic reaction catalyzed by urease. As the result, the formation of hydroxide ions as by-products generally affects the pH value of the system and leads to a change in the potential on the surface of the sensing electrode as shown in Equation 6.

$$NH_2CONH_2 + 3H_2O + urease \rightarrow 2NH_4^+ + HCO_3^- + OH^- \tag{6}$$

Later, the same research group demonstrated urea biosensors with improved sensitivity by modifying the GO/NiO sensing membrane with γ-Fe_2O_3 NPs [32].

It is worth noting that most biomolecule-detecting sensors exploit specific enzymes as receptors and/or oxidizing agents, which leads to a general chemical and thermal instability of the developed devices. Moreover, in some cases, additional chemicals are required as linkers for enzyme immobilization processes. As an alternative to these shortcomings, the method of non-enzymatic oxidation by direct electrocatalytic oxidation of biomolecules with functional nanomaterials, including oxides, is developed. For example, Cu_2O [33,34], Co_3O_4 [35], and $NdNiO_3$ [36] have recently been used as active sensing metal oxides of electrochemical flexible sensors for detecting dopamine, glucose, and ascorbic acid in human sweat, respectively, and opening perspectives for the construction of wearable physiological health monitoring and medical diagnostic systems.

It is well known that transition metal oxides have several oxidation states, which allows them to initiate redox reactions and operate as detector parts of biosensors, catalyzing the electrochemical oxidation of biomolecules. As a rule, the analytical/sensing signal of the target molecules detection by an electrochemical system is determined by the appearance of significant oxidation and reduction peaks on chronoamperometry, cyclic voltammetry, or differential pulse voltammetry. To cite an example, Figure 9.2c illustrates characteristic CV curves obtained by experimental detection of 2 mM glucose in a base electrolyte solution with a pure laser-induced graphene (LIG) electrode and LIG modified with a Co_3O_4 NPs electrode. The study was also conducted in the absence of the analyte. When 2 mM glucose is added, both electrodes show distinct peaks resulting from the redox response signal to glucose [35].

Recently, Manjakkal and his team presented a study testing a portable enzyme-free sensor to detect glucose in sweat [34]. The printing technology was used to transfer graphene paste onto a flexible cellulose cloth and then coated with Cu_2O by drop casting (Figure 9.3). There is no clear understanding of the mechanism of catalytic oxidation of

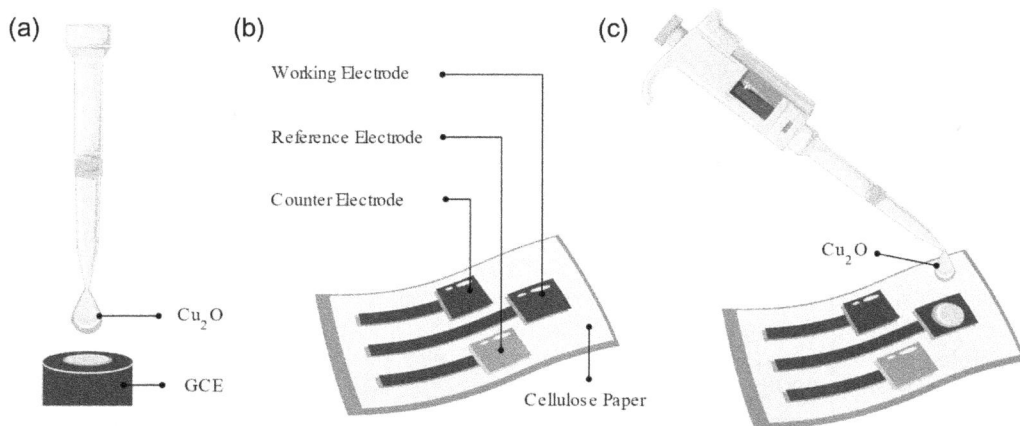

FIGURE 9.3
Schematic fabrication of the glucose sensor. a) Cu_2O nanoclusters drop casted on the GCE for material study. b) Graphene paste printed on cellulose cloth with an Ag/AgCl RE. c) Modified WE with Cu_2O on the graphene-printed cellulose substrate. Adapted with permission from [34], Copyright The Authors, some rights reserved; exclusive licensee MDPI. Distributed under a Creative Commons Attribution License 4.0 (CC BY) https://creativecommons.org/licenses/by/4.0/).

glucose by cuprite, however, some suggestions have been made that CuOOH is responsible for inducing the oxidation of glucose to gluconolactone according to the following chemical reactions:

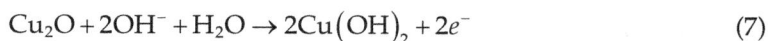

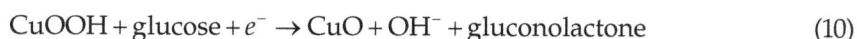

$$Cu_2O + 2OH^- + H_2O \rightarrow 2Cu(OH)_2 + 2e^- \tag{7}$$

$$2Cu(OH)_2 \rightarrow CuO + H_2O \tag{8}$$

$$CuO + OH^- \rightarrow CuOOH + e^- \tag{9}$$

$$CuOOH + glucose + e^- \rightarrow CuO + OH^- + gluconolactone \tag{10}$$

$$2CuO + H_2O + 2e^- \rightarrow Cu_2O + 2OH^- \tag{11}$$

Continuing the development of non-enzymatic flexible biosensors, Rosatto et al. invented an ascorbic acid (AA) electrochemical sensing device based on different-sized $NdNiO_3$ nanotubes. The sensing properties of $NdNiO_3$ nanotube electrodes with two different diameters of 20 nm and 100 nm coated onto laser-induced graphene (GO) were tested via electrochemical characterization. The best performance was obtained for the GO/$NdNiO_3$-100 nm electrode, which was then used to manufacture a flexible wearable electrochemical biosensor with a minimum detectable limit (3.8 μmol/L) and sensitivity (0.031 μA μM^{-1} cm^{-2}) of AA [36].

9.6 Temperature Sensors

Tracking changes in human body temperature is always one of the main indicators for medical assessment of health status. It is well known that healthy people usually have a body temperature in the range of 36 to 37°C. Even slight deviations from normal temperature values can indicate a deterioration in health and sometimes can lead to fatal consequences. Therefore, the development of modern wearable sensor devices for detecting body temperature in real time and in continuous mode is relevant. Currently, thermistors are used for these purposes, which have a high sensitivity to small temperature fluctuations. Among the materials from which thermistors are made, a special place is occupied by metal oxides, since the electrical resistance of some of them varies significantly depending on temperature. Thermistors come in two types: negative temperature coefficient (NTC) and positive temperature coefficient (PTC) [37]. Most oxides are considered NTC thermistors (resistivity decreases with increasing temperature) due to their high thermal constant value (β-value), which is related to temperature and thermistor resistance according to the following formula [38]:

$$R_t = R_0 * \exp\beta\left(\frac{1}{T} - \frac{1}{T_0}\right) \tag{12}$$

where R_t is the resistance at a given temperature, and T and R_0 are the resistance at a reference temperature, T_0. The temperature coefficient of the thermistor, α, is another important parameter that characterizes its sensitivity. The relationship between the values of α and β can be established using Equation 13 [38].

$$\alpha = \frac{\beta}{T^2}\left(\% K^{-1}\right) \tag{13}$$

The temperature constant and coefficient are essential indicators to measure the resistance-based performance of thermistors. High values of β indicate stability at high temperatures, which makes such materials suitable for use in temperature-sensing devices [37].

The material used in commercial thermistors must have high β values above 3500 K ($\alpha \sim$ −4%/K at 20°C). Liao and his team synthesized NiO nanoparticles and inkjet printed them on a flexible polyimide substrate for fabrication thermistors. As a result, the temperature sensing device operated in the range from room temperature to 200°C and maintained great sensitivity due to high β values (~4300 K). The wearability of the thermistor was achieved by depositing oxide on PI films consequently tested by bending and attaching to uneven surfaces [39]. Similarly, Kahn et al. also fabricated a temperature sensor by stencil printing NiO_x nanoparticle inks onto a Kapton PI substrate. Compared with the work of Liao's group, the present study used a lower annealing temperature for the deposited NiO film and used inkjet-printed gold electrodes instead of silver counterparts. After testing, the constructed NTC thermistor showed a relatively higher temperature coefficient $\alpha \approx$ −5.84% K^{-1} and material constant $\beta \approx 4330$ K [38]. Further application of NiO NPs with a new fabrication technique in thermistors was continued by Shin et al. Currently, their results with excellent β-values (~8162 K) acquired near room temperature fabrication by laser sintering techniques led to the highest sensitivity ever reported for thermistor-based temperature sensors [40].

Nakajima and Tsuchiya jointly published their excellent results on preparing a $Mn_{1.56}Co_{0.96}Ni_{0.48}O_4$ (MCN) spinel film on a PET substrate for flexible NTC thermistor applications. Even though the MCN nanoparticles have a high crystallization temperature of 700–1200°C, they were deposited on a flexible substrate by a photoreaction of nanoparticles (PRNP) with pulsed laser irradiation at room temperature. Crystallization of MNCs at 55 mJ/cm^2 under 600 pulses was induced by the optimized photothermal heating effect of the nanoparticles, resulting in well-densified and planarized NTC thermistor films with β -values of 4429 K, superior to commercial counterparts [41].

9.7 Challenges and Future Perspectives

There is an increasing number of applications of metal oxide-based flexible and wearable sensors, however, the majority of designs are quite primitive and often bulky. They still lack applications that can be directly applied to textiles or even attached to the skin. This is due to the inherent crystal structure of metal oxides that are prone to cracking unless proper hybridization is applied in their synthesis step, which needs a higher temperature. Another main challenge is the large band gaps of these metal oxides. Most of the researchers are focusing on the development of binary metal oxides; the researchers need to extend development to ternary metal oxides or introduce complex hybridization to further control electronic properties. Another problem with metal oxide-based flexible and wearable sensors is distinguishing signals from other external and multiple stimuli; researchers have tried complex hybridization to address this, but it still needs to be properly explored to be able to be commercially viable in the future. A further limitation of the flexible and wearable sensors is their substrates, as the majority of the researchers use polyethylene terephthalate or polyimide films, which need to further expand, especially if we are looking for an environmentally sustainable type of solution. With these challenges, there is still a bright future ahead for the development of metal oxide in flexible and wearable sensors due to its low cost, high surface-to-volume ratio, and easy tuning with the use of inorganic and/or organic materials. It is also an avenue for multidisciplinary research that brings all STEM fields together.

References

1. J. Yi, J.M. Lee, W. il Park, Vertically aligned ZnO nanorods and graphene hybrid architectures for high-sensitive flexible gas sensors, *Sensors and Actuators B: Chemical.* 155 (2011) 264–269.
2. M.S. Choi, M.Y. Kim, A. Mirzaei, H.S. Kim, S. il Kim, S.H. Baek, D.W. Chun, C. Jin, K.H. Lee, Selective, sensitive, and stable NO₂ gas sensor based on porous ZnO nanosheets, *Applied Surface Science.* 568 (2021) 150910.
3. M. Tonezzer, T.T. le Dang, N. Bazzanella, V.H. Nguyen, S. Iannotta, Comparative gas-sensing performance of 1D and 2D ZnO nanostructures, *Sensors and Actuators B: Chemical.* 220 (2015) 1152–1160.
4. J.-W. Kim, Y. Porte, K.Y. Ko, H. Kim, J.-M. Myoung, Micropatternable double-faced ZnO nanoflowers for flexible gas sensor, *ACS Applied Materials & Interfaces.* 9 (2017) 32876–32886.
5. P.M. Perillo, D.F. Rodriguez, Low temperature trimethylamine flexible gas sensor based on TiO₂ membrane nanotubes, *Journal of Alloys and Compounds.* 657 (2016) 765–769.
6. Y. Zhou, J. Wang, X. Li, Flexible room-temperature gas sensor based on poly (para-phenylene terephthalamide) fibers substrate coupled with composite NiO@CuO sensing materials for ammonia detection, *Ceramics International.* 46 (2020) 13827–13834.
7. A.U. Alam, Y. Qin, S. Nambiar, J.T.W. Yeow, M.M.R. Howlader, N.X. Hu, M.J. Deen, Polymers and organic materials-based pH sensors for healthcare applications, *Progress in Materials Science.* 96 (2018) 174–216.

8. Y. Tang, L. Zhong, W. Wang, Y. He, T. Han, L. Xu, X. Mo, Z. Liu, Y. Ma, Y. Bao, S. Gan, L. Niu, Recent advances in wearable potentiometric pH sensors, *Membranes (Basel)*. 12 (2022) 504.
9. A. Fog, R.P. Buck, Electronic semiconducting oxides as pH sensors, *Sensors and Actuators*. 5 (1984) 137–146.
10. S. Trasatti, Physical electrochemistry of ceramic oxides, *Electrochimica Acta*. 36 (1991) 225–241.
11. J.A. Mihell, J.K. Atkinson, Planar thick-film pH electrodes based on ruthenium dioxide hydrate, *Sensors and Actuators B: Chemical*. 48 (1998) 505–511.
12. H.N. McMurray, P. Douglas, D. Abbot, Novel thick-film pH sensors based on ruthenium dioxide-glass composites, *Sensors and Actuators B: Chemical*. 28 (1995) 9–15.
13. M.L. Zamora, J.M. Dominguez, R.M. Trujillo, C.B. Goy, M.A. Sánchez, R.E. Madrid, Potentiometric textile-based pH sensor, *Sensors and Actuators B: Chemical*. 260 (2018) 601–608.
14. P. Marsh, S. Member, L. Manjakkal, X. Yang, M. Huerta, T. Le, L. Thiel, J. Chiao, H. Cao, S. Member, R. Dahiya, Flexible iridium oxide based pH sensor integrated with inductively coupled wireless transmission system for wearable applications, *IEEE Sensors Journal*. 20 (2020) 5130–5138.
15. M. Hilal, W. Yang, A dual-functional flexible sensor based on defects-free Co-doped ZnO nanorods decorated with CoO clusters towards pH and glucose monitoring of fruit juices and human fluids, *Nano Convergence*. 9 (2022) 14.
16. N. Liu, L.Q. Zhu, P. Feng, C.J. Wan, Y.H. Liu, Y. Shi, Q. Wan, Flexible sensory platform based on oxide-based neuromorphic transistors, *Scientific Reports*. 5 (2015) 18082.
17. Y.-M. Fu, J.-C. Pan, K.-L. Tsou, Y.-T. Cheng, A flexible WO$_3$-based pH sensor array for 2-D pH monitoring using CPLoP technique, *IEEE Electron Device Letters*. 39 (2018) 881–884.
18. L. Santos, J. P. Neto, A. Crespo, D. Nunes, N. Costa, I. M. Fonseca, P. Barquinha, L. Pereira, J. Silva, R. Martins, E. Fortunato, WO$_3$ nanoparticle-based conformable pH sensor, *ACS Applied Materials & Interfaces*. 6 (2014) 12226–12234.
19. F. Liang, L.B. Luo, C.K. Tsang, L. Zheng, H. Cheng, Y.Y. Li, TiO$_2$ nanotube-based field effect transistors and their application as humidity sensors, *Materials Research Bulletin*. 47 (2012) 54–58.
20. B. Du, D. Yang, X. She, Y. Yuan, D. Mao, Y. Jiang, F. Lu, MoS$_2$-based all-fiber humidity sensor for monitoring human breath with fast response and recovery, *Sensors and Actuators B: Chemical*. 251 (2017) 180–184.
21. Y. Li, C. Deng, M. Yang, A novel surface acoustic wave-impedance humidity sensor based on the composite of polyaniline and poly(vinyl alcohol) with a capability of detecting low humidity, *Sensors and Actuators B: Chemical*. 165 (2012) 7–12.
22. V.K. Tomer, R. Malik, V. Chaudhary, A. Baruah, L. Kienle, Noble metals–metal oxide mesoporous nanohybrids in humidity and gas sensing applications, in S. Mohapatra, T.A. Nguyen, P. Nguyen-Tri, (Eds.), *Noble Metal-Metal Oxide Hybrid Nanoparticles: Fundamentals and Applications*. Woodhead Publishing, Cambridge, MA, (2019) 283–302.
23. J. Yang, R. Shi, Z. Lou, R. Chai, K. Jiang, G. Shen, Flexible smart noncontact control systems with ultrasensitive humidity sensors, *Small*. 15 (2019) 1902801.
24. B. Baptayev, S. Adilov, M.P. Balanay, Surface modification of TiO$_2$ photoanodes with In^{3+} using a simple soaking technique for enhancing the efficiency of dye-sensitized solar cells, *Journal of Photochemistry and Photobiology A: Chemistry*. 394 (2020) 112468.
25. J. Schneider, M. Matsuoka, M. Takeuchi, J. Zhang, Y. Horiuchi, M. Anpo, D.W. Bahnemann, Understanding TiO$_2$ photocatalysis: Mechanisms and materials, *Chemical Reviews*. 114 (2014) 9919–9986.
26. G. Dubourg, A. Segkos, J. Katona, M. Radović, S. Savić, G. Niarchos, C. Tsamis, V. Crnojević-Bengin, Fabrication and characterization of flexible and miniaturized humidity sensors using screen-printed TiO2 nanoparticles as sensitive layer, *Sensors*. 17 (2017) 1854.
27. H. Zhao, R. Su, L. Teng, Q. Tian, F. Han, H. Li, Z. Cao, R. Xie, G. Li, X. Liu, Z. Liu, Recent advances in flexible and wearable sensors for monitoring chemical molecules, *Nanoscale*. 14 (2022) 1653–1669.
28. Y.S. Rim, S.-H. Bae, H. Chen, J.L. Yang, J. Kim, A.M. Andrews, P.S. Weiss, Y. Yang, H.-R. Tseng, Printable ultrathin metal oxide semiconductor-based conformal biosensors, *ACS Nano*. 9 (2015) 12174–12181.

29. W. Han, H. He, L. Zhang, C. Dong, H. Zeng, Y. Dai, L. Xing, Y. Zhang, X. Xue, A self-powered wearable noninvasive electronic-skin for perspiration analysis based on piezo-biosensing unit matrix of enzyme/ZnO nanoarrays, *ACS Applied Materials & Interfaces.* 9 (2017) 29526–29537.

30. R.D. Munje, S. Muthukumar, S. Prasad, Lancet-free and label-free diagnostics of glucose in sweat using zinc oxide based flexible bioelectronics, *Sensors and Actuators B: Chemical.* 238 (2017) 482–490.

31. Y.-H. Nien, T.-Y. Su, J.-C. Chou, C.-H. Lai, P.-Y. Kuo, S.-H. Lin, T.-Y. Lai, M. Rangasamy, Investigation of flexible arrayed urea biosensor based on graphene oxide/nickel oxide films modified by Au nanoparticles, *IEEE Transactions on Instrumentation and Measurement.* 70 (2021) 1–9.

32. Y.-H. Nien, T.-Y. Su, C.-S. Ho, J.-C. Chou, C.-H. Lai, P.-Y. Kuo, Z.-X. Kang, Z.-X. Dong, T.-Y. Lai, C.-H. Wang, The analysis of potentiometric flexible arrayed urea biosensor modified by graphene oxide and γ-Fe_2O_3 nanoparticles, *IEEE Transactions on Electron Devices.* 67 (2020) 5104–5110.

33. Y. Jiang, T. Xia, L. Shen, J. Ma, H. Ma, T. Sun, F. Lv, N. Zhu, Facet-dependent Cu_2O electrocatalysis for wearable enzyme-free smart sensing, *ACS Catalysis.* 11 (2021) 2949–2955.

34. F.F. Franco, R.A. Hogg, L. Manjakkal, Cu_2O-based electrochemical biosensor for non-invasive and portable glucose detection, *Biosensors (Basel).* 12 (2022) 174.

35. J. Zhao, C. Zheng, J. Gao, J. Gui, L. Deng, Y. Wang, R. Xu, Co_3O_4 nanoparticles embedded in laser-induced graphene for a flexible and highly sensitive enzyme-free glucose biosensor, *Sensors and Actuators B: Chemical.* 347 (2021) 130653.

36. J.H.H. Rossato, M.E. Oliveira, B. v. Lopes, B.B. Gallo, A.B. la Rosa, E. Piva, D. Barba, F. Rosei, N.L. v. Carreño, M.T. Escote, A flexible electrochemical biosensor based on NdNiO3 nanotubes for ascorbic acid detection, *ACS Applied Nano Materials.* 5 (2022) 3394–3405.

37. B. Arman Kuzubasoglu, S. Kursun Bahadir, Flexible temperature sensors: A review, *Sensors and Actuators A: Physical.* 315 (2020) 112282.

38. Y. Khan, M. Garg, Q. Gui, M. Schadt, A. Gaikwad, D. Han, N.A.D. Yamamoto, P. Hart, R. Welte, W. Wilson, S. Czarnecki, M. Poliks, Z. Jin, K. Ghose, F. Egitto, J. Turner, A.C. Arias, Flexible hybrid electronics: Direct interfacing of soft and hard electronics for wearable health monitoring, *Advanced Functional Materials.* 26 (2016) 8764–8775.

39. C.-C. Huang, Z.-K. Kao, Y.-C. Liao, Flexible miniaturized nickel oxide thermistor arrays via inkjet printing technology, *ACS Applied Materials & Interfaces.* 5 (2013) 12954–12959.

40. J. Shin, B. Jeong, J. Kim, V.B. Nam, Y. Yoon, J. Jung, S. Hong, H. Lee, H. Eom, J. Yeo, J. Choi, D. Lee, S.H. Ko, Sensitive wearable temperature sensor with seamless monolithic integration, *Advanced Materials.* 32 (2020) 1905527.

41. T. Nakajima, T. Tsuchiya, Flexible thermistors: Pulsed laser-induced liquid-phase sintering of spinel Mn–Co–Ni oxide films on polyethylene terephthalate sheets, *Journal of Materials Chemistry C.* 3 (2015) 3809–3816.

10

Transition Metal Chalcogenide-Based Flexible and Wearable Sensors

G. Subashini and C. Nithya

CONTENTS

10.1 Introduction

The worldwide uncontrollable consumption of non-renewable energy sources encourages us to find greener technologies to meet today's energy demand. The progressive development of safer and cost-effective energy storage technologies is one of the possible solutions to overcome global warming and its consequences. Furthermore, the aforementioned solution leads to a sustainable world and also poses a key challenge simultaneously because these technologies must be affordable to create new pathways for invention. These green energy technologies, such as thermal, solar, wind, mechanical, and biomass, have played a significant role in the invention of new energy harvesting systems. More importantly, while designing new energy storage technologies, we need to keep many factors in mind, for example, high energy and power density, ultrafast charging, and long cycle life according to the application. At the same time, the manufacturing components must be cheap and affordable to every individual. In addition, the newly designed energy storage technologies must be safe for particular applications; therefore, we need proper monitoring devices to ensure the safety of these devices. An effective monitoring component is a sensor, and sensors have played a vital role in human lives. These sensors not only play an important role in energy harvesting systems but also in healthcare monitoring, biomedical fields, and environmental monitoring [1]. Sensors consist of sensing components that are used for detecting and recognizing diverse signals emanating from the physical environment.

Due to the emerging growth of smart internet technologies, the monitoring of any such device is easy and accessible thanks to the expansion of the Internet of Things (IoT). This makes a large demand on the sensors, especially for the market of sensor-enabled devices because these are required for monitoring the newly designed energy technologies as well as in other emerging systems. Errors can occur in conventional rigid electronic sensors because a mismatch occurs between the data collection and motion artifacts during the measurement [2]. However, these kinds of errors are overcome by the invention of flexible electronic devices due to their unique parameters, such as structural conformability and flexibility, which make a natural interface between the human body and electronic devices. Recently, in the field of energy storage technologies, wearable and flexible sensors [3] have played a major role because they are assembled on a flexible substrate that is easily installed. Moreover, these flexible and wearable sensors have also gained attention owing to their prospective applications in the field of healthcare monitoring (including monitoring of blood glucose level, blood pressure, tracking steps and pulse rate, etc.) and biomedical areas including therapeutic/drug delivery monitoring systems. Nowadays, wearable sensors are mostly used in personalized medicine to realize the significant health status changes from time to time for taking necessary medication. Apart from these applications, these sensors sense the environmental conditions/surrounding parameters such as humidity, air pollutant indexes and pathogens present in the air, and so on. Therefore, these flexible and wearable sensors help us to prevent diseases, early diagnoses of epidemic/pandemic diseases, and effective management of natural disasters. The sensing performance of the wearable/flexible sensors is highly dependent on how effectively the functional components detect the variation in signal intensities. This again makes it a challenging issue on designing sensors consisting of functional groups with appropriate physicochemical properties.

The important components to design flexible electronic devices are the substrate, interface layer, and active layer. Among them, the active layer is significant because it determines the performance of the sensors. Generally, it is prepared by the chemical method in nanometer thickness and shows impressive physiochemical properties, enabling the electron/hole transfer process and structural/mechanical strength. Nowadays, layered-type materials with dimensional functionalities have gained a lot of interest in sensing applications due to their distinctive structural, chemical, and electrical properties. In general, two-dimensional materials consist of free-standing individual layers with a thickness of few angstroms. The flexibility of 2D materials can be greatly enhanced by decreasing the thickness of individual layers. This unique structural flexibility makes 2D materials have prospective applications in various fields due to their excellent electrical, mechanical, thermal, and optical properties. One of the best examples of 2D materials is graphene [4], which possesses outstanding electrical and mechanical properties. Due to their structural flexibility, they are best suited for flexible electronic devices. However, the band gap energy of graphene is zero, therefore a lot of leakage issues happen in flexible electronic devices when used as an active material. Therefore, modified graphenes such as graphene oxide and reduced graphene oxide play a vital role in flexible and wearable electronics due to their structural defects. Besides graphene-based analogs, there are abundant varieties of other 2D materials that can be employed as active materials in flexible and wearable electronic devices.

Among the investigated various functional materials, chalcogenides possess considerable attention among researchers owing to their impressive sensing applications [5]. It is a class of 2D functional layered-type materials (e.g., transition metal sulfides and selenides) that is attractive in flexible and wearable sensors due to its unique structural functionalities. For example, MoS_2 in which Mo and S atoms from the molecular layers are stacked in the sequence of covalently bonded S-Mo-S. This unique 2D-type layered structure (2DLM) makes MoS_2 a better candidate for assembling flexible and wearable sensors. In contrast, the S atoms located on the surface are saturated; molecular layers of MoS_2 are pooled by weak van der Waals forces. As per beam theory, at a given bending radius, the deflection of material is indirectly proportional to the thickness [6]. Therefore, bulk van der Waals materials are not suitable for application in flexible and wearable optoelectronic and electronic devices due to their structural instability because their bonds are easily broken.

Among the 2D-type layered materials, transition metal dichalcogenides (e.g., MoS_2, $MoSe_2$, WS_2, WSe_2) have gained much attention in the field of sensors owing to their unique crystal structure, low-cost components, eco-friendliness, and good mechanical and thermal stability [7]. In addition, these 2D-layered chalcogenides have a band gap energy of 1–2 eV, and they exhibit impressive electrical and optical properties created by confined geometry [8]. Apart from the considerable band gap energy, these 2D TMCs possess high carrier charge mobility, high photoconductivity, and thickness-dependent flexibility [9]. This chapter summarizes the recent progress of flexible and wearable sensors employing the 2D transitional metal-based chalcogenides (2D-TMCs). Designing sensors, new material invention, fabrication of wearable sensors for particular applications, and challenges toward the integration of sensors with the human body are also discussed. Moreover, we systematically review the important concepts and various synthesis approaches of 2D-TMCs for designing wearable and flexible sensors. Possible strategies and bottlenecks while incorporating 2D-TMCs into wearable and flexible sensors are also highlighted. Furthermore, this chapter also discusses the challenges and future perspectives for 2D-TMCs employed in flexible and wearable sensors in various fields.

10.2 Copper-Based Chalcogenides

Copper sulfides have attracted the scientific community due to their versatile properties such as their nontoxicity, being environmentally friendly, abundant in nature, and recyclability. Copper sulfides have different crystallographic structures depending on the stoichiometry of copper and sulfur (Cu_2S to CuS_2). From the stability of copper sulfides, five phases of it can exist at room temperature, namely covellite CuS, djurleite $Cu_{1.95}S$, digenite $Cu_{1.8}S$, anilite $Cu_{1.75}S$, and chalcocite Cu_2S. Crystallographic structure and band gap vary depending on the $Cu_{2-x}S$ stoichiometry and their properties can be modulated by varying the composition [10]. Cu_2S is a p-type semiconductor with a band gap of 1.2 eV, which makes it suitable for various applications [11]. Cu_2S possesses lattice parameters close to that of MoS_2, this allows the growth of MoS_2/Cu_2S hybrids. Flexible MoS_2/Cu_2S hybrids are used for sensing humidity, temperature, and breath and to determine ethanol adulteration. The growth of MoS_2/Cu_2S on disposable cellulose paper was integrated into a smartphone for wireless multifunctional sensing. The sensitivity of the humidity sensor was calculated to be 0.6% RH; the TCR of the temperature sensor was found to be 2×10^{-5}°C. The sensor was also applied for real-time monitoring of hydration levels of the lungs [12].

Wearable flexible electronic textiles (E-textiles) are presently attractive due to their important role in the ever-rising era of artificial intelligence [13–14]. Recently, the more common PET fabric has been selected as a flexible substrate owing to its excellent mechanical property, corrosion resistance, fracture resistance, and low cost. The conducting electrodes are prepared by coating carbon nanotube or graphene on the flexible PET fabric. Even though carbon materials have excellent conductivity and electrochemical stability, high cost and complicated procedure are against them. Of all metal sulfides, CuS is of prime importance owing to its excellent electrical conductivity, easy preparation, and low cost. Tailoring CuS with PET for flexible electrodes yields electrodes that have improved properties of flexibility, electrical conductivity, and capacitance [15, 16]. Liu and his group fabricated the flexible PET/CuS/PPy composite electrode that shows high electrical conductivity and large capacitance with an added advantage of rate performance and cycle stability [17].

10.3 Nickel-Based Chalcogenides

Nickel-based chalcogenides are extensively used for energy storage and the generation of devices owing to their redox properties, high conductivity, unique crystal structure, and tunable stoichiometric composition [18, 19]. The nickel atoms are covalently connected to a trigonal arrangement of chalcogenides within the same layer and the individual layers are held together by van der Waals forces [20, 21]. Sulfides of nickel usually exist as NiS stoichiometry but they also exist in non-stoichiometric polymorphs such as N_9S_8 and Ni_3S_2. NiS_2 forms pyrite-type crystals that have disulfide dianion linkage. The naturally occurring NiS (Millerite) and the NiS stoichiometric polymorph are odorless and they are insoluble in water, which makes them functionalization for electrocatalysis. The sulfides of nickel are highly stable, and under alkaline conditions, they form an oxide/hydroxide layer on the sulfide surface that is useful for oxygen evolution reactions and hydrogen evolution reactions (OER/HER) [22]. The property of OER/HER enhances when the sulfides of nickel are doped with Co and Fe. Wearable real-time sensors are of prime importance for healthcare devices, and an increasing focus on sensors paved the way for rapid growth. There are various types of glucose monitoring techniques which are classified into enzymatic and non-enzymatic based on the method. The biocompatibility, comfort, flexibility, and stability of cotton fabric make it a flexible sensor for glucose sensing.

Farzaneh et al. fabricated nickel-cobalt sulfide on CFs through an easy microwave-assisted method for sensing blood serum and saliva glucose levels. The fabricated sensing platform showed high selectivity, good repeatability, and long-term stability (non-enzymatic sensor) for the detection of glucose levels in physiological samples [23]. As repeatability and reproducibility are the two essential criteria for sensing, Ni-Co-S@CFs were evaluated for repeatability for glucose. The amperometry response was investigated after successive additions of standard glucose solution to the electrolyte (0.1 M KOH). Repeatability of 4.74% and reproducibility of 8.41% were obtained on two different days for Ni-Co-S@CFs. Yu et al. fabricated flexible electrodes – polydimethylsiloxane (PDMS) tailored with nickel-cobalt-sulfide nanosheets for high-performance supercapacitors and pressure sensors. The presence of a metallic network enhanced the conductivity and chemical stability, and wrinkled nanostructure provided numerous active sites that increased electron pathways for pressure sensors. Both these properties pave the way for the electrode to withstand large mechanical deformation and tensile strength without any damage to the structure. The fabricated Ag@NiCo NWs-NiCoS [24] electrode exhibited high specific capacitance of 125 mF/cm^2 and capacity retention after 5000 charge-discharge cycles with an energy density of 40.0 Wh/kg with a power density of 1.1 kW/kg. The constructed capacitive pressure sensor exhibited high sensitivity of 0.049 kPa^{-1} in the low detection range of <60 kPa and broad detection range up to 260 kPa with a fast response of approximately 66 ms. The unique Ag@NiCo NWs-NiCoS can be used as a tactile and wearable sensor to monitor physical stimuli and human motions in real-time sensing.

10.4 Molybdenum-Based Chalcogenides

Among the various TMCs, MS2 has gained great attention due to its electronic and chemical properties. Tunability of the bandgap from 1.2 eV to 1.8 eV [25] by thickness modulation of MoS_2 has opened enormous room for its use in electronic applications. The amicability of MoS_2 with flexible substrates makes it a versatile material for flexible and wearable sensors. Fabrication of MoS_2 devices can be brought about by chemical vapor deposition (CVD) and mechanical exfoliation [26, 27]. The CVD method is afflicted by variation in performance caused by the complexity of the transfer process which results in a small lateral size and less yield. Hence direct growth of MoS_2 on flexible substrates remains a problem ultimatum, Ly.

Depending on the mode of coordination and stacking orders, different crystal phases are possible for MoS_2. Usually, MoS_2 exists in any of the three crystalline phases – hexagonal (2H), octahedral (1T), and rhombohedral (3R) – which are vertically stacked layers of MoS_6 units. Thermodynamically stable 2H-MoS_2 has an ABAB stacking pattern and 3R-MoS_2 has an ABCABC stacking pattern with trigonal prismatic coordination; 2H-MoS_6 units are octahedrally coordinated with the AAA pattern. Molybdenum atoms are sandwiched between two sulfur atoms through weak van der Waals force. Due to significant variations in the crystal structure and phases, MoS_2 possesses unique properties of bandgap tunability and electronic flexibility [28, 29]. Vertically aligned molybdenum disulfides (VA-2D MoS_2) possess an enhanced surface area with defective sites that increase stretchability and flexibility. VA-2D MoS_2 layers with a large area >cm^2 were grown by the CVD method and converted into a flexible substrate for NO_2 sensitivity (from 5–30 ppm). They created a serpentine pattern with the ability to withstand a high degree of tensile strength and reversible stretchability of up to 40% [30]. The electrochemical property of MoS_2 can be improved with the dispersion of MoS_2 on a highly conductive carbon surface. Lu et al. reported the fabrication of MoS_2 on carbonized silk fabric (MoS_2/CSilk) by the solvothermal process. Vertically aligned MoS_2 nanosheets grown on CSilk can be used as a pressure sensor. MoS_2/CSilk coaxial fibers are capable of detecting human physiological signals [31].

There is a high need for tactile sensors with high-density arrays and large surface areas in the field of biomedical, robotics, wearable, and electronic skin applications. In a passive matrix (PM), a tactile sensor results in an inaccurate sensing of pressure [32] due to crosstalk among pressure gauges, which always leads to a false result. Hence, for precise determination of shapes and sizes, an active matrix (AM) array is required where the distribution of pressure will be uniform [33]. Park et al. fabricated large-area, flexible MoS_2 tactile sensors by an Am backplane circuitry. The fabricated AM tactile sensor [34] had an isolation value of up to 24.8 dB, whereas it was 6.6 dB for the PM matrix, which was able to sense pressure values of 1 to 120 kPa. This integrated MoS_2-based AM potentially eliminated the crosstalk issues and excellently responded when an object was grasped in the palm. MoS_2 is potentially considered to be an important material for humidity sensing due to the presence of inherent defective sites. Zao et al. developed a humidity sensor with a narrow relative humidity of RH 0–35% [35]. Park et al. developed an rGO/MoS2-based humidity sensor with a long recovery time of 30.8 s [36]. The above-mentioned humidity sensors suffer from low sensitivity, narrow range, and slow response time. To overcome this, Mondal et al. developed a unique resistive type of humidity sensor with an anodic aluminum oxide-assisted MoS_2 honeycomb structure (AMHS) [37] with a quick response to humidity from the atmosphere, human skin, and human breath in the range of 20–85%. AMHS sensor showed high sensitivity of 668 at RH 83% compared with MoS_2 film which was around 80.23% in the same RH condition. The enhanced sensitivity of AMHS is due to the presence of defective sites that can accommodate H_2O molecules. The so-developed AMHS sensor had excellent repeatability with non-degradation, even after 1.5 h.

To improve the thermoelectric property of MoS_2, it was decorated with Au as a hetero-conjugated system by in-situ growth of Au-nanoparticles [38]. An Au-MoS_2 composite was used for the construction of a device in the form of a wristband with five p-type strips. Au-MoS_2 composite films achieved a power factor (σS^2) of 166.3 μ Wm^{-1} K^{-2} at room temperature due to the increased conductivity(σ) and Seebeck coefficient(S). Badhulika and the group reported a multifunctional sensor that reacted to different chemical stimuli. A flexible MoS_2/Cu_2S hybrid grown on disposable cellulose paper was used for the sensing of humidity, temperature, breath, and ethanol adulteration. The sensitivity of the humidity sensor was found to be 2×10^{-5}°C^{-1} with an activation energy of 86.7 meV. The sensor was able to detect 2% of water in ethanol [39].

10.5 Iron-Based Chalcogenides

Iron sulfide-based chalcogenides have gained attention due to energy storage and conversion systems. Iron sulfides exist as FeS (mackinawite), $Fe_{1-x}S$(pyrrhotite), FeS_{2p} (pyrite), FeS_{2m} (marcasite), Fe_3S_4 (greigite), and Fe_9S_{11} (smythite). FeS exists in a tetragonal structure where each iron is coordinated to four sulfurs [40]. $Fe_{1-x}S$ has a monoclinic hexagonal structure, FeS_{2p} shows a cubic structure with a stable iron (II) disulfide, FeS_{2m} exists in a metastable state with orthorhombic lattice, Fe_3S_4 as a cubic metastable state, and Fe_9S_{11} as a hexagonal lattice [41]. Among the various sulfides, FeS, FeS_2, Fe_7S_8, and Fe_3S_4 have extensively attracted researchers for conversion systems.

Sushmitha et al. reported the fabrication of FeS_2 on cellulose paper with silver, and aluminum as top and bottom electrodes for a memristor and a non-contact breath sensor. The decorated FeS_2-based memristor device exhibited an excellent resistive switching behavior with a good data retention capability of 2_*10^4 s, switching endurance of up to 500 cycles, and a decent (R_{off}/R_{on}) ratio of 4_*10^2 with extremely good flexibility [42]. Metal-decorated hydrothermally grown FeS_2 exhibited a distinguished response for oral and nasal breath with a recovery time of 5 s and 2 s, respectively. To demonstrate the practical application of the breath sensor, the data was wirelessly transmitted to a smartphone through a Bluetooth module [43]. The smartphone was used for breath count during various physical activities like walking, jumping, and running.

10.6 Hafnium-Based Chalcogenides

Hafnium-based chalcogenides have gained interest due to their intriguing electronic and optical properties and they have high mobility (1800 to 3500 cm^2 V^{-1} S^{-1}) and large absorption spectral range (1–2 eV). Hafnium sulfide HfS_2 has an octahedral geometry with the P3m1 space group. HfS_2 is a semiconducting material with an indirect band gap of 2.0 eV and single layers of HfS_2 have a band gap of 1.2 eV. HfS_2 finds applications in photocatalysis, photodetectors, and photo sensors, but it is less efficient in broadband detection.

However, a combination of HfS_2 and $HfSe_2$ improves the band gap which in turn creates a promising platform for rapid photoresponse [44, 45]. $Hf(S_xSe_{1-x})_2$ film was prepared through chemical vapor deposition and its optical properties were tailored by altering the S/Se ratio. $Hf(S_xSe_{1-x})_2$ exhibits hexagonal geometry where adjacent layers are bound by van der Waals interaction and they are octahedrally coordinated to Hf. Rajesh Kumar and group synthesized 2D HfSSe by chemical vapor transport method that showed phototransistor performance from visible to the near-infrared region. The synthesized single crystal can be used as a flexible photodetector with a responsivity of up to 1.3 A at 980 nm. The multi-elemental HfSSe chalcogenides pave the way for a vibrant future for wearable electronics [46].

10.7 Tungsten- and Vanadium-Based Chalcogenides

Tungsten-based chalcogenides have attracted researchers owing to their high sensitivity and low detection limit. Tungsten sulfide, WS_2, has a layer-like structure with van der Waals force similar to graphene. WS_2 has been extensively studied for hydrolytic properties and earth-abundance. WS_2 belongs to the P6/3mmc space group with a direct band gap of 2.1 eV and an indirect band gap of 1.4 e V [47]. Due to the reduced mass of electrons in WS_2, theoretically, they have the highest photon-limited electron mobility of over 1000 cm^2 V^{-1} s^{-1}. WS_2 and MoS_2 have hybrid structures and alloying of both improves sensing application at low operating temperatures. The direct synthesis of WS_2 and MoS_2 thin films on Si chips offers a route for NH_3 sensing. Obtained thin films have excellent selectivity to NH_3 with a high sensitivity of 0.10 ± 0.02 ppm^{-1} [48]. Dongzhi et al. synthesized WS_2/MoS_2 nanohybrid for humidity sensing. The inclusion of SnO_2 into c enhances oxygen vacancy and defects, which creates room for the adsorption of water molecules. The so-created hetero junctions at the interfaces of n-type WS_2 and n-type SnO_2 enhance the response of WS_2/SnO_2 film [49].

Vanadium chalcogenides can exist as VS_2, VS_4, and V_5S_8. TMCs like VS_2 have a sandwich structure where the V atom is sandwiched between two sulfur atoms with an interlayer distance of 5.76 Å, which provides room for the intercalation and deintercalation process. V_5S_8 possesses a different crystallographic structure in which V atoms within two VS_2 monolayers occupy a quarter of S-S sites. TMCs of vanadium have high conductivity, high specific area, and variable oxidation states from +2 to +5. Siowwoon et al. demonstrated the functionality of VS_x incorporated graphite film for energy storage and conversion. The flexible VS_x /graphite film has high near-infrared photoresponse and it can be used for infrared absorbance in outer space [50].

10.8 Outlook and Future Perspectives

In this chapter, we have summarized the state-of-the-art research progress on transition metal chalcogenides, particularly focused on metal sulfides (MX/MX_2 in which M = Cu, Fe, Hf, Mo, Ni, V, W, and X = S) and flexible and wearable sensors. Recent advancements in architecture design and synthesis of advanced materials have gained attention in the field of flexible and wearable sensors. The emerging TMCs provide a new platform for flexible and wearable sensing. The atomic-level thickness of mono-layered and multilayered TMCs renders increase flexibility, which is an important parameter for flexibility. In addition, they exhibit more flexibility as they possess the ability for printing, casting, screening, and other 3D constructions. TMCs with unique physical and chemical properties have tunable properties like band gap, high mobility, large surface-to-volume ratio, and mechanical stability. In the fabrication of TMCs as flexible and wearable sensors, the following criteria should be considered: (i) molecular engineering of TMCs and enhancing the potential to increase the active sites for analyte interaction; (ii) reducing the thickness of the layer to improve the surface-to-volume ratio; (iii) improving the charge transfer through doping of heteroatoms with superior electrical properties; and (iv) tailoring the fabricated device with the incorporation of specific analytes like enzymes or ligands to improve selectivity and sensitivity in sensing. Even though TMCs are promising candidates for flexible and wearable sensors, there is an ongoing demand for the implementation of TMCs for analytical applications. The large-scale production of high-quality monolayers of TMCs with enhanced surface area is still a problem in synthesis. Reproducibility in sensing remains a challenge, which includes sensitivity and stability. The presence of structural defects has a considerable influence on the photonics and electronic properties of TMCs, therefore, this requires innovation, research, and further development.

References

1. X. Cao, A. Halder, Y. Tang, C. Hou, H. Wang, J.O. Duus, Q. Chi. Engineering two-dimensional layered nanomaterials for wearable biomedical sensors and power devices. *Mater. Chem. Front.* 2, 2018, 1944–1986.
2. W. Gao, H. Ota, D. Kiriya, K. Takei, A. Javay. Flexible electronics toward wearable sensing. *Acc. Chem. Res.* 52, 2019, 523–533.
3. F. Yi, H. Ren, J. Shan, X. Sun, D. Wei, Z. Liu. Wearable energy sources based on 2D materials. *Chem. Soc. Rev.* 47, 2018, 3152–3188.
4. M. Zou, Y. Ma, X. Yuan, Y. Hu, J. Liu, Z. Jin. Flexible devices: From materials, architectures to applications. *J. Semicond.* 39, 2018, 011010–18.
5. J. Yao, G. Yang. 2D group 6 transition metal dichalcogenides toward wearable electronics and optoelectronics. *J. Appl. Phys.* 127, 2020, 030902–13.
6. F. Wang, Z.Wang, Z. Yin, R. Cheng, J. Wang, Y. Wen, T.A. Shifa, Y. Wang, Y. Zhang, X. Zhan, J. He. 2D library beyond graphene and transition metal dichalcogenides: A focus on photodetection. *Chem. Soc. Rev.* 47, 2018, 6296–6341.
7. M. Samadi, N.Sarikhani, M. Zirak, H. Zhang, H.L. Zhang, A.Z. Moshfegh. Group 6 transition metal dichalcogenide nanomaterials: Synthesis, applications and future perspectives. *Nanoscale Horiz.* 3, 2018, 90–204.
8. X. Li, L. Tao, Z. Chen, H. Fang, X. Li, X. Wang, J.B. Xu, H. Zhu. Graphene and related two-dimensional materials: Structure-property relationships for electronics and optoelectronics. *Appl. Phys. Rev.* 4, 2017, 021306.
9. D. Akinwande, N. Petrone, J. Hone. Two-dimensional flexible nanoelectronics. *Nat Commun.* 5, 2014, 5678.
10. Y. Liu, M. Liu, M.T. Swihart. Plasmonic copper sulfide-based materials: A brief introduction to their synthesis, doping, alloying, and applications. *J. Phys. Chem. C* 121, 2017, 13435–13447.
11. C.Y. Wu, Z.Q. Pan, Z. Liu, X.Y. Wang, F.X. Liang, Y.Q. Yu, L.B. Luo. Controllable synthesis of p-type Cu_2S nanowires for self-driven NIR photodetector application. *J. Nano. Res.* 19, 2017, 35.

12. S. Parikshit, K. Anand, G. Harshit, P. Thanga Gomathi, B. Sushmee. Flexible, disposable cellulose paper-based MoS_2-Cu_2S hybrid for wireless environmental monitoring and multifunctional sensing of chemical stimuli, *ACS Appl. Mater. Interfaces* 10, 2018, 9048–9059.
13. L. Gao, C. Zhu, L. Li, C. Zhang, J. Liu, H.D. Yu, W. Huang. All paper-based flexible and wearable piezoresistive pressure sensor. *Appl. Mater. Interfaces* 11, 2019, 25034–25042.
14. J. Chen, Y. Jiang, J. Yang, Y. Sun, L. Shi, Y. Ran, Q. Zhang, Y. Yi, S. Wang, Y. Guo, Y. Liu. Copolymers of bis-diketopyrrolopyrrole and benzothiadiazole derivatives for high-performance ambipolar field-effect transistors on flexible substrates. *ACS Appl. Mater. Interfaces* 10, 2018, 25858–25865.
15. Z. Pan, F. Cao, X. Hu, X. Ji. A facile method for synthesizing CuS decorated Ti_3C_2 MXene with enhanced performance for asymmetric supercapacitors. *J. Mater. Chem. A* 7, 2019, 10815–10815.
16. H. Peng, G. Ma, K. Sun, J. Mu, H. Wang, Z. Lei. High performance supercapacitor based on multi-structural CuS@polypyrrole composites prepared by in situ oxidative polymerization. *J. Mater. Chem. A* 2, 2014, 3303–3307.
17. S. Parikshit, K. Anand, G. Harshit, P. Thanga Gomathi, B. Sushmee. Synchronous dual roles of copper sulfide on the insulating PET fabric for high-performance portable flexible supercapacitors. *Energy Fuels* 35, 2021, 6880–6891.
18. Y.P. Gao, X. Wu, K.J. Huang, L.L. Xing, Y.Y. Zhang, L. Liu. Two-dimensional transition metal diseleniums for energy storage application: A review of recent developments. *Cryst. Engg. Comm.* 19, 2017, 404–418.
19. H. Zang. Ultrathin two-dimensional nanomaterials. *ACS Nano* 9, 2015, 9451–9469.
20. X. Huang, Z. Zeng, H. Zhang. Metal dichalcogenide nanosheets: Preparation, properties and applications. *Chem. Soc. Rev.* 42, 2013, 1934–1946.
21. X. Rui, H. Tan, Q. Yan. Nanostructured metal sulfides for energy storage. *Nanoscale* 6, 2014, 9889–9924.
22. S. Anantharaj, S. Ede, K. Sakthikumar, K. Karthick, S. Mishra, S. Kundu. Recent trends and perspectives in electrochemical water splitting with an emphasis on sulfide, selenide, and phosphide catalysts of Fe, Co, and Ni: A review. *ACS Catal.* 6, 2016, 8069–8097.
23. H. Farzaneh, E. Milad, S. Saeed, U. Husnu Emrah. Microwave-assisted decoration of cotton fabrics with nickel-cobalt sulfide as a wearable glucose sensing platform. *J. Electroanal. Chem.* 890, 2021, 115244.
24. Z. Yu, L. Jionghong, Z. Bolun, W. Fengming, H. Weiqing, C. Guofa, Z. Chi, X. Yue, C. Bohua, H. Xin. Highly stable, stretchable, and versatile electrodes by coupling of NiCoS nanosheets with metallic networks for flexible electronics. *Nanoscale* 14, 2022, 8172–8182.
25. T. Li, G. Galli. Electronic properties of MoS_2 nanoparticles. *J. Phys. Chem. C* 111, 2017, 16192–16196.
26. A. Sanne, R. Ghosh, A. Rai, M. Nagavalli Yogeesh, S.H. Shen, A. Sharma, K. Jarvis, L. Mathew, R. Rao, D. Akinwande, S. Banerjee. Radiofrequency transistors and circuits based on CVD MoS_2. *Nano Lett.* 15, 2015, 5039–5045.
27. S.S. Chou, B. Kaehr, J. Kim, B.M. Foley, M. De, P.E. Hopkins, V.P. Dravid. Chemically exfoliated MoS_2 as near-infrared photothermal agents. *Angewandte Chemie* 125, 2013, 4254–4258.
28. W. Zhao, J. Pan, Y. Fang, X. Che, D. Wang. Metastable MoS_2: Crystal structure, electronic band structure, synthetic approach, and intriguing physical properties. *Chem. A Eur. J.* 24, 2018, 15942–15954.
29. R.J. Toh, Z. Sofer, J. Luxa, D. Sedmidubsky, M. Pumera. 3R phase of MoS_2 and WS_2 outperforms the corresponding 2H phase for hydrogen evolution. *Chem. Commun.* 53, 2017, 3054–3057.
30. Md. A. Islam, H. Li, S. Moon, S.S. Han, H.S. Chung, J. Ma, C. Yoo, T.J. Ko, K. H. Oh, Y.J. Jung, Y. Jung. Vertically aligned 2D MoS_2 layers with strain-engineered serpentine patterns for high-performance stretchable gas sensors: Experimental and theoretical demonstration. *ACS Appl. Mater. Interfaces* 12, 2020, 53174–53183.
31. W. Lu, P. Yu, M. Jian, H. Wang, H. Wang, X. Liang, Y. Zhang. Molybdenum disulfide nanosheets aligned vertically on carbonized silk fabric as smart textile for wearable pressure-sensing and energy devices. *ACS Appl. Mater. Interfaces* 12, 2020, 11825–11832.

32. J. Kim, M. Lee, H.J. Shim, R. Ghaffari, H.R. Cho, D. Son, Y.H. Jung, M. Soh, C. Choi, S. Jung, K. Chu, D. Jeon, S.T. Lee, J. H. Kim, S.H. Choi, T. Hyeon, D.H. Kim. Stretchable silicon nanoribbon electronics for skin prosthesis. *Nat. Commun.* 5, 2014, 5747.

33. G.F. Cai, J.X. Wang, M.F. Lin, J.W. Chen, M.Q. Cui, K. Qian, S.H. Li, P. Cui, P.S. Lee. A semitransparent snake-like tactile and olfactory bionic sensor with reversibly stretchable properties. *NPG Asia Mater.* 9, 2017, e437.

34. Y.J. Park, B.K. Sharma, S.M. Shinde, M.S. Kim, B. Jang, J.H. Kim, J.H. Ahn. All MoS$_2$-based large area, skin-attachable active-matrix tactile sensor. *ACS Nano*, 13, 2019, 3023–3030.

35. J. Zhao, N. Li, H. Yu, Z. Wei, M. Liao, P. Chen, S. Wang, D. Shi, Q. Sun, G. Zhang. Highly sensitive MoS$_2$ humidity sensors array for noncontact sensation. *Adv. Mater.* 29, 2017, 1702076.

36. S.Y. Park, Y.H. Kim, S.Y. Lee, W. Sohn, J.E. Lee, D.H. Kim, Y.S. Shim, K.C. Kwon, K.S. Choi, H.J. Yoo, J.M. Suh, M. Ko, J.H. Lee, M.J. Lee, S.Y. Kim, M.H. Lee, H.W. Jang. Highly selective and sensitive chemoresistive humidity sensors based on rGO/MoS$_2$ Van Der Waals composites. *J. Mater. Chem. A.* 6, 2018, 5016–5024.

37. S. Mondal, S.J. Kim, C.G. Choi. Honeycomb-like MoS$_2$ nanotube array-based wearable sensors for noninvasive detection of human skin moisture. *ACS Appl. Mater. Interfaces* 12, 2020, 17029–17038.

38. G. Yang, D. Chaochao, P. Xu, L. Peiyun, H. Wenxiao, M. Jiuke, H. Chengyi, A.Corey, H.Q. Zhang, Y. Li, L.C. David, H. Wang. Wearable thermoelectric devices based on Au-decorated two-dimensional MoS$_2$. *ACS Appl. Mater. Interfaces* 10, 2018, 33316–33321.

39. S. Parikshit, K. Anand, G. Harshit, P. Thanga Gomathi, S. Badhulika. Flexible, disposable cellulose-paper-based MoS$_2$/Cu$_2$S hybrid for wireless environmental monitoring and multifunctional sensing of chemical stimuli. *ACS Appl. Mater. Interfaces* 10, 2018, 9048–9059.

40. D. Rickard, G. Luther. Chemistry of iron sulfides. *Chem. Rev.* 107, 2017, 514–562.

41. Y. Ye, W. Liping, G. Lizeng. Nano-sized iron sulfide: Structure, synthesis, properties, and biomedical applications. *Front. Chem.* 8, 2020, 818.

42. V. Sushmitha, K. Shivam, S. Rinky, B. Sushmee. Direct growth of FeS2 on paper: A flexible, multifunctional platform for ultra-low-cost, low power memristor and wearable non-contact breath sensor for activity detection. *Mater. Sci. Semicond. Process.* 108, 2020, 104910.

43. A. Shinde, P. Sahatiya, A.A. Kadu, S. Badhulika. Wireless smartphone assisted personal healthcare monitoring system using MoS$_2$ based flexible, wearable and ultra-low cost functional sensor. *Flex. Print. Electron.* 4, 2019, 025003.

44. L. Li, W. Wang, L. Gan, N. Zhou, X. Zhu, Q. Zhang, H. Li, M. Tian, T. Zhai. Ternary Ta$_2$NiSe$_5$ flakes for a high-performance infrared photodetector. *Adv. Funct. Mater.* 26, 2016, 8281.

45. P. Perumal, R. K. Ulaganathan, R. Sankar, Y.M. Liao, T.M. Sun, M.W. Chu, F. C. Chou, Y.T. Chen, M.H. Shih, Y.F. Chen. Ultra-thin layered ternary single crystals [Sn(Sx Se1−x)2] with bandgap engineering for high performance phototransistors on versatile substrates. *Adv. Funct. Mater.* 26, 2016, 3630.

46. U. Rajesh Kumar, S. Raman, L. Chang-Yu, C.M. Raghavan, T. Kechao, F.C. Chou. High-performance flexible broadband photodetectors based on 2D hafnium selenosulfide nanosheets. *Adv. Electron. Mater.* 2, 2019, 1900794.

47. N. Sakhuja, R. Jha, N. Bhat. Tungsten disulphide nanosheets for high-performance chemiresistive ammonia gas sensor. *IEEE Sens. J.* 19, 2019, 11767–11774.

48. J. Topias, S.L. Gabriela, P. Jani, T. Geza, S. Simo, K.V. Vesa, K. Krisztian. WS2 and MoS2 thin film gas sensors with high response to NH3 in air at low temperature. *Nanotechnology* 30, 2019, 405501.

49. Z. Dongzhi, C. Yuhua, L. Peng, W. Junfeng, Z. Xiaoqi. Humidity-sensing performance of layer-by-layer self-assembled tungsten disulfide/tin dioxide nanocomposite. *Sens. Actuat. B: Chem.* 265, 2018, 529–538.

50. N. Siowwoon, G. Kalyan, V. Jan, P. Martin. Two-dimensional vanadium sulfide flexible graphite/polymer films for near-infrared photoelectron catalysis and electrochemical energy storage, *Chem. Eng. J.* 435, 2022, 135131.

11

Recent Developments in MXenes for Advanced Flexible Sensors

Sagar Sardana and Aman Mahajan

CONTENTS

11.1 Introduction

Today, emerging flexible technology serves as a potential alternative to conventional monitoring devices as a result of developing low-cost, small-size, and wearable sensing devices, which enable real-time applications such as environment monitoring, human physiological activity detection, health monitoring, and disease diagnosis [1]. Consequently, different types of flexible sensing devices, such as pressure sensors, gas sensors, humidity sensors, and wearable sensors, have been widely employed to implement the aforementioned applications. In this context, flexible electronics enable the assembly of a sensor array with electronic circuitry onto flexible substrates, which innovatively contributes to their widespread use as flexible sensors. The other advantage of flexible electronics is skin compatibility, as a flexible sensor can easily fit or be mounted onto a curved surface without significant changes in its sensing performance. Advances in nanotechnology, miniaturization, and the smart sensing network make it possible to develop lightweight, portable, non-invasive, flexible, and inexpensive sensors [2]. Furthermore, material selection can facilitate flexibility as well as sensing properties of devices. Flexible sensors require new materials that possess high stretching ability and can withstand large deformations under mechanical stimuli and external pressure. In addition, there are other characteristics, such

DOI: 10.1201/9781003299455-11

as deformability, mechanical reliability, squeezability, sensibility, good detection limits, durability, and resiliency, which are required for realizing flexible and skin-mounted sensing devices.

In this context, a large set of one-dimensional (1D), two-dimensional (2D), and three-dimensional (3D) structured organic and inorganic materials, polymers, metal oxide semiconductors, graphene-related compounds, and hybrid composite-based materials have been exploited owing to their potential physical and chemical properties [3]. In recent years, MXenes ($M_{n+1}X_nT_x$, where M_{n+1} = transition metal, X_n = carbon or nitrogen, and T_x = termination group), a family of 2D inorganic materials, have gained a great deal of interest since their discovery at Drexel University in 2011, enabling great application prospects in flexible electronic devices such as supercapacitors, gas sensors, humidity sensors, energy storage devices, diodes, transistors, and solar cells [4]. Based on experimental findings, MXenes can be broadly classified into carbide, nitride, and carbonitride families, where 25 are from the carbide family, three of them are based on nitrides, and two are from the carbonitride family. More than 70 different compositions of MXenes are predicted theoretically. MXenes are produced through selective etching of 3D hexagonal $M_{n+1}AX_n$ phases (MAX phase: space group P63/mmc) by removal of an A group element (silicon or aluminum) using strong acidic solutions such as hydrofluoric acid, hydrochloric acid, lithium fluoride solution, hydrofluoric acid and lithium chloride solution, and ammonium hydrogen fluoride at room temperature. The etching process functionalizes the MXenes surface with plentiful termination functional groups (–O, –OH, and –F), which makes them potential candidates for desired applications [5]. Given the advantage of exceptional properties such as high conductivity, anisotropic electronic transportation, abundant surface chemistry, mechanical flexibility, and ultrahigh signal-to-noise ratio, MXene-based materials have gained momentous breakthroughs in wearable and flexible device applications. The structural diversification of MXene provides a great possibility for the exploitation of different nanostructures (1D, 2D, and 3D) through which MXenes can be easily tailored into multifunctional sensing geometries according to particular demands. One-dimensional MXene-based sensors include nanofibrous and interwoven structured networks, exhibiting great application prospects in the development of wearable sensors by the direct weaving of sensing material into clothes. On the other hand, 2D nanosheet structured-MXene has advantages in sensing applications due to its accordion-like structure. Further, properties such as the ease of solution processing, scalability, and the functionalized surface of MXene attribute to its ability to form a flexible and free-standing 2D film on paper, fabric, and textile, offering smart textile-based sensors for wearable applications. Besides these, MXene-based sensors can be constructed into 3D structural morphologies such as aerogel, hydrogel, and sponge. The 3D conductive network results in a high tensile ratio and compression resilience, which enables great customizability to engineer the device structure to explore numerous applications such as flexible electronic devices, gas sensors, and piezoresistive sensors [4].

Several review articles have been reported in the literature detailing the potential aspects of MXene-based flexible devices. In these articles, the researchers have put in a great effort to review the developing field of MXene-based materials from material synthesis to practical applications. However, it is impossible to cover all strands in a single review article due to the wide-ranging subject matter. Herein, we will focus on recent advances in the development of MXene-based flexible sensors, which cover the properties and existing fabrication strategies that implement MXene-based materials to realize flexible sensing devices. The main content of the chapter highlights the advances in flexible sensors derived from MXene composite-based materials, which include piezoresistive sensors and triboelectric,

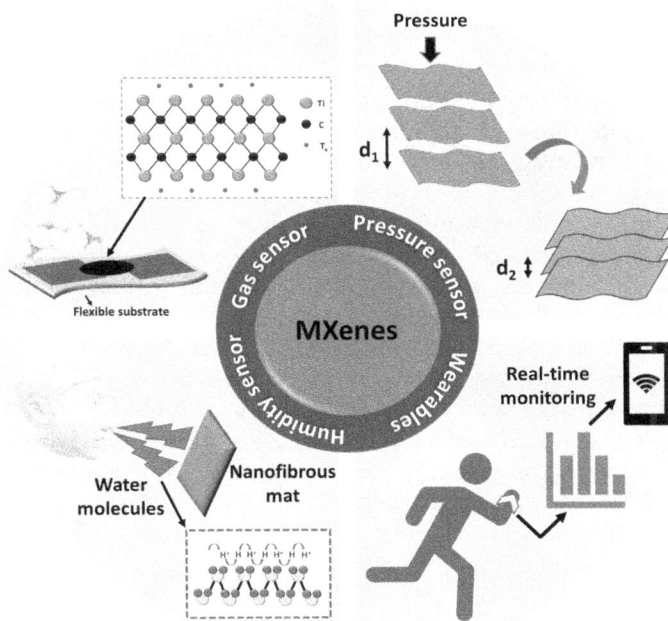

FIGURE 11.1
An overview of the applications of MXene-based flexible sensors.

gas sensors, humidity sensors, and wearable monitoring devices (Figure 11.1). Finally, the concluding remarks provide insights and outlooks on current challenges and the future development of these flexible sensors.

11.2 Synthesis of MXenes

There are several methods reported for the synthesis of 2D materials, which include the wet-etching process, mechanical cleavage exfoliation, the template method, electrochemical metal ion intercalation and exfoliation, and plasma-enhanced pulsed layer deposition, as well as chemical and physical vapor deposition [6]. Owing to the relatively weaker ionic bond strength of the M–A bond than M–X in the MAX phase, it is easy to remove the "A" layers from the commonly used MAX precursors (e.g., Ti_3AlC_2, Ti_2AlC) using the wet-etching method, which is a top-down approach.

In this process, the etching solvent HF acid replaces Al atoms by surface termination groups that include –OH, –O, or –F, from the MAX phase, and separated graphene-like MXene sheets are formed. Technically, the MAX precursor is first chemically treated with etching solvent, the mixture is centrifuged to obtain a solid-liquid separation, a pH value of 4–6 is maintained by employing several washing processes, and then subsequently delaminated into few-layer sheets via sonication. During the etching process, by adjusting the etching time, the concentration of etching solvent, ultrasonication time, and etching temperature, the morphology of MXenes can be controlled for desired sensory functions.

According to the principle, the micro/nano morphology is beneficial for mechanical force-assisted electrical properties while multi-layer morphology enables fast diffusion of target molecules. In addition, the synthesis method offers the possibility of the formation of a different set of MXene combinations with unique properties by varying the element (M and X) ratio. Despite these advantages, the usage of HF leads to organism corrosiveness and toxic treatment owing to its hazardous nature, which hinders the application prospects of MXene in flexible and wearable sensors. Alternatively, mild-etching processes have gained potential over HF etching, which involves fluoride salts such as lithium fluoride, potassium fluoride, and sodium fluoride. On the contrary, the delamination step is not required in the mild-etching process because of the interaction of metal ions between MXene sheets during the etching process.

On the other hand, bottom-up synthesis routes are also beneficial as they can enable the manipulation of morphology, surface termination groups, and size distribution of MXene, which endows MXene with fewer defects and controllable lateral size properties. However, these routes, such as chemical vapor deposition, are limited due to their complex set-up conditions, high-temperature requirements during the nucleation growth process, complex equipment, and the operating principle. In addition, the threshold temperature is required throughout the processing technique to assemble small molecules or atoms into a 2D structure [5].

11.3 Properties of MXenes Over 2D materials

In this section, the significant properties of MXenes on existing 2D materials for application prospects to flexible sensors are described.

11.3.1 Tunable Surface Chemistry

Surface termination groups are the key to tuning the physical and chemical properties of MXenes in compliance with desired applications. For example, −O terminated MXenes exhibit a semiconducting nature, which is beneficial for chemiresistive gas sensors, −OH terminated MXenes are metallic, enabling flexible electronics for the construction of wearable sensors, and MXenes enriched with −F terminational groups are highly electronegative, which hold great application prospects in energy harvesting devices.

11.3.2 Free-Standing Film

The free-standing film capability offered by MXenes endorses them as potential candidates for the construction of flexible sensors. The layered MXene exhibits two to four times stronger interlayer interactions as compared with other 2D materials such as graphite and MoS_2, which enables mechanical flexibility. Further, the interlayer spacing between MXenes sheets can be improved by using polar organic intercalating agents such as hydrazine, dimethyl sulfoxide, or large organic molecules such as tetrabutylammonium hydroxide, or n-butylamine, which assists in the sliding of sheets over each other when subjected to external pressure. The flexible film of MXene "paper" can be easily formed onto filter paper using the vacuum-assisted filtration technique, and their thickness can be readily tailored by adjusting the concentration of the dispersion solution of MXene.

11.3.3 Dispersibility

The hydrophilic surface of MXenes favors the excellent dispersion of MXenes in organic solvents such as ethanol, isopropanol, N, N-dimethylformamide, dimethyl sulfoxide, and propylene carbonate. Consequently, the preparation of MXene-based composite fibers is fairly easy and the key parameters such as flexibility can be easily tuned using an MXene/polymer blending solution for the fabrication of flexible devices. With the possible merits of solution-processing techniques, MXene can be effectively coated on yarns, fibers, fabrics, and so on using spray coating, dip coating, and drop casting methods with a high degree of loading content.

11.3.4 Mechanical Strength

It was found experimentally that single-layered MXene exhibited a tensile strength of 17.3 ± 1.6 GPa, which is in agreement with density functional theory (DFT) studies. Moreover, the presence of transition elements in the MXene structure endows excellent mechanical properties and stability to combine MXenes with flexible and wearable devices. Alexey et al. comprehensively studied the elastic and mechanical properties of MXenes. The comparison of MXene with other 2D materials such as graphene, graphene oxide (GO), reduced graphene oxide (rGO), MoS_2, and so on, was done based on Young's modulus. A force versus deflection (F-δ) study was carried out, and it was found that the occurrence of the fracture point at a load force of 200 nN had a deflection of 38 nm for the case of bi-layered MXene [7]. Further, the extension and retraction curves retrace each other for each loading cycle, indicating the excellent elastic properties of MXene.

11.3.5 Electrical Properties

MXenes offer excellent electrical properties compared with other solution-processed 2D materials. The multi-layered MXene possesses significantly higher electrical conductivity (~20,000 S/cm) than single-layered MXene (~4600 S/cm) due to a large number of interlayers present in the MXene structure [1]. Additionally, the surface termination groups can control the electrical conductivity of the MXene structure. Moreover, MXene coatings used in flexible sensors retain their conductivity while being operated under different mechanical deformations such as twisting, bending, and so on, which is not possible in the case of other 2D materials, for example, graphene, metal nanowires, carbon nanotubes, and organic materials.

11.4 Processing Techniques

The different synthesis routes such as vacuum-assisted filtration, spin coating, solution blending, wet-spinning, and electrospinning can be considered for the processibility of MXene-based materials [8]. To date, most of the researchers utilized a solution blending process followed by a vacuum filtration method to obtain a uniformly distributed wrinkle-like structure over the substrate surface, which resulted in a 2D MXene film with a high specific surface area beneficial for aiming target gas molecules. Nevertheless, the mechanical flexibility of the bare MXene membrane synthesized by the vacuum filtration method is not good enough to be used in the fabrication of wearable devices. In recent years, the reported literature revealed that the blending of MXene with polymer materials facilitates the flexible properties of mechanical stimuli. The blending of polymers not only endows improved flexibility but also increases the interlayer spacing due to cationic intercalation between negative MXene sheets, which results in improved sensing performance. There are several reports based on the MXene/polymer composites (1D nanofibers, 2D composite film, 3D composite aerogel), which were used to fabricate flexible and free-standing film, such as MXene/poly(vinyl alcohol) (PVA), MXene/polyfluorene derivatives, MXene/cellulose, MXene/polyethylenimine, MXene/poly(2,3-dihydrothieno-1,4-dioxin)-poly(styrenesulfonate) (PEDOT:PSS), MXene/natural rubber, MXene/silk, MXene/aramid nanofibers, MXene/ polydiallyldimethylammonium chloride (PDDA), and so on [9].

To produce 1D MXene/polymer composite fibers, wet-spinning and electrospinning methods are used, and functional nanofibers are processed for the fabrication of flexible and wearable sensors. In the wet-spinning technique, the polymeric solution through a spinneret is injected into a coagulation bath, which coagulates to form fibers. Then, the coagulated fibers are removed from the bath and washed several times to remove the trapped solvent. Given the advantage of good dispersion of MXene in polar solvents, polymers such as polycaprolactone, PEDOT:PSS, and polyurethane can be utilized to form MXene/polymer composite fibers through the wet-spinning method. The fiber formation is strongly dependent on spinning device parameters, the coagulation bath, and post-treatment conditions, for example, the coagulation bath of chitosan reduced the exchange rate of solvent, resulting in thick and tightly packed MXene film. On the other hand, ultrathin nanofibers can be synthesized using the electrospinning technique, where the coupling effect of electric field and surface tension produce Taylor's cone of injected polymeric solution at the tip of the needle to extend and stretch fibers at the collector drum. Using the electrospinning technique, nanofibrous films of MXene/PVA, MXene/GO, MXene/polyvinylidene fluoride-trifluoroethylene (PVDF-TrFE), MXene/polydimethylsiloxane (PDMS), and so on were synthesized and found to have potential applications in flexible sensors. Originating from their dimensionality, the high surface-to-volume ratio magnifies exposure surface area to target gases. Specifically, properties such as being soft, tactile, lightweight, and form-fitting combined with electrospun functionalized nanofibers make them potential wearable sensing devices.

In addition to these, there were other fabrication techniques reported for the development of MXene/polymer composite-based devices which includes layer-by-layer (LbL) assembly, electrochemical deposition, and freeze-drying. Using the LbL assembly method, functionalized 2D MXene films with controllable thickness can be produced, where the layers of MXene are connected via electrostatic or hydrogen bond interactions. To prevent the restacking of MXene layers, researchers employed functional materials as spacers between MXene layers, this makes it easy to control the exposed surface area of MXene for sensing applications. On the other hand, the electrodeposition method is widely used in designing free-standing MXene/polymer-based membranes for energy storage applications. The electrochemical polymerization results in the self-assembly of MXene into polymer film to form a 3D porous structure [10]. Efforts in the development and design of new fabrication strategies by researchers are still ongoing to pave the way for fabricating highly flexible and durable sensing devices.

11.5 Piezoelectric and Triboelectric Sensors

Piezoelectric sensors measure the change in electric dipole moment subjected to external mechanical deformation, bending, or any kind of strain on the sensing material. The corresponding changes in the properties of the sensing material can be measured in terms of electric signals such as voltage, current, or resistance. Nowadays, piezoelectric sensors are leading the way in human activity monitoring applications by processing direct information resulting from the mechanical stimulus of different body parts. Originating from the 2D layered structure with ultrathin dimensionality and high surface-to-volume ratio, MXene offers an accordion-like shape with maximum surficial charge density, which makes it a potential candidate for the piezoelectric sensor. Ma Y. et al. [11] fabricated an oxidized MXene-based-flexible pressure sensor onto paper substrates via the dipping-drying method. They compared the output current versus time (I-T) curves for pristine MXene (P-MXene), a lower degree of oxidized MXene (O_1-MXene), and a higher degree of oxidized MXene (O_2-Mxene), where an O_1-MXene-based pressure sensor exhibited partial metallic and semiconducting behavior leading to a larger current output as compared with P-MXene and O_2-MXene-based pressure sensors subjected to the same external pressure (Figure 11.2a). Figure 11.2b exhibits the pressure sensitivity curves that were plotted for P-MXene and O_1-MXene-based pressure sensors under the pressure range of 0–8.00 kPa. Further, the sensor was carefully mounted on the body for monitoring cervical activity (Figure 11.2c), elbow bending (Figure 11.2d), finger bending (Figure 11.2e), and fingertip tapping movements (Figure 11.2f). Further, the different matrix materials were processed to support the MXene active layer, with which the mechanical strength of the MXene layer was improved to sustain the continuous operation of pressure. A flexible pressure sensor based on MXene/reduced graphene oxide (MX/rGO) hybrid structures has been reported where the rGO layer provides a skeleton with a large surface area and high porosity for aerogel formation and MXene contributes its piezoresistive properties [12].

In another study, an MXene/carbon aerogel-based strain sensor having a stable wave-shaped lamellar structure was introduced where cellulose nanocrystals were used as

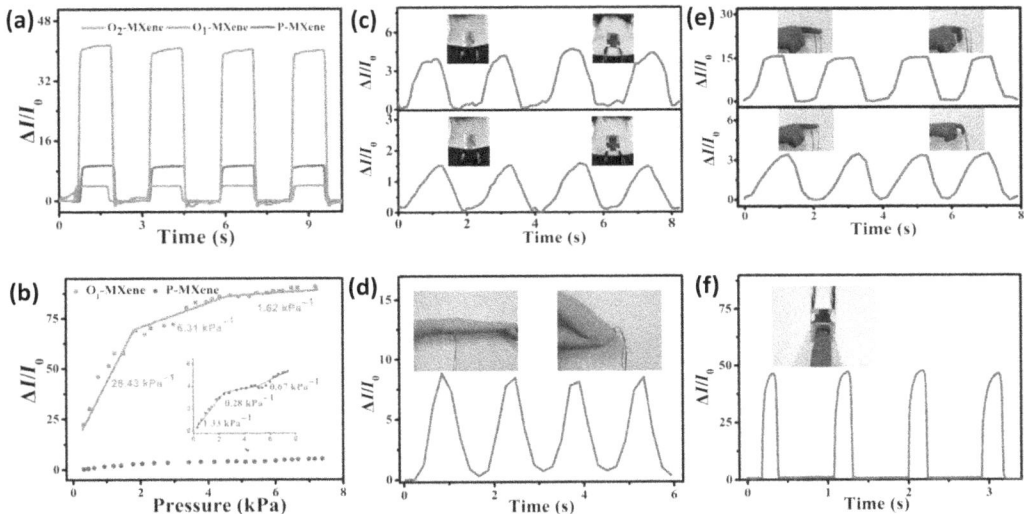

FIGURE 11.2
(a) Current-time plot of P-MXene, O_1-MXene, and O_2-MXene-based pressure sensors at 2.22 kPa. (b) Pressure sensitivity curves of P-MXene and O_1-MXene-based sensor. Applications of pressure sensors for detecting various human motions: (c) cervical vertebra bending signals, (d) elbow movements, (e) finger bending, and (f) fingertip tapping movements. Adapted with permission [11]. Copyright The Authors, some rights reserved; exclusive licensee John Wiley and Sons. Distributed under a Creative Commons Attribution License 4.0 (CC BY).

nano-support to overcome the problem related to the formation of a continuous macro-structure of MXene on carbon aerogel due to its low aspect ratio [13]. The sensor possesses a compression strain of a maximum of 95% and long-term stability with 1000 running cycles at 50% strain. Further, Wang et al. [14] successfully demonstrated the mimicking of human skin micro/nanoscale structure functions by using the interlocked hierarchical structure of a developed flexible pressure sensor based on MXene/natural microcapsule composite. The interlocked hierarchical structure showed a pressure sensitivity of 24.63 kPa^{-1} which is 9.4 times that of the planar structure MXene-based sensor (2.61 kPa^{-1}). Additionally, elastic substrates were also used for obtaining the required flexible geometry of the piezoelectric device. Here, a dip-coating process was utilized to fabricate a highly elastic and mechanically stable MXene-sponge-based piezoresistive sensor using PVA nanowires as an insulating spacer [15]. The researchers studied the potential properties of a sensor in monitoring human activities by integrating a Bluetooth system, which converts the pressure-induced resistance changes into wireless electromagnetic signals. Because of strong van der Waals forces and hydrogen bonds between adjacent MXene layers, which results in a self-restacking phenomenon, the device cannot achieve stable piezoresistive properties induced by the increase/decrease of interlayer spacing of MXene sheets. Su et al. [16] utilized the hydrazine-induced foaming method to tackle the restacking problem of MXene sheets. Further, they combined 1D cellulose nanofibers with MXene sheets as a mechanical support and introduced an MXene/cellulose-foam-based porous structure with enlarged spacing for the flexible piezoresistive sensor.

Triboelectric sensors are based on the principle of electrostatic induction between two frictional surfaces, where one surface acts as an electron donor (triboelectric positive) and the other as an electron acceptor (triboelectric negative). The charge separation between two surfaces results in an electric potential, which can be used as sensing signals to monitor the exerted mechanical forces. MXene can be used as a negative triboelectric layer owing to the electron-withdrawing ability of −F termination groups, which result in intense charge density at its surface. For example, Dong et al. [17] designed a triboelectric nanogenerator (TENG) by successfully integrating MXene thin film-coated glass with ITO-coated polyethylene terephthalate as a triboelectric pair. However, the long-term performance of the triboelectric nanogenerator is hindered due to the cracking of films under the action of continuous tapping, which also results in limited flexibility of the device. This problem was solved by fabricating MXene nanofiber-based TENG, which not only enhanced the flexibility of the device but also improved the output performance of TENG owing to the large surface area of processed nanofibers and improved dielectric properties, resulting in high charge accumulation on their surface [18]. The real-time monitoring of human activities by mounting the device on different human body parts was demonstrated. Another approach followed by Wang et al. [19] was the fabrication of a flexible 3D MXene/PDMS nanocomposite-based TENG, where PDMS was applied as a supporting skeleton for connecting MXene sheets. They also showed the potential applications of prepared nanocomposites in electrical and thermal properties.

Further, Rana et al. [20] utilized PVDF-TrFE/MXene nanocomposite nanofibers with different volume fraction MXene solutions for the fabrication of TENG, where they reported improved dielectric properties and high surface charge density for 10% volume fraction. They demonstrated smart home applications such as controlling fire alarms, fans, other electrical appliances, and smart doors using a self-powered switch. The other group worked on a self-powered and flexible foot detector using MXene/PVDF composite nanofiber-based TENG, which can control the stair-step lights using foot motions assisted tapping [21]. In addition to the triboelectric layer, an electrode material, which is an important component of TENG for the charge transfer process, must possess flexible properties along with high conductivity. In this context, Jia et al. [22] developed flexible and highly conductive Ag microparticle/MXene-based film and paired it with silicone rubber to form a sandwiched TENG structure (STENG) (Figure 11.3a). Figure 11.3b,c exhibits a variation in output voltage as a function of different tapping frequencies and instantaneous power density under different external loads, respectively. The repeatability test of STENG performed over 2000 pressing-releasing cycles and a stable response with no distortion in structure was found (Figure 11.3d). They also demonstrated powering applications such as LED lighting (Figure 11.3e) and charging electronic watches (Figure 11.3f) using STENG.

11.6 Gas Sensors

Gas sensors are devices that can monitor the presence and concentration of target gases. Gas sensors are mainly constituted of two components: receptor and transducer. The receptor part can be considered as a sensing material, which changes its properties such as resistance, electrode potential, dielectric constant, mass, and work function on reacting with the target gas. On the other hand, a transducer can be defined as a device that can convert receptor information into readable signals. The classification of gas sensors can be done based on the transducer used, which includes chemiresistive, electrochemical, capacitive, catalytic, optical, acoustic, and so on gas sensors. In recent years, MXene-based materials possess great potential in chemiresistive gas sensing applications owing to their high specific surface area, ultrahigh signal-to-noise ratio, tunable electronic properties, and excellent flexibility. It has been already proven that the termination functional groups ($-O$, $-OH$, and $-F$) of MXene control the electronic states by providing active sites for the adsorption of target analytes. The researchers theoretically studied the role of functional groups in the adsorption/desorption process of different target gases (H_2, O_2, CO_2, CH_4, NH_3, etc.) by first-principles simulation [23]. They found the preferable selectivity of $-O$ terminated MXene monolayer for NH_3 among other target gases.

Accordingly, Lee et al. [24] fabricated a flexible sensor by integrated MXene film on a polyimide sheet via the solution casting method. Their sensing results showed the highest response and lowest response of fabricated sensor for NH_3 and acetone, respectively, which was explained in terms of how the charge carriers of MXene interacts with gas species. Nevertheless, the synthesis methods such as drop casting and vacuum filtration of MXene film result in a thicker and denser structure, which can block the active sites and hence, greatly limit the performance of the sensing device. In this view, the electrospinning technique was used to develop a high-performing and flexible 3D MXene (3D-M) framework-based volatile organic compound (VOC) sensor [25]. Here, the self-assembling of negatively charged MXene sheets onto positively charged polymer nanofibers was done to form 3D-M, resulting in an interconnected porous structure that enabled the fast adsorption/desorption of gas molecules. The sensor demonstrated good flexibility and stable sensing response even after 1000 bending/unbending cycles. Theoretical studies revealed that the adsorption energy between MXene and NH_3 is higher than other gases, and as a consequence, the surface of MXene cannot be fully recovered after the removal of the target NH_3.

In this context, Zhao et al. [26] introduced flexible sensing devices using a 2D hybrid system based on polyaniline (PANI)/MXene nanocomposites where the resulting heterointerfaces

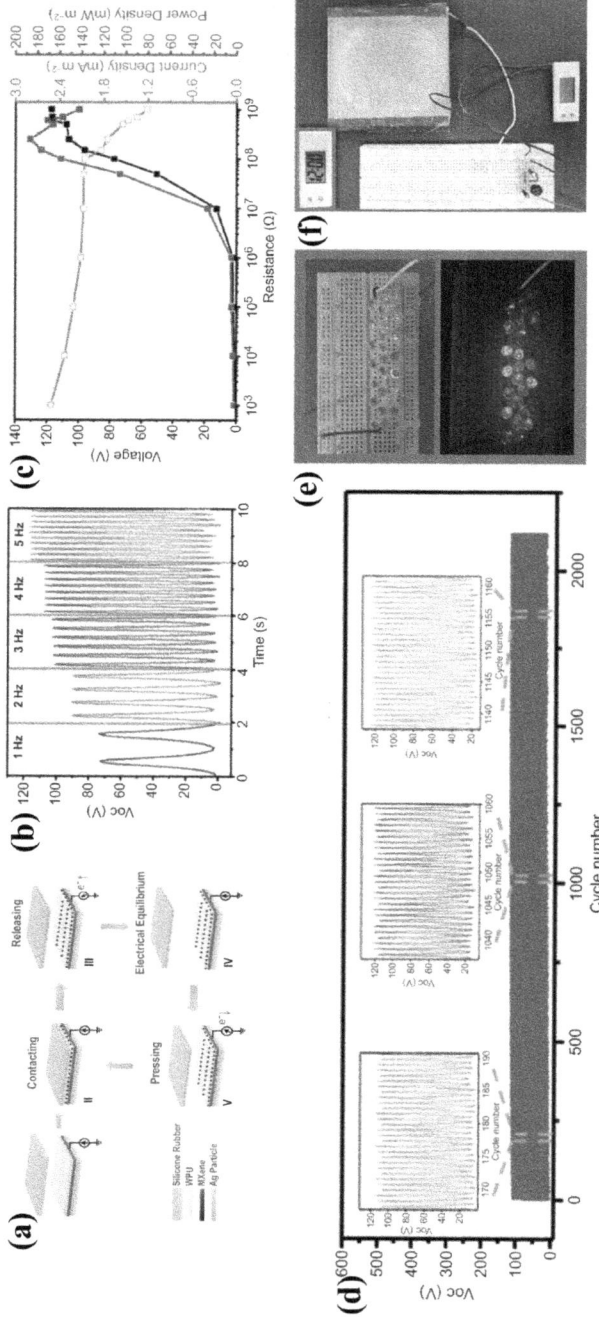

FIGURE 11.3

(a) Schematic illustration of the vertical contact-separation mode of sandwich-structured triboelectric nanogenerator (STENG). (b) Output-voltage waveform of STENG under the operation of different tapping frequencies. (c) The power density of STENG under different external loads (1 kΩ–1 GΩ). (d) Durability test of STENG. Powering (e) LEDs and (f) electronic watch. Adapted with permission [22]. Copyright The authors, some rights reserved; exclusive licensee Springer Nature. Distributed under a Creative Commons Attribution License 4.0 (CC BY).

between PANI and MXene provided efficient interaction between the active material and the target gas molecules. For the flexibility test, they recorded the response under different bending angles (0°–120°) and did not observe any significant change in sensitivity before and after bending. They also performed a bending cycle (140 cycles) test under a bent state of 60° to check the mechanical stability of the sample. The other group developed a flexible NH_3 sensor based on MXene/graphene fibers using a wet-spinning process, which provided low resistance fluctuations (±0.2%) after 2000 bending cycles [27]. Furthermore, they demonstrated potential applications for next-generation portable and wearable devices by integrating NH_3 sensors into a lab coat with a multimeter.

In another study, Li et al. [28] fabricated PANI/MXene hybrid film on a flexible Au electrode deposited PI substrate toward the monitoring of agricultural ammonia volatilization. Here they tested the sensing performance under the operation of different bending angles (20°, 30°, and 40°) and different bending times (100, 300, and 500 times), where they achieved the excellent feasibility of a fabricated device for practical applications of NH_3 monitoring. Further, Tang et al. [29] reported a wearable and stretchable acetone gas sensor based on MXene/polyurethane core-sheath fibers. They integrated a sensor into clothing by knitting core-sheath fibers into fabric and investigated its flexibility and stability properties by carrying out 1000 bending-relaxing cycles of fabric. The sensor-integrated fabric was also examined for a breathability test, where it showed good vapor permeability capable of wearing devices.

Zhang et al. [30] proposed a self-powered gas sensor composed of MXene/Co_3O_4 sensing material and a ZnO/MXene nanowire array-based piezoelectric nanogenerator. The sensor was suited for real-time monitoring without requiring the external power supply, where the energy collected from human movements was used to operate the sensor. Similarly, the MXene/metal-organic framework-derived CuO nanohybrid ammonia sensor was introduced by Wang et al. [31] where TENG based on latex/polytetrafluoroethylene was used for self-operation applications. Our group also worked on an MXene-based self-powered ammonia gas sensor, where TENG based on MXene/Cellulose acetate nanofibers enabled the self-operation of an MXene/TiO_2/cellulose nanofiber-based sensor [32]. Figure 11.4a,b exhibits the output performance of TENG under the application of different frequencies and forces. Gas-sensing studies were carried out where the prepared sensor exhibited a higher response (6.84%) to NH_3 as compared with other target gases (Figure 11.4c) and linear sensitivity (inset: Figure 11.4d). We developed a portable gas monitoring system integrated with an LED circuit that could be used to measure the concentration of ammonia by reading the corresponding LEDs (Figure 11.4e). The practical application of wearable devices was also demonstrated by integrating an as-prepared monitor device into shoe insoles (Figure 11.4f).

11.7 Humidity Sensors

Humidity sensors are electronic devices that sense and measure the content of water molecules in the surroundings by detecting changes in the electrical properties of sensing material. Combining the humidity sensing and flexibility properties of sensing material, practical applications like human respiration/breathing rate and skin moisture monitoring can lead to the prevention of serious health problems. MXene-based sensors are great representatives in the field of humidity sensing owing to their fascinating properties. For

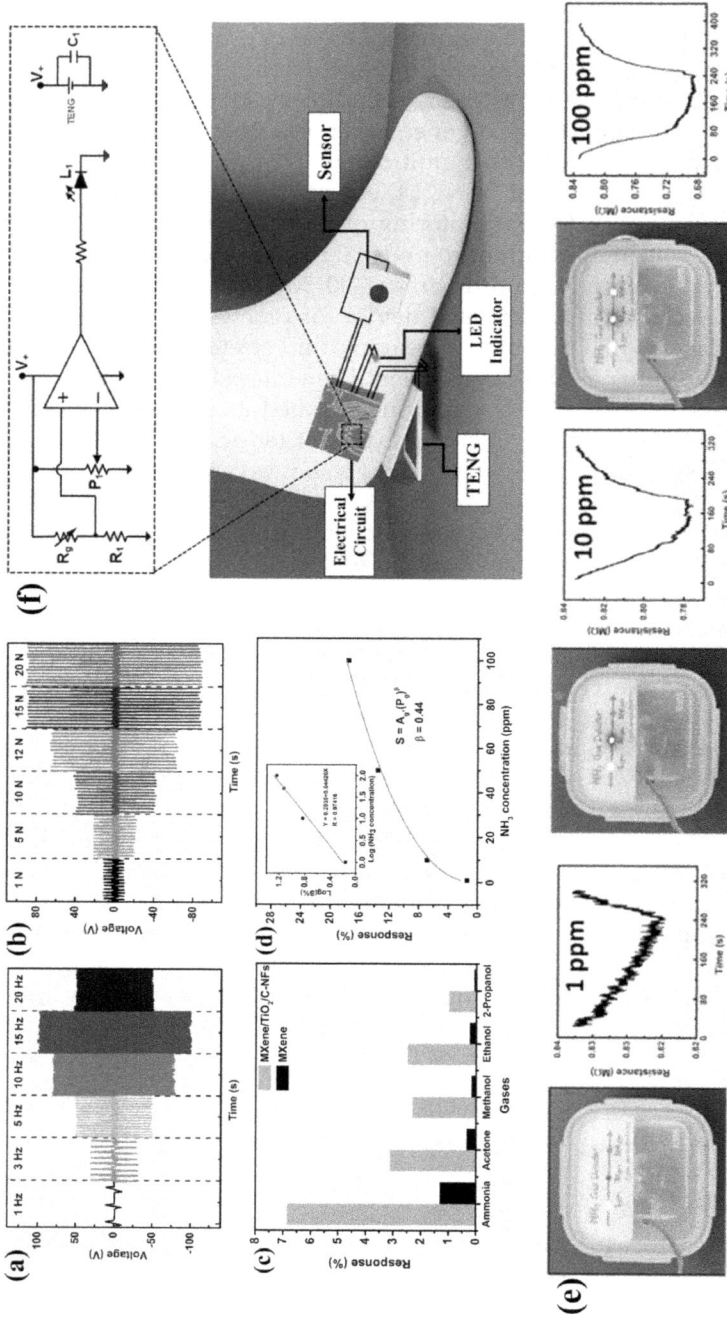

FIGURE 11.4

Voltage-time waveforms of MXene nanofiber-based TENG under (a) tapping frequencies (1–20 Hz) and (b) tapping forces (1–20 N). (c) Histogram showing the comparison of the response of MXene/TiO$_2$/C-NFs heterojunctions and the MXene-based sensor for different gases. (d) Curve-fitting of responses of MXene/TiO$_2$/C-NFs heterojunctions-based sensor for different concentrations of NH$_3$ (inset: response fitting in a logarithmic scale). (e) NH$_3$ monitoring device demonstrating the measurement of concentration (green LED: 1 ppm; green and blue: 10 ppm; green, blue, and red LED: 100 ppm). (f) Schematic illustration demonstrating the shoe insoles integrated with TENG and sensor for real-time monitoring of NH$_3$ without external supply. Adapted with permission [32]. Copyright (2022), American Chemical Society.

example, the −OH termination group of MXene endows the hydrophilic nature conducive to the adsorption of water molecules, which disturbs the charge carrier path of MXene and the resulting disturbance can be measured in terms of electrical signal (resistance). In addition, the interlayer distance between MXene sheets increases after the intercalation of water molecules, which is also a cause of a decrease in the conductivity of MXene. Flexible humidity sensors were reported recently by different research groups by combining these hydrophilic properties of MXene and the flexibility properties of the polymer. For example, Li et al. [33] developed onion-inspired MXene/chitosan–quercetin multilayers (MCQMs) through a layer-by-layer assembling route. The experimental results revealed that the performance of the sensor improved by increasing the number of layers. They demonstrated the practical application of physiological monitoring by integrating a humidity tag into a face mask and successfully recorded different breath patterns such as normal, deep, exercised breath, and apnea. A flexible, biocompatible, highly conductive, and layer-structured humidity sensor based on MXene-graphene oxide film was reported by Jia et al. [34], which was potentially demonstrated to discriminate the respiration patterns along with the monitoring of respiration rates. Another group introduced multifunctional MXene-cellulose-based smart fabric for humidity sensing and real-time temperature monitoring applications [35]. Based on H_2O-induced swelling/contraction of MXene interlayers, they recognized different breath conditions such as light mouth breath, deep mouth breath, and nose breath.

Wang et al. [36] reported a flexible and self-powered humidity sensor based on MXene/PVA nanofibers driven by a monolayer molybdenum diselenide piezoelectric nanogenerator. They demonstrated the potential of a piezoelectric nanogenerator in energy harvesting applications during finger bending, walking, running, and weightlifting (Figure 11.5a–c). Figure 11.5d exhibits the sensing performance of a piezoelectric nanogenerator-powered

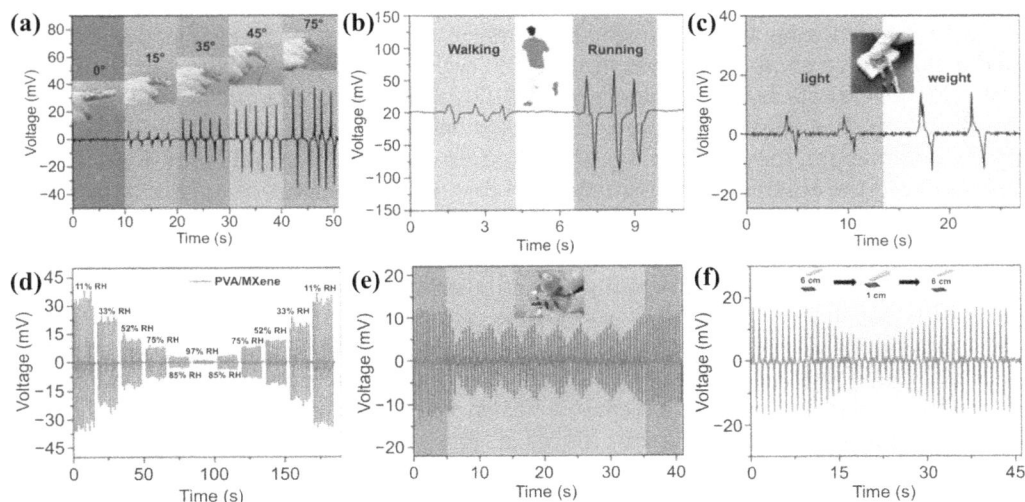

FIGURE 11.5
Energy harvesting performance of a device mounted on different body parts (a–c) for self-operation of a PVA/MXene nanofiber-based humidity sensor. (d) Output-voltage waveform of self-powered humidity sensor upon exposure to different humidity values. Applications of PVA/MXene nanofiber-based humidity sensor for monitoring (e) human breathing rate and (f) finger moisture. Adapted with permission [36]. Copyright The authors, some rights reserved; exclusive licensee Springer Nature. Distributed under a Creative Commons Attribution License 4.0 (CC BY).

MXene/PVA humidity sensor. The fast response/recovery time, flexibility, skin compatibility, and low hysteresis enabled the practical applications of breath monitoring (Figure 11.5e) and finger-moisture detection (Figure 11.5f). Further, Liu et al. [37] utilized a vacuum-assisted layer-by-layer method to fabricate a flexible and multifunctional sensing device based on MXene/AgNW-decorated silk textile. The constructed leaf-like nanostructure improves the conductivity by maintaining the air permeability of silk textile. The as-prepared sensing device was capable to monitor the moisture levels in mouth breath, nose breath, deep mouth breath, and skin surface sweating.

Xing et al. [38] synthesized a mechanical robust MXene/MWCNT electronic fabric-based humidity sensor for real-time respiration monitoring. The experimental results showed that the MXene/MWCNT fabric sensor was more resistant to deformation (7% under stretch) as compared with the pristine MXene fabric sensor (35% under stretch). Further, they identified the human respiration patterns wirelessly using fabric sensor-integrated facemasks. Wang et al. [39] proposed a dual strain/humidity sensor by employing MXene-based core-sheath yarn synthesized using a spinning trial machine, where polyurethane and polyester filament were mixed and twisted to form springlike core-sheath yarn. The unique structure of yarn leads to high flexibility and mechanical strength, which can be employed in clothing for monitoring human activities and exhaled breath.

11.8 Wearable Health Monitors

Human body monitoring has emerged as a key element to provide the necessary information related to human status for the early diagnosis of health issues. In this context, wearable monitors enable the tracking of various biomarkers released from the human body, which provides potential insights into non-invasive diagnostic applications. MXene-based e-skins have gained enormous interest in collecting physiological information such as muscle movements, body temperature, respiration rate, skin dampness, and blood pressure. Based on this, a wearable pressure sensor based on MXene/PVDF-TrFE composite nanofibrous-mat for the monitoring of human physiological signals was reported [40]. They observed the enhanced performance of a device with increased sensitivity by a factor of five and decreased compression modulus of a scaffold. The practical application of the sensor in identifying arterial stiffness and abnormalities for the detection of Parkinson's disease was demonstrated by analyzing the radial artery's pulse waveform, which includes P-wave, T-wave, and D-wave. The nanofibrous-mat-based sensor was integrated into the mask to monitor the alterations in the respiration pattern in the form of an electrical signal, which can show early diagnostic signs for serious respiratory problems. Altogether, this device showed potential as a wearable healthcare device for real-time monitoring and prognosticating the patient's health. Ma et al. [41] fabricated ethanol-dispersed MXene films on PI substrates and demonstrated real-time monitoring applications such as eye blinking, cheek bulging, throat swallowing, and bending movements of the elbow, finger, and knee joints (Figure 11.6a–f).

Wang et al. [42] reported self-assembled and ultra-stretchable MXene composite-based hydrogel for wearable bioelectronics. They proposed an efficient strategy to deal with the restacking problem of MXene and the stretchability of gel in an aqueous environment. The in-situ growth of TiO_2 nanoparticles on MXene sheets and self-assembly with poly-acrylic acid to form hydrogel enabled anti-aggregation of MXene sheets and improved

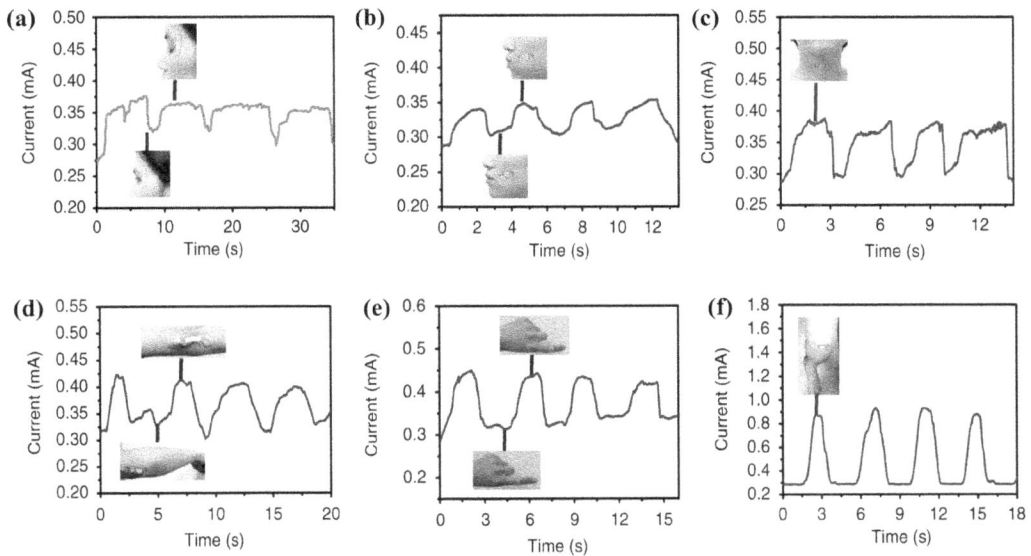

FIGURE 11.6
Human activity monitoring of (a) eye blinking, (b) cheek bulging, (c) throat swallowing, (d) elbow bending, (e) finger bending, and (f) knee bending. Adapted with permission [41]. Copyright The authors, some rights reserved; exclusive licensee Springer Nature. Distributed under a Creative Commons Attribution License 4.0 (CC BY).

crosslinking of hydrogel gel, respectively. Further, they mounted an as-developed soft hydrogel-based MXene bioelectrode on specific parts of human skin and successfully collected information related to electrocardiograms and electrooculography in the form of electrical signals. This kind of information is valuable for predicting health issues such as heart attack, sleep apnea, and musculoskeletal disorders.

Furthermore, Bi et al. [43] developed an AgNWs/MXene composite-based piezoresistive aerogel for e-skin applications. The directional freezing strategy facilitated the interweaving of 1D AgNWs among MXene sheets to minimize stacking. The testing of bending and durability was investigated to confirm the stability of the device, where the piezoresistive response was stable to the investigated bending angle (30° to 90°) and compression cycles (1000 cycles). Similarly, one other group also worked on wearable monitoring for healthcare applications based on multi-layered AgNWs/waterborne polyurethane fiber-structured sensors fabricated using a layer-by-layer self-assembly technique [44]. The prepared sensor accomplished outstanding flexible properties such as tensile strength of ~15 MPa and fracture strain of ~800%, demonstrating prominence in wearable monitoring applications. The practical application of the fabricated device for the surveillance of different body postures such as shoulder imbalance, head forward, and kyphosis was demonstrated, which is valuable for reducing health problems such as back pain, shoulder pain, and neck stiffness. The researchers employed tissue papers to support MXene sheets for the fabrication of flexible and wearable pressure sensors [45]. The fabricated device showed the potential to differentiate respiration patterns such as normal respiration rate and after-exercise respiration rate, which can be synchronized via wireless communication modules for early diagnosis of pulmonary fibrosis, opioid overdose, and cardiopulmonary-related diseases.

11.9 Summary and Future Outlook

Within a decade of emergence, MXenes have demonstrated accelerating potential in the field of sensing applications. As discussed in this chapter, MXenes possess fascinating properties compared with other 2D materials, which have attracted great attention in the development of flexible sensors. Thanks to the versatility of its functional properties, different nanostructures (1D, 2D, and 3D) can be realized using different fabrication techniques, providing enormous access to flexible designs of sensor arrays that are an alternative to bulk and rigid sensors. The recent advancement in the various sensing applications of MXene-based flexible sensors, such as monitoring of body-part movements, gas sensing, humidity sensing, biosensing, and body-released biomarkers monitoring, was also discussed.

However, there are still many problems from synthesis to application parts that limit their long-term performance. For example, the water dispersibility of MXene involves the interaction of hydroxyl groups with dissolved oxygen and water molecules, which leads to oxidation of the MXene surface, and hence, electrical conductivity loses its originality after the long-term operation of the device. The other structural problem is related to interlayer aggregation that hampers the processability and functionality of MXene-based materials. In addition to this, safety problems are a crucial concern in the adaptability of wearables because the synthesis process of MXenes involves the use of hazardous hydrofluoric acid, which affects the biocompatibility of wearable devices. For the consideration of desired applications, properties like flexible structural design, miniaturization, long-term stability, stretchability, skin compatibility, accuracy, integration of the portable sensor array, data collection, and signal processing should be ensured. Therefore, it is necessary to design the MXene structure effectively by coupling and hybridizing with other potential materials, which would enable the realization of smart and self-powered wearable devices for real-time monitoring of health status and the surroundings.

References

1. A. Ahmed, S. Sharma, B. Adak, M.M. Hossain, A.M. LaChance, S. Mukhopadhyay, L. Sun, Two-dimensional MXenes: New frontier of wearable and flexible electronics, *InfoMat.* 4 (2022) e12295.
2. R. He, H. Liu, Y. Niu, H. Zhang, G.M. Genin, F. Xu, Flexible miniaturized sensor technologies for long-term physiological monitoring, *NPJ Flexible Electronics.* 6 (2022) 20.
3. J. Park, J.C. Hwang, G.G. Kim, J.-U. Park, Flexible electronics based on one-dimensional and two-dimensional hybrid nanomaterials, *InfoMat.* 2 (2020) 33–56.
4. L. Wang, M. Zhang, B. Yang, J. Tan, X. Ding, W. Li, Recent advances in multidimensional (1D, 2D, and 3D) composite sensors derived from MXene: Synthesis, structure, application, and perspective, *Small Methods.* 5 (2021) 2100409.
5. S. Zhu, D. Wang, M. Li, C. Zhou, D. Yu, Y. Lin, Recent advances in flexible and wearable chemo- and bio-sensors based on two-dimensional transition metal carbides and nitrides (MXenes), *Journal of Materials Chemistry B.* 10 (2022) 2113–2125.
6. M. Xin, J. Li, Z. Ma, L. Pan, Y. Shi, MXenes and their applications in wearable sensors, *Frontiers in Chemistry.* 8 (2020) 297–311.

7. L. Alexey, L. Haidong, A. Mohamed, A. Babak, G. Alexei, G. Yury, S. Alexander, Elastic properties of 2D Ti$_3$C$_2$T$_x$ MXene monolayers and bilayers, *Science Advances*. 4 (2022) eaat0491.

8. G.R. Koirala, A. Chhetry, Chapter 13 - MXenes and their composites for flexible electronics, in: K.K. Sadasivuni, K. Deshmukh, S.K.K. Pasha, T. Kovářík (Eds.), *Mxenes and Their Composites*, Elsevier, 2022: pp. 423–447.

9. R. Bhardwaj, A. Hazra, MXene-based gas sensors, *Journal of Materials Chemistry C*. 9 (2021) 15735–15754.

10. D. Lei, N. Liu, T. Su, L. Wang, J. Su, Z. Zhang, Y. Gao, Research progress of MXenes-based wearable pressure sensors, *APL Materials*. 8 (2020) 110702.

11. Y. Ma, Y. Cheng, J. Wang, S. Fu, M. Zhou, Y. Yang, B. Li, X. Zhang, C.-W. Nan, Flexible and highly-sensitive pressure sensor based on controllably oxidized MXene, *InfoMat*. 4 (2022) e12328.

12. Y. Ma, Y. Yue, H. Zhang, F. Cheng, W. Zhao, J. Rao, S. Luo, J. Wang, X. Jiang, Z. Liu, N. Liu, Y. Gao, 3D synergistical MXene/reduced graphene oxide aerogel for a piezoresistive sensor, *ACS Nano*. 12 (2018) 3209–3216.

13. H. Zhuo, Y. Hu, Z. Chen, X. Peng, L. Liu, Q. Luo, J. Yi, C. Liu, L. Zhong, A carbon aerogel with super mechanical and sensing performances for wearable piezoresistive sensors, *Journal of Materials Chemistry A*. 7 (2019) 8092–8100.

14. K. Wang, Z. Lou, L. Wang, L. Zhao, S. Zhao, D. Wang, W. Han, K. Jiang, G. Shen, Bioinspired interlocked structure-induced high deformability for two-dimensional titanium carbide (MXene)/natural microcapsule-based flexible pressure sensors, *ACS Nano*. 13 (2019) 9139–9147.

15. Y. Yue, N. Liu, W. Liu, M. Li, Y. Ma, C. Luo, S. Wang, J. Rao, X. Hu, J. Su, Z. Zhang, Q. Huang, Y. Gao, 3D hybrid porous MXene-sponge network and its application in piezoresistive sensor, *Nano Energy*. 50 (2018) 79–87.

16. T. Su, N. Liu, Y. Gao, D. Lei, L. Wang, Z. Ren, Q. Zhang, J. Su, Z. Zhang, MXene/cellulose nano-fiber-foam based high performance degradable piezoresistive sensor with greatly expanded interlayer distances, *Nano Energy*. 87 (2021) 106151.

17. Y. Dong, S.S.K. Mallineni, K. Maleski, H. Behlow, V.N. Mochalin, A.M. Rao, Y. Gogotsi, R. Podila, Metallic MXenes: A new family of materials for flexible triboelectric nanogenerators, *Nano Energy*. 44 (2018) 103–110.

18. C. Jiang, C. Wu, X. Li, Y. Yao, L. Lan, F. Zhao, Z. Ye, Y. Ying, J. Ping, All-electrospun flexible triboelectric nanogenerator based on metallic MXene nanosheets, *Nano Energy*. 59 (2019) 268–276.

19. D. Wang, Y. Lin, D. Hu, P. Jiang, X. Huang, Multifunctional 3D-MXene/PDMS nanocomposites for electrical, thermal and triboelectric applications, *Composites Part A: Applied Science and Manufacturing*. 130 (2020) 105754.

20. S.M.S. Rana, M.T. Rahman, M. Salauddin, S. Sharma, P. Maharjan, T. Bhatta, H. Cho, C. Park, J.Y. Park, Electrospun PVDF-TrFE/MXene nanofiber Mat-based triboelectric nanogenerator for smart home appliances, *ACS Applied Materials & Interfaces*. 13 (2021) 4955–4967.

21. T. Bhatta, P. Maharjan, H. Cho, C. Park, S.H. Yoon, S. Sharma, M. Salauddin, M.T. Rahman, S.M.S. Rana, J.Y. Park, High-performance triboelectric nanogenerator based on MXene functionalized polyvinylidene fluoride composite nanofibers, *Nano Energy*. 81 (2021) 105670.

22. Y. Jia, Y. Pan, C. Wang, C. Liu, C. Shen, C. Pan, Z. Guo, X. Liu, Flexible Ag microparticle/MXene-based film for energy harvesting, *Nano-Micro Letters*. 13 (2021) 201.

23. X. Yu, Y. Li, J. Cheng, Z. Liu, Q. Li, W. Li, X. Yang, B. Xiao, Monolayer Ti$_2$CO$_2$: A promising candidate for NH3 sensor or capturer with high sensitivity and selectivity, *ACS Applied Materials & Interfaces*. 7 (2015) 13707–13713.

24. E. Lee, A. VahidMohammadi, B.C. Prorok, Y.S. Yoon, M. Beidaghi, D.-J. Kim, Room temperature gas sensing of two-dimensional titanium carbide (MXene), *ACS Applied Materials & Interfaces*. 9 (2017) 37184–37190.

25. W. Yuan, K. Yang, H. Peng, F. Li, F. Yin, A flexible VOCs sensor based on a 3D Mxene framework with a high sensing performance, *Journal of Materials Chemistry A*. 6 (2018) 18116–18124.

26. L. Zhao, K. Wang, W. Wei, L. Wang, W. Han, High-performance flexible sensing devices based on polyaniline/MXene nanocomposites, *InfoMat*. 1 (2019) 407–416.

27. S.H. Lee, W. Eom, H. Shin, R.B. Ambade, J.H. Bang, H.W. Kim, T.H. Han, Room-temperature, highly durable $Ti_3C_2T_x$ MXene/graphene hybrid fibers for NH3 gas sensing, *ACS Applied Materials & Interfaces*. 12 (2020) 10434–10442.

28. X. Li, J. Xu, Y. Jiang, Z. He, B. Liu, H. Xie, H. Li, Z. Li, Y. Wang, H. Tai, Toward agricultural ammonia volatilization monitoring: A flexible polyaniline/$Ti_3C_2T_x$ hybrid sensitive films based gas sensor, *Sensors and Actuators B: Chemical*. 316 (2020) 128144.

29. Y. Tang, Y. Xu, J. Yang, Y. Song, F. Yin, W. Yuan, Stretchable and wearable conductometric VOC sensors based on microstructured MXene/polyurethane core-sheath fibers, *Sensors and Actuators B: Chemical*. 346 (2021) 130500.

30. D. Zhang, Q. Mi, D. Wang, T. Li, MXene/Co_3O_4 composite based formaldehyde sensor driven by ZnO/MXene nanowire arrays piezoelectric nanogenerator, *Sensors and Actuators B: Chemical*. 339 (2021) 129923.

31. D. Wang, D. Zhang, Y. Yang, Q. Mi, J. Zhang, L. Yu, Multifunctional latex/polytetrafluoro-ethylene-based triboelectric nanogenerator for self-powered organ-like MXene/metal–organic framework-derived CuO nanohybrid ammonia sensor, *ACS Nano*. 15 (2021) 2911–2919.

32. S. Sardana, H. Kaur, B. Arora, D.K. Aswal, A. Mahajan, Self-powered monitoring of ammonia using an MXene/TiO_2/cellulose nanofiber heterojunction-based sensor driven by an electrospun triboelectric nanogenerator, *ACS Sensors*. 7 (2022) 312–321.

33. X. Li, Y. Lu, Z. Shi, G. Liu, G. Xu, Z. An, H. Xing, Q. Chen, R.P.S. Han, Q. Liu, Onion-inspired MXene/chitosan-quercetin multilayers: Enhanced response to H_2O molecules for wearable human physiological monitoring, *Sensors and Actuators B: Chemical*. 329 (2021) 129209.

34. G. Jia, A. Zheng, X. Wang, L. Zhang, L. Li, C. Li, Y. Zhang, L. Cao, Flexible, biocompatible and highly conductive MXene-graphene oxide film for smart actuator and humidity sensor, *Sensors and Actuators B: Chemical*. 346 (2021) 130507.

35. X. Zhao, L.-Y. Wang, C.-Y. Tang, X.-J. Zha, Y. Liu, B.-H. Su, K. Ke, R.-Y. Bao, M.-B. Yang, W. Yang, Smart $Ti_3C_2T_x$ MXene fabric with fast humidity response and joule heating for healthcare and medical therapy applications, *ACS Nano*. 14 (2020) 8793–8805.

36. D. Wang, D. Zhang, P. Li, Z. Yang, Q. Mi, L. Yu, Electrospinning of flexible poly(vinyl alcohol)/MXene nanofiber-based humidity sensor self-powered by monolayer molybdenum diselenide piezoelectric nanogenerator, *Nano-Micro Letters*. 13 (2021) 57.

37. L.-X. Liu, W. Chen, H.-B. Zhang, Q.-W. Wang, F. Guan, Z.-Z. Yu, Flexible and multifunctional silk textiles with biomimetic leaf-like MXene/silver nanowire nanostructures for electromagnetic interference shielding, humidity monitoring, and self-derived hydrophobicity, *Advanced Functional Materials*. 29 (2019) 1905197.

38. H. Xing, X. Li, Y. Lu, Y. Wu, Y. He, Q. Chen, Q. Liu, R.P.S. Han, MXene/MWCNT electronic fabric with enhanced mechanical robustness on humidity sensing for real-time respiration monitoring, *Sensors and Actuators B: Chemical*. 361 (2022) 131704.

39. L. Wang, M. Tian, Y. Zhang, F. Sun, X. Qi, Y. Liu, L. Qu, Helical core-sheath elastic yarn-based dual strain/humidity sensors with MXene sensing layer, *Journal of Materials Science*. 55 (2020) 6187–6194.

40. S. Sharma, A. Chhetry, M. Sharifuzzaman, H. Yoon, J.Y. Park, Wearable capacitive pressure sensor based on MXene composite nanofibrous scaffolds for reliable human physiological signal acquisition, *ACS Applied Materials & Interfaces*. 12 (2020) 22212–22224.

41. Y. Ma, N. Liu, L. Li, X. Hu, Z. Zou, J. Wang, S. Luo, Y. Gao, A highly flexible and sensitive piezoresistive sensor based on MXene with greatly changed interlayer distances, *Nature Communications*. 8 (2017) 1207.

42. Q. Wang, X. Pan, C. Lin, H. Gao, S. Cao, Y. Ni, X. Ma, Modified $Ti_3C_2T_X$ (MXene) nanosheet-catalyzed self-assembled, anti-aggregated, ultra-stretchable, conductive hydrogels for wearable bioelectronics, *Chemical Engineering Journal*. 401 (2020) 126129.

43. L. Bi, Z. Yang, L. Chen, Z. Wu, C. Ye, Compressible AgNWs/$Ti_3C_2T_x$ MXene aerogel-based highly sensitive piezoresistive pressure sensor as versatile electronic skins, *Journal of Materials Chemistry A*. 8 (2020) 20030–20036.

44. J.-H. Pu, X. Zhao, X.-J. Zha, L. Bai, K. Ke, R.-Y. Bao, Z.-Y. Liu, M.-B. Yang, W. Yang, Multilayer structured AgNW/WPU-MXene fiber strain sensors with ultrahigh sensitivity and a wide operating range for wearable monitoring and healthcare, *Journal of Materials Chemistry A*. 7 (2019) 15913–15923.

45. L. Yang, H. Wang, W. Yuan, Y. Li, P. Gao, N. Tiwari, X. Chen, Z. Wang, G. Niu, H. Cheng, Wearable pressure sensors based on MXene/tissue papers for wireless human health monitoring, *ACS Applied Materials & Interfaces*. 13 (2021) 60531–60543.

12

Hydrogel-Based Flexible and Wearable Sensors

Neda Zalpour and Mahmoud Roushani

CONTENTS

DOI: 10.1201/9781003299455-12

12.1 Introduction

The rate at which technology advances to satisfy human beings' daily requirements and facilitate and improve their quality of life is expanding exponentially in the twenty-first century. As discussed extensively in earlier chapters, a variety of methods based on the use of different materials, such as carbon-based nanomaterial, metal oxide-based nanomaterials, chalcogenides, MXene, and conductive polymers, are used to develop traditional flexible and wearable sensors (FWSs). Although all of the elastomer-based wearable and flexible sensors constructed from the aforementioned materials have all the indispensable sensor requirements, such as high conductivity and conversion of mechanical deformations to measurable electrical signals via record changes in capacity or resistance, due to the growing demand for health and movement monitoring, and resolving the existing human-machine interface, there is still a need to manufacture and design sensors with extremely high flexibility, moisture compatibly and mechanical homogeneity with human tissue, electrical conductivity, stretchability, and biocompatibility [1]. Hydrogels are physically or chemically crosslinked hydrophilic polymers that have a highly porous three-dimensional network that can retain 90% or more of the water in their structure. They have unparalleled mechanical performance such as low elastic modulus, tunable toughening flexibility, self-repair, outstanding stretchability, and imitation of inherent skin attributes such as self-healing and stretchability. Hence, they can be used very effectively in the construction of FWSs for prosthetic software robotics and human motion monitoring [2].

Hydrogel's first publication, and attention in various fields of research, began in 1888. as seen in Figure 12.1. The number of articles published on hydrogels has increased steadily based on the Scopus data source search for "Hydrogel". One hundred years after the invention of hydrogels, the first research publication on FWSs was published. The combination of these two approaches and the use of hydrogels in the field of FWSs took about ten years. Figure 12.1a–c shows the result of a search for the terms "Hydrogel", "Flexible and Wearable sensor", and "Hydrogel and Flexible and Wearable sensor" in the Scopus data source, which shows the widespread attention to the use of hydrogels in the field of FWSs in 2020 and the unparalleled growth of this research field to the present.

This chapter provides a broad perspective on the varieties of hydrogels used in the preparation of flexible and wearable sensors, as well as general and detailed design and production procedures, potential applications, performance, and current advances.

12.2 Features

The most important options that a great FWS should have are strain sensitivity, stretchability, linearity, fatigue resistance, compressibility, hysteresis, recovery time, and response time. Flexible substrates traditionally constructed from petroleum-derived polymers were chemically crosslinked weak and fragile inhomogeneous networks and so possessed fewer advantages and application possibilities compared with hydrogels [4,5]. When traditional polymers are doped with conductive substances, they suffer from efficient and adequate interface flexibility and stretchability, resulting in unstable conductivity. Hydrogels, on the other hand, are made up of continuous crosslinked polymeric networks capable of storing substantial amounts of water. Due to the mechanical flexibility provided by

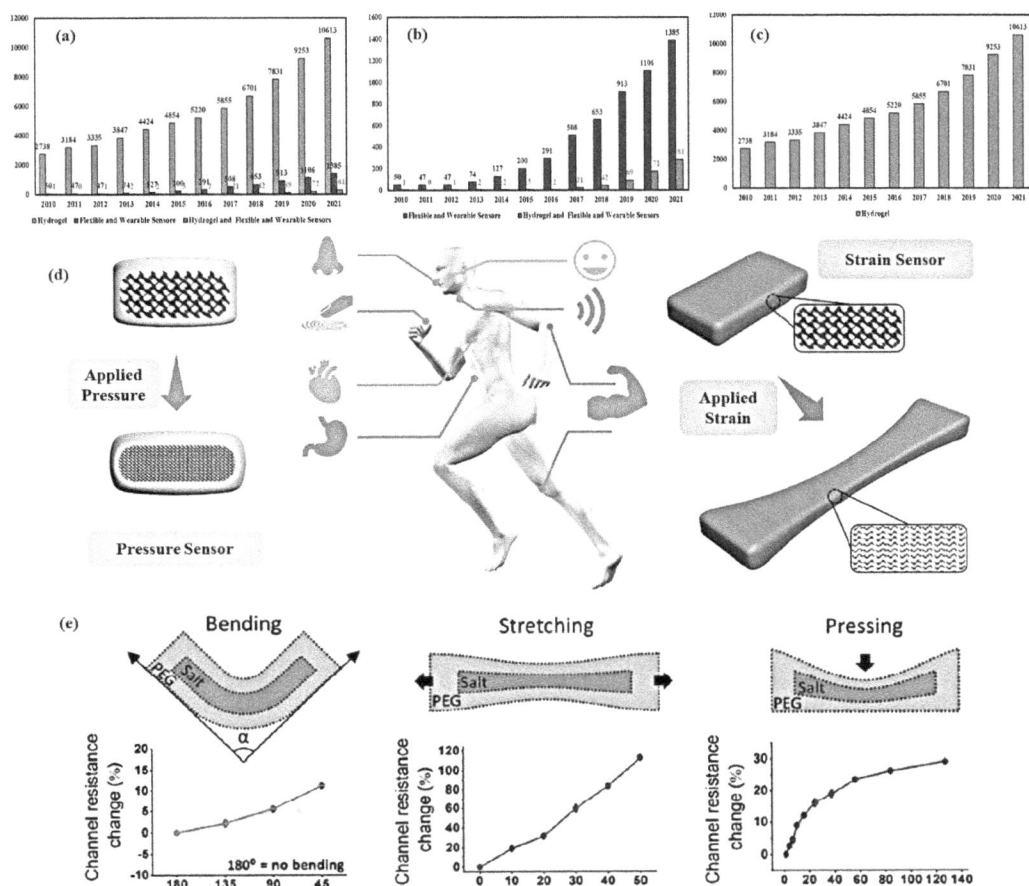

FIGURE 12.1
A bibliometric analysis of international scientific publication trends in (a) hydrogels, (b) flexible and wearable sensors, and (c) hydrogel and flexible and wearable sensor research (2010–2021). (d) Schematic illustration of the implementable devices and in vivo potential application of FWSs. Adapted with permission [2]. Copyright (2020), Journal of Materials Chemistry B. (e) Response schematic pattern of bending, stretching, and pressing patterns of a PEG hydrogel matrix. Adapted with permission [3]. Copyright (2020), Materials.

polymeric networks and the ability to consecutively convey ions by the water contained in them, they can overcome the conflict between stretchability and conductivity available in traditional cases. Other advantages of hydrogels that make them one of the best candidates for constructing FWSs include their inherent properties such as long-lasting stability, injectability, antibacterial effect, cost-effective manufacturing, transparency, ionic and electronic conductivity, high biocompatibility and remarkable renewability, self-healing, self-adhesiveness, multi-functionality, elimination of harmful organic solvents due to biopolymer solubility in water media, excellent biodegradability, non-toxicity, and so on. Petromaterial-derived polymers lack these properties [6]. Due to innovations in science, and the progress made to improve and enhance the relationship between man and machine, different categories of hydrogels with different physicochemical properties and applications in the field of FWS manufacturing have been used, such as artificial skins, prostheses, and tracking and monitoring of human signs and movement (Figure 12.1d)

[2]. The bending, stretching, and pressing properties of FWSs to make them appropriate for these applications are shown in Figure 12.1e [3]. Various approaches have been investigated to overcome the aforementioned constraints and improve the capabilities of hydrogels for usage in FWSs, which will be discussed in closer detail later in this chapter.

12.3 Double Network Hydrogel-Based FWSs

Na et al. initially proposed the notion of "double network (DN) hydrogels" in 2004 [7]. DN hydrogels are new innovative materials that feature two interconnected networks: a tightly crosslinked sacrificial brittle network for dissipating energy by the breaking of irreversible sacrificial bonds and a loosely crosslinked soft and ductile second network capable of maintaining network elasticity [8]. At high pressures, the brittle network plays the role of being sacrificed and, by absorbing and dissipating high energies, preserves the ductile network. This feature has resulted in excellent mechanical strength and toughness and a unique network structure in these hydrogels. In prototypes of DN hydrogels, both brittle and ductile networks were chemically crosslinked to each other. Therefore, when the sample is exposed to external forces, the chemical sacrificial bonds break, the internal network fractures, and irreversible network deformation occurs, which leads to poor recovery and fatigue resistance. To produce DN hydrogel networks capable of recovering and dealing with the issue of conventional synthetic hydrogels with high mechanical strength, studies to promote toughness by developing reversible non-covalent interactions were performed, such as the use of metal-coordination bonds, hydrogen bonds, and hydrophobic associations to physically crosslink with the first network (Figure 12.2a–c) [9]. Recent advancements in upgrading these systems to meet the needs of FWSs are the use of a synthesized DN Ca^{+2}/alginate/polyacrylamide (PAAM) hydrogel, which combines an ionically and covalently crosslinked alginate network, as well as a DN agar/PAAM hydrogel, which combines a ductile network crosslinked with a brittle agar network via a covalent bond (Figure 12.2f) [2]. Creating ductile network hydrogels with high self-recoverability through dual crosslinking, which have strong binding energy and stimuli-responsive functionalities, is another strategy.

12.4 Nanocomposite Hydrogel-Based FWSs

Designing nanocomposite hydrogels (NCs) has involved a thorough investigation of nanocomposite structures because of their exceptional ability to convert external stimuli to electrical signals, high mechanical strength, and excellent sensitivity. Different types of nanofillers such as nanoparticles, nanosheets, nanodots, nanotubes, and nanowires can be either chemically crosslinked or physically added with a uniform distribution to prevent aggregation into hydrogel networks. Crosslinking the nanofillers in the adsorption process into a hydrogel network to construct a hydrogel nanocomposite is based on hydrogen bonding, hydrophobic association, hydrophilic adsorption, chemical bonds, and ionic interactions between the nanoparticles and the polymer chain. Because most nanoparticles are inherently hydrophobic and accumulate in hydrogel hydrophilic networks, thereby

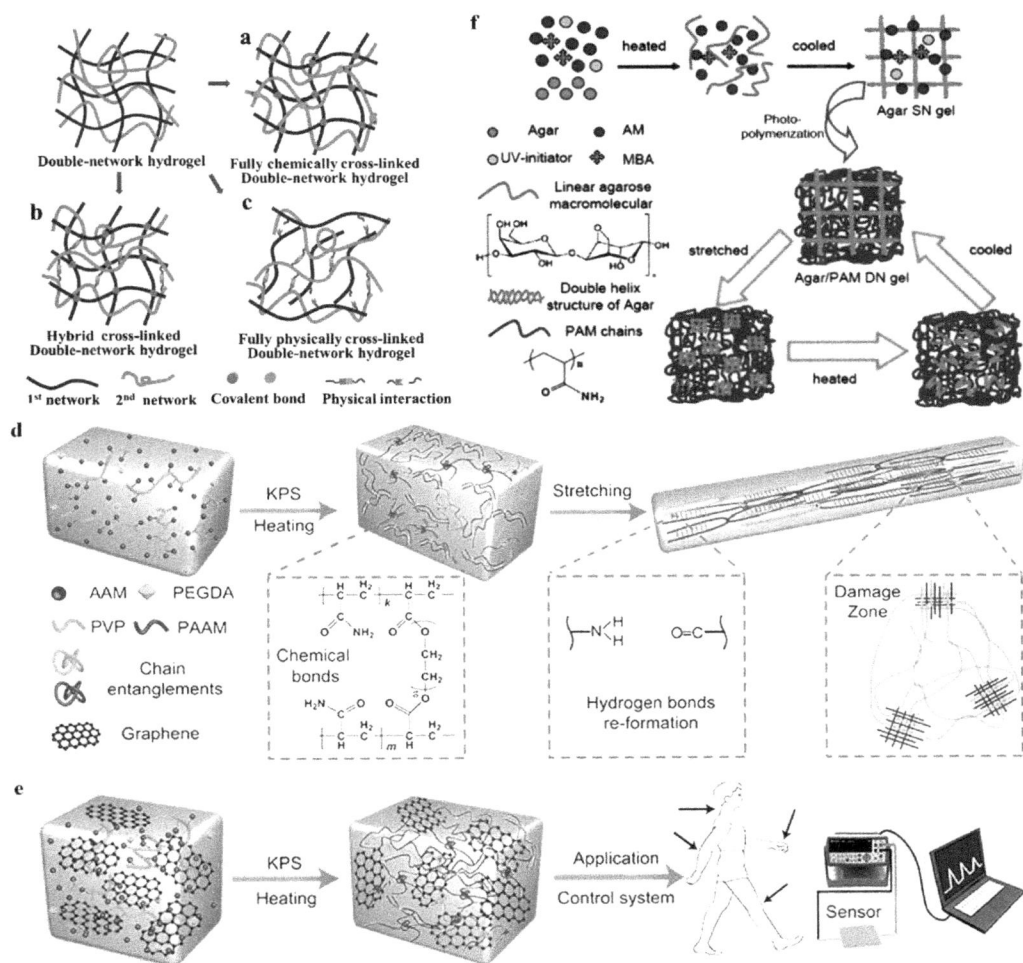

FIGURE 12.2
Schematic diagram of DN hydrogels with different crosslinking methods. (a) Fully chemically, (b) hybrid, (c), and fully physically crosslinked DN hydrogel. Schematic diagram of PAAm/PVP/EG DN hydrogel synthesis. (d) Preparation, molecular structures, and tensile fracture mechanism of PAAm/PVP/EG DN hydrogels. (e) FWS preparation and sensing application. Adapted with permission [9]. Copyright (2020), Materials Today Communications. (f) DN hydrogels based on agar and polysaccharide. Adapted with permission [2]. Copyright (2020), Journal of Materials Chemistry B.

causing a diminishing in the mechanical properties of nanocomposite-based hydrogel networks, various chemical modifications or physical interactions and the in-situ formation of nanofillers have been used to improve their mechanical performance [10]. Chemical modification is done on the surface or edges of nanofillers by oxidization or etching processes by adding some hydrophilic groups (e.g., –COOH, –OH, –O⁻, and –F). In addition to chemical crosslinks, these modifier groups complement and improve the nanocomposite-based hydrogel network features through physical interactions. Modifications of some nanofillers such as gold and silver nanoparticles and nanotubes are difficult and aggregation behaviors occur in the hydrogel precursor solution [11]. One of the unique strategies to decrease the aggregation of insoluble nanofillers and their uniform and tight distribution is to produce conductive hydrogel networks assembled as FWSs as an in-situ nanofiller formation. In the case of gold nanoparticles, for example, this is done through a reduction process of $HAuCl_4$ [10].

12.5 Conductive Hydrogel-Based FWSs

CHs are very attractive candidates for FWS construction due to the compatibility between conductive agents and flexible substrates, which leads to stable conductivity and excellent strain. The main platform of these hydrogels is based on the development of a composite between conventional hydrogels and conductive nanofillers for structural support and provide electrical conductivity, respectively. There have been four basic approaches for their synthesis so far, which are the introduction of ionic salts, in-situ polymerization of conductive polymers within a pre-formed hydrogel, a combination of hydrogel precursors and conductive fillers, and direct gelation of conductive materials. The four mentioned methods are generally divided into two main categories, which are ionically conductive hydrogels and electronically conductive (Figure 12.3) [2,12].

12.5.1 Electronically Conductive Hydrogel-Based FWSs

The main framework for manufacturing ECHs is the usage of conductive polymers or compositing or hybridizing conductive fillers into the hydrogel matrix. Utilizing some conductive polymers in the synthesis of hydrogels and gluing conductive particles to uniquely boost the hydrogels' conductivity is the most conventional manufacturing method for these hydrogels. ECHs are typically composed of a conductive network providing conductors (e.g., conductive fillers or conductive polymer hydrogel networks serving as templates or matrices for monomers in in-situ polymerization in order to fabricate a conductive network through the hydrogel) and a network of hydrogels as a flexible and deformable matrix. These ECHs are scratch resistant, fatigue resistant, have great conductivity, a high degree of sensitivity, and can withstand thousands of cycles. These characteristics make them an excellent choice for use in FWSs. Detailed instructions for producing these hydrogels are provided in reference [13].

12.5.1.1 Interpenetrating Network Conductive Hydrogel-Based FWSs

To produce an effective actuator or biological FWS, it is important to combine a hydrogel's high electric field sensitivity factor and great mechanical strength factor. It is expected that the development of an interpenetrating polymer network (IPN) could generate relatively robust mechanical properties, making it a good choice for meeting these requirements. Typically, ECHs consist of high filler content and have inhomogeneous filler distribution, and filler aggression occurs within the hydrogel matrix. Therefore, to improve weak interfacial interactions of most rigid fillers with the polymer matrix, surface modification is an essential requirement. Due to the difficulty of rapid transmission of external stress and strain into hydrogel networks, especially at low levels of strain and stress, the rapid reaction in hydrogel networks is diminished. Hence, the creation of a homogeneous conduciveness network in which both networks are interconnected is of special importance. An efficient approach to fabricating an IPNCH matrix is the penetration of conductive

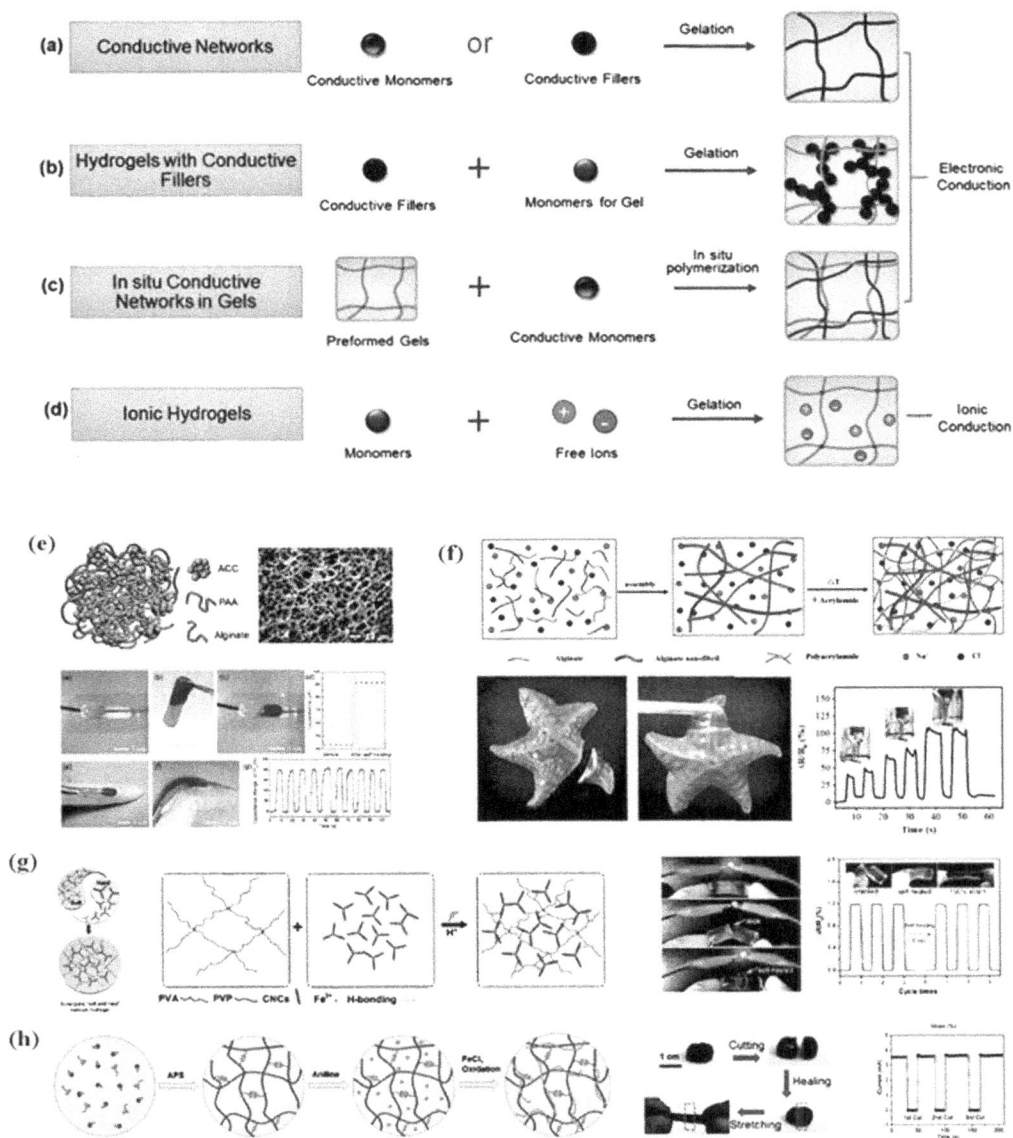

FIGURE 12.3
Main strategies to prepare CHs (a–d). Adapted with permission from [2] Copyright (2020) Journal of Materials Chemistry B. Self-healing CHs-based FWSs. (e) Self-healable ionic hydrogel to finger motion sense. (f) A self-healing supramolecular sodium alginate nanofibrillar DN hydrogel. (g) Conductive and self-healing FWS-based hydrogels made of synergistic "soft and hard" structures. (h) A rapid high conductive self-healing UPy/ PANI/PSS hydrogel FWS by multiple hydrogen bonding. Adapted with permission [12]. Copyright (2020), Trends in Analytical Chemistry.

polymer processors into a hydrogel network and then an in-situ polymerization process. Conventionally, 3,4-ethylenedioxythiophen/poly(sodium 4-styrenesulphate (EDOT: NaPSS) is firstly dissolved in agarose solution, afterward poly EDOT (PEDOT) is produced via an electropolymerization process in the agarose gel (Figure 12.4a) [2]. Using electrode

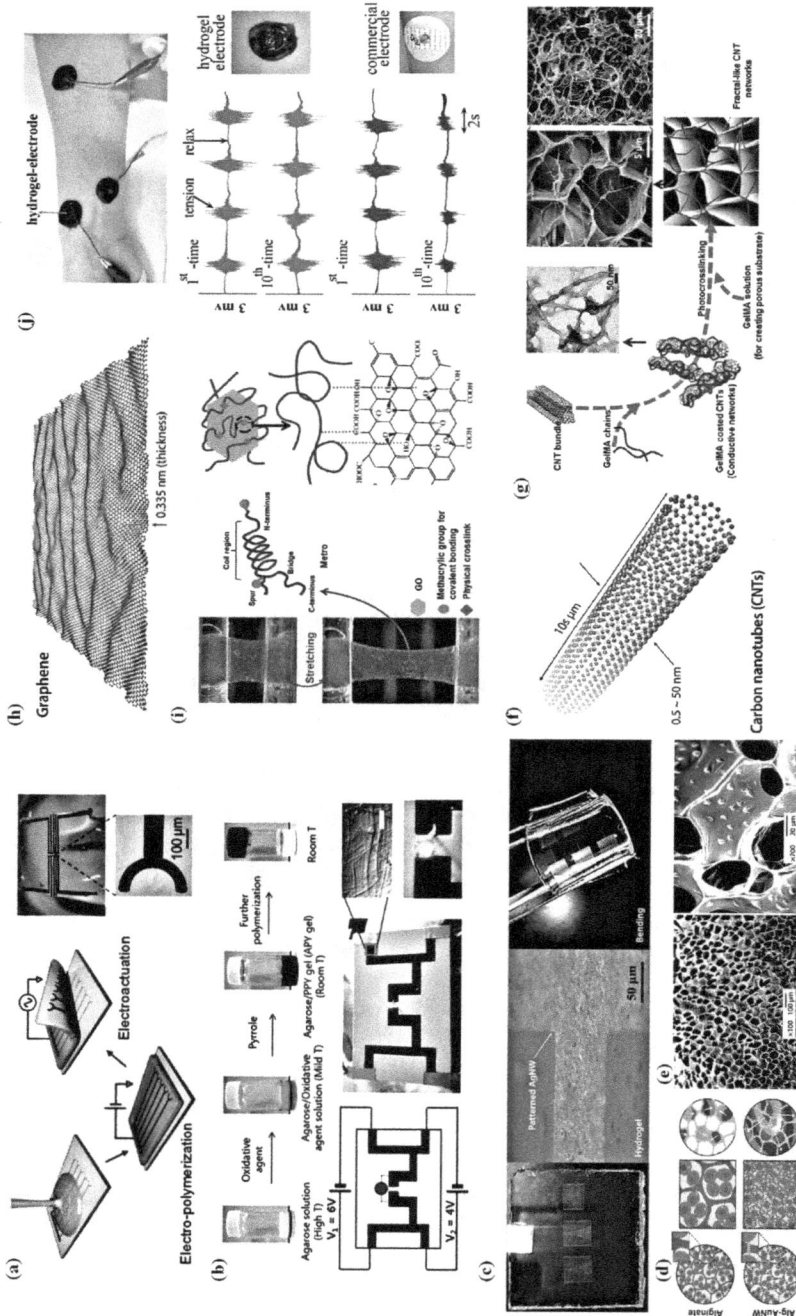

FIGURE 12.4

IPNCH. (a) Electropolymerization of an IPNCH-PEDOT network in an agarose hydrogel matrix by the in-situ electropolymerization method. Adapted with permission [2]. Copyright (2020), Journal of Materials Chemistry B. Conductive metal nanocomposite hydrogels. (c) A highly conductive and flexible microelectrode. (d) The conductive alginate–Au nanowire NC-based FWS. (e) SEM images of conductive alginate–Au nanowire NC-based FWS. (f) Carbon nanotubes for developing conductive carbon nanocomposite hydrogels. (g) A CH composite is created by incorporating the fractal-like carbon nanotube network into the GelMA hydrogel. (h) Conductive carbon nanocomposite hydrogels based on graphene. (i) The hybrid nanocomposite between graphene and proteins derived from humans facilitates the creation of highly elastic CHs. (j) Combining graphene with polydopamine hydrogel results in the production of a hydrogel that is not only conductive but also self-adhesive and self-healing. Adapted with permission [15]. Copyright (2019), Chemical Society Reviews.

patterns or arrays, the distribution patterning of EPDOT in the agarose hydrogel can be accurately defined. The IPNCH produced by in-situ oxidation polymerization of pyrrole is an agarose hydrogel with stretchability of up to 35% and self-healing properties (Figure 12.4b) [2]. These hydrogels acquired conductivities range from 20 to 70 s.m^{-1}. Direct painting and patterning of the polypyrrole/agarose hydrogels for bioelectronics are possible on biological surfaces. Lately, numerous conductive polymers have been incorporated into various hydrogel networks, hence creating a continuous electronic transmission network. These hydrogels have weak mechanical characteristics or limited sensitivity, which can be caused by a poor interaction between the conductive polymer hydrogel network and the conductive polymer [14].

12.5.1.2 Conductive Composite Hydrogel-Based FWSs

As mentioned in previous sections, to eliminate the insulating properties of hydrogels, ionic currents are generated through water-soluble ions and electrical conductivity is amplified by various approaches. The conductor hydrogels have weak biomechanical interactions. Hence, for high electrical, biomechanical performance, and further improve tissue–electrode interfaces, nanocomposite hydrogels are used to improve the electrical properties of FWSs (such as increased charge injection capacity and lower interfacial impedance) without sacrificing their desired biomechanical characteristics (e.g., low biocompatibility and mechanical modulus). Numerous varieties of electronically conductive nanoparticles, such as graphene, carbon nanotubes, metal nanoparticles, and nanowires, have been incorporated into the hydrogel matrix in order to maintain the hydrogel's ionic and electronic conductivity qualities as well as its distinct biomechanical advantages (Figure 12.4c) [15]. Despite the superior electrical properties of metal nanocomposites, they are generally susceptible to degradation and tissue damage in wet environments and have a reduction in electrical performance due to corrosion (except for noble metals like platinum and gold, which exhibit greater resilience to electrochemical degradation in physiological environments). As an illustration, nanogold nanowires (AuNWs) supply high electrical conductivity. They have been composited with the alginate hydrogel, which provides an extracellular matrix-like three-dimensional tissue scaffold that is biocompatible for developing a bioelectronic cardiac patch. They are capable of synchronously stimulating embedded cardiomyocytes (Figure 12.4d,e) [15].

Carbon-based nanofibers, such as carbon nanotubes (CNTs) and graphene, are promising candidates for nanocomposite CHs due to their abundance, excellent morphology, high electrical properties, extraordinary mechanical and optical properties, and lack of corrosion-related issues observed in metal nanomaterials. Due to various crosslinking approaches (such as polymer grafting, covalent or physical crosslinking) and extraordinary chemical functionalization of conductive CNT–hydrogel nanocomposites, a diverse range of biocompatible hydrogels (e.g., alginate, polyethylene ethylene glycol, collagen, chitosan, poly (vinyl alcohol), gelatin, and guar-gum) have been utilized to construct FWSs (Figure 12.4f,g) [15].

Graphene is one of the most significant structures of nanomaterials made of carbon. It has sheets capable of self-assembly and network formation via electrostatic interactions or chemical bonds with other polymers to develop nanocomposite CHs usable in FWSs. Graphene-based nanocomposite hydrogels exhibit simultaneously high electrical conductivity, favorable mechanical properties (such as elasticity and low modulus), and superior in vitro and in vivo biocompatibility (Figure 12.4h–j) [15].

12.5.2 Ionically Conductive Hydrogel-Based FWSs

The conduction through free ions transportation is of considerable importance in biological systems in which FWSs are used. Typically, large numbers of free ions are produced by dissolving ionic salts (e.g., NaCl, LiCl) in three-dimensional ionic conductive polymer networks. As a result of the large concentration of free ions, these unique hydrogels have outstanding ionic conductive properties (>10 S m^{-1}), in addition to their elastic (>10 times the original length) and tissue-like softness (E < 100 kPa) capabilities, making them widely used in FWSs. Due to the high salt concentration required for ICHs (>1 M) compared with physiological ionic concentrations (<300 mM), ICHs have been widely used in non-invasive bioelectronic applications such as ICH-based FWSs and epidermal electrodes. Due to the fact that hydrogels can interchange dissolved elements like ions with the tissue around them through the diffusion mechanism, high ion concentrations in ICHs may result in unstable bioelectronic performance and issues with biocompatibility in more unstable bioelectronic applications (e.g., implantable devices). Recently, modified hydrogel methods to address this issue based on phase separation between polyethylene glycol hydrogels and aqueous salt solution were devised (Figure 12.5a) [15]. Phase-separated aqueous ionic solutions can create soft, flexible, and highly ionically conductive circuits inside the biocompatible polyethylene glycol hydrogels without unpleasant ion diffusion into the surrounding tissue medium (Figure 12.5b,c) [15]. Because of the greater stability in physiological conditions, in vivo bioelectronic stimulation of muscles using ICH circuits in direct contact with tissues can be performed for longer periods of time [12].

12.5.3 Tough Conductive Hydrogel-Based FWSs

The skin is the largest organ in the human body and it enables protection from external injury and conveys a range of mechanical sensations to the brain. Therefore, skin-inspired hydrogels have the potential to be used in numerous domains, such as FWSs and soft robots. There are two general methodologies for synthesizing and designing an FWS hydrogel network structure to meet the necessary needs for an efficient sensor, such as achieving superior conductivity and high mechanical strength. In one approach, tough hydrogels are employed as the guest matrix for host conductive fillers or polymer networks to attain the highest level of strength and toughness. A conductive network is employed in another method to obtain maximal tensile strength (Figure 12.5d–i) [17].

12.5.3.1 Double Network Conductive Hydrogel-Based FWSs

In a type of double network conductive hydrogel (DNCH)-based FWS, one of the networks is replaced by conductive filler materials, and the in-situ polymerization of the conductive network is performed in a bed of a host tough or a flexible network. Throughout most cases, the second network has a uniform distribution in the first network bed, and the interactions between them improve the mechanical properties of the resulting hydrogel. As a result, these hydrogels are particularly sensitive due to the rapid response of the conductive network, in addition to reducing the number of conductive polymers and fillers applied in the production process. In another platform for manufacturing DNCHs, the conductive polymer is one-pot synthesized in a hydrogel matrix [18].

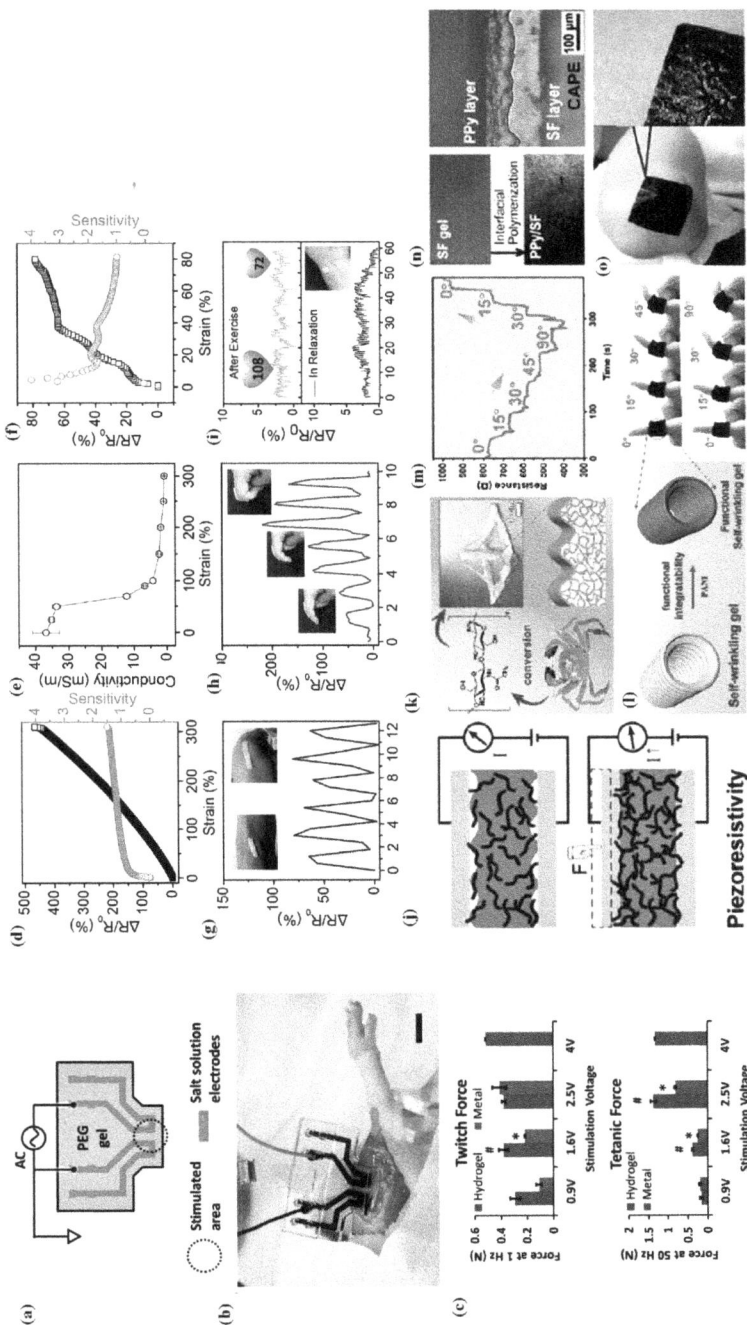

FIGURE 12.5

Hydrogel-based ionic circuits for in vivo electrical stimulation of muscle tissue. (a) An ICH based on the phase-separation method. (b and c) With external electronic connections, the ICH stimulating electrode is implanted in the rat muscle. Adapted with permission [15]. Copyright (2019), Chemical Society Reviews. (d) FWS based on PVA/PSBMA hydrogel electronic properties. The dependence of sensitivity and resistance change ratio, (e) conductivity of hydrogels on tensile strain, (f) and resistance change ratio of ICH on compression strain. Changes in relative resistance over time as measured by ICH sensors. (g) Cyclically bending on the elbow, (h) based on the finger bending at multiple angles. (i) Wrist pulse prior to and following activity. Biopolymer-based FWS-based hydrogel with piezoresistive transduction methods. Adapted with permission [13]. Copyright (2019), Journal of Material Chemistry B. (j) Diagram illustrating the operating mechanism of piezoresistive hydrogel (PH) sensors. (k) Self-wrinkling tubular chitosan/PANI hydrogel as an electronic skin attached to the finger to measure finger joint bending action, and (m) its associated fluctuations in resistance. (n) Through interfacial polymerization, a biocomposite conformal and adhesive polymer electrode (CAPE) with interlocking structures was created. (o) CAPE attached to the human elbow demonstrated remarkable conformability. Adapted with permission [16]. Copyright (2019), ACS Applied Bio Materials.

12.6 Piezoresistive Hydrogel-Based FWSs

When mechanical deformation occurs in piezoresistive sensors, the resistance changes, resulting in this force conversion converting to electrical signals. Despite their relatively ordinary architectures, the piezoresistive hydrogel (PH) sensors that use free ions such as electrolytes as sensing elements and/or conducting fillers, such as conductive polymers, metal nanowires, and carbon nanotubes, exhibit sensory and unparalleled tissue-like qualities. In reaction to an external stimulus, the resistive-type sensing instrument is often determined by changes in geometrical structure, tunneling resistance, or contact resistance (e.g., strain or pressure) (Figure 12.5j–n) [16].

12.7 Capacitive Hydrogel-Based FWSs

Typically, in this sort of sensor, a layer of dielectric hydrogel is integrated with two layers of capacitive hydrogel (CH), thereby being able to respond to external stimuli such as strain and pressure by sensing changes in the dielectric capacity of the parallel-plate capacitor. According to the formula presented in Figure 12.6a [16], for capacitors in a parallel-plate configuration, $C = \varepsilon S/4\pi kd$ (C, ε, k, S, and d, are the capacitance, the dielectric constant, the electrostatic constant, the effective area of the conductive layer, and the thickness of the dielectric layer, respectively), an increasing area (S) or a decreasing distance (d) results in a rise in capacitance (C). Consequently, CH-based FWSs offer unmatched pressure and strain sensitivity, as well as the ability to be precisely customized by evaluating the governing equation. A typical transistor structure includes a semiconductor, source-drain

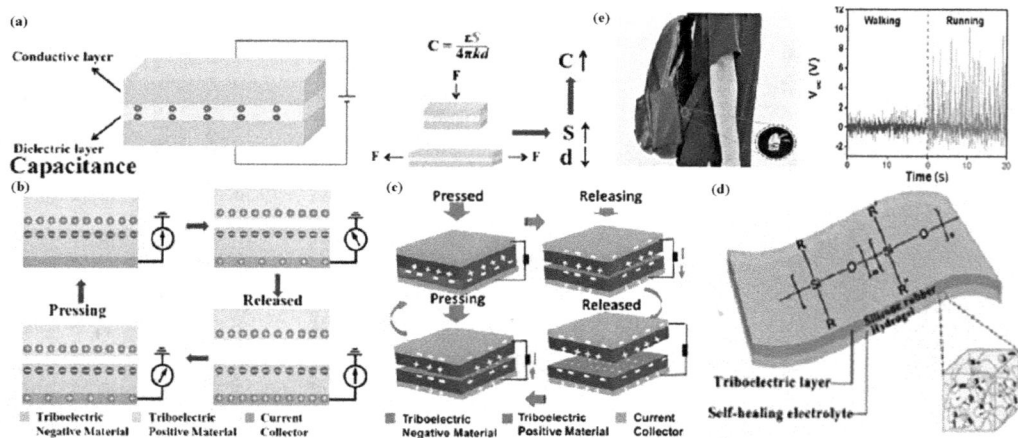

FIGURE 12.6
(a) Working principle and schematic design of CH sensor. (b) FWSs based on biopolymers that use piezoelectric and transduction triboelectric techniques, single-electrode mode. (c) Mode of vertical contact separation nanogenerator with triboelectricity. (d) Layered structure of the hydrogel-based self-healed TENG (HS-TENG). (e) With a foldable FG-TENG, which is attached to the back of a bag, there is self-powered sensing of specific bodily movements (running and walking). Adapted with permission [16]. Copyright (2020), ACS Applied Bio Materials.

electrodes, a gate electrode, and a gate dielectric. If micropatterned elastic ionic hydrogels have been used as a dielectric layer, due to the formation of a dual electrical layer in them, the capacitance changes drastically after the applied pressure, so these sensors not only are very sensitive but also greatly reduce the operating voltage [19].

12.8 Triboelectric Hydrogel-Based FWSs

When friction between two materials causes electrostatic changes, the triboelectric effect occurs. Therefore, triboelectric nanogenerators as a superior technology for collecting ambient energy, based on the coupling between triboelectric effect and electrostatic induction, can convert mechanical energy into electrical energy. Self-powered biopolymer-based hydrogel sensors that utilize triboelectric effects enable the integration of ultrathin devices on flexible textiles or skin for FWS applications and have therefore received increased interest in wearable applications, such as artificial skin, because they can receive energy from the environment and work without an external energy source (Figure 12.6b–e) [16].

12.9 Polysaccharide-Based Hydrogel-Based FWSs

Numerous intrinsic functional groups in the repeating units of polysaccharides, such as monosaccharides through different kinds of O-glycosidic linkage, are capable of creating covalent and non-covalent crosslinking, as well as the production of adjustable architectural biocompatible and biodegradable hydrogels. Due to the fact that different polysaccharides have different intrinsic and structural properties (e.g., one-dimensional, three-dimensional, cyclic), chemical and physical modifications can be used to achieve properties such as improved solubility, as well as incorporate functional moieties for crosslinking. By mixing nanoparticles with polymers to create polysaccharide composites, as the crosslinking or flexibility of the matrix is improved, the mechanical processibilities or performances of the resulting hydrogels are significantly increased. Due to the remarkable tunability of the crosslinking points (such as chemical and physical crosslinking including host-guest, hydrophobic, metal coordination, and hydrogen bonding interactions) and compositions, polysaccharide-based hydrogels can offer self-healing, excellent flexibility, high strength, and enhanced ionic conductivity (e.g., alginate, cellulose, and chitosan). Polysaccharide-based CHs can be developed by adding electronic or ionic conductors to the hydrogel's hydrogel substrate. Metallic nanoparticles/nanowires (such as Ag and Cu), carbon nanomaterials (such as CNTs) and reduced graphene oxide (RGO), and conductive polymers (such as polyaniline (PANI) and polypyrrole (PPy)) are the three main types of conductors found in electronic devices. Alkalis (such as NaOH), metallic salts (such as NaCl, KCl, LiCl, $FeCl_3$, and $CaCl_2$), and ionic liquids (such as benzyltrimethylammonium hydroxide ($BzMe_3NOH$)) are the three primary groups of substances that can produce free ions in water (Table 12.1) [20]. Therefore, electronic conductive and ICHs are the two broad categories for polysaccharide-based CHs (PCHs). The classification of these hydrogels is described in the following sections [21].

TABLE 12.1

Selective Representation of Polysaccharide-Based Hydrogels [20]

Materials	Functional moieties	Modification methods	Crosslinkers	Crosslinking mechanism
CNC	–COOH	TEMPO oxidation	-	Hydrogen bonding
CNC	–NH$_2$	Amidation reaction	-	Hydrogen bonding
CNC	–SO$_3$H	Sulfuric acid hydrolysis	-	Electrostatic interaction
CMC	–COOH	Etherification	Fe^{3+}and Ru^{3+}	Metal coordination
CNC, PVA	–OH	-	Borax	Hydrogen bonding, boronate ester bonds
CNC, PVA	Viologen, naphthyl	ATRP	CB [8]	Host–guest
CNC, PAM	–OH	-	MBA	Covalent bonds
CNC	Dihydrazide	Amidation reaction	PEG	Acylhydrazone, disulfide bonds
Alginate	–COOH	-	Ca^{2+}	Metal coordination
Alginate	–COOH	-	Chitosan	Electrostatic interaction
Alginate	Furan	Amidation reaction	Bismaleimide	Diels–Alder addition
Alginate	Phenylboronic acid	Amidation reaction	PVA	Metal coordination, boronate ester bonds
Alginate, PAM	–COOH	-	MBA	Metal coordination, covalent bonds
Chitosan	–NH$_2$	-	-	Hydrogen bonding
Chitosan	–NH$_2$	-	Metal ions	Metal coordination
Chitosan	–NH$_2$	-	TPP	Electrostatic interaction
Chitosan	–NH$_2$	-	SDS	Electrostatic interaction
Chitosan	–NH$_2$	-	Cellulose	Imine bonds
Chitosan	Double bond, catechol	Amidation reaction	PEG	Electrostatic interaction, covalent bonds
Cyclodextrin (CD)	-	-	-	Inclusion complexation
CD	-	-	Ad	Host–guest interaction
CD	-	-	PNIPAM	Host–guest interaction

12.9.1 Cellulose-Based Hydrogel-Based FWSs

Due to its inflexible backbone, cellulose – the most common polysaccharide biomass resource in nature – displays a remarkable 1D shape. The presence of many hydroxyl groups along the backbone of cellulose further contributes to its poor dispersibility in common solvents due to the creation of intra- and intermolecular hydrogen bonds between cellulose backbones. Finding suitable solvents to dissolve cellulose is therefore the fundamental challenge in the synthesis of cellulose-based hydrogels because the fabrication of hydrogels involves two steps: dissolution and crosslinking. To dissolve cellulose, two strategies are used: raw cellulose can be dissolved using base/urea aqueous solutions or ionic liquids, and (ii) cellulose can be modified to make it more soluble in typical solvents. It is important to note that there are commercial cellulose derivatives that can dissolve in a common solvent, such as hydroxypropyl cellulose (HPC) and carboxymethyl cellulose (CMC). These techniques offer fantastic prospects for creating cellulose-based hydrogels with specialized properties. Due to their excellent appealing features, such as biocompatibility and biodegradability, customizable physiochemical features, and renewability, cellulose-based CHs as reliable substrates for sensor-related applications have been developed (e.g., tunable mechanical and electrical properties, superior flexibility). This section provides a summary of various common FWSs based on cellulose-based CHs, covering structural designs and material, as well as characteristics to produce high-performance FWSs, such as strain, pressure, temperature, and multifunctional sensors [22].

12.9.2 Electronic Conductive Cellulose-Based Hydrogel-Based FWSs

Nanocellulose made from bacteria or plants is typically split into three types: cellulose nanocrystals (CNCs), cellulose nanofibrils (CNFs), and bacterial cellulose (BC). Nanocellulose can promote the dispersion of electron conductors in solutions and significantly increase the mechanical properties of the resulting hydrogels owing to its distinctive qualities, including high mechanical strength, large specific surface area, and numerous functional groups. Similarly, nanocellulose can stop the agglomeration of metal nanoparticles or nanowires during the sensing process. A subclass of electronic conductive cellulose-based hydrogels is electronic conductive cellulose nanocomposite hydrogels (ECCHs). They are typically made by mixing a homogenous dispersion of nanocellulose and electronic conductors with a hydrogel matrix. Stretchability, strength, toughness, and modulus may all be improved by synergistic interactions between the hydrogel network and the nanocellulose/electronic conductor network, as well as the development of a conductive network with highly effective percolation. The cellulose nanocomposite hydrogels should also be endowed with distinctive traits such as shape recovery, plasticity, self-healing, high sensitivity, self-adhesion, conformability, and multisensory capabilities [13].

12.9.3 Ionic Conductive Cellulose-Based Hydrogel-Based FWSs

Because of their rationally designed porous structures, the three-dimensional cellulose-based hydrogel networks are composed of a large number of free ions to produce ionic conductive cellulose-based hydrogels (ICCHs), which maintain high conductivity via ion transport and permit the free movement of water molecules in the hydrogel networks that are perfect for soft FWSs. There are typically two ways to increase conductivity in ICCHs: (1) preparing a polyelectrolyte network to encourage free ion transport and (2) constructing a large number of ion channels in hydrogel networks composed of cellulose to facilitate ion transport. In contrast to ECCHs, ICCHs offer a unique combination of benefits and features, which expands their applicability in FWSs. ICCHs frequently exhibit greater transparency compared with the majority of black electron-CHs, which provides competitive advantages for applications needing excellent transparency. Additionally, some ICCHs have integrated special properties for water retention, anti-drying, and anti-freezing, which significantly increases the robustness and environmental adaptability of FWSs. By chemically or physically crosslinking cellulose solutions, it is simple to create pure cellulose hydrogels. However, due to the hydrogen-bonded network and inflexible crystalline structure, water and common organic solvents do not readily dissolve cellulose. The main obstacle to creating cellulose hydrogels is the lack of an appropriate dissolving system. There have recently been efforts to construct single network ICCHs using solvent systems including inorganic salt, alkali/urea, and ionic liquids [12].

12.9.4 Chitosan-Based Hydrogel-Based FWSs

Chitosan is a random copolymer of D-glucosamine and N-acetyl glucosamine with good hydrogel-forming capabilities that is formed via the full or partial deacetylation of chitin. It is a biocompatible, biodegradable, non-toxic, and amine-rich biopolymer. When it comes to chitosan hydrogels, the copious free ions that result from the acid chitosan solution are typically present in chitosan hydrogels, and enough water contained in the hydrogel matrix assures adequate ion transfer, resulting in an inherent current. Additionally, conductive polymers (such as polyaniline, polythiophene, and polypyrrole) and nanofillers (nanosilver, carbon nanotube, and graphene) are frequently incorporated into the hydrogel matrix to further improve conductivity. Mechanical performance is the key area of concern because the low strength and tensile strain of conventional chitosan hydrogels prevent their usage as FWSs. To address this issue, DN hydrogels – which consist of two different kinds of polymer networks – are frequently used. In the first network, which is made up of rigid and brittle polymers, sacrificial bonds are formed to dissipate energy, and the second network is made up of soft polymers to maintain the integrity of the hydrogel during deformation. Due to highly reactive protonated amine groups that can be crosslinked by negatively charged groups such as $SO4^{-2}$ and Cit^{-3}, its rigid chain structure is an ideal material for constructing the first network [23].

12.9.5 Chitosan-Based Hydrogels for Stimuli-Response Sensors

Specially constructed structures of many chitosan hydrogels can respond to particular stimuli with distinct behaviors, such as photoreception, shape memory, and resistance change, which can be identified and produced as electrical signals. These behaviors can occur in response to stimuli like relative humidity light, temperature, or changes in the external environment. To create a self-powered humidity sensor, chitosan hydrogels were employed as the insulator inside a tunable Fabry–Pérot resonator with a metal-insulator-metal structure. Changes in relative humidity caused the chitosan hydrogel film to swell, which changed the multilayer structure's transmissive structural color [24].

12.10 Protein-Based Hydrogel-Based FWSs

Protein engineering is a cutting-edge technology that plays a vital part in biomedical applications. The building blocks of proteins are amino acids joined by peptide bonds in a condensation process. When combined with polymeric hydrogels with dynamic covalent bonds in the field of biomedical engineering, proteins' physicochemical properties, particularly their mechanical properties, create protein-based hydrogels (PBHs) that have an appropriate structure, stability, and strength, as well as many distinct qualities like stimuli responsiveness and self-healing properties. Covalent (chemical) and non-covalent (physical) crosslinking can be used to create the gel matrix of PBHs. Over the years, several functional proteins have been created and integrated into PBHs for a variety of FWSs [25,26].

Table 12.2 [26] summarizes developments in the field of PBHs, including hydrogel kinds, components, and their effects, with a particular emphasis on PBH applications in FWSs.

TABLE 12.2

Summary of Tissue Engineering-Related Applications of PBHs [26]

Proteins	Crosslinking	Hydrogel components	Effects of protein component	Target tissue
Collagen	Physical crosslinking Adjustment of pH (50 mM HEPES), employment of $CaCl_2$ and thrombin, and UV exposure	-	Showed rapid mouse myoblast cell infiltration and micro-vascularization	Heart
	Thermally crosslinking Incubation for 30 min	-	Formed a lattice pattern for cornea structure	Cornea
	-	Alginate	Increased chondrocyte cell viability (up to 90%)	Cartilage
	Thermally crosslinking Incubation at 37°C	-	Displayed significant osteogenic differentiation	Bone
Gelatin	Physical crosslinking UV exposure, Irgacure 2959 (0.5% w/v)	GelMA	Produced endothelial cell-responsive tissues	Blood vessel
	Chemical crosslinking Borax (0.1 M, 30 s)	Alginate	Promoted mouse chondrocyte adhesion, viability, and proliferation	Cartilage
	-	PEI-Ppy	Developed antibacterial properties	Skin
	Physical crosslinking UV exposure, Irgacure 2959 (0.5% w/v)	GelMA	Aided keratinocyte proliferation and differentiation	Skin
Serum albumin	Chemical crosslinking Adjustment of pH (NaH_2PO_4 and Na_2HPO_4)	PEG-SS2- bioglass	Accelerated the wound healing process	Skin
	Ionic crosslinking (Ag^+)	-	Significantly increased osteogenesis differentiation	Bone
	Ionic crosslinking ($CaCl_2$)	Sodium alginate Hydroxyapatite	Affected the differentiation and proliferation of human bone marrow-derived mesenchymal stem cells	Cartilage
Elastin	Thermally crosslinking Incubation at 75–80°C	Fibroin	Created contractile heart tissue	Heart
	Thermally crosslinking incubation at 37°C	Collagen	Created contractile heart tissue	Heart
	Modification of SKS concentration	Plasma	Accelerated heart valve endothelial cell growth	Heart
	Chemical crosslinking N-hydroxysuccinimide (NHS)-1-ethyl-3-(3- dimethylaminopropyl (EDC)	Collagen	Improved mechanical characteristics and biological capabilities	Skin
	-	-	Tackled bacterial infection	Bone

(Continued)

TABLE 12.2 (CONTINUED)

Summary of Tissue Engineering-Related Applications of PBHs [26]

Proteins	Crosslinking	Hydrogel components	Effects of protein component	Target tissue
Keratin	Disulfide crosslinking	-	Showed rapid penetration, propagation, and differentiation of MSCs	Cartilage
	Chemical crosslinking Sodium trimetaphosphate	Konjac glucomannan, Oat	Aided collagen formation	Skin
	Disulfide crosslinking	Glucose-triggered	Decreased gel formation time	Skin
	Disulfide crosslinking	-	Developed hydrogel biocompatibility	Bone and skin
Resilin	Chemical crosslinking Tris(hydroxymethyl phosphine)	Fibronectin	Increased human MSC proliferation	Cartilage
	Chemical crosslinking PEG macromers	PEG-vinyl sulfone	Increased aortic cell viability	Cardiovascular
	Chemical crosslinking Tris(hydroxymethyl phosphine)	-	Increased hydrogel flexibility and bioactivity	Vocal fold
	Chemical crosslinking 3,3'dithiobis(sulfosuccinimidy lpropionate)	-	Displayed remarkable NIH/3T3 fibroblast growth in a day (>95%)	-
Silk	-	Fibroin	Improved rat cardiomyocytes cell attachment and activities	Heart
	Enzyme-mediated crosslinking	-	Provided the repair of osteochondral tissue	Bone and cartilage
	Physical crosslinking UV exposure, LAP (0.6% w/v)	Glycidyl methacrylate	Displayed proliferation and viability of chondrocytes cell after four weeks	Cartilage
	Thermally crosslinking at physiological temperature	Chitosan	Positively impacted MC3T3-E1 cell adhesion and proliferation	Bone

12.10.1 Collagen-Based Hydrogel-Based FWSs

The most prevalent fibrous protein, collagen, is found in both hard and soft tissues. As a result, it has received considerable research to be used as a biocompatible material in a variety of sectors including FWSs. Physicomechanical properties of thermo-reversible, physically created collagen hydrogels are inferior to those of covalently crosslinked ones produced by glutaraldehyde or diphenyl phosphorylate azide. Additionally, gelation conditions have a substantial impact on the creation of the hydrogel; to be more precise, pH and temperature can be precisely controlled to tailor the characteristics of the collagen hydrogel's fibrous matrix. Furthermore, collagen-based hydrogels generated at lower temperatures show greater pore diameters and improved osteoblast-like cell responsiveness, whereas those made at higher temperatures (e.g., 37°C) encourage fibrillogenesis and have smaller pore diameters. Collagen's natural bonding is destroyed by chemical functionalization during the extraction process; as a result, biomaterials containing collagen require inter- and intramolecular crosslinking to enhance their mechanical properties [27].

12.10.2 Gelatin-Based Hydrogel-based FWSs

Gelatin, a hydrophilic biopolymer derived from denatured collagen, has been utilized extensively in FWS applications due to its biocompatibility, wide availability, non-immunogenicity, biodegradability, high carbon content, low cost, and so on. At 37°C, it exhibits amphoteric behavior and is easily soluble. For example, because of their high fluid absorption capacity, gelatin-based hydrogels (GBHs) can be exploited as wound dressings. When developing gelatin-based formulations, it is important to consider the different isoelectric point (pI) values of cationic and anionic gelatin (pI equal to pH 7–9 for type A and pH 4.7–5.1 for type B), as pI can alter the retention of gelatin's active components. GBHs are widely utilized to manufacture contact lenses, FWS scaffolds, and drug delivery devices due to these properties. Gelatin's functional groups, which include primary amine, carboxyl, and hydroxyl, allow it to be modified with a variety of crosslinkers and medicinal compounds, making it a perfect alternative for tissue regeneration. Gelatin can display a reversible sol-gel transition feature due to its thermo-responsive features. When the temperature drops, the solution transforms into a gel, and this transformation can be reversed by restoring the temperature of the mixture to the physiological temperature [28].

12.10.3 Resilin-Based Hydrogel-Based FWSs

Resilin, found by Torkel Weis-Fogh in the hinges of locust wings, belongs to the elastomeric protein category. Resilin has a remarkable rubber-like elasticity with a Young's modulus of 50–300 kPa and a tensile strength of 60–300 kPa. This is due to its low stiffness, great resilience, and effective energy storage (492%) [29]. The structural properties of resilin have been used to engineer resilin hydrogels that are responsive to environmental stimuli like temperature or pH. According to Elvin et al., rec-resilin hydrogel exhibits thermo-responsive behavior, transitioning between protein-rich and protein-poor phases at specific temperatures. Zeta-potential measurements were used to determine the rec1-resilin's pH-dependent properties in an aqueous solution and showed that the resilin-like polypeptides (RLPs) had varying surface charges at various pH levels [30]. Moreover, in its usage in the fields of cell culture and tissue engineering, rec1-resilin was manufactured with nanometer-scale gold nanoparticles (AuNPs) through a one-step covalent binding procedure. The fluorescence quenching of AuNP by the fluorophore moiety of rec1-resilin has potential uses in fluorescence-based detection FWSs and nanoparticle-based electronics. Recently, Hu et al. produced graphene-conjugated RLP hydrogel with glycerol that is extensible to more than four times its original length and has an adhesion strength of up to 24 kPa, allowing the gel to stick to a variety of substrate surfaces (i.e., skin, rubber glove, metal, cellulosic products, and glass). Interestingly, the graphene-conjugated RLP hydrogel demonstrated exceptional electroconductivity under mechanical deformations, with a gauge factor of 3.4 at 200% strain, and was effectively applied to real-time monitoring of swallowing, finger bending, and phonating. These hybrid RLP hydrogels with electroconductivity, stretchability, and adhesion may offer a promising material for application in FWSs [31].

12.10.4 Silk-Based Hydrogel-Based FWSs

A product of the silkworm *Bombyx mori*, silk fibroin has been used for thousands of years because of its exceptional strength, flexibility, and toughness. Flexible and wearable silk fibroin-based sensors have excellent flexibility, great sensitivity, and a wide range of practical strains, allowing them to be easily incorporated into textiles or installed on human skin to detect mechanical stimuli. Two typical methods have so far been employed to raise the conductivity of the silk fibroin-based sensor. First, a sensor with exceptional elasticity and high sensitivity can be created directly by integrating conductive elements (such as metal nanowires, carbon nanotubes, and graphene oxide) onto soft substrates (such as hydrogel, polydimethylsiloxane (PDMS), and rubber). On the other hand, accurate modifications to the original silk strands can give the fibers conductive functionality or improve their intrinsic qualities [16,32].

12.10.4.1 Silk Fibroin FWSs Based on Conductive Components

Silk fibroin wearable sensors have good mechanical properties but are not enough for real-world applications. Therefore, multiple alternatives must be provided. For instance, reports on carbon materials including graphene, carbon aerogels, carbon fiber, amorphous carbon, CNTs, carbon black, and their hybrids are numerous. Recent research has revealed that a wearable sensor constructed of silk fibroin and graphene-based nanoparticles (GBNs) can increase biomedical conductivity and mechanical properties. In addition to their superior mechanical properties, silk fibroin and graphene-based sensors with multiple sensing functions must be developed to monitor compression and tensile deformation and simultaneously identify different body signals such as joint movement, facial gestures, and vocal vibration. By proportionately mixing polyacrylamide (PAM), graphene oxide (GO), silk fibroin (SF), and poly(3,4-ethylenedioxythiophene): poly(styrenesulfonate) (PEDOT:PSS), Cui et al. produced a soft CH with high elasticity and the ability to simultaneously monitor strain/pressure data (Figure 12.7a) [16]. The resulting sensor has significant stretchability and compressibility, which allowed for a wide sensing range (a strain of 2600%; a pressure of 0.5119.4 kPa), rapid recovery from massive deformations, and stable performance. Besides this, the equivalent sensor could simultaneously recognize many bodily signals, such as joint movement, pulse, facial gesture, and respiration. Moreover, CNTs as a conductive component have been widely incorporated into FWSs for a variety of applications due to their superior electrical characteristics and thermal stability. In comparison with graphene and CNTs, carbon-based materials derived from natural biomaterials have garnered extensive interest due to their superior electrical conductivity, environmental friendliness, and scalability. The practical applications of soft electronics,

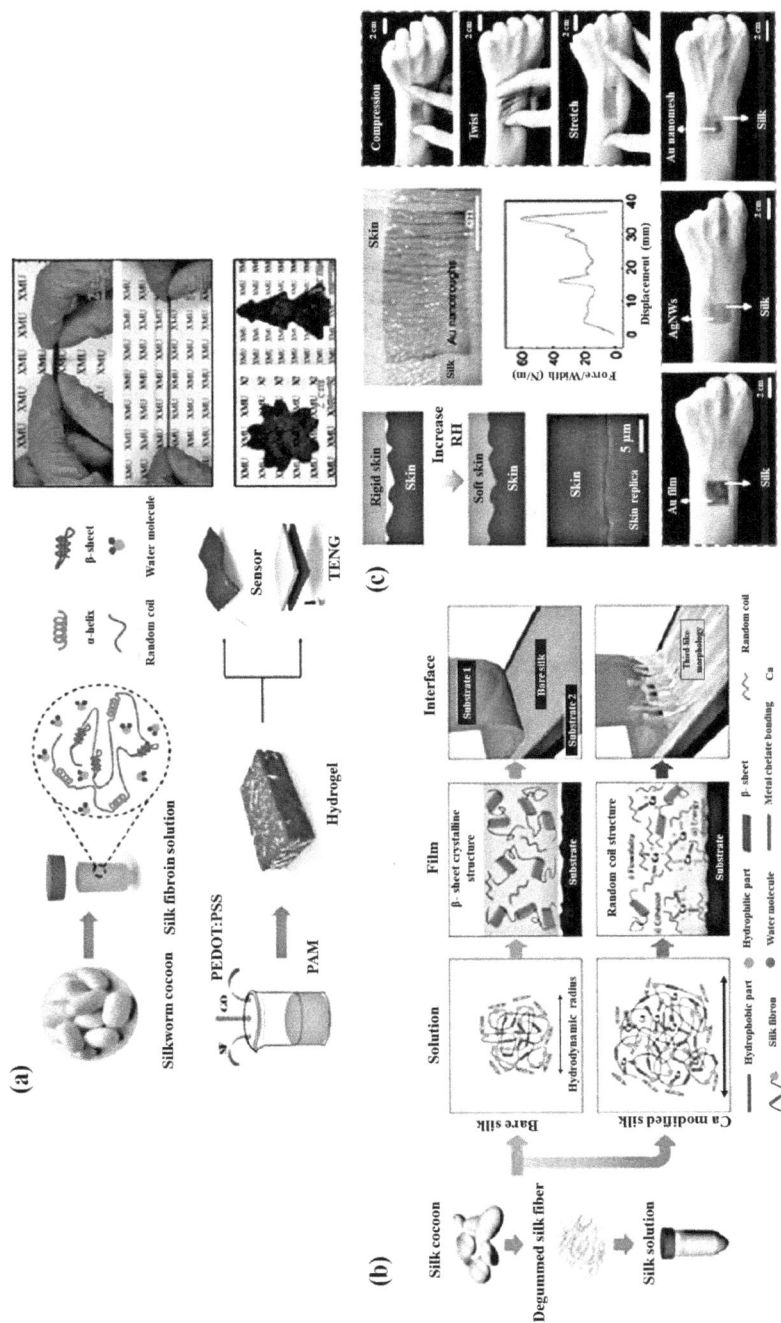

FIGURE 12.7

Composites made of silk fibroin and nanoparticles based on graphene. (a) PAM/SF/GO/PEDOT:PSS (PSGP) hydrogel, triboelectric nanogenerator schematic illustration. (b) Adhesive mechanism of Ca-modified silk. (c) Electrical and mechanical stability of the silk epidermal electrode. Adapted with permission [16]. Copyright (2020), ACS Applied Bio Materials.

including muscles, triboelectric generators, and artificial skins have increased significantly in recent years as a result of the combination of smart materials with the benefits of hydrogels and ionic conductors [33].

12.10.4.2 Calcium-Modified Silk Fibroin Electronic Skin

The human sensor based on ultrathin and flexible silk fibroin electronics is a potential next-generation FWS that could provide real-time monitoring of physiological and physical signals (temperature, mechanical forces, and humidity). In comparison with conventional FWSs, the epidermal system offers competitive advantages in terms of improved user comfort and maintaining the physiological equilibrium between the body and the surrounding environment via different physiological signals. To provide accurate measurements with a high signal-to-noise ratio, it is necessary to develop a non-allergic, stable, and high adhesion epidermal system on the skin's surface, thereby preventing inevitable physical interaction with the environment. To date, various methods have been proposed to produce a durable, non-allergic, and highly adhering interface for epidermal systems, such as the chemical alteration of the silk fibroin surface and the development of biocompatible adhesives. As an adhesive for epidermal devices, calcium (Ca)-modified silk fibroin was produced, which was a possible candidate for the next-generation adhesive for epidermal FWSs. Ca-modified silk displayed outstanding physical adhesive properties due to the crosslinking of Ca ions and random coil chains of silk through water-capturing sites and metal chelate complexes. The Ca ions in a Ca-modified silk film generated a flexible and fluidic thin film by absorbing water molecules from the environment, allowing for a high correlation between the connected epidermal FWS and the adhesive/substrate interface of Ca-modified silk. Furthermore, metal chelation and water absorption of Ca ions by increasing viscoelasticity and dissipating substantial amounts of energy at the interface caused the peel strength of Ca-modified silk and high correlation with the interface. These properties permitted a vast variety of applications as epidermal electronic FWS systems for effective monitoring of electrophysiological signals, including hydrogel-based drug delivery, electrocardiogram monitoring sensors, resistive strain sensors, and capacitive touch sensors [34].

12.11 Summary, Challenges, and Outlook

This chapter attempted to introduce the various types of hydrogel-based FWSs as well as their synthesizing techniques, applications, and modifications. As explained in detail in the chapter, hydrogels have unique inherent properties such as biodegradability, abundant and renewable supply, antimicrobial activity, non-toxicity, adsorption, low cost, renewability, and abundant functional groups (e.g., carboxyl, hydroxyl, amino, and amine groups). The polymer molecular chains endow biopolymers with great hydrophilicity and hence are the best candidates for FWS architecting. In contrast, due to the strong reactivity of these functional groups and to fulfill the demands of diverse specific functionalities, modification is achievable by modifying electron transfer rates, mechanical characteristics, solubility, and crystal structure.

Even though significant progress has been made, several challenges still need to be resolved. The spread of biopolymer applications will be facilitated by demystifying the fundamental concepts of the structure-property connection of biopolymers such as molecular-level organization, physicochemical properties, and material structures. More crucially, it is essential to avoid both the freezing of water under subzero temperatures and the evaporation of water from hydrogels under ambient conditions for practical applications. In general, despite several challenges, it is thought that natural biopolymer-based hydrogels still have great potential for FWS development.

References

1. Y. Wang, H. Liu, X. Ji, Q. Wang, Z. Tian, S. Liu, Recent advances in lignosulfonate filled hydrogel for flexible wearable electronics: A mini review, *Int. J. Biol. Macromol.* 212 (2022) 393–401.
2. Z. Wang, Y. Cong, J. Fu, Stretchable and tough conductive hydrogels for flexible pressure and strain sensors, *J. Mater. Chem. B.* 8 (2020) 3437–3459.
3. L. Tang, S. Wu, J. Qu, L. Gong, J. Tang, A review of conductive hydrogel used in flexible strain sensor, *Materials (Basel).* 13 (2020) 3947.
4. C. Ma, M. Ma, C. Si, X. Ji, P. Wan, Flexible MXene-based composites for wearable devices, *Adv. Funct. Mater.* 31 (2021) 2009524.
5. H. Wang, J. Xiang, X. Wen, X. Du, Y. Wang, Z. Du, X. Cheng, S. Wang, Multifunctional skin-inspired resilient MXene-embedded nanocomposite hydrogels for wireless wearable electronics, *Compos. Part A Appl. Sci. Manuf.* 155 (2022) 106835.
6. T. Qin, W. Liao, L. Yu, J. Zhu, M. Wu, Q. Peng, L. Han, H. Zeng, Recent progress in conductive self-healing hydrogels for flexible sensors, *J. Polym. Sci.* 60 (2022) 2607–2634.
7. Y.-H. Na, T. Kurokawa, Y. Katsuyama, H. Tsukeshiba, J.P. Gong, Y. Osada, S. Okabe, T. Karino, M. Shibayama, Structural characteristics of double network gels with extremely high mechanical strength, *Macromolecules.* 37 (2004) 5370–5374.
8. S. Xia, S. Song, G. Gao, Robust and flexible strain sensors based on dual physically cross-linked double network hydrogels for monitoring human-motion, *Chem. Eng. J.* 354 (2018) 817–824.
9. X. Huang, J. Li, J. Luo, Q. Gao, A. Mao, J. Li, Research progress on double-network hydrogels, *Mater. Today Commun.* 29 (2021) 102757.
10. X. Sun, F. Yao, J. Li, Nanocomposite hydrogel-based strain and pressure sensors: A review, *J. Mater. Chem. A.* 8 (2020) 18605–18623.
11. Y. Zhang, B. Liang, Q. Jiang, Y. Li, Y. Feng, L. Zhang, Y. Zhao, X. Xiong, Flexible and wearable sensor based on graphene nanocomposite hydrogels, *Smart Mater. Struct.* 29 (2020) 75027.
12. L. Wang, T. Xu, X. Zhang, Multifunctional conductive hydrogel-based flexible wearable sensors, *TrAC Trends Anal. Chem.* 134 (2021) 116130.
13. W. Zhang, P. Feng, J. Chen, B. Zhao, Flexible energy storage systems based on electrically conductive hydrogels, *Prog. Polym. Sci.* 88 (2019) 220–240.
14. S. Sekine, Y. Ido, T. Miyake, K. Nagamine, M. Nishizawa, Conducting polymer electrodes printed on hydrogel, *J. Am. Chem. Soc.* 132 (2010) 13174–13175.
15. H. Yuk, B. Lu, X. Zhao, Hydrogel bioelectronics, *Chem. Soc. Rev.* 48 (2019) 1642–1667.
16. C. Cui, Q. Fu, L. Meng, S. Hao, R. Dai, J. Yang, Recent progress in natural biopolymers conductive hydrogels for flexible wearable sensors and energy devices: Materials, structures, and performance, *ACS Appl. Bio Mater.* 4 (2020) 85–121.
17. Z. Wang, J. Chen, L. Wang, G. Gao, Y. Zhou, R. Wang, T. Xu, J. Yin, J. Fu, Flexible and wearable strain sensors based on tough and self-adhesive ion conducting hydrogels, *J. Mater. Chem. B.* 7 (2019) 24–29.
18. T. Wang, J. Wang, Z. Li, M. Yue, X. Qing, P. Zhang, X. Liao, Z. Fan, S. Yang, PVA/SA/MXene dual-network conductive hydrogel for wearable sensor to monitor human motions, *J. Appl. Polym. Sci.* 139 (2022) 51627.

19. G. Wu, M. Panahi-Sarmad, X. Xiao, F. Ding, K. Dong, X. Hou, Fabrication of capacitive pressure sensor with extraordinary sensitivity and wide sensing range using PAM/BIS/GO nanocomposite hydrogel and conductive fabric, *Compos. Part A Appl. Sci. Manuf.* 145 (2021) 106373.

20. Z. Li, Z. Lin, Recent advances in polysaccharide-based hydrogels for synthesis and applications, *Aggregate.* 2 (2021) e21.

21. P. Heidarian, H. Yousefi, A. Kaynak, M. Paulino, S. Gharaie, R.J. Varley, A.Z. Kouzani, Dynamic nanohybrid-polysaccharide hydrogels for soft wearable strain sensing, *Sensors.* 21 (2021) 3574.

22. R. Tong, G. Chen, J. Tian, M. He, Highly stretchable, strain-sensitive, and ionic-conductive cellulose-based hydrogels for wearable sensors, *Polymers (Basel).* 11 (2019) 2067.

23. L. Quan, J. Tie, Y. Wang, Z. Mao, L. Zhang, Y. Zhong, X. Sui, X. Feng, H. Xu, Mussel-inspired chitosan-based hydrogel sensor with pH-responsive and adjustable adhesion, toughness and self-healing capability, *Polym. Adv. Technol.* 33 (2022) 1867–1880.

24. S. Murugesan, T. Scheibel, Chitosan-based nanocomposites for medical applications, *J. Polym. Sci.* 59 (2021) 1610–1642.

25. Y. Tang, X. Zhang, X. Li, C. Ma, X. Chu, L. Wang, W. Xu, A review on recent advances of Protein-polymer hydrogels, *Eur. Polym. J.* 162 (2022) 110881.

26. N. Davari, N. Bakhtiary, M. Khajehmohammadi, S. Sarkari, H. Tolabi, F. Ghorbani, B. Ghalandari, Protein-based hydrogels: Promising materials for tissue engineering, *Polymers (Basel).* 14 (2022) 986.

27. M. Zhang, Q. Yang, T. Hu, L. Tang, Y. Ni, L. Chen, H. Wu, L. Huang, C. Ding, Adhesive, antibacterial, conductive, anti-UV, self-healing, and tough collagen-based hydrogels from a pyrogallol-Ag self-catalysis system, *ACS Appl. Mater. Interfaces.* 14 (2022) 8728–8742.

28. X. Wang, Z. Bai, M. Zheng, O. Yue, M. Hou, B. Cui, R. Su, C. Wei, X. Liu, Engineered gelatin-based conductive hydrogels for flexible wearable electronic devices: Fundamentals and recent advances, *J. Sci. Adv. Mater. Devices.* 7 (2022) 100451.

29. W. Ahn, J.-H. Lee, S.R. Kim, J. Lee, E.J. Lee, Designed protein-and peptide-based hydrogels for biomedical sciences, *J. Mater. Chem. B.* 9 (2021) 1919–1940.

30. C.M. Elvin, A.G. Carr, M.G. Huson, J.M. Maxwell, R.D. Pearson, T. Vuocolo, N.E. Liyou, D.C.C. Wong, D.J. Merritt, N.E. Dixon, Synthesis and properties of crosslinked recombinant pro-resilin, *Nature* 437(7061) (2005) 999–1002.

31. X. Hu, X.-X. Xia, S.-C. Huang, Z.-G. Qian, Development of adhesive and conductive resilin-based hydrogels for wearable sensors, *Biomacromolecules.* 20(9) (2019) 3283–3293.

32. Y. Han, L. Sun, C. Wen, Z. Wang, J. Dai, L. Shi, Flexible conductive silk-PPy hydrogel toward wearable electronic strain sensors, *Biomed. Mater.* 17 (2022) 024107.

33. F. He, X. You, H. Gong, Y. Yang, T. Bai, W. Wang, W. Guo, X. Liu, M. Ye, Stretchable, biocompatible, and multifunctional silk fibroin-based hydrogels toward wearable strain/pressure sensors and triboelectric nanogenerators, *ACS Appl. Mater. Interfaces.* 12 (2020) 6442–6450.

34. J. Seo, H. Kim, K. Kim, S.Q. Choi, H.J. Lee, Calcium-modified silk as a biocompatible and strong adhesive for epidermal electronics, *Adv. Funct. Mater.* 28 (2018) 1800802.

13

Multifunctional Flexible and Wearable Sensors

Naveen Kumar, Jyoti Rani, and Rajnish Kurchania

CONTENTS

13.1 Introduction

The recent progress in multifunctional flexible and wearable sensors has gained attention in the fields of healthcare, biomedical, robotics, and many others where high sensitivity, accuracy, flexibility, and low cost are required. Multifunctionality is an important aspect in the field of flexible, wearable, and stretchable electronics [1]. These multifunctional sensors require the integration of three important components: electrical conductivity, sensing performance, and heating management. Flexible and wearable sensors that can monitor human activities and health conditions have attracted significant attention [2]. Smartwatches and wristbands can track human physical activities and provide body health conditions such as heart rate, blood pressure, and calories burned are already available on the market. However, these commercially available devices are not fully considered flexible and wearable devices. Figure 13.1 illustrates the broad range of applications of multifunctional sensors [3].

Conventional sensors present today suffer from poor signal transduction, however, flexible sensors obtain data efficiently and convert it to high-quality signals. Although rigorous efforts have been made in the development and fabrication of multifunctional sensing sensors, the integration of such sensors into a stretchable form restricts the wearability of that sensor. To date, several reports have been reported regarding flexible sensors like

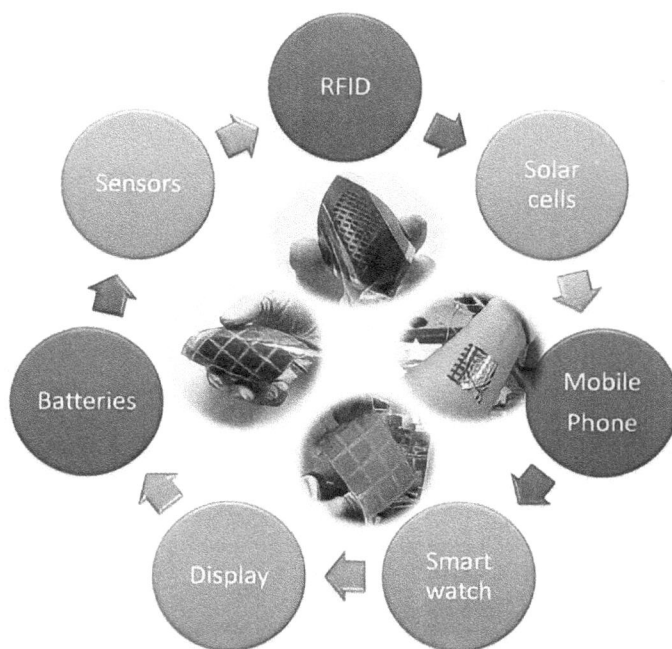

FIGURE 13.1
Schematics of the broad range of applications of multifunctional sensors. Adapted with permission [3]. Copyright (2017), Wiley-VCH.

strain, pressure, and temperature sensors using nanowires, nanotubes, nanosheets, carbon nanotubes (CNTs), and graphene. The fabrication of flexible sensors requires innovative methods in design, a selection of active materials, and a flexible substrate. A flexible electronic device generally comprises multiple components such as the substrate, the active layer, and the interface layer. The active layer is normally made up of inorganic nanomaterials synthesized by the physical transfer or solution process methods [4]. These inorganic nanomaterials possess superior electron-hole mobility, physicochemical properties, and mechanical strength. Despite the rapid growth in flexible and wearable technologies, and although some of the results are impressive, the crucial performing parameters or aspects of such sensors, like high sensitivity, are difficult to attain simultaneously with desired working range [5]. This is hindered because of specific structures such as 0D, 1D, and 2D structures of the functional materials. In the case of 0D materials like nanoparticles, their sensing range is limited because of their short aspect ratio caused by the separation of particles from each other during straining; 1D materials like nanowires and nanotubes have a high aspect ratio and a wide sensing range but their high aspect ratio makes them entangle with each other resulting in low sensitivity. Moreover, in 2D materials, large cracks are formed under the tensile stress that breaks the conductive pathway, restricting the sensitivity of the sensor. However, these 2D sensors can be used where a narrow sensing range is required [6].

The approach of integrating the multiple parameters of the sensor into a common substrate is a basic principle in developing and fabricating multifunctional flexible and wearable devices. These sensors can easily overcome existing limitations such as the detection of more than one parameter without interference simultaneously [6]. Although intensive research has been carried out in the field of multifunctional sensors, the union or integration of several parameters into a particular substrate leads to an increase in the size of the sensors affecting the wearability of the device. Considering this, an ultrathin, lightweight, high mechanical strength, flexible, wearable, and stretchable multifunctional sensor with a wide sensing range is a key strategy for fabricating multifunctional sensors. In this chapter, we will discuss multifunctional flexible and wearable sensors, their applications, and their challenges [7].

13.2 Various Multifunctional Flexible and Wearable Sensors

13.2.1 Multifunctional Temperature Sensors

Body temperature is an indicator of many warning signs, such as insomnia, fever, and hypothermia. Body temperature provides beneficial medical diagnostic information about the disease. Real-time and accurate determination of localized body temperature changes is the key parameter in monitoring the health of a patient. Traditional medical examination practice normally relies on the regular measurement of an individual's body temperature by thermometers. Typically, rigid temperature detectors, such as thermometers, are used to measure the temperature of the body by contacting the skin [8, 9]. However, measuring the temperature in covered conditions, such as inside the clothing or prosthetic area, is the main problem involved in the use of a rigid sensor. Moreover, this methodology is also not reliable for children, as the placement of thermometers is not very promising because direct human body and thermometer contact is preferred for high-precision temperature measurements. Moreover, traditional rigid temperature detectors are unable to provide a comfortable touching base with irregular surfaces. This factor plays an important role in the rise of flexible and wearable temperature sensors [10–12]. Soft, flexible, biocompatible, lightweight, durable, and non-irritating temperature sensors are key to considering wearable requirements. Physical stimuli that are associated with temperature such as resistance, volumetric expansion, and vapor pressure are used for the development of temperature measurement systems. Resistance temperature detectors (RTDs), thermocouples, and thermistors are the most common type of temperature sensors. Figure 13.2 shows the schematics of different temperature sensors [13].

13.2.1.1 Resistance Temperature Detectors

RTDs use electrical resistance as a function of temperature. The electron vibrations at high temperatures obstruct the free flow of the electron in the conductive materials. This obstruction in the flow of the electron results in increments of electrical resistance of the material. The higher the temperature, the higher the resistance [2]. These RTDs have a high degree of accuracy and the response time is also very low. Moreover, RTDs can also be fabricated by putting a thin layer or thin film of conductive materials over the substrate. The temperature sensitivity or thermal response of an RTD can be defined by the temperature coefficient of resistance (TCR) defined by Eq. (1).

$$TCR = \frac{R(T) - R(To)}{R(To)(T - To)} \tag{1}$$

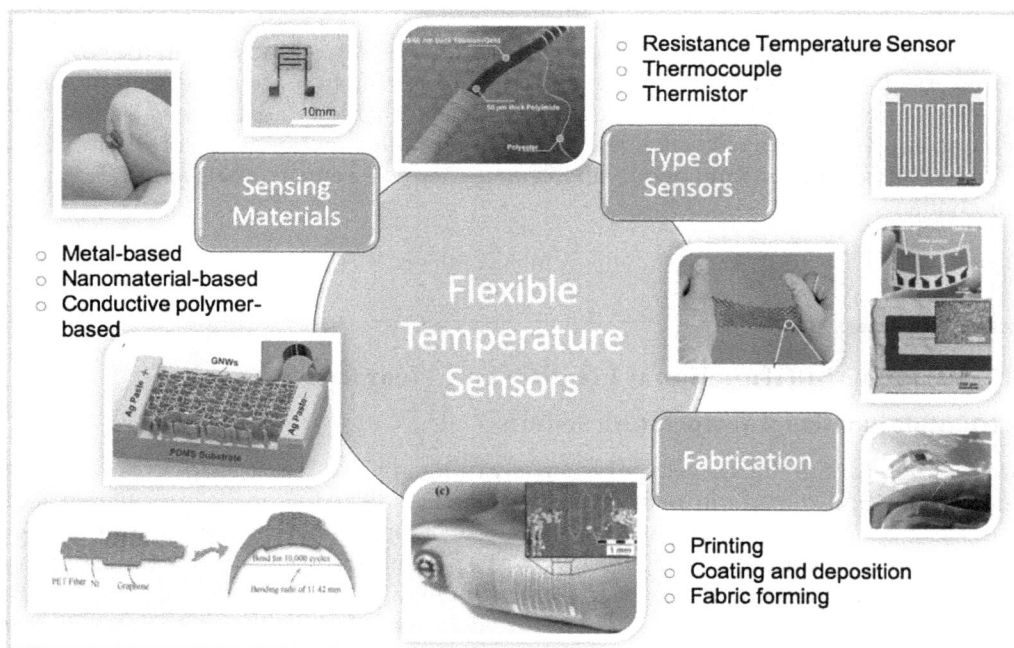

FIGURE 13.2
Schematics of different types of temperature sensors. Adapted with permission [13]. Copyright (2020), Elsevier.

Where $R(T)$ is the resistance measured at temperature T and $R(T_o)$ is the initial resistance measured at temperature T_o. A high value of TCR indicates high precision in temperature measurement. Depending on the resistive wire arrangements, RTDs can be fabricated in two forms: thin film and wire wound. For thin film RTDs, a thin film of active or sensing material is deposited on a substrate, whereas in wire wound RTDs, a fine conducting wire is draped over the ceramic rod, which is further concealed with an insulating material [13].

13.2.1.2 Thermocouple

A thermocouple is a kind of temperature sensor that is fabricated by combining two different thermoelements. It is based on thermoelectric phenomena also called the Seebeck effect. In the Seebeck effect, thermal voltage or an electromotive force (emf) is developed across the terminals when heat is applied to the different terminals of the conductor. The application of heat causes the movement of electrons to form the hot terminal to the cold terminal, causing a force in the reverse direction of the electron movement. This force is called Seebeck's electromotive force and the phenomenon is called Seebeck's thermoelectric effect. The thermal voltage is calculated by using Eq. 2

$$V = \propto \times \Delta T \tag{2}$$

Where $\propto$ is the difference between the Seebeck coefficient of two different thermoelectric elements and ΔT is the temperature difference between two ends [13–16].

13.2.1.3 Thermistor

A thermistor also called a resistance thermometer and is an instrument to measure temperature. It is a temperature-dependent resistor made of metal oxides and pressed into desired shapes like discs or cylinders. The temperature is measured with respect to the change in resistance of the resistor as the thermistor starts self-heating its element when the ambient temperature is changed. These are thermally sensitive resistors whose temperature coefficient can be negative as well as positive. The temperature measurement of the thermistor is determined by measuring the thermal index (β) as in Eq. 3.

$$\beta = \frac{\ln\left(\frac{R(To)}{R(T)}\right)}{\frac{1}{To} - \frac{1}{T}} \tag{3}$$

Where, $R(T)$ is the resistance measured at temperature T and $R(T_o)$ is the initial resistance measured at temperature T_o [13, 17–19].

13.2.2 Pressure or Strain Sensors

In recent years, pressure sensors, or strain sensors, have gained attention for wearable and flexible devices specifically for skin-stimulated electronic devices. Possessing awesome properties of flexibility, biocompatibility, and durability in addition to being lightweight, pressure and strain sensors can be easily stuck to the human skin for the real-time and immediate tracking of physiological conditions of health, inclusive of coronary heart and respiration rates. To comprehend the high overall performance of skin-stimulated sensors, a range of touchable sensor structures has been suggested via rational engineering consisting of nanomaterials and hybrid micro- or nanostructures primarily based on powerful transduction mechanisms through changing outside stimuli into electric signals. These transduction techniques commonly encompass piezoresistivity, piezoelectricity, and capacitance. Among them, piezoresistivity is taken into consideration because of its multifunctionality and high-performing capacity. Piezoelectric materials show a property where they can be polarized by the application of mechanical force or they show deformation or strain when an electric field is applied across the material. Piezoelectric materials fall under the category of dielectric materials that may be polarized via the utility of outside stimuli, that is, mechanical, electrical, and thermal. Piezoelectric materials have a subcategory of "pyroelectrics" and "ferroelectrics". Pyroelectrics show an electric polarization that may be managed by the application of a thermal field or heat energy and vice-versa, whereas ferroelectrics can be electrically polarized when subjected to an electric field and vice-versa. All ferroelectric materials can be considered both pyroelectric and piezoelectric and all pyroelectrics are piezoelectric materials. However, all piezoelectrics cannot be considered pyroelectric as quartz can only be polarized by applying mechanical stress. Therefore, it can be said that the maximum of the piezoelectrics is intrinsic multifunctional materials that can impart a minimum of capability for a multifunctional flexible and wearable device [20–24].

13.2.3 Chemical Sensors

Flexible chemical sensors can track human fitness through the fast detection of biomarkers from the human body. Chemical flexible sensors are different from many sensors that aim to monitor the human body's health, fitness level, and physical activities. Chemical sensors offer a greater reliable way for non-invasive fitness tracking on the molecular level. In recent years, many body fluids such as saliva, blood, sweat, and so on have been used for healthcare diagnostics with the aid of flexible wearable chemical sensors [25–27]. For example, a bio-interfaced sensor fabricated of hybrid graphene/electrode/silk shape can be used in chemical and organic sensing. A flexible chemical sensor is capable of apprehending H. pylori cells in human saliva. They can detect up to 100 cells when the sensors are placed onto the enamel. Regarding breath analyses, volatile organic compounds (VOCs) are contained in exhaled breath, which gives a powerful direction for security screening. Additionally, blood diagnostics may be carried out with the aid of using implantable sensors, however, it is harder to recognize non-invasive size primarily based on wearable systems [28, 29]. However, sweat-based sensors have an exceptional advantage over different body bio-fluid sensing due to their continuous tracking capability of human health as they analyze different components like glucose, sodium, and potassium present in sweat. The timely and continuous analysis of sweat guarantees early detection and diagnosis of related diseases. Various wearable systems have been established such as patches and wristbands for effective sweat sampling and sensing. Among all wearable sweat sensors, microfluidic-based sweat sensors are one of the most promising techniques for shorter sweat sampling and sensing times. Koh et al. developed a stretchable microfluidic device with the functionality to be intimately and robustly bonded to human skin. The sweat can be harvested from pores present in the skin, which is further introduced to distinctive networks for the analysis and monitoring of pH, chloride, and glucose concentrations [30]. However, some wearable sweat sensors cannot extract enough sweat from the skin, stopping their application. An electrochemically improved iontophoresis interface may overcome this difficulty through incorporation into a wearable sweat-sensing machine as it can locally stimulate sweating. This process can allow health monitoring even in low sweat-producing environments. Although advances have been made in flexible wearable sweat sensors, only small molecules can be extracted, whereas the extraction of large biomolecules such as proteins, DNA, RNA, and so on from sweat has not been clearly understood [31–33].

13.3 Fabrication of Multifunctional Flexible and Wearable Sensors

The development of flexible and wearable sensors needs improvements in each material technology and fabrication strategy. Generally, the fabrication of traditional sensors is completed by photolithography. However, this strategy is not properly compatible with flexible substrates, making it unsuitable for the fabrication of flexible and wearable devices. Therefore, distinct fabrication strategies are needed for the manufacturing of flexible sensing platforms such as printed electronics, electrospinning, and additive manufacturing. Among such strategies, functional printing strategies are the most desirable fabrication method. Printing techniques require conductive ink, which normally incorporates a liquid suspension of metal or inorganic materials [34]. This particular composition lets in the uniform deposition of the conductive ink at the preferred polymeric substrates and the next curing of the ink-coated substrates at excessive temperatures. The requirement of high sintering temperature in printing technologies restricts the application of this method to a small variety of thermoplastics such as polyimide, moreover, polyimide is an expensive polymer which further makes the fabrication costlier. However, improvements in materials synthesis and fabrication have eased the requirement for higher sintering temperatures [35]. This enables the use of inexpensive thermoplastic alternatives like polyethylene naphthalate and decreases the fabrication cost as well as the product cost. Conductive ink printing can be done using many different approaches. These strategies include display screen printing, gravure printing, roll-to-roll printing, and inkjet printing. These strategies generally consist of display screen masks, nozzles, or patterned drums for the deposition of the conductive ink on the desired substrate. Moreover, large-scale roll-to-roll printing provides an opportunity to manufacture many multiarray microsensors over a large substrate sheet [36]. However, despite their many advantages, the use of thermoplastics is restricted to their stretchable property because of their incompatibility with three-dimensional contours. However, tender elastomeric substrates, which have comparable similarities to skin, are not restricted by those issues. The fabrication of those tender elastomeric substrates entails a resin aggregate followed by a curing procedure [37].

Adjustments have been made considering these fabrication strategies, such as material selection and synthesis simultaneously with development in design. Moreover, weaving is also an old fabrication procedure that requires the integration of conductive fibers with a substrate. These fibers may be woven, knitted, or sewn into textiles to create consumable clothing. They can maintain excessive stress without mechanical breakage. Currently, conductive textiles are found in apparel such as gloves, shirts, and socks. These can show the opportunity of integrating flexible sensors into fashionwear for numerous healthcare and clinical engineering applications [1, 38]. Figure 13.3 illustrates the schematics of the fabrication technologies [39].

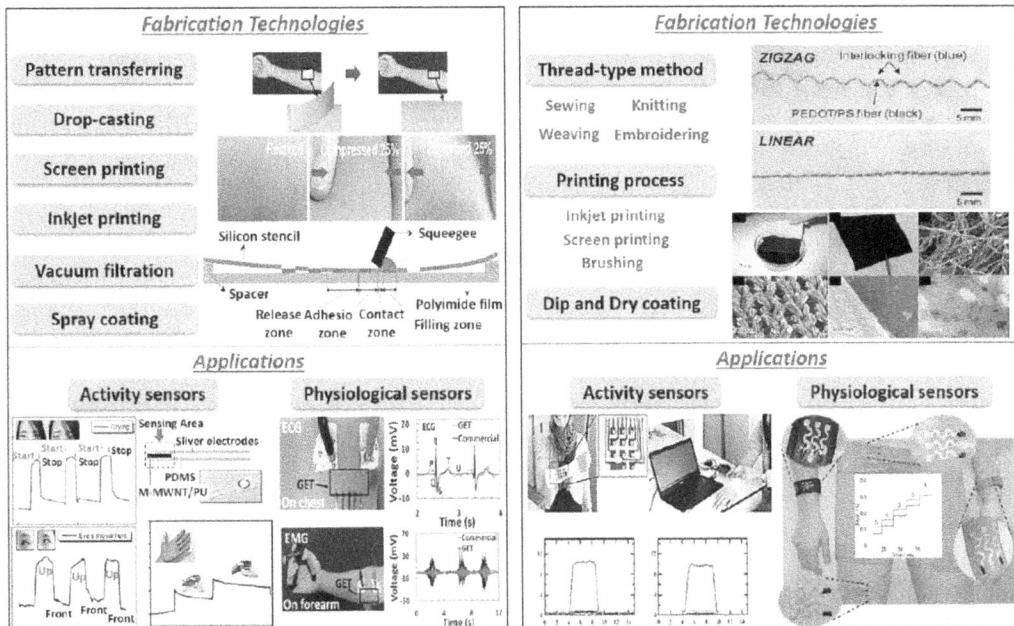

FIGURE 13.3
Schematics of the fabrication technologies for multifunctional sensors. Adapted with permission [40]. Copyright (2020), MDPI.

13.4 Applications of Multifunctional Flexible and Wearable Sensors

The rise of human-machine interaction has led to the emergence of new materials and devices, particularly sensing devices. These sensors find their application in touch-sensing devices, monitoring human health and activities, temperature-sensing devices, flexible LEDs, and many more [40].

13.4.1 Applications in Human Health

The improvements in wearable and flexible multifunctional sensors have attracted enormous attention. Particularly in the field of smart biomedical and healthcare, multifunctional sensors can be an excellent chance for curing and preventing diseases. The real-time monitoring of an individual's physiological status can assist clinical personnel in the early prediction and diagnosis of heart ailments or atypical body conditions. These multifunctional sensors can monitor physiological parameters, such as pulse rate and body temperature, and can also be used as an electrocardiogram (ECG). As a core factor of wearable devices, flexible and stretchable multifunctional sensors have an important role in the health monitoring system and the human-machine interface. Although extensive research efforts have been accomplished in the development of multifunctional sensors, the difficulty of seamless and effortless incorporation of such sensors into a flexible platform prevents the wearability and selectivity of some stimuli. Moreover, developing a frequent functional material that is responsive to many stimuli is a major limiting factor for multisensing wearable electronics. Considerable efforts have been committed to the production and development of smart wearable electronics and interesting improvements have been made in novel materials, manufacturing, and sensing mechanisms in the last few years [40].

Figure 13.4 shows the working principle of flexible sensors in human health monitoring. Silver chloride (AgCl) gel electrodes are predominantly used in a scientific setting for obtaining skin biopotentials, however, signal sensing capability in the long term is degraded due to the volatile nature of the liquid in the gel electrolyte, motives pores, and skin rashes. Metal films of gold (Au) possess low contact impedance that can result in the easy observation of high noise on biopotential signals caused by curved surfaces and/ or due to body movements. Due to the easy handling and biocompatibility of poly(3,4-ethylenedioxythiophene) polystyrene sulfonate (PEDOT: PSS) electrodes, they have been actively studied in recent years in bioelectronics. However, because of the limitations of the substrate materials in terms of flexibility, greater contact impedances restrict its sensing ability. Therefore, selecting a suitable functional material, design engineering, and an environmentally friendly fabrication technique is critical for performance improvements of the emerging type of flexible and wearable electronics. The composite of laser-induced graphene (LIG) and polymerized MXene-$Ti_3C_2T_x$ with 3,4-ethylenedioxythiophene (EDOT) fabricated by Zhang et al. showed higher pressure sensitivity, a wider operating range, and long-term durability and stability. Deformations to larger body joint bending motions caused by the arterial pulse and the vibration of Adam's apples can be easily measured [6]. The high TCR of LIG/MXene-$Ti_3C_2T_x$@EDOT was once attributed to the aid of charge carriers with temperature loading and hopping transport. The material possesses excessive conductivity, mechanical flexibility, and biocompatibility. These properties favor its suitability for epidermal biopotential measurement. Overall, the advanced multifunctional sensors based on the LIG/MXene-$Ti_3C_2T_x$@EDOT composite have laid the foundation for a

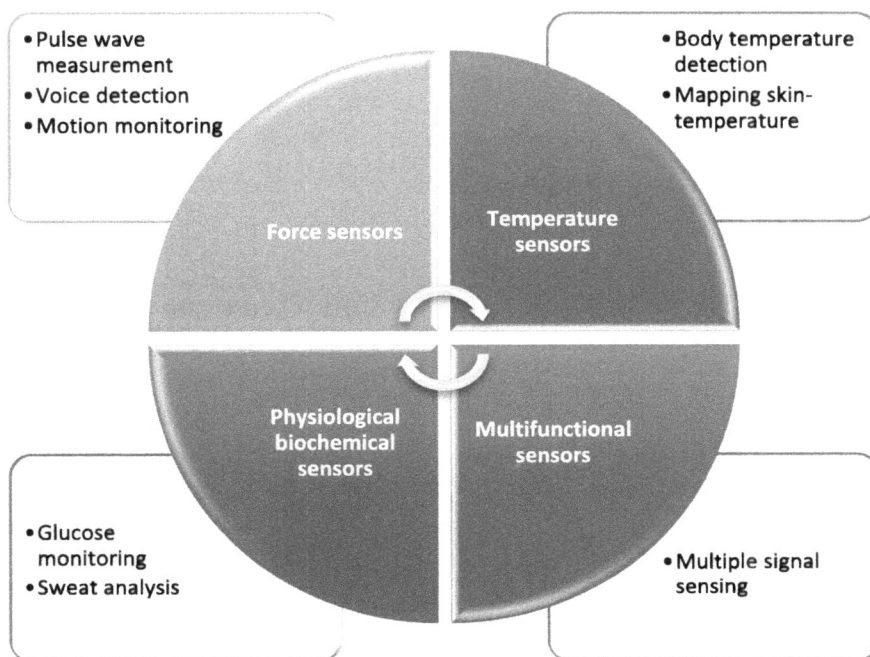

FIGURE 13.4
Working principle of flexible sensors in human health monitoring. Adapted with permission [40]. Copyright (2019), Springer.

promising route of development in forthcoming wearable devices by the incorporation of artificial intelligence and the complete tracking capability of instantaneous physiological observing indicators for e-skin and smart healthcare applications [40].

13.4.2 Artificial Electronic Skins

Biological skin-based sensory receptors (like mechanics and thermoreceptors) collect and transmit streams of physical variables from the exterior environment to the body. Despite significant developments in mechanical and thermal sensations, the replication of biological skin sensory capabilities in artificial skins and prosthetics is still an elusive goal. Consequently, as a functional substitute for serving as an alternative for natural limbs, prosthetics and synthetic skins are regularly worn merely as supplementary motion aids or for cosmetic utility. Recent changes in the layout of sensor-laden prosthetics and synthetic skins for the integration with rigid and semi-flexible sensing platforms have delivered promising alternatives with restricted spatiotemporal resolution, stretchability, and conformability. However, extreme mechanical incompatibility between soft biological tissues and the electronic parts of biomedical devices further obstructs the utility and performance of these systems. In recent years, stretchable e-skins with high sensitivity have gathered significant attention due to the fact that they are capable of matching human pore and skin functionality in detecting understated changes in external stimuli, such as stress, strain, shear, temperature, vibration, and pain, and converting these to electronic signals. They are also conformable and curved to complicated surfaces. These advances make flexible e-skin feasible. Until now, composite films of CNTs and polydimethylsiloxane (PDMS)

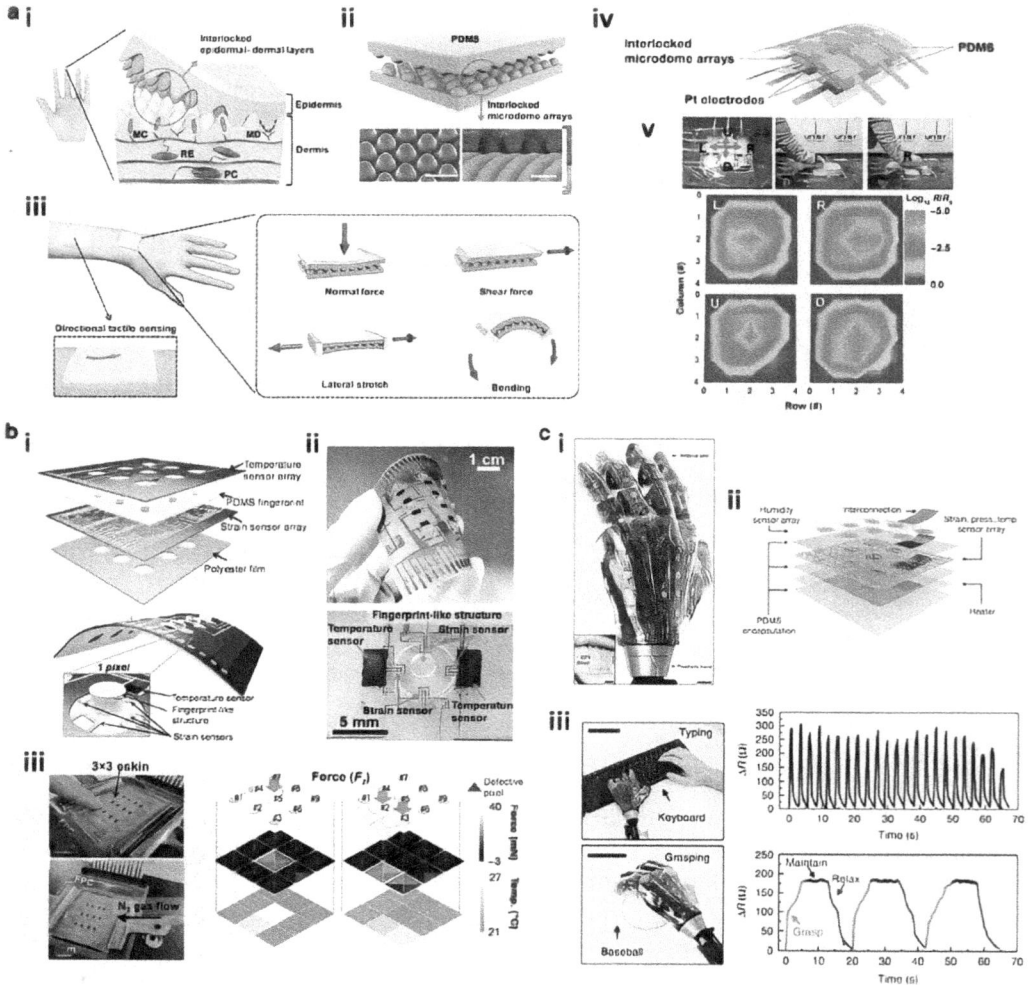

FIGURE 13.5
Performances of the CNT-PDMS e-skin. Adapted with permission [40]. Copyright (2016), Springer Nature.

are the artificial e-skins that are closest to the human skin. Figure 13.5 shows the performance of the CNT-PDMS e-skin [41].

13.4.3 Opto-Electronic Applications

Flexible photodetectors have proved to be essential in the application prospects of wearable devices. Flexible photodetectors can replicate and simulate the function of real human eyes and notice light signal information. The device can have many applications including as a wi-fi signal receiver to attain information for unique functions. Conventional photodetectors regularly use semiconductor materials such as silicon and germanium to discover optical signals. However, the hard and brittle nature of these semiconductors makes them tough to build with wearable devices. Graphene has exquisite flexibility and optoelectronic properties which supports its application potential in flexible photodetectors.

Moreover, graphene has a wide light-absorbing band and it also has high carrier mobility. However, the photo-generated carrier lifetime of graphene is brief and the dark current is also high, which will restrict its photoelectric performance as a photodetector. Schneider et al. fabricated a photodetector using graphene and amorphous silicon [41]. Compared with regular devices (coated via aluminum-doped zinc oxide (ZnO:Al) electrodes), the photodetector fabricated with graphene-amorphous silicon acquires 440% enhancement of photo-response in the 320 nm ultraviolet (UV) region. Due to the precise gasoline impermeability of graphene, it can properly prevent the oxygen desorption of AuO_x, thereby achieving the top steadiness of AuO_x. In this structure, the warm electrons generated by using graphene are separated at the AuO_x/graphene heterojunction, which results in the high responsivity of 3100 A/W, 58 A/W, and 9 A/W at 310 nm, 500 nm, and 1550 nm, respectively, below a very low utilized bias of 0.1 V of the photodetector. Moreover, the reduced graphene oxide (rGO) can additionally be used to fabricate flexible photodetectors. Chang et al. stated that rGO can exhibit the infrared light responsivity of ~0.7 A/W with the help of oxygen defect engineering [42]. Accordingly, sensors and positioners can be extended as unique applications of flexible photodetectors. Kang et al. reported a crumpled graphene-based stretchable photodetector [43]. This crumpled graphene-based photodetector exhibited improvement in photo responsivity of up to 400% compared with the flat one. Furthermore, the stretchable structure of graphene/gold can be built-in over the contact lens surface, exhibiting an excessive signal-to-noise ratio beneath laser irradiation, which is anticipated to be used in wearable optical detection devices. Moreover, graphene-based heterostructures can additionally be constructed as photograph sensors and memories. For example, Jang et al. presented a photodetector comprised of a non-volatile reminiscence function, which is built of three materials: organic semiconductors, gold nanoparticles (AuNPs), and graphene. This structure features a light-sensitive material such as an organic semiconductor [44]. The graphene present plays the role of absorbing photo-generated holes, and the AuNPs create plasmonic enhancement effects by enticing the carriers. The photocurrent can be retained under exceptional mild intensity with pulse light. Moreover, the related structural structures can also recognize picture sensing and memory functions and are predicted to improve the processing effectivity and image recognition rate [45].

13.5 Challenges and Future Perspectives

The rapid improvement in material science and engineering technology has resulted in the development and improvement of flexible and wearable devices. Despite the development of wearable sensing devices, they still face significant demanding situations that restrict their proper application. The biocompatibility of active materials to human bodies or their organs, mainly for invasive applications, is still a major concern. Integrating sensors into wearable structures and the cross-sensitivity of various sensors are major challenges in flexible devices. Multifunctional sensors should be able to sense and detect different parameters such as pressure, stress, temperature, humidity, and gas atmosphere simultaneously and avoid any crosstalk or interference between them. One sensor communicates with a couple of stimuli, making it hard to discover the precise form of stimulus and the depth of every stimulation. Therefore, it is vital to have a look at new sensing mechanisms with low cross-sensitivity and strong integration. In the case of temperature sensors, it is harder to gain excessive stretchability, sensitivity, and strain adaptability concurrently. Improving the overall sensing performance and putting off the effect of the elastic deformation of the sensor on temperature detection remain major research topics. Wearable sensors nonetheless still require development before they can be applied for practical applications such as monitoring human activities.

Researchers are consequently constantly searching for higher sensitive, selective detection techniques, measuring concepts, and new analytical methodologies to expand present-day sensing devices and instruments. Improvements in sensing devices' overall performance can be executed via the exploitation of 0D, 1D, and 2D materials collectively with flexible and bio-stimulated concepts. Biomaterials can be used for their healing and therapeutic applications as they have proved their usability in implantable healthcare devices. Moreover, they are biocompatible and biodegradable. They are much more suitable for primarily skin-based wearable devices. The substitution of a regularly used polymeric elastomer with greater biocompatible biomaterial-based substrates, such as protein films, can enhance the wearability of flexible devices. Moreover, multifunctional sensors also have abilities such as electricity harvesting, self-recovery, and stimuli-responsivity, or multistimuli-responsivity. This will have a probable effect on decreasing the size, cost, and complexity of fabricating flexible devices. Considering the distinguished purpose of wearable sensors for human movement tracking, researchers should be cognizant of leveraging low-cost, biocompatible, and environmentally pleasant materials with feasible fabrication methods.

13.6 Conclusions

This chapter contains an overview of multifunctional flexible and wearable devices and their applications. They have gained attention in the fields of healthcare, biomedicine, robotics, and many others in recent years. These multifunctional sensors require the integration of three important parameters: electrical conductivity, sensing performance, and heating management. These sensors are based on self-healing materials and self-powered devices. Skin-inspired flexible wearable electronic devices have a high potential in the coming generation of smart, compact, and portable electronic devices. These sensors demand a high degree of sensitivity, accuracy, precision, reproducibility, mechanical flexibility, and low cost. Carbon-based materials such as 1D and 2D carbon nanotubes and graphene, silver fractal dendrites, and hydrogel-based materials are commonly used as active materials for the fabrication of flexible sensors. To date, there have been several reports regarding flexible sensors, such as strain and temperature sensors, using nanowires, nanotubes, nanosheets, CNTs, and graphene. Fabrication of flexible sensors requires innovative methods in design, selection of active materials, and a flexible substrate. A flexible electronic device generally comprises several components such as a substrate, an active layer, and an interface layer. The approach of integrating the multiple parameters of the sensor into a common substrate is a basic principle in developing and fabricating multifunctional flexible and wearable devices. Printed electronics, electrospinning, and additive manufacturing are fabrication strategies for manufacturing flexible and wearable devices.

Wearable and flexible multifunctional sensors in biomedical and healthcare applications can help in monitoring the instantaneous physiological status of an individual, such as pulse rate and body temperature, and can be used as an ECG, as well as assisting personnel in the early detection and diagnosis of any atypical body conditions. Overall, advanced multifunctional sensors based on LIG/ MXene-Ti$_3$C$_2$T$_x$@EDOT composite laid the foundations for the development of forthcoming wearable devices with the incorporation of artificial intelligence and the complete tracking capability of instantaneous physiological observing indicators for e-skin and smart healthcare applications. Composite films of carbon nanotubes and polydimethylsiloxane are suitable for artificial e-skins. The rapid improvement in material science and engineering technology has resulted in the development and improvement of flexible and wearable devices. However, they still require development before they can be applied in practical applications such as monitoring human activities.

References

1. S. Patra, R. Choudhary, R. Madhuri, P. K. Sharma, "Graphene-based portable, flexible, and wearable sensing platforms: An emerging trend for health care and biomedical surveillance," *Graphene Bioelectronics* 2018, 2018, 307–338.
2. B. Tian, L. Qun, L. Chengsheng, F. Yu, W. Wei, "Multifunctional ultrastretchable printed soft electronic devices for wearable applications," *Advanced Electronic Materials* 6, 2, 2020, 1900922.
3. S. Han, H. Peng, Q. Sun, S. Venkatesh, K. Chung, S. Chuen Lau, Y. Zhou, V. A. L. Roy, "An overview of the development of flexible sensors," *Advanced Materials* 29, 33, 2017, 1700375.
4. C. Tong, *Advanced Materials for Printed Flexible Electronics*, Springer Science and Business Media LLC, 2022.
5. W. Gao, O. Hiroki, K. Daisuke, T. Kuniharu, J. Ali, "Flexible electronics toward wearable sensing," *Accounts of Chemical Research* 52, 3, 2019, 523–533.
6. S. Zhang, A. Chhetry, Md. Zahed, S. Sharma, C. Park, S. Yoon, J. Y. Park, "On-skin ultrathin and stretchable multifunctional sensor for smart healthcare wearables," *NPJ Flexible Electronics* 6, 1, 2022, 1–12.
7. H. Joh, S. Lee, M. Seong, W. Lee, S. Oh, "Engineering the charge transport of Ag nanocrystals for highly accurate, wearable temperature sensors through all-solution processes," *Small* 13, 24, 2017, 1700247.
8. Q. Li, L. Zhang, X. Tao, X. Ding, "Review of flexible temperature sensing networks for wearable physiological monitoring," *Advanced Healthcare Materials* 6, 12, 2017, 1601371.
9. P. Descent, R. Izquierdo, C. Fayomi, "Printing of temperature and humidity sensors on flexible substrates for biomedical applications," in *2018 IEEE International Symposium on Circuits and Systems (ISCAS)*, 2018, 1–4.
10. E. F. J. Ring, "Progress in the measurement of human body temperature," *IEEE Engineering in Medicine and Biology Magazine* 17, 4, 1998, 19–24.
11. S. Bielska, M. Sibinski, A. Lukasik, "Polymer temperature sensor for textronic applications," *Materials Science and Engineering: B* 165, 1–2, 2009, 50–52.
12. N. J. Blasdel, E. K. Wujcik, J. E. Carletta, K. Lee, C. N. Monty, "Fabric nanocomposite resistance temperature detector," *IEEE Sensors Journal* 15, 1, 2014, 300–306.
13. B. A. Kuzubasoglu, S. K. Bahadir, "Flexible temperature sensors: A review," *Sensors and Actuators A: Physical* 315, 2020, 112282.
14. R. N. Peter, C. J. R. Greenwood, C. A. Long, "Review of temperature measurement," *Review of Scientific Instruments* 71, 8, 2000, 2959–2978.
15. A. Graf, M. Arndt, G. Gerlach, "Seebeck's effect in micromachined thermopiles for infrared detection. A review," *Estonian Journal of Engineering* 13, 4, 2007, 338–353.
16. A. Davidson, A. Buis, I. Glesk, "Toward novel wearable pyroelectric temperature sensor for medical applications," *IEEE Sensors Journal* 17, 20, 2017, 6682–6689.

17. R. E. Tressler, G. L. Messing, C. G. Pantano, R. E. Newnham, *Tailoring Multiphase and Composite Ceramics*, Springer Science & Business Media, 20, 2012.
18. A. Tong, "Improving the accuracy of temperature measurements," *Sensor Review* 21, 3, 2001, 193–198.
19. J. P. Bentley, *Principles of Measurement Systems*, Pearson Education, 2005.
20. Y. Wu, Y. Ma, H. Zheng, S. Ramakrishna, "Piezoelectric materials for flexible and wearable electronics: A review," *Materials & Design* 211, 2021, 110164.
21. Y. Ding, T. Xu, O. Onyilagha, H. Fong, Z. Zhu, "Recent advances in flexible and wearable pressure sensors based on piezoresistive 3D monolithic conductive sponges," *ACS Applied Materials & Interfaces* 11, 7, 2019, 6685–6704.
22. Y. Zhang, H. Kim, Q. Wang, W. Jo, A. I. Kingon, S. Kim, C. Jeong, "Progress in lead-free piezoelectric nanofiller materials and related composite nanogenerator devices," *Nanoscale Advances*, 2, 8, 2020, 3131–3149.
23. B. An, H. Cho, Y. Kim, "Fabrication of planar and curved polyimide membranes with a pattern transfer method using ZnO nanowire arrays as templates," *Materials Letters*, 149, 2015, 109–112.
24. Z. A. Alhasssan, Y. S. Burezq, R. Nair, N. Shehata, "Polyvinylidene difluoride piezoelectric electrospun nanofibers: Review in synthesis, fabrication, characterizations, and applications," *Journal of Nanomaterials*, 2018, 2018, 1–12.
25. A. J. Bandodkar, D. Molinnus, O. Mirza, T. Guinovart, J. R. Windmiller, G. Valdés-Ramírez, F. J. Andrade, M. J. Schöning, J. Wang, "Epidermal tattoo potentiometric sodium sensors with wireless signal transduction for continuous non-invasive sweat monitoring," *Biosensors and Bioelectronics*, 54, 2014, 603–609.
26. M. Caldara, C. Colleoni, E. Guido, V. Re, G. Rosace, "Optical monitoring of sweat pH by a textile fabric wearable sensor based on covalently bonded litmus-3-glycidoxypropyltrimethoxysilane coating," *Sensors and Actuators B: Chemical*, 222, 2016, 213–220.
27. M. Parrilla, J. Ferré, T. Guinovart, F. J. Andrade, "Wearable potentiometric sensors based on commercial carbon fibres for monitoring sodium in sweat," *Electroanalysis* 28, 6, 2016, 1267–1275.
28. W. Jia, A. J. Bandodkar, G. Valdés-Ramírez, J. R. Windmiller, Z. Yang, J. Ramírez, G Chan, J. Wang, "Electrochemical tattoo biosensors for real-time noninvasive lactate monitoring in human perspiration," *Analytical Chemistry* 85, 14, 2013, 6553–6560.
29. J. Kim, S. Imani, W. R. de Araujo, J. Warchall, G. Valdés-Ramírez, T. R. L. C. Paixão, P. P. Mercier, J. Wang, "Wearable salivary uric acid mouthguard biosensor with integrated wireless electronics," *Biosensors and Bioelectronics*, 74, 2015, 1061–1068.
30. A. Koh, D. Kang, Y. Xue, S. Lee, R. M. Pielak, J. Kim, T. Hwang, S. Min, A. Banks, P. Bastien, M. C. Manco, "A soft, wearable microfluidic device for the capture, storage, and colorimetric sensing of sweat," *Science Translational Medicine*, 8, 366, 2016, 366ra165.
31. C. Chen, Q. Xie, D. Yang, H. Xiao, Y. Fu, Y. Tan, S. Yao, "Recent advances in electrochemical glucose biosensors: A review," *Rsc Advances* 3, 14, 2013, 4473–4491.
32. S. T. Gaylord, T. L. Dinh, E. R. Goldman, G. P. Anderson, K. C. Ngan, D. R. Walt, "Ultrasensitive detection of ricin toxin in multiple sample matrixes using single-domain antibodies," *Analytical Chemistry* 87, 13, 2015, 6570–6577.
33. A. J. Bandodkar, I. Jeerapan, J. Wang, "Wearable chemical sensors: Present challenges and future prospects," *ACS Sens*, 1, 2016, 464–482.
34. S. Khan, L. Lorenzelli, R. S. Dahiya, "Technologies for printing sensors and electronics over large flexible substrates: A review," *IEEE Sensors Journal* 15, 6, 2014, 3164–3185.
35. C. L. Bauer, and R. J. Farris, "Determination of poisson's ratio for polyimide films," *Polymer Engineering & Science* 29, 16, 1989, 1107–1110.
36. C. Yeom, K. Chen, D. Kiriya, Z. Yu, G. Cho, A. Javey, "Large-area compliant tactile sensors using printed carbon nanotube active-matrix backplanes," *Advanced Materials* 27, 9, 2015, 1561–1566.
37. M. Melzer, D. Karnaushenko, G. Lin, S. Baunack, D. Makarov, O. G. Schmidt, "Direct transfer of magnetic sensor devices to elastomeric supports for stretchable electronics," *Advanced Materials* 27, 8, 2015, 1333–1338.

38. G. H. Büscher, R. Kõiva, C. Schürmann, R. Haschke, H. J. Ritter, "Flexible and stretchable fabric-based tactile sensor," *Robotics and Autonomous Systems*, 63, 2015, 244–252.
39. J. Heo, Md. F. Hossain, I. Kim, "Challenges in design and fabrication of flexible/stretchable carbon-and textile-based wearable sensors for health monitoring: A critical review," *Sensors* 20, 14, 2020, 3927.
40. Y. Gu, T. Zhang, H. Chen, F. Wang, Y. Pu, C. Gao, S. Li, "Mini review on flexible and wearable electronics for monitoring human health information," *Nanoscale Research Letters* 14, 1, 2019, 1–15.
41. D. S. Schneider, A. Bablich, M. C. Lemme, "Flexible hybrid graphene/a-Si: H multispectral photodetectors," *Nanoscale* 9, 25, 2017, 8573–8579.
42. H. Chang, Z. Sun, Q. Yuan, F. Ding, X. Tao, F. Yan, Z. Zheng, "Thin film field-effect phototransistors from bandgap-tunable, solution-processed," *Advanced Materials* 22, 43, 2010, 4872–4876.
43. P. Kang, M. C. Wang, P. M. Knapp, S. Nam, "Crumpled graphene photodetector with enhanced, strain-tunable, and wavelength-selective photoresponsivity," *Advanced Materials* 28, 23, 2016, 4639–4645.
44. S. Jang, E. Hwang, Y. Lee S. Lee J. H. Cho, "Multifunctional graphene optoelectronic devices capable of detecting and storing photonic signals," *Nano letters* 15, 4, 2015, 2542–2547.
45. Y. J. Chuan, C. T. Lim, "Emerging flexible and wearable physical sensing platforms for healthcare and biomedical applications," *Microsystems & Nanoengineering* 2, 1, 2016, 1–19.

14

Nanocomposite-Based Flexible and Wearable Sensors

Soheil Jalali, Sadaf Mehrasa, Shadi Fathi, and Fahimeh Hooriabad Saboor

CONTENTS

14.1 Introduction

Throughout history, various methods have been used to diagnose health, physical, and environmental conditions. Many of these diagnostic methods are complex and in some cases use specialized devices. These methods require time and money to check health-related parameters, which, in most cases, make people suffer [1]. One of the newest areas of medical research is wearable sensors. Wearable skin-like devices have recently attracted the attention of many researchers and have presented significant capabilities in a wide range of functions for human-interactive approaches. The reason is that these devices can supply the basis for the next era of artificial intelligence [2,3].

Flexible wearable sensors are one of the developing areas in detecting physiological and biochemical parameters, which can achieve outstanding achievements due to their advantages over conventional complex methods. Wearable sensors measure the biological, chemical, and physical health conditions of humans using different methods and provide them to the individual. Using these sensors, it is possible to get accurate and useful information about the health conditions of the body, and in addition, the problems of traditional methods are eliminated. From a commercial point of view, many of the produced flexible sensors have made significant progress. Due to the increase in demand, the production of more accurate sensors with diverse diagnostic capabilities is steadily increasing [4].

The most essential things in the production of wearable sensors are the materials used, the fabrication technologies, and the methods used to increase their efficiency. The materials used in the wearable sensors must be such that they meet the expectations of these sensors, such as desired toughness, flexibility, and conductivity. Moreover, in various environmental conditions, there should be no disruption in the sensor's work. Therefore, composite materials need to be used to overcome the drawbacks of single materials through the synergetic effects of composite materials. Among various materials, hydrogel and graphene-based materials have shown good capabilities for wearable sensors, such as stretchability, conductivity, and toughness [5,6].

In addition, the materials chosen to make these sensors must be made with the right technology so that the final product can provide the best performance in different conditions. Also, these technologies should be a way to improve the performance of the device and be up to date and economical. One of the modern technologies with many applications in this field is 3D printing, which can be used to make these sensors with many advantages. Despite the remarkable progress of flexible wearable sensors, the manufacturing of these sensors is facing many challenges. To obtain a device with the desired capabilities, all upcoming refinements must be checked and addressed [7,8].

In this chapter, we review hydrogel and graphene-based materials and their composites for use in flexible wearable sensors. Moreover, we have tried to reach an acceptable point of view about flexible wearable sensors by examining the technologies that can be used in fabrication and we discuss the ongoing challenges.

14.2 Materials

Making flexible wearable sensors requires the use of materials that have desirable features. Since there is no material with all the required properties, the materials must be used in the form of composites. In this section, we will discuss hydrogel- and graphene-based materials and explain their main characteristics applicable to wearable sensors.

14.2.1 Hydrogel-Based Materials

Hydrogel is a three-dimensional (3D) hydrophilic polymer network composed primarily of water, produced by crosslinking hydrophilic polymer chains or by polymerizing water-soluble monomers with crosslinkers. The first hydrogel was created and reported in 1960. Hydrogels can be used as biological materials due to their high similarity to the soft tissues of plants and animals and their suitable biocompatibility [9–11].

Soft tissues have achieved advanced functions due to their structures and water content. An example is the human body, which contains natural hydrogels in some of its tough and fragile tissues [9,12,13]. The mechanical, electrical, and chemical properties of this biocompatible material are modifiable, and its behavior is unique, like fluids and solids. Hydrogels have low conductivity, which limits their performance. Also, many hydrogels become dry and stiff over time. Composite materials consisting of hydrogels and nanomaterials are commonly used for biocompatible materials to enhance their performance, efficiency, and durability [14].

14.2.1.1 Which Applications Can Hydrogel Be Used For?

Hydrogels can respond to environmental changes due to temperature, light, magnetic and electric fields, or pH changes. These environmental changes can be responded to via hydrogel swelling and/or shrinking, which can be used for stimulation and sensation [15–19]. One of the applications of hydrogels is absorbing sweat with high capacity and swelling in response to it. This can be a sensory response of this substance and prevent dehydration in the body due to the quantity of sweat absorbed [20]. Moreover, hydrogels can be used as sensors to detect glucose because of the changes in its frequency and swelling when it comes in contact with glucose. This glucose detection can be ongoing in the range of 0–50 mM [21]. In other applications, this material can be used in pressure sensors, flexible electrodes, and touch screens [22].

14.2.1.2 The Structure and Properties of Hydrogel for Use in Wearable Sensors

Among all materials, hydrogel-based composites are one of the best candidates for use in flexible and wearable sensors if we could tune the essential properties such as toughness, elasticity, stretchability, conductivity, anti-freezing, and self-healing in these structures. It is possible to make high-performance wearable sensors using appropriate methods to increase conductivity, produce hydrogels with self-healing and anti-freezing features, and select hydrogels with good toughness and elasticity [12].

14.2.1.2.1 Structure

Conventional hydrogels, such as single-network hydrogels (SN), are typically soft, feeble, and fragile and have finite stretchability (<100%), low elastic modulus, and little fracture energy (0.1–10 J/m^2) [14,23,24]. In recent years, scientists have made many efforts to improve the solidity and toughness of hydrogels to use this material in various cases [25].

Double-network hydrogel (DN) is one of the best examples in terms of high toughness due to its performance. DN hydrogel with a fracture toughness of more than 1000 J/m^2 was first developed in 2003 and has attracted a great deal of attention. DN hydrogel has two strong polymer networks, and its opposite properties include its strength, network density, crosslink density, and molecular weight. This material is a kind of gel with a special structure consisting of two interpenetrating polymer networks that differ from each other with special properties. In a DN structure, the molar concentration of the first network is generally one-tenth of the second one [26–28]. The mechanical properties of DN hydrogel are listed in Table 14.1. As shown in Figure 14.1, there are various methods to synthesize DN hydrogels, such as combining two different hydrogel networks, the classical two-step polymerization method, preparation of the physically–chemically crosslinked alginate-PAAm DN hydrogel, and the composition of a typical physically–chemically crosslinked alginate-PAAm hydrogel. Practical methods include two-step polymerization or 3D printing [27,29].

14.2.1.2.2 Conductivity

Hydrogels are considered almost insulating materials due to their structure and ingredients. This feature has changed with great effort in recent decades, and researchers are trying to improve the conductivity of hydrogels by using different materials and methods

TABLE 14.1

Mechanical Properties of DN hydrogel [5]

Properties	Range
Water content	90%
Tensile fracture strain	1000–2000%
Elastic modulus	0.1–10 MPa
Fracture toughness	100–16000 J/m^2
Failure compressive nominal stress	20–60 MPa
Compressive fracture stain	90–95%

Adapted with permission [5]. Copyright (2021), Elsevier.

[12,30]. For example, metal nanomaterials, carbon nanotubes, and graphene can be used as conductive fillers for building conductive nanocomposite hydrogels [22]. Electrolyte-based and polyelectrolyte-based hydrogels and conductive polymer-based hydrogels are the other conductive hydrogels. Most conductive hydrogels use ions to transmit signals through their network, but in some cases, such as nanomaterial-based conductive hydrogels, there is a different approach [5,22,31,32].

14.2.1.2.3 Anti-Freezing Properties

One of the challenges facing hydrogels is freezing. Because it consists of a large amount of water, the hydrogel freezes at sub-zero temperatures and its stretchability and conductivity will be affected, which causes the sensors to malfunction. Therefore, one of the essential factors in the design of hydrogels is their anti-freezing properties. By using various methods and materials, it is possible to build hydrogels with a proper performance at sub-zero temperatures. For example, by adding glycerol to the hydrogel, we can create a polyacrylamide network in the hydrogel [4,22].

14.2.1.2.4 Self-Healing Properties

One of the exciting topics concerning hydrogels is the self-healing properties of these materials. This feature has led to a longer lifetime and more applications of self-healing hydrogels than conventional ones [33].

14.2.2 Graphene-Based Materials

Graphene-based materials are usually classified into three groups based on their structure, including graphene (G), graphene oxide (GO), and reduced graphene oxide (rGO). GO can be obtained from graphite by applying ultrasonic or mechanical separation by increasing the interlayer distance. While rGO and G can be obtained through the reduction of GO, GO is hydrophilic and has oxygen-rich functional groups (e.g., –C=O, –OH, –COOH) to absorb H_2O molecules from the environment, making it ideal for perceiving moisture based on changing resistance [34,35]. rGO and G have high electronic conductivity (graphene paper is about 3.51 S/m). As shown in Figure 14.2(A), it can be used in different wearable sensors to transform body signals into electrical signals to monitor human conditions.

Graphene-based materials, as sensing materials, can be combined with different substrates such as SiO_2, cellulose, PDMS, PEN, PET, PI, Al_2O_3, and PVC. Also, these composites can be used inside enclosures such as nylon, epoxy resin, PAE, EVA, and PVA to make wearable sensors for monitoring human health [36,37].

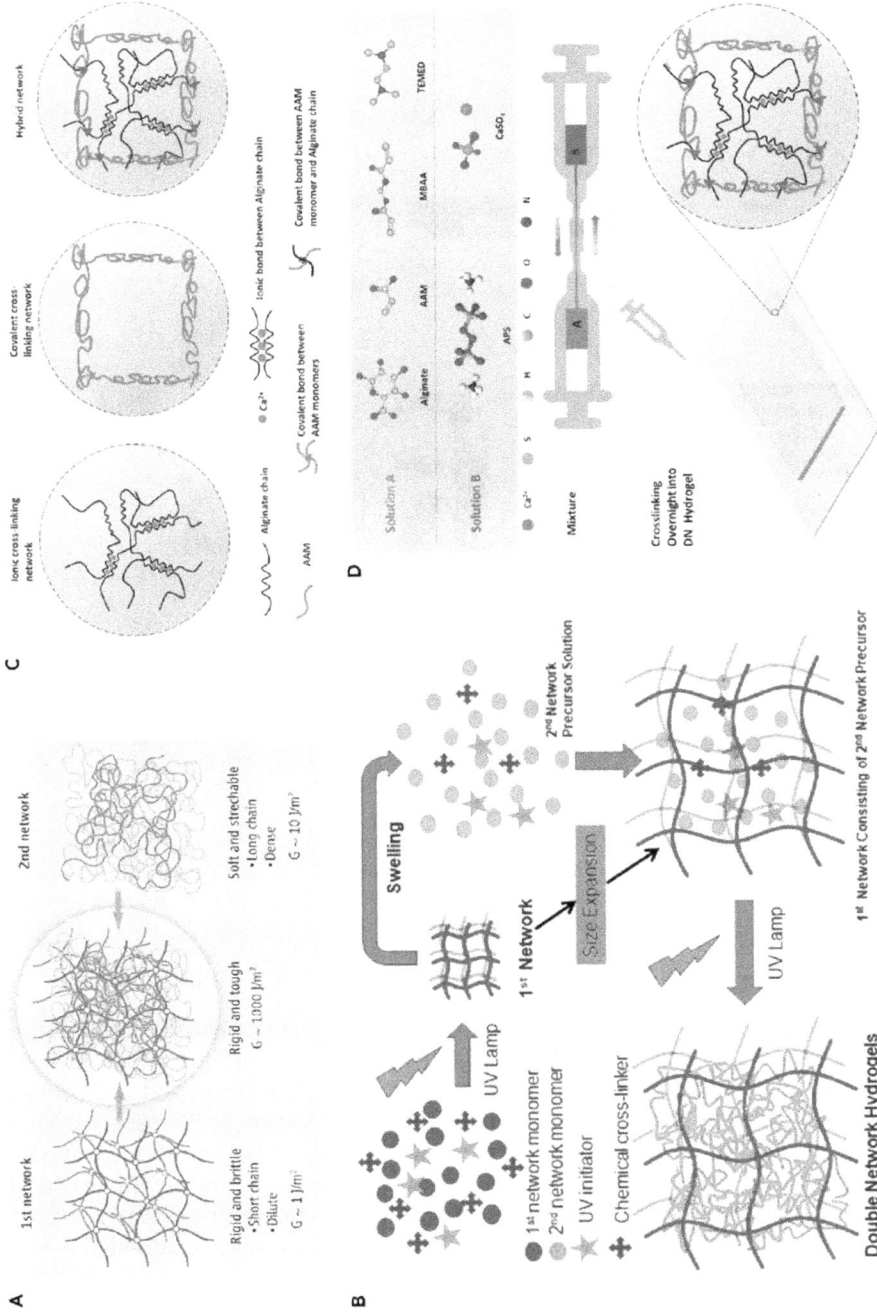

FIGURE 14.1

Two typical DN hydrogel structures and their synthesis strategies, including (A) combining two different hydrogel networks, tough DN hydrogels can be created. (B) Classical two-step polymerization method to prepare chemically–chemically crosslinked DN hydrogels. (C) The composition of a typical physically–chemically crosslinked alginate-PAAm hydrogel. (D) Preparation of the physically–chemically crosslinked alginate-PAAm DN hydrogel. Adapted with permission [5]. Copyright (2021), Elsevier.

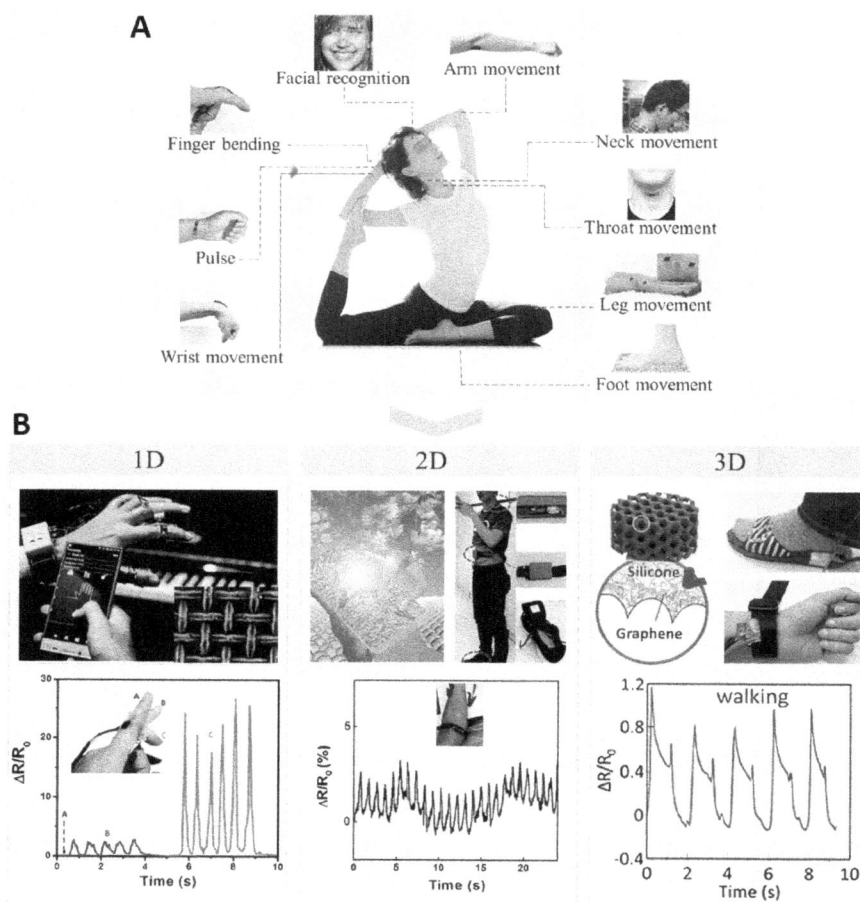

FIGURE 14.2
(A) Graphene-based wearable sensors for healthcare monitoring of wrist, pulse, finger, face, arm, neck, throat, leg, and foot. (B) One-dimensional fiber-based, 2D film-based, and 3D foam-based, graphene-based wearable sensors. Adapted with permission [34]. Copyright (2022), Elsevier.

14.2.2.1 Structural Design of Graphene-Based Materials for Use in Wearable Sensors

Structural design is required to improve the detection sensitivity and mechanical flexibility of graphene-based sensors. As shown in Figure 14.2(B), popular strategies for structural design are classified into 1D, 2D, and 3D structures with different fabricated techniques. One-dimensional fibers can be modeled on textiles for the fabrication of 3D sensors. Two-dimensional films can lead to a similar connection with the human body and improve the accuracy of sensing data. Using 3D foams, the adsorption capability of fluids or gases with biomarkers can improve due to sufficient space in the porous structure of foams.

Various forms of graphene have been extensively studied in wearable piezoresistive sensors because of the high surface area, easy large-scale preparation, and superior electrical and mechanical properties. The use of traditional strain sensors based on sheet semiconductors and metals is restrained due to the very low gauge coefficient (G=1 at 30% strain), narrow working range (less than 5% strain), and poor flexibility. Lightweight, flexible graphene fibers have been promoted as stimulus-sensitive materials for developing flexible piezoresistive sensors, which can provide benefits such as good washing speed, the ability to follow many complex human body movements, easy integration into clothing, and wearable comfort [38].

Besides 1D fiber sensors, 2D or thin-layer piezoresistive sensors have also been extensively studied due to their easy connection to human skin surfaces, ease of fabrication, low weight, and high flexibility. As a typical 2D conductive carbon substance, graphene can be easily adapted to work with metal nanowires, other 2D materials, and conductive polymers to achieve high compressibility sensitivity and mechanical elasticity [38,39]. Zhou et al. fabricated graphene-woven fabrics by direct graphene growth on copper grids with the chemical vapor deposition (CVD) technique and then engraved them with an aqueous solution of HCl and $FeCl_3$. When small vibrations and strains are applied, a high density of random cracks appears in the built-in sensor network, which destroys the suitable path and increases the resistance [40,41].

Extensive research has been conducted on developing 3D graphene-based piezoresistive sensors; 3D graphene-based materials with unique geometries offer higher compressibility and excellent mechanical elasticity, which can adequately prevent the destruction of sensor configurations during external stimuli. Three-dimensional conductive graphene networks can be classified into sponges, foams, aerogels, and hydrogels based on the construction mechanism. Graphene hydrogels, usually synthesized by hydrothermal or chemical methods, are widely studied due to the combination of the dual benefits of a 3D lattice structure and good conductivity [42]. Sun et al. developed vanadium nitride-graphene constructions to simulate a pressure sensor using the spray-printing method for suitable skin compatibility, which also showed outstanding stability and high sensitivity (40 kPa^{-1}) [38].

14.2.2.2 *Various Stimulants Affecting Graphene-Based Materials*

As shown in Figure 14.3(A, B), graphene-based materials are an ideal candidate for fabricating wearable sensors to monitor biophysical signals such as mechanical, thermal, and electrophysical signals for healthcare functions [34]. Graphene-based materials can reply to physical stimuli such as heat, light, and mechanics. By converting these signals into electrical signals, different sensory operations can be accomplished. Temperature is an essential indicator of health. Graphene-based thermal sensors can receive body temperature signs through resistance changes due to transferring electron-hole pairs. As illustrated in Figure 14.3, in addition to physical signals, graphene-based sensors can reply to chemical signals, inclusive of biomolecules, ions, and gas.

Therefore, the development of wearable sensors could pave the way for a breakthrough in medical science. Using these sensors, we can measure our vital signals. Even in sports, these sensors can conduct measurements, for example, blood pressure. Applying graphene-based materials and graphene-based composites can improve the performance of these sensors and ultimately gives us hope that the use of these sensors will be implemented soon.

14.2.2.3 *Application of Graphene-Based Materials in Wearable and Flexible Sensors*

Graphene and graphene-based composites have many applications for use in wearable and flexible sensors. One of these applications is making sensors capable of detecting

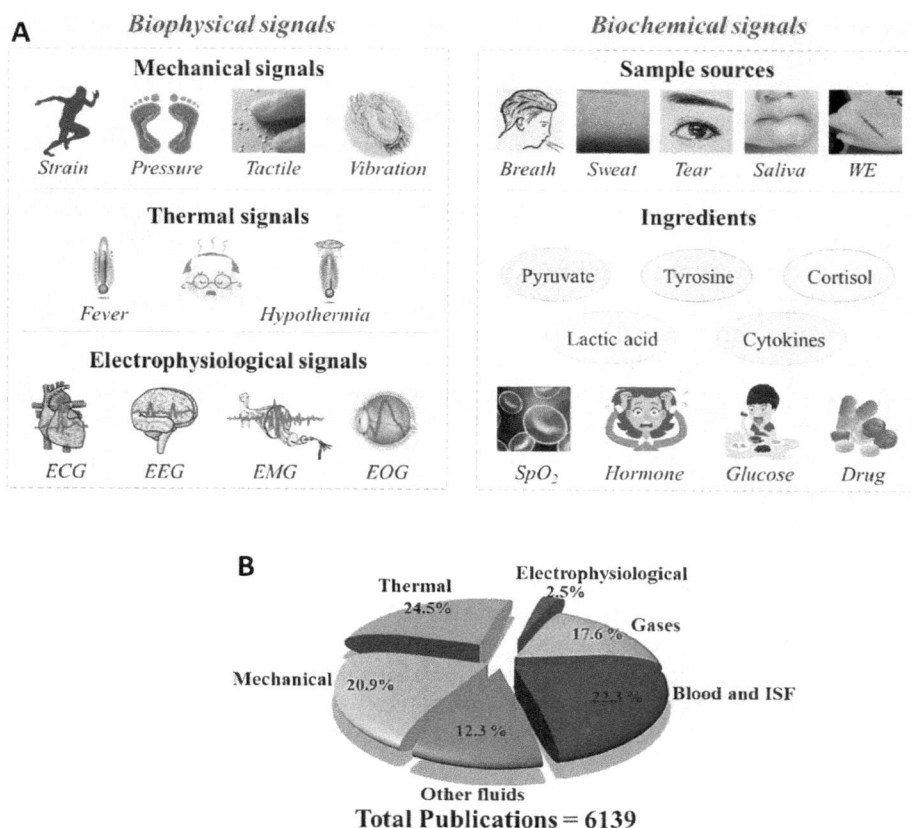

FIGURE 14.3
(A) Graphene-based wearable sensors for monitoring biophysical and biochemical signals such as mechanical, thermal, and electrophysiological signals based on sample sources and ingredients. (B) Statistical data analysis on various signals and sample sources in graphene-based wearable sensors in the Web of Science from 2010 to 2020. Note: WE means wound exudate. Adapted with permission [34]. Copyright (2022), Elsevier.

health-related signals. Wearable graphene sensors can monitor health-related biological signals and help users know about their physical condition [34]. Graphene biochemical sensors can detect cortisol and glucose, and the location of these signals can be in wound secretions and mouth saliva. Making these sensors requires the use of unique technology [34]. The temperature and pH have strangely affected the performance of enzymatic biosensors. Regarding the problems caused by enzymatic biosensors, researchers have introduced non-enzymatic graphene biosensors with biological and electrocatalytic activity. A copper/graphene electrode with copper placed on graphene was designed to measure glucose, which shows a detection range of 0.5 μM to 4.45 mM with a response time of 2 s [34].

As shown in Figure 14.4(A, B), biomarkers related to tear health can be detected by integrating a graphene sensing material into a soft contact lens to diagnose eye health. To detect health-related biomarkers, a graphene-based saliva sensor mounted on a tooth or dental guard can be used for healthcare monitoring. For example, a graphene-based saliva sensor is synthesized by printing graphene on a silk layer, contacting it with an antenna,

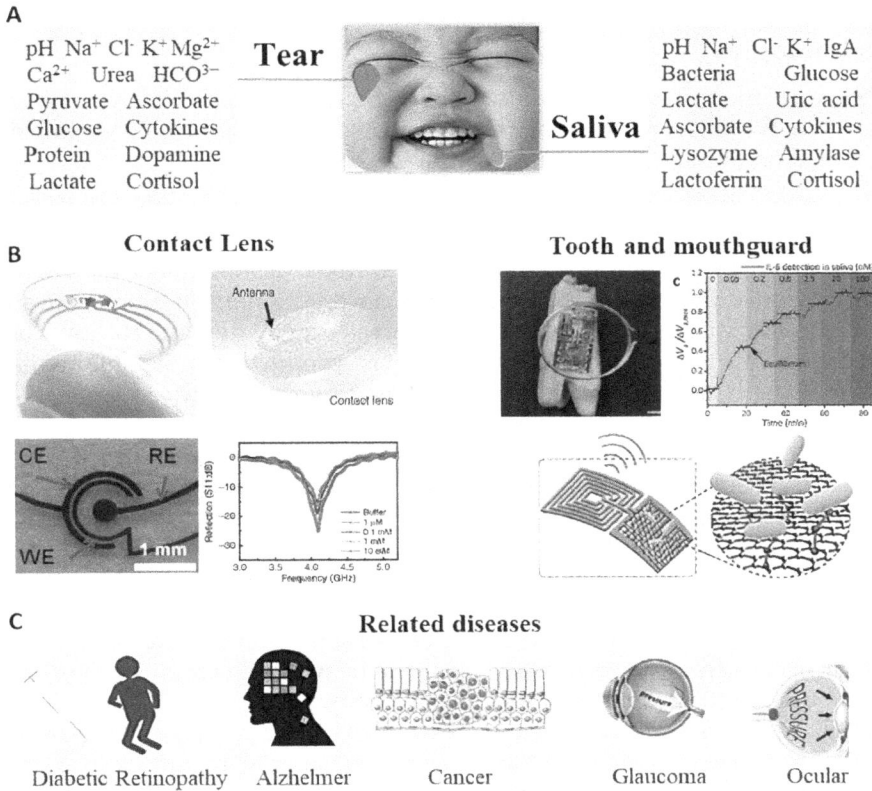

FIGURE 14.4
(A) Health-related signals in tears and saliva. (B) Graphene-based wearable sensors for tears and saliva monitoring. (C) Early diagnosis of related diseases based on tears and saliva monitoring. Adapted with permission [34]. Copyright (2022), Elsevier.

and transferring it to tooth enamel or a tooth guard to monitor respiration and bacteria in saliva. A portable, graphene-based nanosensor has been proposed to make it easier to detect cytokine biomarkers in high-sensitivity saliva. Using on-site custom signal processing circuits, this handheld device can provide information on accurately detecting cytokine levels in saliva on the embedded liquid crystal display. The device can also wirelessly transmit information to the smartphone via a module on the board to visualize cytokine concentration changes online. These components in tears or saliva are also essential biomarkers related to various diseases such as diabetes, Alzheimer's, cancer, glaucoma, and eye diseases (see Figure 14.4) [34]. Moreover, wearable sensors based on graphene can be used to detect cancer biomarkers, and this technology can lead to significant advances and achievements in the field of medicine. In particular, using epitaxial graphene on silicon carbide, they could repeatedly detect 8-hydroxydeoxyguanosine (8-OHdG), a biomarker of DNA damage [34].

Moreover, graphene-based biofuels have shown great potential in automatic biosensors as wearable sensors by extracting energy from human metabolites (such as glucose in saliva or sweat) and converting chemical energy into electricity [34].

14.3 Technologies

There are several modern and economical technologies to make flexible wearable sensors. In this part, we will discuss methods that can be used to make flexible wearable sensors using hydrogel or graphene-based composites.

14.3.1 3D Printing Technology

Three-dimensional printing technology is one of the most up-to-date sciences in the world, which has expanded in various fields. Due to the unconditional and unique advantages of this new technology, it has become more widespread day by day, and its applications are also increasing. This technology is a new era in the production of flexible wearable sensors, and by using it, it is possible to produce sensors with practical advantages by using suitable inks and proper manufacturing methods.

Three-dimensional printing technology can cause many changes in biomedical wearable devices and sensors. In the field of health monitoring, biological wearable sensors, including wearable electronic sensors, pressure sensors, artificial skin, sensors that are used on the skin as tattoos, glucose detection sensors, and strain sensors, have attracted much attention and are continuously improving. To make these widely used wearable sensors, advanced 3D printing technology can be used, which has an outstanding contribution to the development of wearable biological sensors. Using this technology, wearable devices or sensors can be used directly on the skin with heterogeneous soft materials, according to the unique geometry of the patient's body.

In this regard, 3D-printed hydrogels are extremely attractive among researchers. This attractiveness is due to the convenient features of this material, regarding its methods and preparation materials, as well as its chemical and physical properties. These properties are improving day by day and expanding 3D-printed hydrogel applications. For example, hydrogels can be used as inks in this technology and then used to develop wearable sensors for various applications with high accuracy. The hydrogel used as ink can be combined with other materials, and using appropriate methods, can lead to the production of a 3D-printed hydrogel with the desired capabilities [43].

Using a digital light processing (DLP) 3D printer and PAAm-PEGDA hydrogel, a super sensitive and flexible ionic conductive hydrogel was made as a wearable device for strain sensing. Moreover, for sensing dynamic and static pressures with high accuracy and sensitivity, a capacitive sensor was made by including a VHB™ dielectric strip between the hydrogel layers [44].

Regarding glucose detection, various methods have been developed for glucose monitoring such as electrochemical and optical spectroscopy. Also, recently, 3D printing methods have been used to make lightweight wearable biosensors to measure various quantities, as the sensor has been made to determine the quantity of glucose [7].

Another application of 3D printing technology is making wearable oximeters that were urgently needed in the Covid-19 pandemic to detect and determine the amount of oxygen in the blood and also heart rate. A wearable pulse oximeter was made using a reversible 3D printing method to make a PDMS elastomeric cuff using a sacrificial layer of hydrogel to shape the final PDMS geometry. These oximeters can work in different conditions [45].

14.3.2 Nanogenerators in Flexible Wearable Sensors

One of the evolving technologies in the field of flexible wearable sensors is the construction of self-powered sensors. One way to achieve this capability is to use nanogenerators in wearable and flexible sensors. Piezoelectric nanogenerators (PENG) and triboelectric nanogenerators (TENG) are two general nanogenerators used in this field. Piezoelectric nanogenerators emit electrical signals when mechanical force is applied to them, which makes them suitable for use as self-powered sensors. The piezoelectric sensor can be used for detecting mechanical forces such as pressure. There are different ways to make piezoelectric sensors [46,47].

Microfabrication processing has been implemented to achieve these requirements through many technologies. Some studies have shown the ability to use graphene (nanowires/graphene) as a piezoelectric base material to produce a pressure sensor. Accordingly, new technology can be achieved by producing touch screens and wearable sensor monitors [48–50].

PENG can be used in heart rate sensors to monitor heart rate according to the said properties and with the help of a suitable production method. One of the most popular sensors is the respiratory system monitoring sensor, and many technologies have been developed in this field. However, these sensors fail in the wrong conditions. One of the uses of PENG is in producing respiration monitoring sensors. For example, the fabrication of polyvinylidene fluoride (PVDF) PNEG is a promising technology and can be used to build a self-powered wearable respiratory sensor [51].

Nowadays, the importance of energy generation in wearable sensors by using body motions has become noticeable. Researchers are trying to convert mechanical energy from body motions to a suitable energy source for wearable sensors. In this regard, PENG is suitable for using this energy from body motions. For example, hybrid piezoelectric structures or SE nanowires were used to produce wearable nanogenerators [52,53].

On the other hand, there are triboelectric nanogenerators that can generate energy by friction and electric induction, which have lower manufacturing costs and a variety of materials for their construction. Also, their structure is simple, and the resulting energy is high. This has made them suitable for use as self-powered nanogenerators in wearable sensors that can be used to generate energy using human activities. Many studies have been conducted to make pulse monitoring sensors based on TENG, and according to the results, TENG is one of the suitable candidates for this type of sensor. TENG can be used for building sensors to monitor heart rate, and due to its self-power capability, it can also be made as a wireless sensor [47,54]. TENG has also been used in the fabrication of self-powered sensors to monitor respiration, which has led to the development of many applications. A TENG-based sensor has been reported for detecting the concentration of NH_3 in human breath [47,55]. TENGs are applicable in body motion monitoring sensors. For example, the use of a sensor on the insole, which can detect walking, or its use in sensors that can control a robotic hand using a human hand or human touch receivers [47,56–58].

14.3.3 Methods to Improve Conductivity

Typically, conductive hydrogels are structurally designed as composites that can be filled with various fillers to provide conductivity. In general, conductive hydrogels can be included in two main groups: ionically conducting hydrogels (ICH) and electrically conductive hydrogels (ECH). Figure 14.5 shows different methods that have been applied for the fabrication of conductive hydrogels. Direct gelling of conductive materials such as conductive polymers, graphene, carbon nanotubes, and graphene oxide has been applied to prepare conductive hydrogels. The other methods include the dispersion of ions in the hydrogel structure, the addition of fillers in precursor hydrogels (such as graphene and carbon nanotubes), and the polymerization of conductive polymers in prefabricated hydrogels.

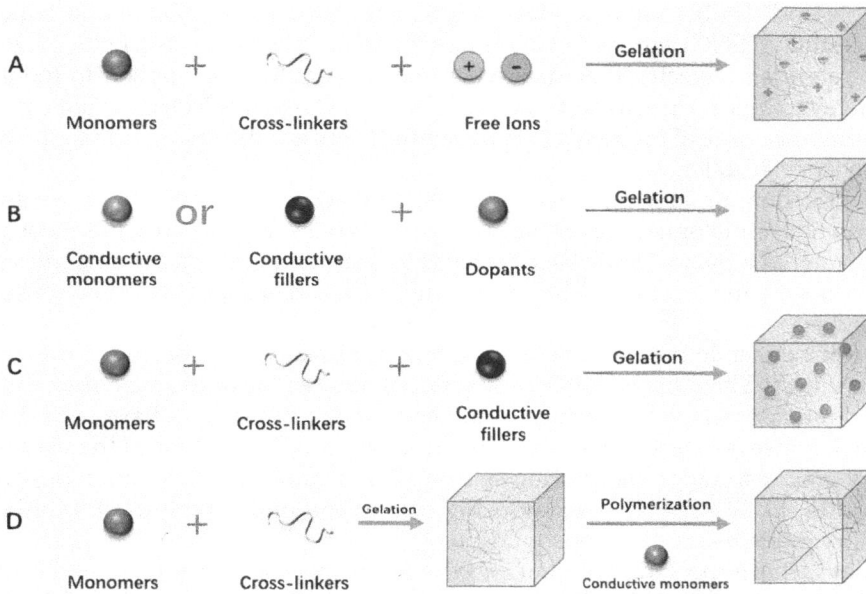

FIGURE 14.5
Various methods to prepare conductive hydrogels. Adapted with permission [22]. Copyright (2021), Elsevier.

Graphene is one of the suitable materials that can be combined with hydrogels using the mentioned methods to make a hydrogel with suitable conductivity. Due to its flexible and conductive 2D structure, this material can fit in the hydrogel's 3D structure and lead to graphene hydrogel production [22,30,59,60,61].

14.4 Challenges

In addition to the advantages mentioned earlier of hydrogel and graphene in flexible wearable sensors based on nanocomposites, it is worth mentioning the disadvantages and challenges of these cases to conclude whether it is cost-effective to use these materials and technologies. The challenges that flexible wearable sensors face can be listed as follows:

- The materials on which these sensors are made
- The manufacturing technology of the sensors
- The sensing performance
- The energy required by these sensors and how to supply them in different ways
- Time to receive a response from these sensors
- The construction of sensors with memory
- The lifetime, security, and safety of these sensors
- Ease and convenience of using these sensors
- Not being harmful to human health [62]

One of the most critical challenges facing flexible wearable sensors is the materials on which these sensors are made. These materials must have the desired toughness, flexibility, conductivity, and so on, to be used as the base structure of the sensor. Also, the selected materials should be appropriately integrated with electronic sensors. Moreover, in different situations and different environmental conditions, there should be no problems in the connection between these materials and the sensors [22]. Therefore, it is difficult to find a material with these features in its original form without modification or integration with other materials. Therefore, the utilization of composite materials can meet these needs. Another problem is the adhesion between materials and sensors. Other challenges include the declining electrochemical performance of wearable sensors [4].

Pressure sensors act by receiving mechanical energy and converting it into electrically recognizable signals. They are very efficient in sports activities, health monitoring, robotics, and so on. To make wearable and flexible pressure sensors, materials are needed that can withstand the desired strain and have good stretchability. However, even the best-selected materials cannot withstand strains that are too high. So, making sensors with high strain tolerance, good stretchability, and sensitivity in detection is always challenging [63–65].

In 3D printing technology, one of the challenges is to provide ink with good performance and strong printing. It is possible to use hydrogel inks with a high forming speed before printing, which is always challenging when producing these inks and using the appropriate materials [66]. One of the main challenges is energy management in these sensors. Due to the dimensions of the sensors, high-capacity batteries cannot be used and energy will be consumed quickly, and the device will turn off. However, with proper planning for energy consumption in these sensors and the optimal use of energy, the hours of use can be increased [62]. In addition to the energy management challenge, we can discuss the use of nanogenerators in these sensors, which can provide the energy required by these sensors to some extent. Many resources in the human body can be used, and nanogenerators can address the energy problems in these sensors.

A good response time for the wearable sensor is the wish of every consumer, especially in sensors with medical uses, where the response time can be critical. So, accurate and rapid detection of the desired substances in the body, on the skin, and the detection of somebody's behaviors is one of the critical challenges in this field. In this regard, accurate, easy, and rapid detection of the substances in body fluids has remained a challenge due to their low concentration [67].

Stability, good service lifetime, and reliability of wearable sensors are other challenges. The performance of wearable sensors should be stable in different environmental conditions [22]. Another challenge that can be mentioned is the ease of use of wearable sensors. These sensors should be suitable and comfortable, especially for patients [68]. In addition to the many applications and excellent benefits that wearable sensors bring to us, we must answer the question of whether these sensors can be commercialized or not. After using the proper methods and technology to build a sensor with the best efficiency at the lowest possible cost, these flexible wearable sensors should be commercialized and sent to the market for public consumption.

14.5 Conclusion

Wearable sensors face many challenges in various fields, including materials and technology. It can be concluded that two materials, i.e., hydrogel, graphene, and their composite, have a high potential for use in flexible wearable sensors. High-performance sensors can be made from these materials using 3D printing technology. Also, by using nanogenerators in wearable sensors, one of the main challenges ahead, i.e., the energy problem of these sensors, can be addressed.

Although flexible wearable sensors have many challenges, due to their advantages such as short response time, ease of use, carrying capacity, and cost-effectiveness compared with traditional methods, it can be concluded that these sensors have a high potential to become the main tool for detecting health and environmental conditions.

References

1. Z. Chu, W. Zhang, Q. You, X. Yao, T. Liu, G. Liu, G. Zhang, X. Gu, Z. Ma, W. Jin, A separation-sensing membrane performing precise real-time serum analysis during blood drawing, *Angewandte Chemie - International Edition.* 59 (2020) 18701–18708.
2. M. Bariya, H.Y.Y. Nyein, A. Javey, Wearable sweat sensors, *Nature Electronics.* 1 (2018) 160–171.
3. W.A.D.M. Jayathilaka, K. Qi, Y. Qin, A. Chinnappan, W. Serrano-García, C. Baskar, H. Wang, J. He, S. Cui, S.W. Thomas, S. Ramakrishna, Significance of nanomaterials in wearables: A review on wearable actuators and sensors, *Advanced Materials.* 31 (7) (2019) 1805921.
4. H. Luo, B. Gao, Development of smart wearable sensors for life healthcare, *Engineered Regeneration.* 2 (2021) 163–170.
5. B. Ying, X. Liu, Skin-like hydrogel devices for wearable sensing, soft robotics and beyond, *IScience.* 24 (2021) 103174.
6. B. Purohit, B. Divya, N.P. Shetti, P. Chandra, Materials for wearable sensors, in *Wearable Physical, Chemical and Biological Sensors: Fundamentals, Materials and Applications.* (2022) 5–40.
7. A. Kalkal, S. Kumar, P. Kumar, R. Pradhan, M. Willander, G. Packirisamy, S. Kumar, B.D. Malhotra, Recent advances in 3D printing technologies for wearable (bio)sensors, *Additive Manufacturing.* 46 (2021) 102088.
8. M. Abshirini, M. Charara, P. Marashizadeh, M.C. Saha, M.C. Altan, Y. Liu, Functional nanocomposites for 3D printing of stretchable and wearable sensors, *Applied Nanoscience.* 8 (2019) 2071–2083.
9. H. Fan, J.P. Gong, Fabrication of bioinspired hydrogels: Challenges and opportunities, *Macromolecules.* 53 (8) (2020) 2769–2782.
10. C. Xu, Y. Yang, W. Gao, Skin-interfaced sensors in digital medicine: From materials to applications, *Matter.* 2 (2020) 1414–1445.
11. O. Wichterle, D. Lím, Hydrophilic gels for biological use, *Nature.* 185 (1960) 117–118.
12. X. Liu, J. Liu, S. Lin, X. Zhao, Hydrogel machines, *Materials Today.* 36 (2020) 102–124.
13. X. Zhao, Multi-scale multi-mechanism design of tough hydrogels: Building dissipation into stretchy networks, *Soft Matter.* 10 (2014) 672–687.
14. Y.S. Zhang, A. Khademhosseini, Advances in engineering hydrogels, *Science (1979).* 356 (2017) 6337.
15. T.H. Kang, H. Chang, D. Choi, S. Kim, J. Moon, J.A. Lim, K.Y. Lee, H. Yi, Hydrogel-templated transfer-printing of conductive nanonetworks for wearable sensors on topographic flexible substrates, *Nano Letters.* 19 (2019) 3684–3691.
16. Y. Piao, H. You, T. Xu, H.P. Bei, I.Z. Piwko, Y.Y. Kwan, X. Zhao, Biomedical applications of gelatin methacryloyl hydrogels, *Engineered Regeneration.* 2 (2021) 47–56.
17. Z. Dong, X. Zhao, Application of TPMS structure in bone regeneration, *Engineered Regeneration.* 2 (2021) 154–162.
18. R. Yoshida, K. Uchida, Y. Kaneko, K. Sakai, A. Kikuchi, Y. Sakurai, T. Okano, Comb-type grafted hydrogels with rapid deswelling response to temperature changes, *Nature.* 374 (1995) 240–242.

19. P. Calvert, Hydrogels for soft machines, *Advanced Materials*. 21 (2009) 743–756.
20. L. Wang, T. Xu, C. Fan, X. Zhang, Wearable strain sensor for real-time sweat volume monitoring, *IScience*. 24 (2021) 102028.
21. M. Elsherif, M.U. Hassan, A.K. Yetisen, H. Butt, Wearable contact lens biosensors for continuous glucose monitoring using smartphones, *ACS Nano*. 12 (2018) 5452–5462.
22. L. Wang, T. Xu, X. Zhang, Multifunctional conductive hydrogel-based flexible wearable sensors, *TrAC - Trends in Analytical Chemistry*. 134 (2021) 116130.
23. Q. Chen, H. Chen, L. Zhu, J. Zheng, Fundamentals of double network hydrogels, *Journal of Materials Chemistry B*. 3 (2015) 3654–3676.
24. J.P. Gong, Why are double network hydrogels so tough?, *Soft Matter*. 6 (2010) 2583–2590.
25. J.P. Gong, Materials science. Materials both tough and soft, *Science*. 344 (2014) 161–162.
26. M.A. Haque, T. Kurokawa, J.P. Gong, Super tough double network hydrogels and their application as biomaterials, *Polymer (Guildf)*. 53 (2012) 1805–1822.
27. J.P. Gong, Y. Katsuyama, T. Kurokawa, Y. Osada, Double-network hydrogels with extremely high mechanical strength, *Advanced Materials*. 15 (2003) 1155–1158.
28. T. Nonoyama, J.P. Gong, Double-network hydrogel and its potential biomedical application: A review, *Proceedings of the Institution of Mechanical Engineers. Part H*. 229 (2015) 853–863.
29. S.E. Bakarich, R. Gorkin, M. Panhuis, G.M. Spinks, Three-dimensional printing fiber reinforced hydrogel composites, *ACS Applied Materials & Interfaces*. 6 (2014) 15998–16006.
30. H. Yuk, B. Lu, X. Zhao, Hydrogel bioelectronics, *Chemical Society Reviews*. 48 (2019) 1642–1667.
31. Q. Rong, W. Lei, M. Liu, Conductive hydrogels as smart materials for flexible electronic devices, *Chemistry – A European Journal*. 24 (2018) 16930–16943.
32. C. Yang, Z. Suo, Hydrogel ionotronics, *Nature Reviews Materials*. 3 (2018) 125–142.
33. D.L. Taylor, M. In Het Panhuis, Self-healing hydrogels, *Advanced Materials*. 28 (2016) 9060–9093.
34. H. Zhang, R. He, Y. Niu, F. Han, J. Li, X. Zhang, F. Xu, Graphene-enabled wearable sensors for healthcare monitoring, *Biosensors and Bioelectronics*. 197 (2022) 113777.
35. H. Chen, M.B. Müller, K.J. Gilmore, G.G. Wallace, D. Li, Mechanically strong, electrically conductive, and biocompatible graphene, *Faculty of Science - Papers (Archive)*. 20 (2008) 3557.
36. X. Cai, L. Lai, Z. Shen, J. Lin, Graphene and graphene-based composites as Li-ion battery electrode materials and their application in full cells, *Journal of Materials Chemistry A*. 5 (2017) 15423–15446.
37. S.K. Singh, M.K. Singh, M.K. Nayak, S. Kumari, S. Shrivastava, J.J.A. Grácio, D. Dash, Thrombus inducing property of atomically thin graphene oxide sheets, *ACS Nano*. 5 (2011) 4987–4996.
38. K.Y. Chen, Y.T. Xu, Y. Zhao, J.K. Li, X.P. Wang, L.T. Qu, Recent progress in graphene-based wearable piezoresistive sensors: From 1D to 3D device geometries, *Nano Materials Science*. 28 (2022) 1561–1580.
39. D. Niu, W. Jiang, G. Ye, K. Wang, L. Yin, Y. Shi, B. Chen, F. Luo, H. Liu, Graphene-elastomer nanocomposites based flexible piezoresistive sensors for strain and pressure detection, *Materials Research Bulletin*. 102 (2018) 92–99.
40. Y. Wang, T. Yang, J. Lao, R. Zhang, Y. Zhang, M. Zhu, X. Li, X. Zang, K. Wang, W. Yu, H. Jin, L. Wang, H. Zhu, Ultra-sensitive graphene strain sensor for sound signal acquisition and recognition, *Nano Research*. 8 (2015) 1627–1636.
41. M. Zhou, Y. Zhai, S. Dong, Electrochemical sensing and biosensing platform based on chemically reduced graphene oxide, *Analytical Chemistry*. 81 (2009) 5603–5613.
42. Q. Wang, X. Pan, C. Lin, H. Gao, S. Cao, Y. Ni, X. Ma, Modified Ti3C2TX (MXene) nanosheet-catalyzed self-assembled, anti-aggregated, ultra-stretchable, conductive hydrogels for wearable bioelectronics, *Chemical Engineering Journal*. 401 (2020) 126129.
43. C. Liu, N. Xu, Q. Zong, J. Yu, P. Zhang, Hydrogel prepared by 3D printing technology and its applications in the medical field, *Colloids and Interface Science Communications*. 44 (2021) 100498.
44. X.Y. Yin, Y. Zhang, X. Cai, Q. Guo, J. Yang, Z.L. Wang, 3D printing of ionic conductors for high-sensitivity wearable sensors, *Materials Horizons*. 6 (2019) 767–780.
45. S. Abdollahi, E.J. Markvicka, C. Majidi, A.W. Feinberg, 3D printing silicone elastomer for patient-specific wearable pulse oximeter, *Advanced Healthcare Materials*. 9 (2020) 1901735.

46. D.Y. Park, D.J. Joe, D.H. Kim, H. Park, J.H. Han, C.K. Jeong, H. Park, J.G. Park, B. Joung, K.J. Lee, Self-powered real-time arterial pulse monitoring using ultrathin epidermal piezoelectric sensors, *Advanced Materials*. 29 (37) (2017) 1702308.

47. Y. Mao, Nanogenerators in wearable sensors, in *Nanobatteries and Nanogenerators: Materials, Technologies and Applications: A Volume in Micro and Nano Technologies.* (2020) 587–616.

48. K. Takei, T. Takahashi, J.C. Ho, H. Ko, A.G. Gillies, P.W. Leu, R.S. Fearing, A. Javey, Nanowire active-matrix circuitry for low-voltage macroscale artificial skin, *Nature Materials*. 9 (2010) 821–826.

49. G. Schwartz, B.C.K. Tee, J. Mei, A.L. Appleton, D.H. Kim, H. Wang, Z. Bao, Flexible polymer transistors with high pressure sensitivity for application in electronic skin and health monitoring, *Nature Communications*. 4 (2013) 1–8.

50. Z. Chen, Z. Wang, X. Li, Y. Lin, N. Luo, M. Long, N. Zhao, J. bin Xu, Flexible piezoelectric-induced pressure sensors for static measurements based on nanowires/graphene heterostructures, *ACS Nano*. 11 (2017) 4507–4513.

51. Z. Liu, S. Zhang, Y.M. Jin, H. Ouyang, Y. Zou, X.X. Wang, L.X. Xie, Z. Li, Flexible piezoelectric nanogenerator in wearable self-powered active sensor for respiration and healthcare monitoring, *SeScT*. 32 (2017) 064004.

52. M. Lee, C.Y. Chen, S. Wang, S.N. Cha, Y.J. Park, J.M. Kim, L.J. Chou, Z.L. Wang, A hybrid piezoelectric structure for wearable nanogenerators, *Advanced Materials*. 24 (2012) 1759–1764.

53. M. Wu, Y. Wang, S. Gao, R. Wang, C. Ma, Z. Tang, N. Bao, W. Wu, F. Fan, W. Wu, Solution-synthesized chiral piezoelectric selenium nanowires for wearable self-powered human-integrated monitoring, *Nano Energy*. 56 (2019) 693–699.

54. Z. Lin, J. Chen, X. Li, Z. Zhou, K. Meng, W. Wei, J. Yang, Z.L. Wang, Triboelectric nanogenerator enabled body sensor network for self-powered human heart-rate monitoring, *ACS Nano*. 11 (2017) 8830–8837.

55. S. Wang, Y. Jiang, H. Tai, B. Liu, Z. Duan, Z. Yuan, H. Pan, G. Xie, X. Du, Y. Su, An integrated flexible self-powered wearable respiration sensor, *Nano Energy*. 63 (2019) 103829.

56. Z. Lin, Z. Wu, B. Zhang, Y.C. Wang, H. Guo, G. Liu, C. Chen, Y. Chen, J. Yang, Z.L. Wang, A triboelectric nanogenerator-based smart insole for multifunctional gait monitoring, *Advanced Materials Technologies*. 4 (2019) 1800360.

57. X. Pu, H. Guo, Q. Tang, J. Chen, L. Feng, G. Liu, X. Wang, Y. Xi, C. Hu, Z.L. Wang, Rotation sensing and gesture control of a robot joint via triboelectric quantization sensor, *Nano Energy (Print)*. 54 (2018) 453–460.

58. W. Choi, I. Yun, J. Jeung, Y.S. Park, S. Cho, D.W. Kim, I.S. Kang, Y. Chung, U. Jeong, Stretchable triboelectric multimodal tactile interface simultaneously recognizing various dynamic body motions, *Nano Energy*. 56 (2019) 347–356.

59. S. Seyedin, P. Zhang, M. Naebe, S. Qin, J. Chen, X. Wang, J.M. Razal, Textile strain sensors: A review of the fabrication technologies, performance evaluation and applications, *Materials Horizons*. 6 (2019) 219–249.

60. W. Zhang, P. Feng, J. Chen, Z. Sun, B. Zhao, Electrically conductive hydrogels for flexible energy storage systems, *Progress in Polymer Science*. 88 (2019) 220–240.

61. Z. Wang, Y. Cong, J. Fu, Stretchable and tough conductive hydrogels for flexible pressure and strain sensors, *Journal of Materials Chemistry B*. 8 (2020) 3437–3459.

62. M.R. Yuce, Implementation of wireless body area networks for healthcare systems, *Sensors and Actuators, A: Physical*. 162 (2010) 116–129.

63. J. Lee, S. Kim, J. Lee, D. Yang, B.C. Park, S. Ryu, I. Park, A stretchable strain sensor based on a metal nanoparticle thin film for human motion detection, *Nanoscale*. 6 (2014) 11932–11939.

64. T.Q. Trung, N.E. Lee, Flexible and stretchable physical sensor integrated platforms for wearable human-activity monitoringand personal healthcare, *Advanced Materials*. 28 (2016) 4338–4372.

65. S.H. Zhang, F.X. Wang, J.J. Li, H.D. Peng, J.H. Yan, G.B. Pan, Wearable wide-range strain sensors based on ionic liquids and monitoring of human activities, *Sensors*. 17 (2017) 2621.

66. M. Nadgorny, J. Collins, Z. Xiao, P.J. Scales, L.A. Connal, 3D-printing of dynamic self-healing cryogels with tuneable properties, *Polymer Chemistry*. 9 (2018) 1684–1692.
67. W. Gao, H.Y.Y. Nyein, Z. Shahpar, H.M. Fahad, K. Chen, S. Emaminejad, Y. Gao, L.C. Tai, H. Ota, E. Wu, J. Bullock, Y. Zeng, D.H. Lien, A. Javey, Wearable microsensor array for multiplexed heavy metal monitoring of body fluids, *ACS Sensors*. 1 (2016) 866–874.
68. A. Banerjee, K. Venkatasubramanian, S.K.S. Gupta, Challenges of implementing cyber-physical security solutions in body area networks, *BODYNETS 2009 - 4th International ICST Conference on Body Area Networks*. 18 (2011) 1–8.

15

Electrochemical Wearable Sensors Based on Laser-Induced Graphene for Health Monitoring

Eider Aparicio-Martínez and Rocio B. Dominguez

CONTENTS

15.1 Introduction

Electrochemical sensors and biosensors have played a crucial role in the detection and periodical monitoring of healthcare biomarkers associated with diabetes mellitus (DM), cancer, renal function, cardiac conditions, and inflammatory disease [1]. Through the incorporation of nanomaterials combined with highly selective biomolecules, notable advances have been made in terms of sensitivity of detection, mainly operated in a fixed point of care (PoC) format. These devices usually operate with bulky electrode materials that register the target biomarker information through non-continuous measurements. However, new healthcare directions demand the development of on-body devices for real-time monitoring and acquiring information from non-invasive samples, with high integration of functions and optimized for mobile operation [2]. Healthcare wearable sensors have been increasingly reported to fulfill these needs, and there is a high demand due to factors like the aging of society, the concerns of novel pathogens and viruses, and the need of tracking continuous physiological data for personalized medicine [2,3]. However, to provide on-body real-time information, it is mandatory to replace the conventional format of rigid electrodes with novel flexible transducers of lightweight materials that conform mechanically to the body curvatures during detection without being delaminated or removed. Unlike the recording of physical parameters (e.g., blood pressure, electrocardiogram, temperature), chemical detection possesses the additional challenge of continuous interaction between the receptor and the target biomarker on a confined surface to produce analyte content information [4].

Figure 15.1 shows the schematic interaction of a user with a wearable system for health monitoring. First, the wearable sensor performs the chemical or biochemical interaction with the target biomarker; then, the produced chemical information is processed by electronic circuits into an electrical signal that can be stored or transmitted. The recorded value can be presented to the user, transmitted to a physician facility, or analyzed for an outcome with proper treatment or medication for the user. In a wearable scenario, the described procedure is continuously performed for tracking the dynamic evolution of the analyzed biomarkers. The potential applications for these on-body devices in monitoring imply sampling information from biological fluids other than blood, such as sweat, saliva, urine, tears, and interstitial fluid [2]. This could potentially lead to novel non-invasive detection systems for key biomarkers like glucose [5], uric acid (UA) [6], dopamine (DA) [7], and cortisol [8] to name a few.

A confluence of disciplines is required for the development of wearable systems such as materials science, chemistry, biochemistry, microelectronics, engineering, data science, and a broad range of medical specialties. The rapid evolution of wearable systems has been encouraged by the development of materials that can fulfill the needs of translation from typical rigid detection to on-body monitoring [4]. For electrochemical systems, this implies materials that guarantee good conductivity as electrodes but are also lightweight for mobile operation and conformability to curvatures of the body. So far, manufacturing of flexible

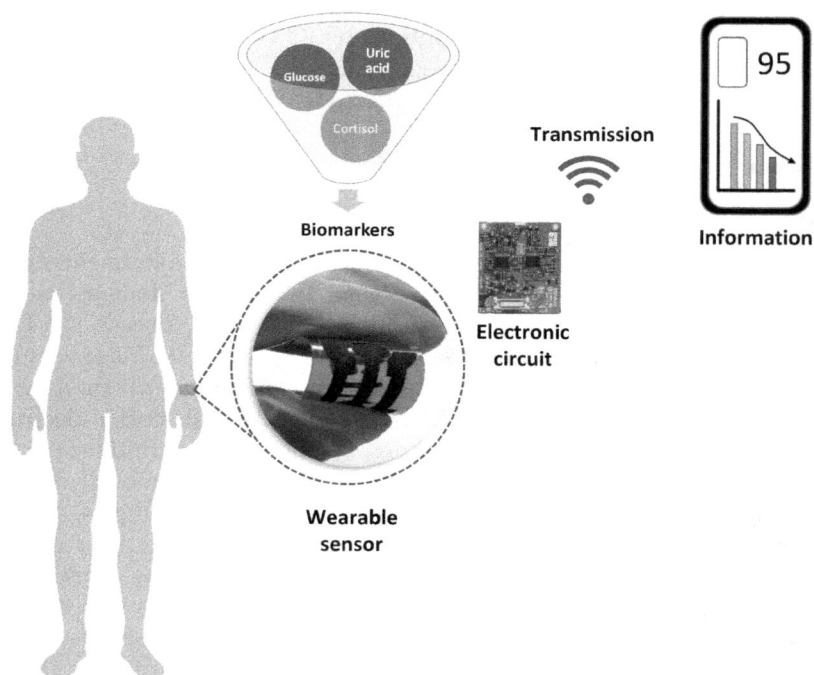

FIGURE 15.1
Schematic representation of a wearable system and its components: biomarker content detected by a wearable sensor attached to the body, recorded signals by an electronic circuit, transmission of data, and interaction with the final user.

electrodes relies on techniques like screen printing, modified textiles, or processing of polymers, but laser-induced graphene (LIG) has steadily gained popularity as a suitable, cost-effective, scalable, and flexible electrochemical platform since its introduction in 2014 by Prof. James Tour's group [9]. Previously, graphene had been largely proposed as an excellent material for flexible applications, but the difficulties associated with its large-scale production limited the manufacturing of wearable graphene-based electrochemical sensors. LIG broke into the scientific literature as a graphene-based material derived from the irradiation of commercial polyimide (PI) without previous synthesis steps. The facilities in LIG production contributed to its fast growth as a transducer; for example, PI was already frequently used in flexible sensor fabrication, the CO_2 laser source for writing was also commonly found in research facilities, and there was no need for a controlled atmosphere to induce the procedure. Thus, the fabrication of flexible electrochemical systems based on LIG has been increasingly reported for environmental, security, and energy fields, with special emphasis on healthcare applications [10].

This chapter summarizes the state-of-the-art advancements of LIG-based wearable electrochemical detection for biomedical analytes, even though it is not intended to be an exhaustive search. First, the background of laser-assisted fabrication of LIG is covered, exploring the unique structural characteristics and morphology of the material and its suitability as a template for electrochemical wearable sensors. Then, we classify the current LIG-based wearable sensors according to the target biomarker, while discussing the role of LIG in the system either as bare electrode material, in combination with active/ hybrid nanomaterials like multiwalled carbon nanotubes (MWCNT) [11], MXenes [12] or metallic nanoparticles [7,13] to improve its inherent performance, or as immobilization support of biomolecules to create highly selective layers. We conclude with the challenges associated with the implementation of wearable LIG in real-life biomedical systems and perspectives on the evolution of the field.

15.1.1 Laser-Induced Graphene for Electrochemical Sensors

Laser-assisted methods have been largely reported for the synthesis, reduction, thinning, and patterning of graphene structures [14]. Commonly, these methods depart from graphene oxide (GO) as a precursor to produce photothermal reduction and rearrange the sp^3 hybridization, resulting in reduced graphene oxide (RGO) with sp^2 configuration. LIG stands out among laser-assisted methods by the absence of GO precursors to directly produce graphene formation on the surface of commercial polymer films without a controlled atmosphere. The most studied polymeric precursor for LIG is polyimide (PI) sheets but aromatic polymers like polyether imide (PEI), polyether ether ketone (PEEK), and polysulfone (PSU), along with untreated carbon-based sources like cellulose, lignin, cloth, bread, and coconut husk, have been successfully applied as precursors [10]. CO_2 lasers from commercial engraver machines were first reported as radiation sources, but the methodology has been successfully expanded to UV lasers [10].

Although the mechanism of LIG formation is not fully described, it is accepted that the photothermal pyrolysis process drives the induction of graphene over the PI surface through a high localized rising temperature (> 2500°C), producing the broken C–O, C=O, and C–N bonds of the PI network [14]. This process rearranges the aromatic compounds into graphitic structures with five or seven carbon rings, differing from the original six carbon ring of pristine graphene. The final material exhibits a surface morphology characterized by a highly porous structure, which is embedded over the surface of PI as the procedure does not impact the entire precursor. These characteristics of laser induction allow the uniting of synthetic and manufacturing procedures within a single step, as previous synthetic steps can be avoided to proceed directly to scribing customized LIG patterns, which are embedded over inherent flexible substrates. For LIG-based electrochemical systems, geometries like that presented in Figure 15.2a can be directly written, either as a single working electrode (WE) or as an integrated electrochemical cell of three electrodes. In both cases, the active area and the conductive paths are completely manufactured of LIG, and the active electrode and clamp area are delimited by an insulating layer (Figure 15.2a,b). Although visually the patterns resemble the configuration obtained in screen-printed electrodes, laser induction can be performed without using any mask and LIG exhibits superior characteristics such as high surface area compared with carbon ink.

Even though there are no standardized conditions for the production of LIG, the morphology, conductivity, surface area, and electrochemical performance are directly related to the processing conditions applied to the precursor. However, as observed in Figure 15.2c, the parameters of the laser cutting machine as scan speed, laser power, beam size, and ppi density should be carefully studied and selected as the combination of values can

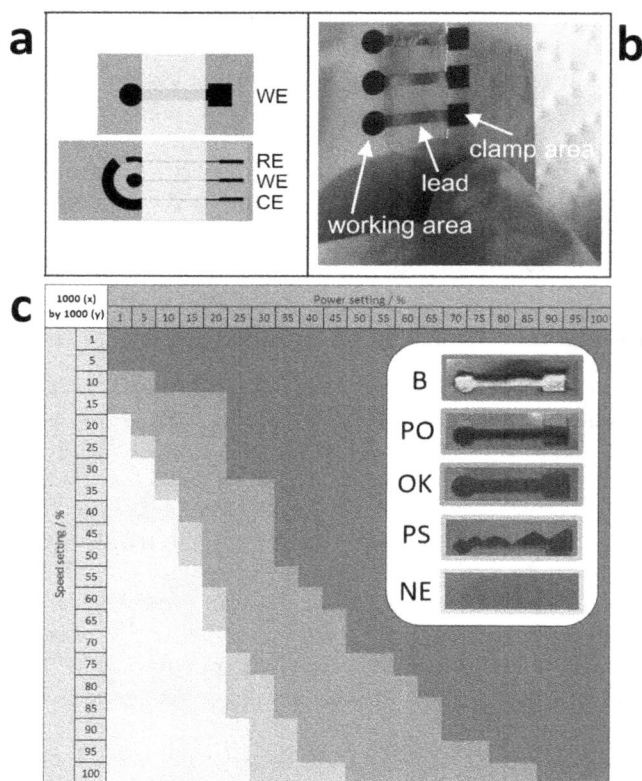

FIGURE 15.2
(a) Common geometries evaluated in LIG-based electrochemical sensors either as single WE or as an integrated three-electrode cell. (b) Example of LIG-WE with defined working and clamp areas isolated. (c) Experimental settings for speed and laser power parameters to produce no effect (NE), partial scribing (PS), LIG (OK), peeled off material (PO). and burned substrate (B). Adapted with permission [15]. Copyright The Authors, some rights reserved; exclusive licensee [Springer]. Distributed under a Creative Commons Attribution License 4.0 (CC BY) https://creativecommons.org/licenses/by/4.0/.

produce no induction at all to complete burning of the PI [15]. Additionally, fluence has been presented as a more convenient and quantitative parameter to study LIG formation rather than, for example, laser power with a fluence of at least 5 J/cm^2 required to ensure the formation of graphitic structures [16,17]. The most common morphologies are three-dimensional foam [18], cellular networks [17], lamellar [19], and fibers [16]. They are obtained by increasing the energy flow applied during the induction process, specifically to form fibers; a lower ppi density is required to reduce the overlap of the laser beam during treatment. LIG obtained from PI has a conductivity from 2 to 35 S/cm [20,21], and sheet resistance from 17 to 58 Ω/sq [9,22], while LIG fibers increase up to 200 Ω/sq [16]. When precursors such as paper or a composite of Kraft lignin and cellulose nanofibers are used, they can present a sheet resistance of 56 Ω/sq and 2000 Ω/sq, respectively [23,24]. Pristine monolayer graphene has a heterogeneous electron transfer rate constant (k_{eff}^0) with a value of 1.11×10^{-3} cm/s [25]. On the other hand, LIG presents a value greater than 3×10^{-3} cm/s up to 1.46×10^{-2} cm/s, representing an improvement between 2.7 and 13 times. The increase in the physicochemical properties is due to a greater number of exposed edges, planes and microstructural defects in LIG. The surface area obtained by the BET technique for LIG reaches up to 340 m^2/g [9] and 106 m^2/g [26] when induced in air and nitrogen atmospheres, respectively. Electrochemically active areas between 0.057 cm^2 [27] and 1.3 cm^2 [28] have been calculated using electrochemical techniques for LIG obtained from PI.

As LIG has been increasingly used as an electrochemical template, additional procedures have been incorporated to confer novel properties than that obtained during fabrication. For example, to add stretchability to the electrodes, LIG has been transferred to polydimethylsiloxane (PDMS) substrates [7], which also allows the incorporation of patterned microfluidic systems for the sampling of biofluids [12]. The casting of PI with precursors for metallic nanoparticles takes advantage of the laser processing for growing gold (Au) [6], silver (Ag) [19], and Cu-Ru [29] nanoparticles, creating selective layers over the LIG surface within a single step. Additional pretreatments over PI substrate include UV radiation for improving the roughness and wettability of the sheet [19] and ozone treatment to increase hydrophilic and consequently better permeation of polymeric ion-selective membrane [30]. Prepared LIG has been subjected to acetic acid treatment to increase C–C bonds and enhance the conductivity of the electrodes [5] and to 1H-pyrrole propionic acid (PAA) modification to create moieties for covalent antibody immobilization [8].

15.2 Applications of Laser-Induced Graphene Sensors in Health Monitoring Biomarkers

15.2.1 Glucose

Glucose is by far the most reported analyte for chemical sensors and wearable systems. This trend has been motivated by the rising number of cases of DM registered worldwide, which has resulted in a highly active market for glucose biosensors. According to the 2021 report of the International Diabetes Federation, there are 537 million people with DM and about 240 million without diagnosis [31]. Since the most significant number of cases is concentrated in developing countries, it is crucial to find new methods of diagnosis and monitoring of glucose levels that are less traumatic, cheaper, and provide continuous tracking of the dynamic levels of glucose during the day. Selective glucose detection has largely been dependent on the glucose oxidase (GOx) enzyme as a biocatalyst, but the design of a novel nanostructured surface for enzyme-less glucose oxidation is rising great interest. Even the most common method of measuring glucose levels implies taking a blood sample; current trends are to look for alternative, less invasive ways by measuring glucose levels in other body fluids such as saliva, urine, tears, and sweat [32]. Wearable systems are particularly well suited for such analysis, either as enzymatic or non-enzymatic detectors; consequently, a significant amount of research has been conducted on LIG-based wearable sensors for glucose.

For example, Yoon H. et al. manufactured LIG-based electrodes that undergo an acetic acid treatment by dip coating. The procedure provided an additional reduction to the LIG structure, increasing the C–C bonds by 15% for the resulting material and improving the electrical properties compared with untreated LIG. The selectivity toward glucose of the acetic acid-treated LIG (AT-LIG) system was achieved with platinum nanoparticles (PtNPs) electrodeposited on the WE followed by an immobilized GOx enzyme. The AT-LIG/PtNPs/GOx biosensor exhibited a working range from 0.3 μM to 2.1 mM, a detection limit of 300 nM, and a sensitivity of 4622 μA/mM [5]. With the obtained analytical parameters, glucose detection on collected sweat samples was evaluated and compared with concentrations in blood glucose in three different scenarios: before meal, right after meal, and two hours after meal, registering similar trends in the recorded glucose levels after each scenario.

On the other hand, although the detection of glucose based on non-enzymatic methods is highly desirable, the main disadvantage of this approach is the need for an alkaline solution to produce glucose oxidation. In an interesting work presented by Zhu J. et al., the elaboration of LIG foam and LIG fibers (LIGF) decorated with bimetallic nickel (Ni) and Au layers was proposed as a non-enzymatic wearable sensor for glucose (Figure 15.3a) [33]. To allow the sensor to work in a wearable fashion, the Au/Ni/LIG was encapsulated with a reaction cavity that serves as a container for alkali solutions of 0.05 M NaOH (Figure 15.3b). The skin-interfaced component allowed on-body non-enzymatic glucose detection for sweat samples. Moreover, the sensor performance was evaluated in bending

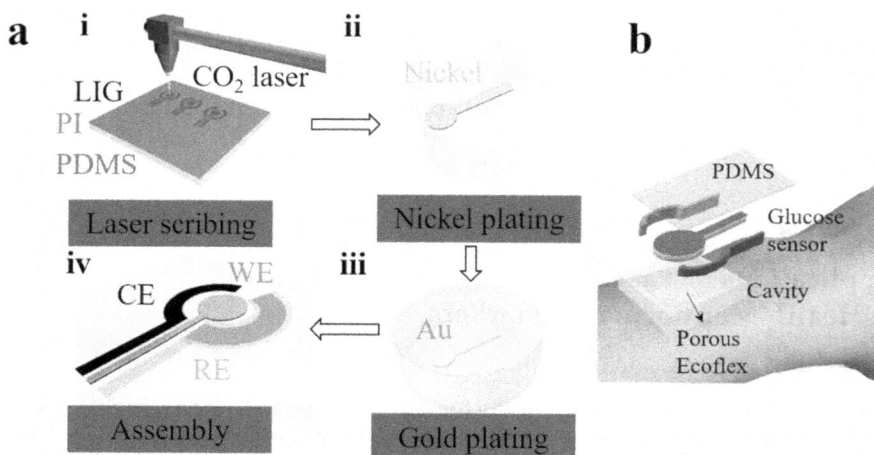

FIGURE 15.3
(a) Manufacturing of a non-enzymatic LIG-based wearable sensor for glucose: (i) LIG preparation by CO_2 laser writing, (ii) electroplating of Ni over WE, (iii) electroplating of Au over LIG/Ni/WE, (iv) assembly of three-electrode system with non-enzymatic Ni-Au catalyst. (b) Incorporation of a LIG sensor with a microfluidic system for sample collection and an alkaline reservoir designed for on-body detection. Adapted with permission [33]. Copyright (2021), Elsevier

conditions, showing only a low 2% variation in sensitivity. Other nanoparticles studied as non-enzymatic catalytic nanomaterials were Au [13], cobalt oxide (Co_3O_4) [34], and copper (Cu) nanostructures [35]. Table 15.1 summarizes the reported LIG-based sensors with their detailed analytical parameters classified according to the strategy for glucose detection, either enzymatic or non-enzymatic means. Notably, the presented LIG-based enzymatic glucose systems presented a more comprehensive performance, as they operate for glucose sweat detection at physiological conditions and the measurements have been contrasted with blood glucose measurements; even though glucose detection in sweat shows promising results, additional evaluations in long-term performance need to be carefully studied in these systems. As for non-enzymatic detection, the dependence of alkaline detection and the lack of evaluation in a real sample matrix that is more suitable for wearable operation are two missing points that require further development.

15.2.2 Hydrogen Peroxide

Hydrogen peroxide (H_2O_2) is a molecule present in human tissue as a product of mitochondrial respiration; when it is generated in abnormal amounts, it induces oxidative damage in cells, causing cardiovascular problems, brain damage, the appearance of tumors, Alzheimer's, and Parkinson's disease [40]. In addition, H_2O_2 is a well-known subproduct of oxidoreductase enzymatic reactions for lactic acid, glucose, cholesterol, and so on. Therefore, continuous measurement of H_2O_2 levels with sensors at low cost, that are less sensitive to environmental conditions, and are easy to manufacture is highly desirable for diagnosing or monitoring these diseases. In this sense, Zhao G. et al proposed anchoring silver nanoparticles (AgNPs) to LIG electrodes for highly sensitive non-enzymatic H_2O_2 detection. The AgNPs were fabricated jointly with LIG (Figure 15.4) by casting $AgNO_3$ solution over PI films and subsequently irradiated with a defocused laser at a 10 mm distance. To ensure the wettability of PI to the $AgNO_3$ solution, the substrate was pretreated with UV lasing. The morphology analysis showed that AgNPs were homogenously

TABLE 15.1

LIG-Based Sensors for Glucose Detection

Detection	Material	Linear range (mM)	Sensitivity (µA mM^{-1} cm^{-2})	LOD (µM)	Electrolyte	Sample	Ref.
Non-enzymatic	LIG/Ni/Au and LIGF	0–4(0.5V), 0–30 (0.1V)	1200 (0.5V), 3500 (0.1V)	1.5	0.05 M NaOH	Sweat	[33]
	LIG/Co_3O_4	0.001–9	214	0.41	0.1 M NaOH	Human serum	[34]
	LIG	0–1, 1–15	27.24, 5.42 µA/mM	130	PBS	-	[23]
	LIG	0–8	43.15	431	PBS	-	[36]
	LIG/AuNP	0.01–10	0.044±0.005 mA/log(M)	6.3	0.5M KOH	-	[13]
	LIG	0.001–0.1	-	0.138	-	Human serum	[37]
	LIG/CuNPs	0–6	495	0.39	0.1 M NaOH	Human serum	[35]
	LIG/CuNCs	0.025–4.5	4532.2	0.25	0.1 M NaOH	-	[38]
Enzymatic	LIG/ PEDOT:PSS/ Pt@PdNPS/GOx	0.003–9.2	247.3	3	0.1 M PBS	Sweat	[39]
	LIG/PtNPs/GOx	0.0003–2.1	4.622 µA/mM, 65.6	0.3	50 mM PBS	Sweat	[5]

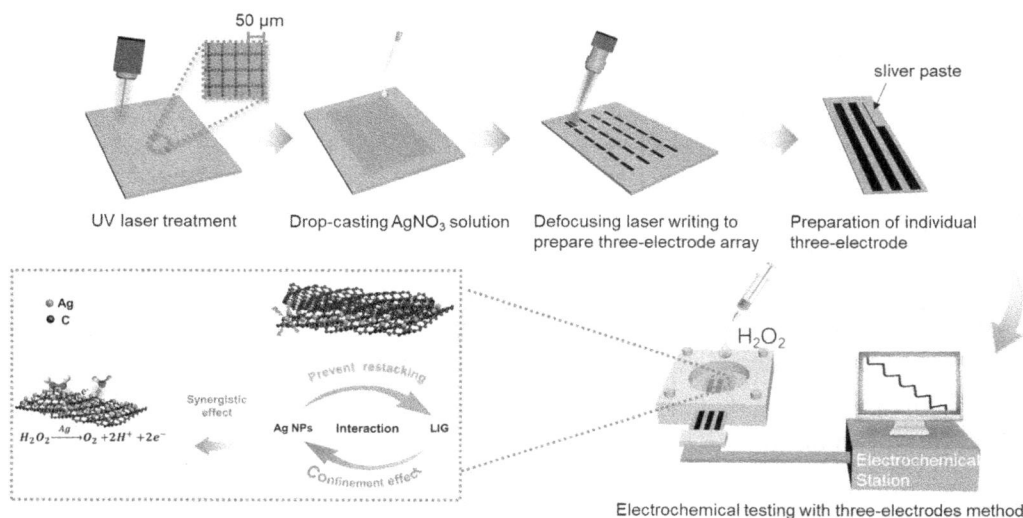

FIGURE 15.4

Development of a LIG/AuNPs sensor: UV pretreatment of PI, drop casting of Ag precursor, laser writing by defocusing, preparation of three-electrode cell, electrochemical evaluation of the H_2O_2 LIG sensor, and obtained morphology of the material. Adapted with permission [19]. Copyright (2021), Elsevier.

TABLE 15.2

LIG-Based Electrochemical Sensors for H_2O_2

Material	Linear range (mM)	Sensitivity (μA mM^{-1} cm^{-2})	Limit of detection (μM)	Ref.
LIG@Ag	0.01–0.55, 0.55–2.61	28.6	2.8	[19]
B-LIG, MWCNT-LIG	2–12	−6.25, −8.63 μA/mM	–	[11]
LIG/Cu-Ru	0.01–4.32	136.7	1.8	[29]
LIG/PtNPs	0–5	248.4	0.2	[41]

dispersed on LIG sheets without excessive growing, which resulted in sensing of H_2O_2 with a low limit of detection (2.8 μM), fast response (3 s), and long-term stability [19]. In a different study, the Thirumalai group synthesized copper–ruthenium bimetallic nanoparticles (Cu-Ru NPs) to mimic peroxidase enzyme in situ by laser treatment [29]. In a first step, LIG electrodes were manufactured by irradiation of PI, then precursors for Cu and Ru nanoparticles were cast over LIG and adsorbed due to the porous structure of the material. Finally, the prepared LIG was irradiated again to produce the growth of the bimetallic nanoparticles. As a result, the LIG/Cu-Ru NPs achieved detection of H_2O_2 in a wide linear range of 0.01–4.32 mM with a detection limit of 1.8 μM, high sensitivity of 136.7 μA mM^{-1} cm^{-2}, and good selectivity. The additional reported work for H_2O_2 detection with LIG systems is presented in Table 15.2.

15.2.3 Dopamine

Dopamine (DA) is the main neurotransmitter of the central nervous system, and abnormal levels are linked to neurological conditions like Alzheimer's, schizophrenia, and Parkinson's disease. DA's crucial importance has motivated a plethora of electrochemical sensors, either using enzymatic methods with tyrosinase enzyme or nanostructured surfaces for detection. Electrochemical detection based on platinum nanoparticles (PtNPs) anchored on LIG electrodes has been evaluated for DA detection [42]. In the first work, the PtNPs were fabricated by electrodeposition, achieving a high sensitivity of 6995.6 μA mM^{-1} cm^{-2} for DA. Considering this approach, Hui et al updated the procedure by transferring the induced LIG into PDMS to improve the mechanical properties of the system, enhancing flexibility and adding stretchability to the system (Figure 15.5) [7]. Additionally, the bimetallic Pt-AuNPs were added as a catalyst; the PDMS/LIG/Pt-AuNPs sensor showed a high sensitivity of 865.80 μA mM^{-1} cm^{-2} and a wide working range with a low LOD of 75 nM. The change in resistance of the WE was evaluated under increasing elongation, resulting in a maximum allowance of 10% of elongation. Noticeably, the PDMS/LIG/Pt-AuNPs sensor successfully performed in the presence of interferences like ascorbic acid (AA) and UA, even at high concentrations without affecting DA detection. The efficacy of the sensor in the real sample matrix was evaluated in human urine, showing potential application for the detection of DA in biofluids. The additional work for DA detection based on LIG systems along with the proposed modifiers is listed in Table 15.3.

15.2.4 Ions

Detection of K$^+$, Na$^+$, and NH$_4^+$ has implications for physiological conditions like hypokalemia, heat stroke, and dehydration, which can be especially severe in infants and the elderly. The level of ions can be estimated by monitoring through potentiometric detection with ion-selective electrodes (ISE). Typically, a body with an inner filling solution to maintain the ionic solution between the electrode and the selective membrane is needed, although all-solid-state ion-selective graphene electrodes have largely been an attractive

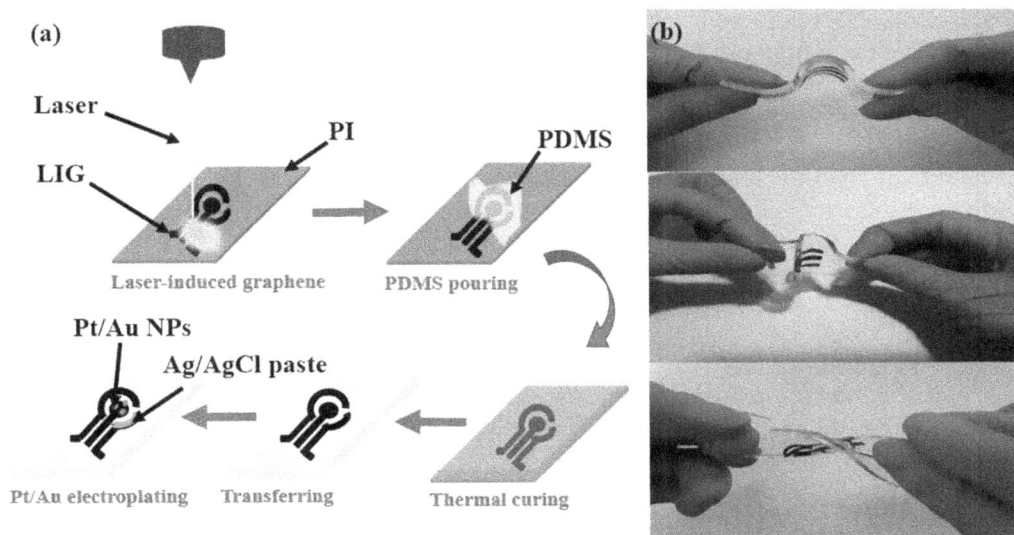

FIGURE 15.5
(a) Manufacturing and transferring of a LIG-based sensor for DA to a PDMS polymer: writing the LIG electrodes, then pouring PDMS over induced material, thermal curing of PDMS-LIG-PI substrate, transferring LIG to PDMS, and decoration of WE with Pt-Au NPs. (b) Evaluation of flexibility and stretchability of the transferred LIG sensor for DA. Adapted with permission [7]. Copyright (2019), Elsevier.

TABLE 15.3

LIG-Based Sensors for DA

Material	Linear range (μM)	Sensitivity (μA mM^{-1} cm^{-2})	Limit of detection (μM)	Sample	Ref.
LIG (Kraft lignin and CNFs)	5–40	4.39	3.4	-	[24]
PDMS/LIG/Pt-AuNPs	0.95–30	865.8	0.075	Human urine	[7]
LIG/ITO	5–100, 0.2–24	-	5, 0.2	-	[43]
LIG-PEDOT	1–150	0.220 μA· μM^{-1}	0.33	-	[44]
LIG/PtNPs	0–30	6995.6	0.07	-	[42]

CNFs, cellulose nanofibers.

TABLE 15.4

LIG-ISE Applications for Ions

Material	Analyte	Linear range (mM)	Sensitivity (mV/dec)	Limit of detection (mM)	Sample	Ref.
PDMS/lignin LIG	Na$^+$, K$^+$	0.0001–100, 0.00001–100	63.6 (Na$^+$), 59.2 (K$^+$)	-	Sweat	[46]
LIG	NH^{4+}, K$^+$	0.1–150, 0.1–150	51 (NH^{4+}), 53 (K$^+$)	0.03 (NH4+), 0.1 (K+)	Urine	[47]
O-LIG	Na$^+$	10–1000	60.2	0.001	Sweat	[30]

option but with a complex fabrication process. LIG was applied to simplify the fabrication process of all-solid-state ion-selective electrodes (ISEs) for K$^+$ and NH$^+_4$ [45]. The LIG electrodes were modified with polymeric membranes containing ionophores valinomycin and nonactin. LIG-ISE effectively detected the variation of K$^+$ and NH$^+_4$, showing an average of 54 mV/dec for both ions. The response was evaluated at different pH and in the presence of different ions to evaluate cross-selectivity. Additionally, LIG-ISEs were evaluated in the clinical range to monitor hydration in urine samples of two volunteers engaged in physical activity to induce dehydration with the recorded levels of K$^+$ and NH$^+_4$ effectively correlated with hydration status. Other approaches in Table 15.4 include inducing LIG over PDMS/lignin composite [46] and pretreatment of PI with ozone (OLIG) for improving the adhesion of the polymeric selective membrane [30].

15.2.5 Cortisol

Cortisol is a steroid hormone commonly known as the "stress hormone" but it has a wide variety of functions involving systems such as musculoskeletal, cardiovascular, respiratory, endocrine, and nervous. When chronically elevated, cortisol can have adverse effects on weight, immune function, and chronic disease risk [48]. Because its balance is essential for human health, the continuous monitoring of this hormone by wearable devices is of high interest. However, the LIG surface needs to be modified to create selective biolayers that allow cortisol detection through affinity interactions since it is a small molecule. In this sense, a fully integrated wearable system was presented by Torrente-Rodríguez et al. to study the dynamic nature of sweat cortisol to correlate with serum cortisol [8]. The LIG electrodes were modified with PPA for subsequent anti-cortisol mAb antibody immobilization by covalent bonding. The competitive immunoassay is multi-step, which

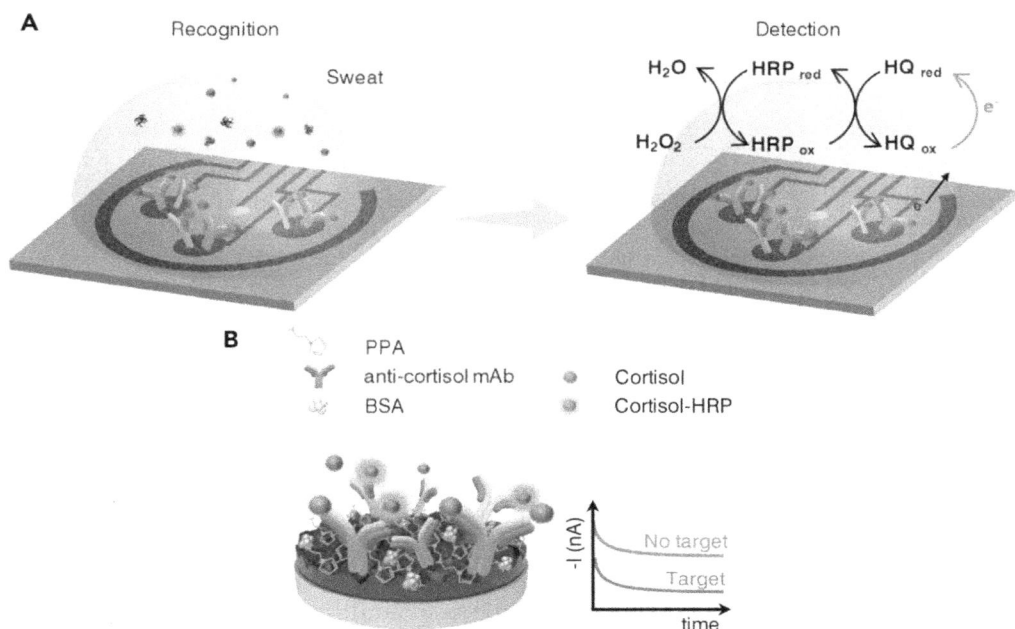

FIGURE 15.6
(a) Interaction and the principle of detection of a LIG-based immunosensor for cortisol detection in sweat. (b) Assembly of PPA and anti-cortisol mAb over a LIG electrode for detection. Adapted with permission [8]. Copyright (2020), Elsevier.

was further optimized to a short detection time (Figure 15.6). The affinity assay took 15 minutes, but it could be as low as 1 minute to emulate real-time monitoring. The LIG/PPA/anti-cortisol mAb biosensor showed a linear range from 0.43–50.2 ng/dL and a LOD of 0.08 ng/dL. Notably, this sensor was coupled with a miniaturized potentiostatic circuit to handle information and applied to study the sweat cortisol dynamic in a pilot study for the first time in a wearable fashion. On the other hand, San Nah et al. developed a wearable sensor for cortisol detection using a LIG transferred to PDMS [12]. To compensate for the surface reduction of LIG during the transference, the surface was enhanced with MXenes, which increased the surface area and provided additional moieties for antibody immobilization. To operate as an on-body system, a microfluidic system handled the sampling of sweat when the system was attached to the skin. The patch PDMS/LIG/MXene immunosensor showed linear relation from 0.01 to 100 nM and a LOD of 88 pM.

15.2.6 SARS-CoV-2

After the appearance of the new SARS-CoV-2 in Wuhan in December 2019, the COVID-19 disease produced by this virus rapidly became a pandemic. To help to control the rapid transmission of this new coronavirus, there is an urgent need to develop accurate quick tests that allow the population to be diagnosed. Fast results would allow the subjects to take healthcare actions and isolate them to limit the chains of contagion. Potentially, wearable SARS-CoV-2 sensors would allow helping persons to identify the risk of infection in their daily routine. For this purpose, a group led by Cui developed a LIG field-effect transistor to detect the SARS-CoV-2 spike protein [49]. This method can detect a virus

TABLE 15.5

LIG-Based Immunosensors for SARS-CoV-2

Material	Linear range		Limit of detection	Sample	Ref.
LIG-FET	-		1 pg/mL	Human serum	[49]
LSG/AuNS	5.0–500 ng/mL		2.9 ng/mL	Blood serum	[27]
	Serum (µg/mL)	Saliva (µg/mL)	-	Serum, saliva	[50]
LIG/NP,	0.1–0.8	0.5–2.0 ng/mL			
LIG/Anti-IgG,	20–40	0.2–0.5			
LIG/Anti-IgM,	20–50	0.6–5.0			
LIG/Anti-CRP	10–20	0.1–0.5			

concentration of up to 1 pg/mL in human serum with high specificity. On the other hand, in 2021, Beduk et al. reported the development of an electrochemical immunosensor made up of LIG, Au nanostructures, a cross-linking agent, and the SARS-CoV-2 spike antibody to detect up to 2.9 ng/mL with a linear range of 5 to 500 ng/mL. Although the device was conceived as a PoC detector, the miniaturized potentiostat in combination with a flexible sensor could be translated into a wearable scenario. In addition, the system was evaluated in a clinical trial with 23 patients, and the results achieved a high correlation with the results obtained by the RT-PCR test [27]. On the other hand, the multianalyte detection platform SARS-CoV-2 RapidPlex includes sensing of nucleocapsid protein (NP), IgG and IgM immunoglobulins, and the inflammatory C-reactive protein (CRP) biomarker within a single device [50]. The prototype aimed for a more comprehensive diagnosis of COVID-19, including parallel detection of viral infection, immune response, and severity of the disease. Table 15.5 shows the current LIG-based immunosensors for SARS-CoV-2 along with their exhibited analytical parameters and the evaluated samples during the assay.

15.3 Challenges for On-Body Measurements with LIG Wearable Sensors

The revised literature gives an outlook on the current efforts to develop wearable LIG-based sensors for healthcare biomarkers. Although extensive progress has been made in terms of chemical detection, some major challenges prevent the implementation of LIG-based wearable systems for real-life applications. First, most of the presented systems were focused on the design of a sensitive layer with LIG as a base, but flexibility and possible deflaking of LIG layers during continuous operation were only studied in a minor set of works [7,42]. Mechanical characterization of the sensor performance after continuous bending is commonly included for physical LIG-based sensors like gauges, but for further LIG-based chemical systems, this characterization should be included and considered as important as structural and electrochemical evaluation. Another key point is the on-body sample collection for continuous operation, which needs to be solved for operation in a wearable fashion. A major portion of the revised literature was focused on evaluating detection in a non-continuous mode in a phosphate buffered saline (PBS) electrolyte, which is more suitable for a fixed PoC format rather than wearable operation. Solutions like chambers and integrated microfluidic systems for sampling biofluids are promising

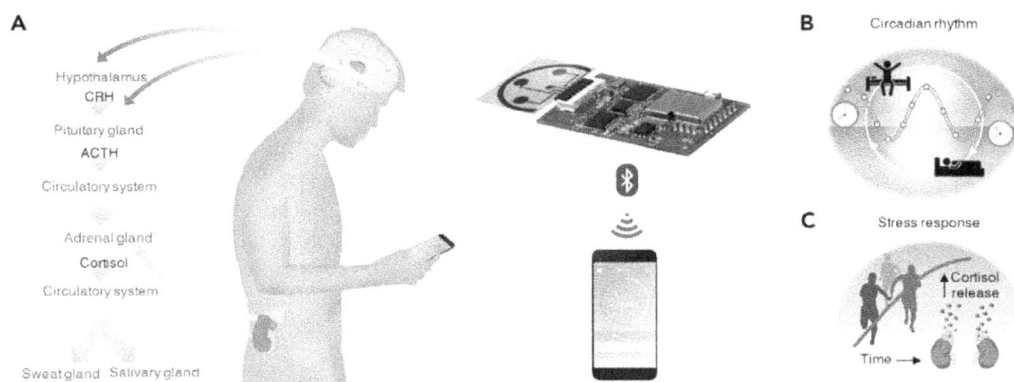

FIGURE 15.7
(a) Cortisol segregation mechanism for sweat and saliva along with the fully integrated wearable system based on LIG electrodes for detection, including electronic circuits for acquisition and data transmission to a smartphone. (b,c) Dynamic cortisol pattern study with the LIG-based cortisol system. Adapted with permission [8]. Copyright (2020), Elsevier.

and have already been successfully evaluated in sweat, but for additional samples like urine or saliva, additional strategies should be considered. Chemical detection needs to be evaluated in conditions as close as possible to real body fluid characteristics, such as concentration range and adequate pH. This is a pending task, especially for non-enzymatic glucose detection, where the alkaline electrolyte is needed during measurement adding and additional processing of the sample. LIG-based wearable systems should operate as reagent-less detectors or, in the case of immunoassays, to optimize measurements as close as possible to emulate real-time monitoring.

Noticeably, progress has been made in the fully wearable implementation of healthcare LIG sensors like the cortisol system presented in Figure 15.7 or the wearable sensors for sweat glucose detection; however, there are still challenges that need to be fully addressed. Fully operational wearable systems require a multidisciplinary approach that includes electronic engineers and technologists to develop adequate instrumentation for acquiring continuous electrochemical signals and the tools for wireless communication to transmit the information to mobile devices. Only a minor quantity of reports includes electronic circuits that are suitable for wearable operation, as most of the sensors were evaluated with benchtop potentiostats. The current electronic systems include custom instrumentation [8,27,50] or Arduino-based platforms [46], but further contributions could be conceived as a collaborative effort under the open-source hardware scheme to boost current sensing instrumentation. This necessity includes appropriate storage modulus and energy sources for continuous operation. The full integration of all parts will allow the evaluation of wearable systems in pilot studies, which can potentially validate the collected real-time information against reference methods. The final goal for these systems would be to operate in an outside-of-lab environment bringing information about healthcare biomarkers on a daily basis.

15.4 Conclusions

In this chapter, state-of-the-art wearable electrochemical systems based on LIG electrodes were presented. The background of LIG formation as a combination of laser engraving parameters, the importance of precursors, and the structural and electrochemical characteristics were presented. LIG-based systems for key biomedical analytes were revised; even though most works were focused on glucose and H_2O_2, sensing systems for DA, K^+, NH_4^+, and Na^+ ions, cortisol, and SARS-CoV-2 were also developed. To create sensitive layers, the growing of metallic nanoparticles, casting of ionophores, and the immobilization of antibodies were the most reported strategies. Although some key challenges remain, fully integrated LIG-based wearable systems have been evaluated for glucose, ion detection, and cortisol. The promising results encourage a multidisciplinary approach to the remaining challenges to create fully operational devices.

References

1. K.K. Reddy, H. Bandal, M. Satyanarayana, K.Y. Goud, K.V. Gobi, T. Jayaramudu, J. Amalraj and H. Kim, Recent trends in electrochemical sensors for vital biomedical markers using hybrid nanostructured materials, *Adv. Sci.* 7 (2020), pp. 1902980.
2. A. Yang and F. Yan, Flexible electrochemical biosensors for health monitoring, *ACS Appl. Electron. Mater.* 3 (2021), pp. 53–67.
3. H. Liu and C. Zhao, Wearable electrochemical sensors for noninvasive monitoring of health—A perspective, *Curr. Opin. Electrochem.* 23 (2020), pp. 42–46.
4. M. Xu, D. Obodo and V.K.V.K. Yadavalli, The design, fabrication, and applications of flexible biosensing devices, *Biosens. Bioelectron.* 124–125 (2019), pp. 96–114.
5. H. Yoon, J. Nah, H. Kim, S. Ko, M. Sharifuzzaman, S.C. Barman, X. Xuan, J. Kim and J.Y. Park, A chemically modified laser-induced porous graphene based flexible and ultrasensitive electrochemical biosensor for sweat glucose detection, *Sensors Actuators, B Chem.* 311 (2020), pp. 127866.
6. K. Samoson, A. Soleh, K. Saisahas, K. Promsuwan, J. Saichanapan, P. Kanatharana, P. Thavarungkul, K.H. Chang, A.F. Lim Abdullah, K. Tayayuth and W. Limbut, Facile fabrication of a flexible laser induced gold nanoparticle/chitosan/porous graphene electrode for uric acid detection, *Talanta* 243 (2022), pp. 123319.
7. X. Hui, X. Xuan, J. Kim and J.Y. Park, A highly flexible and selective dopamine sensor based on Pt-Au nanoparticle-modified laser-induced graphene, *Electrochim. Acta* 328 (2019), pp. 135066.
8. R.M. Torrente-Rodríguez, J. Tu, Y. Yang, J. Min, M. Wang, Y. Song, Y. Yu, C. Xu, C. Ye, W.W. IsHak and W. Gao, Investigation of cortisol dynamics in human sweat using a graphene-based wireless mHealth system, *Matter* 2 (2020), pp. 921–937.
9. J. Lin, Z. Peng, Y. Liu, F. Ruiz-Zepeda, R. Ye, E.L.G. Samuel, M.J. Yacaman, B.I. Yakobson and J.M. Tour, Laser-induced porous graphene films from commercial polymers, *Nat. Commun.* 5 (2014), pp. 5714.
10. F.M. Vivaldi, A. Dallinger, A. Bonini, N. Poma, L. Sembranti, D. Biagini, P. Salvo, F. Greco and F. Di Francesco, Three-dimensional (3D) laser-induced graphene: Structure, properties, and application to chemical sensing, *ACS Appl. Mater. Interfaces* 13 (2021), pp. 30245–30260.
11. K. Settu, Y.C. Lai and C.T. Liao, Carbon nanotube modified laser-induced graphene electrode for hydrogen peroxide sensing, *Mater. Lett.* 300 (2021), pp. 130106.
12. J.S. Nah, S.C. Barman, M.A. Zahed, M. Sharifuzzaman, H. Yoon, C. Park, S. Yoon, S. Zhang and J.Y. Park, A wearable microfluidics-integrated impedimetric immunosensor based on Ti3C2Tx MXene incorporated laser-burned graphene for noninvasive sweat cortisol detection, *Sensors Actuators, B Chem.* 329 (2021), pp. 129206.
13. B. Zhu, L. Yu, S. Beikzadeh, S. Zhang, P. Zhang, L. Wang and J. Travas-Sejdic, Disposable and portable gold nanoparticles modified - laser-scribed graphene sensing strips for electrochemical, non-enzymatic detection of glucose, *Electrochim. Acta* 378 (2021), pp. 138132.

14. R. You, Y.Q. Liu, Y.L. Hao, D.D. Han, Y.L. Zhang and Z. You, Laser fabrication of graphene-based flexible electronics, *Adv. Mater.* 1901981 (2019), pp. 1–22.

15. A. Behrent, C. Griesche, P. Sippel and A.J. Baeumner, Process-property correlations in laser-induced graphene electrodes for electrochemical sensing, *Microchim. Acta* 188 (2021), pp. 159.

16. L.X. Duy, Z. Peng, Y. Li, J. Zhang, Y. Ji and J.M. Tour, Laser-induced graphene fibers, *Carbon N. Y.* 126 (2018), pp. 472–479.

17. M. Abdulhafez, G.N. Tomaraei and M. Bedewy, Fluence-dependent morphological transitions in laser-induced graphene electrodes on polyimide substrates for flexible devices, *ACS Appl. Nano Mater.* 4 (2021), pp. 2973–2986.

18. D.X. Luong, A.K. Subramanian, G.A.L. Silva, J. Yoon, S. Cofer, K. Yang et al., Laminated object manufacturing of 3D-printed laser-induced graphene foams, *Adv. Mater.* 30 (2018), pp. 1707416.

19. G. Zhao, F. Wang, Y. Zhang, Y. Sui, P. Liu, Z. Zhang, C. Xu and C. Yang, High-performance hydrogen peroxide micro-sensors based on laser-induced fabrication of graphene@Ag electrodes, *Appl. Surf. Sci.* 565 (2021), pp. 150565.

20. S. Kaur, D. Mager, J.G. Korvink and M. Islam, Unraveling the dependency on multiple passes in laser-induced graphene electrodes for supercapacitor and H2O2 sensing, *Mater. Sci. Energy Technol.* 4 (2021), pp. 407–412.

21. L. Meng, A.P.F. Turner and W.C. Mak, Conducting polymer-reinforced laser-irradiated graphene as a heterostructured 3D transducer for flexible skin patch biosensors, *ACS Appl. Mater. Interfaces* 13 (2021), pp. 54456–54465.

22. S. Rauf, A.A. Lahcen, A. Aljedaibi, T. Beduk, J. Ilton de Oliveira Filho and K.N. Salama, Gold nanostructured laser-scribed graphene: A new electrochemical biosensing platform for potential point-of-care testing of disease biomarkers, *Biosens. Bioelectron.* 180 (2021), pp. 113116.

23. T. Pinheiro, S. Silvestre, J. Coelho, A.C. Marques, R. Martins, M.G.F. Sales and E. Fortunato, Laser-induced graphene on paper toward efficient fabrication of flexible, planar electrodes for electrochemical sensing, *Adv. Mater. Interfaces* 8 (2021), pp. 2101502.

24. F. Mahmood, Y. Sun and C. Wan, Biomass-derived porous graphene for electrochemical sensing of dopamine, *RSC Adv.* 11 (2021), pp. 15410–15415.

25. D.A.C. Brownson, A. Garcia-Miranda Ferrari, S. Ghosh, M. Kamruddin, J. Iniesta and C.E. Banks, Electrochemical properties of vertically aligned graphenes: Tailoring heterogeneous electron transfer through manipulation of the carbon microstructure, *Nanoscale Adv.* 2 (2020), pp. 5319–5328.

26. E.R. Mamleyev, S. Heissler, A. Nefedov, P.G. Weidler, N. Nordin, V.V. Kudryashov, K. Länge, N. MacKinnon and S. Sharma, Laser-induced hierarchical carbon patterns on polyimide substrates for flexible urea sensors, *NPJ Flex. Electron.* 3 (2019), pp. 1–11.

27. T. Beduk, D. Beduk, J.I. de Oliveira Filho, F. Zihnioglu, C. Cicek, R. Sertoz, B. Arda, T. Goksel, K. Turhan, K.N. Salama and S. Timur, Rapid point-of-care COVID-19 diagnosis with a gold-nanoarchitecture-assisted laser-scribed graphene biosensor, *Anal. Chem.* 93 (2021), pp. 8585–8594.

28. C. Fenzl, P. Nayak, T. Hirsch, O.S. Wolfbeis, H.N. Alshareef and A.J. Baeumner, Laser-scribed graphene electrodes for aptamer-based biosensing, *ACS Sensors* 2 (2017), pp. 616–620.

29. D. Thirumalai, J.U. Lee, H. Choi, M. Kim, J. Lee, S. Kim, B.S. Shin and S.C. Chang, In situ synthesis of copper-ruthenium bimetallic nanoparticles on laser-induced graphene as a peroxidase mimic, *Chem. Commun.* 57 (2021), pp. 1947–1950.

30. S. Choudhury, S. Roy, G. Bhattacharya, S. Fishlock, S. Deshmukh, S. Bhowmick, J. McLaughlign and S.S. Roy, Potentiometric ion-selective sensors based on UV-ozone irradiated laser-induced graphene electrode, *Electrochim. Acta* 387 (2021), pp. 138341.

31. IDF, *IDF Diabetes Atlas*, 10th ed. Brussels, Belgium, 2021.

32. P. Makaram, D. Owens and J. Aceros, Trends in nanomaterial-based non-invasive diabetes sensing technologies, *Diagnostics* 4 (2014), pp. 27–46.

33. J. Zhu, S. Liu, Z. Hu, X. Zhang, N. Yi, K. Tang, M. Gregory Dexheimer, X. Lian, Q. Wang, J. Yang, J. Gray and H. Cheng, Laser-induced graphene non-enzymatic glucose sensors for on-body measurements, *Biosens. Bioelectron.* 193 (2021), pp. 113606.

34. J. Zhao, C. Zheng, J. Gao, J. Gui, L. Deng, Y. Wang and R. Xu, Co3O4 nanoparticles embedded in laser-induced graphene for a flexible and highly sensitive enzyme-free glucose biosensor, *Sensors Actuators B Chem.* 347 (2021), pp. 130653.

35. Y. Zhang, N. Li, Y. Xiang, D. Wang, P. Zhang, Y. Wang, S. Lu, R. Xu and J. Zhao, A flexible non-enzymatic glucose sensor based on copper nanoparticles anchored on laser-induced graphene, *Carbon N. Y.* 156 (2020), pp. 506–513.

36. K. Settu, P.T. Chiu and Y.M. Huang, Laser-induced graphene-based enzymatic biosensor for glucose detection, *Polymers (Basel)* 13 (2021), pp. 2795.

37. M. Bhaiyya, P. Rewatkar, M. Salve, P.K. Pattnaik and S. Goel, Miniaturized electrochemiluminescence platform with laser-induced graphene electrodes for multiple biosensing, *IEEE Trans. Nanobiosci.* 20 (2021), pp. 79–85.

38. F. Tehrani and B. Bavarian, Facile and scalable disposable sensor based on laser engraved graphene for electrochemical detection of glucose, *Sci. Rep.* 6 (2016), pp. 27975.

39. M.A. Zahed, S.C. Barman, P.S. Das, M. Sharifuzzaman, H.S. Yoon, S.H. Yoon and J.Y. Park, Highly flexible and conductive poly (3, 4-ethylene dioxythiophene)-poly (styrene sulfonate) anchored 3-dimensional porous graphene network-based electrochemical biosensor for glucose and pH detection in human perspiration, *Biosens. Bioelectron.* 160 (2020), pp. 112220.

40. A. Nandi, L.-J. Yan, C.K. Jana and N. Das, Role of catalase in oxidative stress- and age-associated degenerative diseases, *Oxid. Med. Cell. Longev.* 2019 (2019), pp. 1–19.

41. Y. Zhang, H. Zhu, P. Sun, C.K. Sun, H. Huang, S. Guan, H. Liu, H. Zhang, C. Zhang and K.R. Qin, Laser-induced graphene-based non-enzymatic sensor for detection of hydrogen peroxide, *Electroanalysis* 31 (2019), pp. 1334–1341.

42. P. Nayak, N. Kurra, C. Xia and H.N. Alshareef, Highly efficient laser scribed graphene electrodes for on-chip electrochemical sensing applications, *Adv. Electron. Mater.* 2 (2016), pp. 1–11.

43. Q. Hong, L. Yang, L. Ge, Z. Liu and F. Li, Direct-laser-writing of three-dimensional porous graphene frameworks on indium-tin oxide for sensitive electrochemical biosensing, *Analyst* 143 (2018), pp. 3327–3334.

44. G. Xu, Z.A. Jarjes, V. Desprez, P.A. Kilmartin and J. Travas-Sejdic, Sensitive, selective, disposable electrochemical dopamine sensor based on PEDOT-modified laser scribed graphene, *Biosens. Bioelectron.* 107 (2018), pp. 184–191.

45. I.S. Kucherenko, D. Sanborn, B. Chen, N. Garland, M. Serhan, E. Forzani, C. Gomes and J.C. Claussen, Ion-selective sensors based on laser-induced graphene for evaluating human hydration levels using urine samples, *Adv. Mater. Technol.* 5 (2020), pp. 1901037.

46. C.W. Lee, S.Y. Jeong, Y.W. Kwon, J.U. Lee, S.C. Cho and B.S. Shin, Fabrication of laser-induced graphene-based multifunctional sensing platform for sweat ion and human motion monitoring, *Sensors Actuators A Phys.* 334 (2022), pp. 113320.

47. I.S. Kucherenko, D. Sanborn, B. Chen, N. Garland, M. Serhan, E. Forzani, C. Gomes and J.C. Claussen, Ion-selective sensors based on laser-induced graphene for evaluating human hydration levels using urine samples, *Adv. Mater. Technol.* 5 (2020), pp. 1901037.

48. M. Zea, F.G. Bellagambi, H. Ben Halima, N. Zine, N. Jaffrezic-Renault, R. Villa, G. Gabriel and A. Errachid, Electrochemical sensors for cortisol detections: Almost there, *TrAC - Trends Anal. Chem.* 132 (2020), pp. 116058.

49. T. Cui, Y. Qiao, J. Gao, C. Wang, Y. Zhang, L. Han, Y. Yang and T. Ren, Ultrasensitive detection of COVID-19 causative virus (SARS-CoV-2) spike protein using laser induced graphene field-effect transistor, *Molecules* 26 (2021), pp. 6947.

50. R.M. Torrente-Rodríguez, H. Lukas, J. Tu, J. Min, Y. Yang, C. Xu, H.B. Rossiter and W. Gao, SARS-CoV-2 RapidPlex: A graphene-based multiplexed telemedicine platform for rapid and low-cost COVID-19 diagnosis and monitoring, *Matter* 3 (2020), pp. 1981–1998.

16

Textile-Based Flexible and Wearable Sensors

Fatma S. Erdonmez and Ayşe Çelik Bedeloğlu

CONTENTS

16.1 Introduction

As the needs and demands of human beings evolved over time, scientists and researchers have looked for ways to deliver to those demands. People live faster lives, and demand mobility, real-time data transfer, quick response times, and remote access, as well as comfort, multifunctionality, cost-effectiveness, and sustainability. In regard to textile materials, fibers, and polymers, there are numerous possible ways to enhance their functionalities by adding properties such as biocompatibility and conductivity, as well as antibacterial and

DOI: 10.1201/9781003299455-16

antistatic properties. New application areas are also introduced with the development of multifunctional textile materials, and one of these new areas is textile-based flexible and wearable sensors. Textile materials are sometimes used as basic textile structures and also as smart textiles applications in almost all sub-industry groups such as automotive, health, agriculture, environment, and military.

Advancements in current technology have enabled electronics to become more user-friendly by making them lighter, more flexible, more compatible, and minimized in size, while also increasing their sensitivity and accuracy. Flexible sensors and wearable electronics have increased the mobility of users by making these devices available for use on the go. They have also increased the number of potential applications for sensors by giving the user access to real-time data. Wearable electronics collect and transfer data, monitor activities, and detect changes in the norm, thus making the lives of professionals easier in a wide variety of sectors. They can be custom designed according to the area of application and consumer specifications. They can be integrated in different ways into various parts of objects such as rings, smart glasses, smart watches, wristbands, and bracelets. Sensors can also be integrated with textiles to form textile-based sensors. Textile-based sensors provide advantages to the user in several ways. They are flexible, breathable, comfortable, can have full contact with body contours, and can stretch or be tailored to fit different body sizes. Textile-based materials are generally more cost-effective than other materials used in flexible sensors such as carbon-based materials.

Flexible sensors are a vital component of flexible and wearable electronics. For application in wearable electronics, the technical requirements are to have lightweight and flexible materials that can withstand mechanical deformation such as bending, twisting, and elongation, and also be integrated into different surfaces while providing comfort to the user. Three different methods can be followed to fabricate electronic textiles. In the first method, an electronic device can be attached to a textile material such as a waistband with a pulsometer attachment or a wearable camera [1]. The textile is only used as a carrier and both components are independent of each other. In the second method, electronic components are integrated on or into the textile material and there are interconnections such as metal push buttons between the two [2]. The third method is the textile-based method where the textile material is part of the electronics. Figure 16.1 describes the relationship between textiles and flexible sensors. This chapter will analyze the materials used, fabrication methods, applications, challenges, and the future of textile-based flexible sensors for wearable applications.

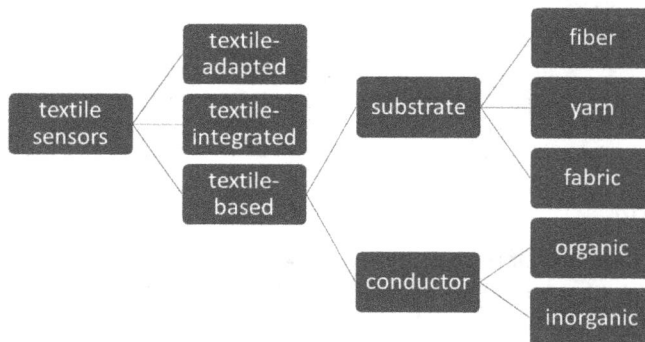

FIGURE 16.1
The relationship between textiles and flexible sensors.

16.2 Materials for Textile-Based Flexible and Wearable Sensors

Physical, mechanical, chemical, and electrical properties of flexible sensors are influential in the decision-making process for material choice, therefore, they should be taken into consideration when choosing the materials and designing the flexible sensor, keeping in mind the fabrication process, application area, and required parameters of the end product. Glass transition temperature and melting temperature should be compatible with processing temperatures. The sensors may require a structure that can conform and adapt to irregular surfaces, depending on the application area of the end product. Elasticity, permeability, tensile strength, surface roughness, adhesion, chemical stability, and electrical conductivity are some of the properties taken into consideration for flexible sensors. Some of the most common materials used for designing textile-based flexible sensors are conventional natural materials like wool, cotton, and paper, as well as polymeric materials such as polyester, polyurethane, polypropylene, polylactic acid, polycarbonate; conductive polymers polyaniline (PANi), polypyrrole (PPy), polyimide (PI) and poly(3,4-ethylenedio xythiophene):poly(styrenesulfonate) (PEDOT:PSS), and metal-based materials, Ag, Au, stainless steel, Cu, and aluminum. Materials for flexible sensors can be examined in two different categories: substrates and conducting medium [3].

16.2.1 Substrates

Substrates for textile-based sensors can be in fiber or filament form, yarn form, fabric, and nonwoven form. Using textile materials as substrates provides breathability, flexibility, comfort for the user, and variability of applications. With advances in technologies concerning textiles and new materials introduced to the industry, nanotechnology and small-scale electronics will make wearable sensors more available. Textile materials can be chosen based on the end use of the flexible sensor. The variety of materials available gives options to the manufacturers. Some of these requirements can be hydrophobicity, air permeability, flame retardancy, mildew resistance, antibacterial, antistatic, and quick drying. In each case scenario, the textile material can be manufactured according to the needs of the product. Textile-based sensors are manufactured in fiber/filament, yarn/thread, or fabric form.

16.2.1.1 Fibers

Fibers are the building blocks of textile materials and can be used to produce a variety of patterns. Fibers are lightweight, flexible, and durable and can be manufactured to carry the required mechanical and electrical properties. Conductive fibers can be used to design flexible sensors. Conductive fibers can be classified as intrinsically or extrinsically conductive. Metal-based filaments such as aluminum, nickel, copper, and stainless steel are intrinsically conductive materials and have been in circulation since the beginning of wearable electronics. Conductive polymers were discovered in the late 1970s by Heeger, Shirakawa, and Macdiarmid. Extrinsically conductive fibers can be manufactured by coating the fibers with a polymeric, carbon-based, or metal-based conductive layer [4, 5].

16.2.1.2 Yarns

Yarns are manufactured from filaments or fibers and they can be a single component as well as a bi-component or a multi-component. They carry the properties of the components as well as the mechanical properties given by the fabrication method, such as the frequency and direction of the twists. If the fibers/filaments forming the yarn have intrinsic electrical properties, the yarn also has intrinsic conductivity. If the yarn is made up of nonconductive materials, it can be coated with a conductive layer to introduce electrical properties to its structure.

16.2.1.3 Fabrics

Fabrics are manufactured by weaving, knitting, and nonwoven technologies. They can be manufactured according to the specifications of the wearable sensor. Breathability, body conformity, air and water permeability, skin compatibility, elasticity, flexibility, and durability are some of the benefits of using textile materials for flexible sensors.

16.2.2 Conducting Medium

A conducting medium can be applied in solution form, melt form, or nanoparticle form as ink or coating material, as well as in fiber, filament, or thread form for knitting, woven, and embroidery applications.

16.2.2.1 Polymers

Conductive polymers, first discovered by Heeger, Shirakawa, and Macdiarmid in the 1970s, have characteristics such as easy applicability, flexibility, biocompatibility, cost-effectiveness, and are lightweight, as well as mechanical properties compatible with flexible surfaces, which gives them an important place in the field of wearable electronics. Due to the doping process during polymerization, conductive polymers display relatively high levels of adjustable electrical conductivity. They are also easy to process. Conductive polymers contain monomers that can acquire positive or negative charges by oxidation or reduction that contribute to the electrical conductivity of intrinsically conductive polymers. Some examples of intrinsically conductive polymers are polyacetylene (PA), polyfuran (PF), PANi, poly(phenylenevinylene) (PPV), PPy, polythiophene (PT), and poly(*para*-phenylene) (PPP). The chemical structures of some of these polymers are shown in Figure 16.2 [4–7].

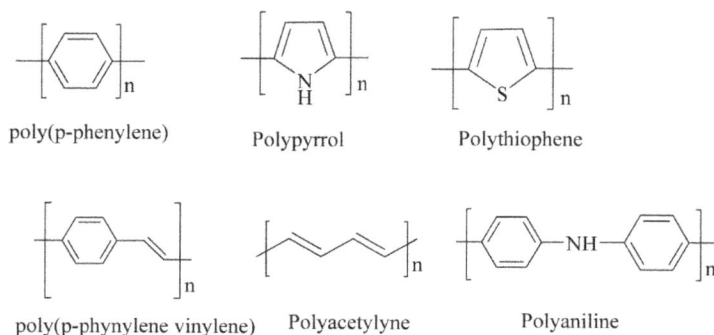

poly(p-phenylene)   Polypyrrol   Polythiophene

poly(p-phynylene vinylene)   Polyacetylyne   Polyaniline

FIGURE 16.2
Conductive polymer structures. Adapted with permission [6]. Copyright The Authors, some rights reserved; exclusive licensee Royal Society of Chemistry. Distributed under a Creative Commons Attribution License 3.0 (CC BY).

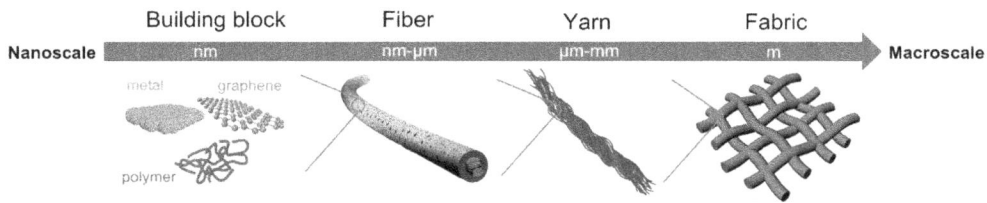

FIGURE 16.3
An overview of textile-based materials used in flexible wearable sensors. Adapted with permission [10]. Copyright The Authors, some rights reserved; exclusive licensee MDPI. Distributed under a Creative Commons Attribution License 4.0 (CC BY).

16.2.2.2 Metal-Based

Metal-based conducting mediums are commonly prepared as conductive inks. Conductive inks are prepared by suspension of metal nanoparticles or metal-organic decomposition (MOD). The process of producing metal nanoparticles is labor intense and energy-consuming, however, they are easy to disperse in inks for application in printed electronics. Stabilizers are added to the ink-metal nanoparticle solution to prevent the agglomeration of nanoparticles. Metals can also be used in filament or thread form and woven, knitted, or embroidered onto textile surfaces [8–13].

16.2.2.3 Carbon-Based

Carbon, one of the most prolific elements found in nature, has various allotropic forms. Graphite is one of these allotropes and is favored in electronic applications for its conductivity and thermodynamic stability. The graphite structure is multiple layers of graphene, which is a single-layered two-dimensional honeycomb structured allotrope of carbon atoms piled together. Other carbon-based materials used in the fabrication of flexible electronics are graphene oxide (GO) and reduced graphene oxide (rGO) derived from graphene: single and multi-wall carbon nanotube (SWCNT, MWCNT), carbon black, and activated carbon. Since graphene is a single-layer structure, it can transmit light, is flexible, and has mechanical strength but conductivity increases as the number of layers goes up. Carbon-based materials are used as fillers in composites. They are also highly favored in conductive inks [8–13].

16.2.2.4 Composite

To maximize the desired properties or to functionalize new properties for materials, they can be combined in several ways to form a composite structure. Carbon-polymer composites and organometallic conductive inks are an example of conductive composites [10,11]. Figure 16.3 presents an overview of textile-based materials used in flexible wearable sensors

16.3 Fabrication Methods for Textile-Based Flexible and Wearable Sensors

Textile-based sensors are expected to replace conventional sensors in wearable electronics because they carry the unique properties of textile materials while they function as sensors.

One advantage of textile-based sensors is that they can be manufactured using state-of-art fabrication methods in the place of conventional textile materials and do not need new investments in expensive high-tech production lines. Although it should also be considered that to improve the properties of textile-based sensors such as precision, accuracy, sensitivity, and durability; adjustments to fabrication methods need to be further studied.

Textile-based sensors can manifest in fiber, yarn/thread, or fabric forms, therefore, fabrication technologies include fiber and yarn spinning, weaving, knitting, and embroidering with conductive materials. In cases where the substrate is not conductive, coating and printing processes are applied. The coating is an efficient method for large-area applications [14,15].

Fabrication technologies are summarized in the following sections.

16.3.1 Coating Processes

Some coating methods are spray coating, in-situ polymerization, vapor-phase polymerization, dip coating, spray coating, roller coating, chemical vapor deposition, and vacuum filtration. They vary depending on the type of material and the expected outcome.

Dip coating is one of the most used techniques because it is a very basic and low-cost process for large-area applications. Dipping the material in a conductive solution and drying after the surface is coated is a simple process but one of the major challenges of dip coating is the weak adhesion of the layer of coating onto the substrate. Several researchers addressed this challenge through the interaction of functional groups on the substrate and the negatively charged coating solution. Others treated the sensors with acid and thermal reactions to increase adhesion as well as the mechanical and electrical properties of the sensors. Dip coating is favorable for large-area and high throughput but it is not good for designing a certain pattern on textile-based flexible sensors [14–16].

16.3.2 Printing Process

Like dip coating, printing methods are also uncomplicated and low cost but unlike dip coating, patterns can be applied to textile surfaces through printing technologies. Screen printing and stencil printing are two printing methods where conductive ink or paste is applied to a textile substrate to transfer a certain pattern onto the surface of the substrate. There are two screen-printing methods, flat bed and rotary screen. In the flat-bed method, the screen with a design mask is placed on top of the substrate and the conductive paste/ink is poured onto the screen and scraped with an applicator. The thickness of the film applied can be adjusted by using a film applicator with thickness options. A flat-bed method is usually applied to smaller surface areas whereas the rotary screen-printing method is a continuous process. In rotary printing, the substrate is passed between two rolling cylinders, one of which acts as a screen and feeds ink from its center. The applicator is also inside the screen cylinder and scrapes ink onto the substrate surface. The second cylinder is called an impression cylinder. Challenges faced in screen and stencil printing are the adjustments to the viscosity of the paste or ink, the thickness of the film applied on the substrate, exposure to air during air-drying, and mask deterioration due to the drying of the ink on the mask surface. With rotary screen printing, the challenge is the difficulty of cleaning the rotors and screens. The ink/paste may dry up and cause malfunctions if not cleaned properly [8, 17].

Inkjet printing is another method used to transfer conductive ink onto substrates. It is preferred for its flexibility, sustainability, and ease of set-up. It requires low material consumption and is user-friendly. Viscosity and particle size of the conductive ink is the focal point of this method. To obtain satisfactory results regarding conductivity and mechanical durability, the printed substrate ink solution should have optimal viscosity, a smaller particle size of conductive materials, and a homogenous dispersion. The working principle of inkjet printing is that ink is deposited via droplets from a nozzle onto a substrate and, as in everyday inkjet printers, any pattern can be transferred onto the substrate surface. The correlation between particle size, viscosity, and nozzle diameter is important because if the viscosity is too high or the particle size is too large for the nozzle diameter, clogging will occur and disrupt the printing process. Flaws or breaks in the pattern will increase resistivity and negatively affect conductivity [8, 16, 18, 19, 20]. Figure 16.4 shows the schematics of screen printing and inkjet printing processes.

A major challenge for printing processes is the mechanical deformations that may form on the coated surface after the ink/paste/solution dries. The textile substrate absorbs some of the coating material but because of the nonlinear structure of textiles, the coating may also be nonuniform. This can be overcome by applying a sublayer onto the textile surface preprinting to enhance uniformity and prevent mechanical deformations such as cracks and delamination on the surface.

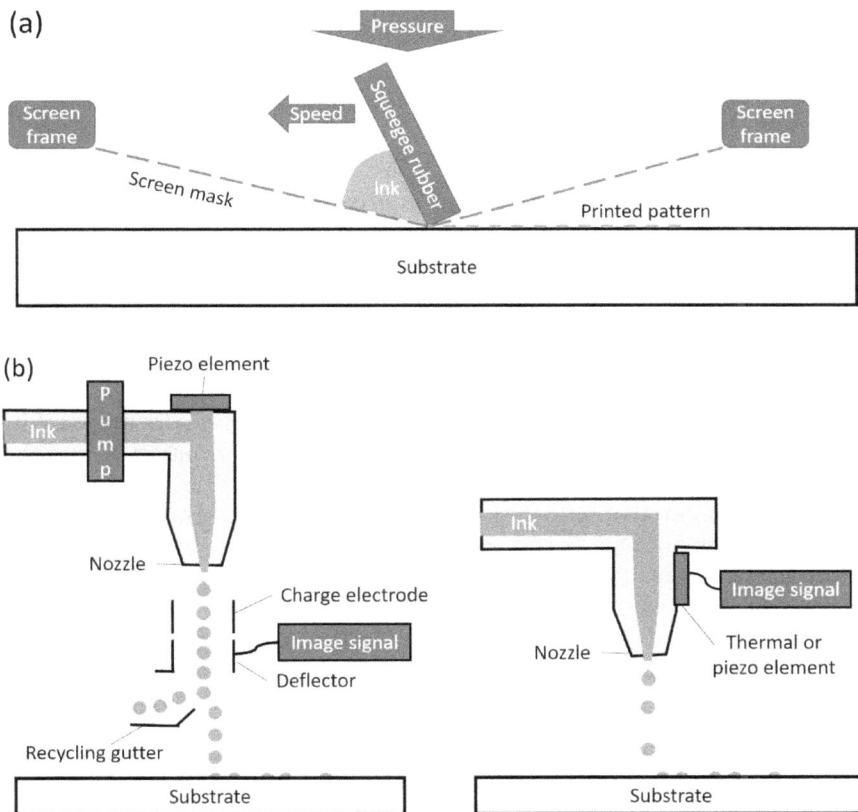

FIGURE 16.4
Schematics of (a) screen printing and (b) inkjet printing. Adapted with permission [8]. Copyright The Authors, some rights reserved; exclusive licensee MDPI. Distributed under a Creative Commons Attribution License 4.0 (CC BY).

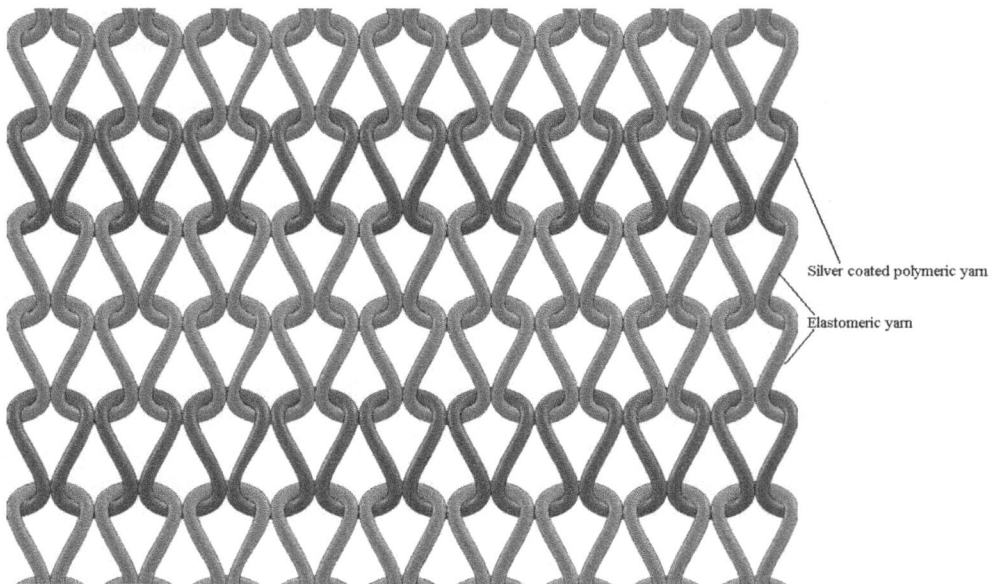

FIGURE 16.5
Knitted conductive fabric structure. Adapted with permission [21]. Copyright The Authors, some rights reserved; exclusive licensee MDPI. Distributed under a Creative Commons Attribution License 3.0 (CC BY).

16.3.3 Yarn-Type Processes: Knitting, Weaving, Sewing, and Embroidering

Conductive threads have been used as fashion accessories for over 1000 years. Artisans have been wrapping fine gold and silver foils around yarns used in embroidery for centuries. Their motivation was fashion but the same method can be applied to fabricate conductive patterns of textile substrates. Conductive filaments and yarns can be knitted, woven, sewn into, or embroidered onto a substrate to form flexible sensors in any specific pattern or shape necessary for the design of the sensor. Their flexible structure, small dimensions, and use of conventional fabrication methods make them good candidates for wearable flexible sensor applications. Woven, knitted, or embroidered sensors are well integrated with the textile interface but they can undergo mechanical deformation during the fabrication process and lose some of their functional characteristics. To avoid this problem, it is important to select materials that can withstand mechanical degradation during fabrication and that presents another challenge: limitation caused by the number of materials available that fit these specifications. Some materials such as silver-coated fibers used in yarn-type processes like embroidering can cause stiffness in the textile surface, therefore, this method is not applied in large-scale areas. Figure 16.5 shows a knitted silver-coated polymeric yarn and elastomeric yarn structure [17, 21].

16.4 Applications for Textile-Based Flexible and Wearable Sensors

16.4.1 Types of Sensors and Areas of Use

Sensors for flexible textile-based applications can be categorized as pressure, strain, temperature, and electrochemical sensors. Figure 16.6 illustrates the classification of types of fibers and areas of use.

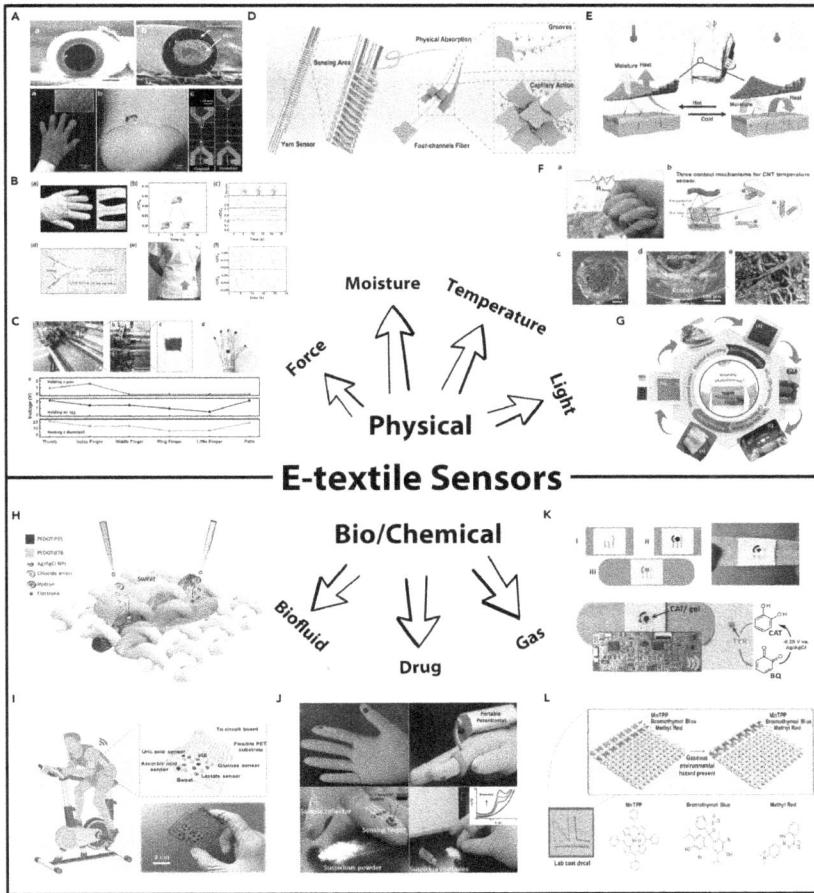

FIGURE 16.6
Types of sensors and areas of use. Adapted with permission [31]. Copyright The Authors, some rights reserved; exclusive licensee MDPI. Distributed under a Creative Commons Attribution License 4.0 (CC BY).

16.4.1.1 Pressure Sensors

Piezoresistive, piezoelectric, and capacitive pressure sensors are three different types of pressure sensors. As understood from their name, pressure sensors generate an electrical signal when they are under mechanical pressure. This signal can be a change in capacitance, resistivity, or electrical charge [2, 15, 22].

16.4.1.1.1 *Piezoresistive Pressure Sensor*

When mechanical stress is applied to a piezoresistive pressure sensor, the contact area of the conductive material alters, thus altering the resistance of the material. The formula for resistance equals resistivity times the length of material divided by cross-sectional area [$R = \rho L/S$]. Any change in these variables causes a change in resistance. Piezoresistive sensors have high sensitivity and precision and are easy to fabricate because of their basic uncomplicated structure. They are used in a wide spectrum of applications in the health and sports industries. However, they have some shortcomings such as low durability and hysteresis. Hysteresis is the expected margin of error when the output value is not the same as the input stimulus [17,23, 24, 25, 26].

16.4.1.1.2 Piezoelectric Sensors

Flexible piezoelectric sensors use materials with piezoelectric characteristics to generate a piezoelectric effect under pressure. These materials convert the mechanical energy formed under pressure into electrical energy. When pressure is applied, the deformation in the material causes positive and negative charges to shift and form a potential difference based on the degree of pressure applied. Piezoelectric materials have piezoelectric constants and this constant determines the performance of the sensor [17, 23, 24, 25].

16.4.1.1.3 Capacitive Sensors

Capacitive sensors are formed from conductive plates that are parallel to each other with a nonconductive dielectric material in between. When pressure is applied and the distance between the plates decreases, a change in capacitance occurs. The variables that affect the performance of capacitive sensors are the thickness and surface area of the sensor and the properties of the dielectric material in between. Conductive plates can be in the fiber, yarn, fabric, or film structure. The dielectric layer or spacer layer is generally composed of foam or rubber. The most common form of the capacitive sensor is the sandwich type with two parallel plates with a nonconductive layer in between. There is another type – a core/sheath structure where the core is dielectric and the sheath is conductive. A core/sheath structure can be manufactured by coaxial spinning or coating. This type of capacitive sensor is usually smaller in size and easier to apply in textile-based electronics [17, 23, 24, 25].

16.4.1.2 Strain Sensors

Similar to pressure sensors, strain sensors sense the physical deformation in their structure and transmit that deformation as an electrical signal. Most strain sensors used for textile-based flexible electronics are resistive and capacitive strain sensors. They are used to monitor motion and muscle movements such as respiration and heart rate [17, 27, 25, 28, 29, 22, 30, 31].

16.4.1.2.1 Resistive Strain Sensor

The difference between a piezoresistive pressure sensor and a resistive strain sensor is that in a pressure sensor, compression is applied, and in a strain sensor, the sensor stretches or contracts. Resistance of the sensor changes with the change in the sensor's length or area during exposure to strain.

16.4.1.2.2 Capacitive Strain Sensor

A capacitive strain sensor is also similar to a capacitive pressure center in its structure of two parallel conductive plates with a dielectric material sandwiched in between. When strain is applied to the sensor and the sensor stretches, the difference between the area or length between the parallel plates shifts and changes capacitance values.

16.4.1.3 Temperature Sensors

Temperature sensors can be electronic, electromechanical, or resistive. Bimetal thermostats are an example of electromechanical temperature sensors. They consist of two different metal types that have different reactions to temperature changes merged into a single strip. Different types of metal have different expansion behaviors and the difference between the rates of expansion of these materials can be measured as a change in temperature. Thermocouples belong to the electronic temperature sensors group. Like bimetal thermostats, they have two different metals in their structure. For thermocouples, different types of metal electrical conductors or semiconductors are brought together on one side of a circuit. Their working principle involves the Seebeck phenomenon where a thermal voltage otherwise known as electromotive force is created between the endpoints of two different conductors and the electron flow is reversed in the direction of a cold endpoint. Resistance temperature detectors (RTDs) are also widely used in temperature sensing. As temperature increases, electrons have higher vibration rates and the resistance of the material increases, restricting the flow of electrons in the sensor. Resistance temperature detectors use this property to determine the temperature. Thermistors are also resistive temperature sensors. They consist of different types of metal oxides attached to a semiconductor [25, 31, 32].

16.4.1.4 Electrochemical/Biosensors

Electrochemical or biosensors are functional in detecting biomolecules or chemical substances in the medium. They can be used to detect body fluids, nutrients, biological samples, and environmental samples. They consist of a single unit sensor with two components: receptors and transducers. The bioreceptor is a biomolecule that attaches itself to a specific molecule in the medium and starts a case-specific biological phenomenon. The transducer recognizes this phenomenon and turns it into a measurable signal [25, 33, 31, 34, 35].

16.4.1.5 Humidity Sensors

Humidity sensors are used to measure water content in the atmosphere. They can measure absolute humidity and relative humidity. Relative humidity is the amount of moisture in air or gas compared with the maximum amount measured at that specific temperature. To measure relative humidity, the sensor also needs to measure temperature. Humidity sensors can be capacitive, resistive, and thermal conductivity based. Textile-based humidity sensors are made with electrically conductive materials that are sensitive to water molecules [25, 31, 36, 37].

16.4.2 Applications

Textile-based wearable flexible sensors open new fields of applications for flexible sensors. As mentioned previously, different types of sensors measure a variety of parameters and these parameters enable sensors to be used for a wide range of applications such as healthcare, defense and military applications, automotive, aerospace, agriculture, sports, and artificial intelligence.

A textile-based biosensor used in military uniforms to detect exposure to biological or chemical agents during wartime can also be applied to other industries where employees are at risk of exposure to chemical and biological agents. They can be applied in agriculture to detect pesticides and chemical fertilizers. Sensor yarns are integrated into fibrous reinforcements (such as bulletproof vests) to investigate the behavior of the textile structure during high-velocity impact. In military applications, textile-based sensors can be used to detect if a soldier has been wounded, what part of their body they were hit, and the extent of their injuries. When real-time data is transmitted back to the command center, the necessary steps can be taken to provide urgent intervention to reach the wounded soldier.

In healthcare, some patients, especially those suffering from chronic illnesses, need round-the-clock close monitoring. It is impossible to provide constant care for these patients through conventional methods due to a lack of resources, the high cost of patient care, and the poor quality of life that such close monitoring in a medical institution would cause to the patient. With the assistance of textile-based flexible sensors, these patients can lead relatively normal lives while the wearable sensor monitors their parameters such as heart rate, blood glucose levels, and blood tension. Textile-based flexible sensors can also transmit these data to the patient's healthcare provider. Scientists are working on models that combine more than one mode of sensor into garments to measure multiple parameters such as heart rate, respiration, pH values, and body temperature [32]. Integrating physical and chemical sensor arrays with textile materials saves time and resources for both the patient and the healthcare providers, while also providing comfort to the user and collecting a continuous flow of data for the patient's and doctor's files.

In a study, a flexible strain-sensing fabric was applied to a finger and was used to monitor the motion of the finger in real time [27]. In another study, a smart glove was designed using piezoresistive sensors to recognize the gestures of the fingers, and the glove was used in laparoscopic surgery. The glove is supposed to recognize and mimic the surgeon's hand gestures to guide a robot during surgery [38]. Figure 16.7a shows an image of the hand gesture sensing glove and Figure 16.7b shows an image of the electronic suit for multiple parameter sensing. A textile-based foot pressure sensor designed as a sandwich structure was used in connection with a supercapacitor to store energy for wearable applications. Graphene-coated textile electrodes manufactured by screen printing provided good results for detecting electrocardiogram (ECG) signals.

Sports and wellness applications are other areas where textile-based flexible sensors are prevalent. Other sensors are also manufactured to monitor joint movement. Especially in sports applications, textile-based flexible wearable sensors can be used to monitor hydration/dehydration rates, heart rate and breathing patterns, and posture analysis. To improve performance and endurance based on custom-made training programs, particularly with professional athletes, these sensors play an important role. Textile-based physiological sensors can also be used for physical rehabilitation patients to monitor the patients' needs and changes in their condition [22, 30, 26, 39, 38, 40, 33, 41, 42, 31, 32].

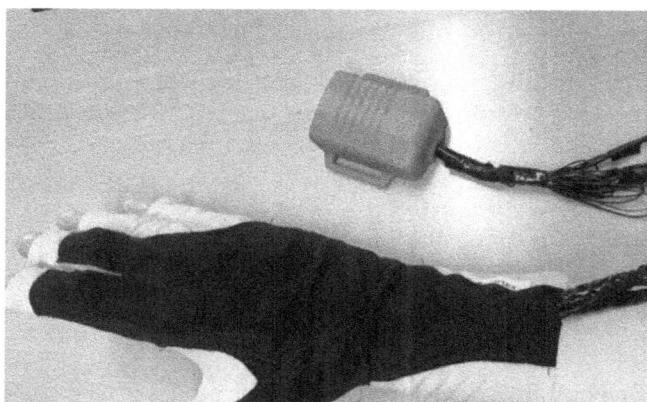

FIGURE 16.7(A)
(a) Smart glove for dynamic gesture recognition. Adapted with permission [38]. Copyright The Authors, some rights reserved; exclusive licensee MDPI. Distributed under a Creative Commons Attribution License 4.0 (CC BY).

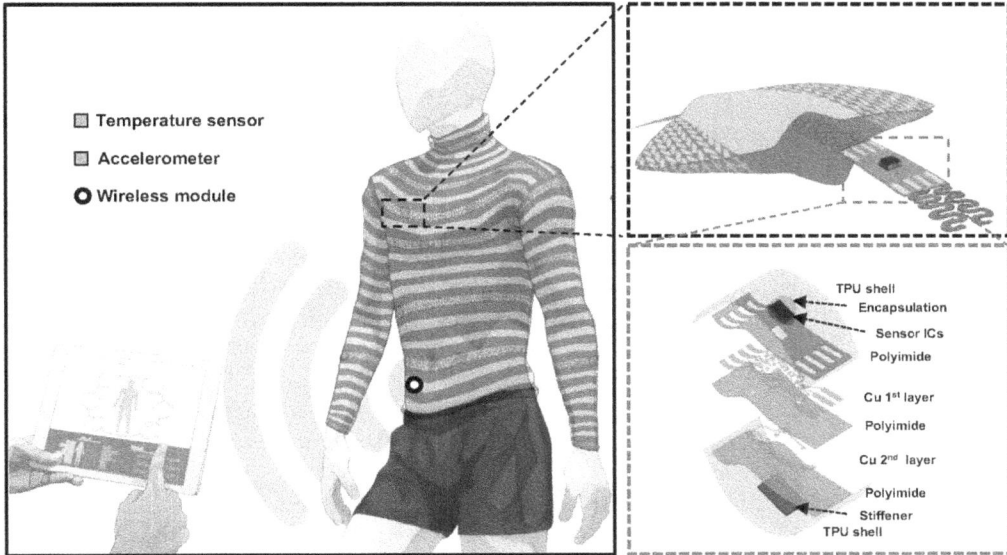

FIGURE 16.7(B)
A tailored, electronic textile conformable suit. Adapted with permission [32]. Copyright The Authors, some rights reserved; exclusive licensee Nature Partner Journals. Distributed under a Creative Commons Attribution License 4.0 (CC BY).

Applications for textile-based flexible sensors are not limited to the above-mentioned areas. They can be applied in various diverse industries and products to measure and monitor a wide range of parameters. Scientists are studying different sensor designs and materials to achieve optimum performance for each application area. The focus is to design textile-based flexible sensors that have high sensitivity, accuracy, and durability and mass-produce standardized commercial products.

16.5 Challenges, Opportunities, and the Future of Textile-Based Flexible and Wearable Sensors

Although advancements in current technology and new materials available to researchers have improved the status of textile-based flexible sensors, when compared with conventional sensors, they still lack accuracy and sensitivity as well as standards for fabrication and data reliability. Researchers have shown great interest in developing and improving textile-based flexible sensors because they portray the potential to make up for all the shortcomings of conventional sensors. Whereas conventional sensors are bulky, heavy, and uncomfortable, textile-based flexible sensors are lightweight, smaller in size, and provide comfort to the user. However, there is still a lot of room to improve.

Sensitivity, reliability, response time, data accuracy, durability, the difficulty of mass manufacturing, and cost are some of the main areas that need to be developed. The structure of the textile material that provides comfort to the user also provides a handicap for the sensor. The flexible, soft, breathable materials cause extra stress to the sensor and decrease its sensitivity and accuracy. In conventional sensors, because of their rigid structure, there is no extra bending and twisting, therefore, they have higher sensitivity and accuracy, and the results can be repeated and are more reliable.

The complex structure of textiles sometimes requires additional treatments for designing sensors, like coating with water repellants or conductive materials, which causes the textile material to lose some of its desirable characteristics such as making it bulkier and less flexible. The design and fabrication of textile-based sensors require a consortium of researchers from multiple disciplines including, but not limited to, textile and polymer engineers, chemists, physicists, computer and software scientists, mechatronics engineers, medical professionals, and fashion designers. This consortium can work together to overcome design flaws that cause mechanical stress and degradation on the sensor or develop sturdier sensors that can withstand the additional stress caused by the textile structure. Another disadvantage of textile-based sensors is that they need to be washed for re-use and the washing cycles shorten the lifespan of the sensor. If a sensor is not a single-use sensor, it needs to provide the same data accuracy and reliability after several cycles of washing as it did pre-washing.

The technology for the fabrication of textile-based sensors is mainly restricted to research laboratories and lacks the necessary standards for mass fabrication. This and the cost of chemicals, polymers, metal-based compounds, carbon-based compounds, and composites used in manufacturing textile-based flexible sensors increase the cost of the end product, making it unavailable to a great percentage of potential end users.

Another challenge of textile-based flexible sensors is the smooth transition between the textile-based sensor and the electronic components. Their structures and levels of flexibility differ and the contact points need to be secured [43, 44]. Since the fabrication of textile-based sensors is a relatively new technology, it is difficult to compare the performance of sensors due to a lack of methods and standards. Most of the testing and standards are developed for conventional sensors or conventional textile materials, therefore, the results are not conclusive [7, 14].

Based on market research reports, in recent years, flexible sensor potential for wearable electronics and Internet of Things applications has attracted researchers' interest, thus increasing their growth pace over the forecast period. One report predicts an annual growth rate of 6.8% for the forecast period of 2022–2029, which brings the market value of US\$ 5.34 billion in 2021 to approximately US\$ 9.04 billion in 2029. The application of flexible sensors in the healthcare industry is one of the drivers of the market. Manufacturers are showing a tendency toward flexible sensor technologies over conventional sensors. Another driving factor is an escalation in the use of smart wearable electronic devices. Developments in technologies such as artificial intelligence and automation will further expand the future growth of the flexible sensors market [45].

The textile-based flexible sensors industry is rapidly growing, however, future research is needed to resolve all technical limitations. Since the flexible wearable textile sensors industry is still young, most of the methods and materials used in fabrication need improvement and verification. The data from wearable textile sensors should be accurate and reproducible under similar conditions. The noise generated by the sensors should be reduced by using artificial intelligence or increasing the sensitivity and accuracy of the sensors. Flexible wearable electronic devices should be self-powered or energy efficient. Complex systems with several connections and arrays should be minimized and simplified and made lighter for a more user-friendly experience. Also, for a user-friendly experience, these products should be easy to apply and be used unaided by the target group without the need for expert intervention. These products will not be cost-effective if they are single-use disposable products, therefore, durability and long-term usage are also areas that need improvement in future research. They should withstand mechanical stress caused by conditions during wearing and washing and the accuracy of the sensor needs to keep at a reasonable level during the lifetime of the product. Comfort and aesthetic levels should increase to meet the demands of potential end users. Researchers should also address the standardization of sensor fabrication and data measurement methods, as well as methods to secure the data collected by these sensors [44].

Acknowledgement

This chapter is based upon work from COST Action "European Network to connect research and innovation efforts on advanced Smart Textiles" (CONTEXT, Ref. CA17107, https://www.context-cost.eu) supported by COST (European Cooperation in Science and Technology, https://www.cost.eu.

References

1. S. Mann, (1997). "Smart clothing": Wearable multimedia computing and "personal imaging" to restore the technological balance between people and their environments, in *MULTIMEDIA '96: Proceedings of the Fourth ACM International Conference on Multimedia*, 163–174.
2. P. Bosowski, M. Hoerr, V. Mecnika, T. Gries, S. Jockenhovel, & T. Dias (eds.), (2015). Chapter 4: Design and manufacture of textile-based sensors, in *Electronic Textiles*, Woodhead Publishing, Boston, 83–115.
3. L.M. Castano, & A.B. Flatau, (2014). Smart fabric sensors and e-textile technologies: A review, *Smart Materials and Structures*, 23(5), 053001.
4. A.M. Grancarić, I. Jerković, V. Koncar, C. Cochrane, F.M. Kelly, D. Soulat, & X. Legrand, (2018). Conductive polymers for smart textile applications. *Journal of Industrial Textiles*, 48(3), 612–642.
5. J. Ouyang, (2018). Recent advances of intrinsically conductive polymers. *Acta Physicochimica Sinica*, 34(11), 1211–1220.
6. K. Namsheer, & C.S. Rout, (2021). Conducting polymers: A comprehensive review on recent advances in synthesis, properties and applications. *RSC Advances*, 11(10), 5659–5697.
7. J.S. Heo, J. Eom, Y.H. Kim, & S.K. Park, (2018). Recent progress of textile-based wearable electronics: A comprehensive review of materials, devices, and applications. *Small*, 14(3), 1–16.
8. J. Wiklund, A. Karakoç, T. Palko, H. Yiğitler, K. Ruttik, R. Jäntti, & J. Paltakari, (2021). A review on printed electronics: Fabrication methods, inks, substrates, applications and environmental impacts. *Journal of Manufacturing and Materials Processing*, 5(3), 89.
9. R. Islam, N. Khair, D.M. Ahmed, & H. Shahariar, (2019). Fabrication of low cost and scalable carbon-based conductive ink for E-textile applications. *Materials Today Communications*, 19, 32–38.
10. S. Wu, (2022). An overview of hierarchical design of textile-based sensor in wearable electronics. *Crystals*, 12(4), 555.
11. İ. Üner, & B.H. Gürcüm, (2019). Elektronik tekstillerde iletken mürekkep uygulamaları. *Pamukkale Üniversitesi Mühendislik Bilimleri Dergisi*, 25(7), 794–804.
12. M. Krifa, (2021). Electrically conductive textile materials—Application in flexible sensors and antennas. *Textiles*, 1(2), 239–257.

13. A. Hatamie, S. Angizi, S. Kumar, C.M. Pandey, A. Simchi, M. Willander, & B.D. Malhotra, (2020). Review—Textile based chemical and physical sensors for healthcare monitoring. *Journal of the Electrochemical Society*, 167(3), 037546.

14. M.R. Miah, M. Yang, M.M. Hossain, S. Khandaker, & M.R. Awual, (2022). Textile-based flexible and printable sensors for next generation uses and their contemporary challenges: A critical review. *Sensors and Actuators A: Physical*, 344, 113696.

15. J.W. Zhang, Y. Zhang, Y.Y. Li, & P. Wang, (2022). Textile-based flexible pressure sensors: A review. *Polymer Reviews*, 62(1), 65–94.

16. Y.S. Rim, S.H. Bae, H. Chen, N. De Marco, & Y. Yang, (2016). Recent progress in materials and devices toward printable and flexible sensors. *Advanced Materials*, 28(22), 4415–4440.

17. N.A. Choudhry, L. Arnold, A. Rasheed, I.A. Khan, & L. Wang, (2021). Textronics—A review of textile-based wearable electronics. *Advanced Engineering Materials*, 23(12), 1–19.

18. W. Zeng, L. Shu, Q. Li, S. Chen, F. Wang, & X.M. Tao, (2014). Fiber-based wearable electronics: A review of materials, fabrication, devices, and applications. *Advanced Materials*, 26(31), 5310–5336.

19. G.M. Islam, A. Ali, & S. Collie, (2020). Textile sensors for wearable applications: A comprehensive review. *Cellulose*, 27(11), 6103–6131.

20. S.T. Han, H. Peng, Q. Sun, S. Venkatesh, K.S. Chung, S.C. Lau, & V.A.L. Roy, (2017). An overview of the development of flexible sensors. *Advanced Materials*, 29(33), 1700375.

21. S.T.A. Hamdani, P. Potluri, & A. Fernando, (2013). Thermo-mechanical behavior of textile heating fabric based on silver coated polymeric yarn. *Materials*, 6(3), 1072–1089.

22. J.C. Costa, F. Spina, P. Lugoda, L. Garcia-Garcia, D. Roggen, & N. Münzenrieder, (2019). Flexible sensors—from materials to applications. *Technologies*, 7(2), 35.

23. W. Wen, & F. Fang, (2022). Flexible sensors in smart textiles and their applications. *Wearable Technology*, 2(2), 83–91.

24. Z. Zhou, N. Chen, H. Zhong, W. Zhang, Y. Zhang, X. Yin, & B. He, (2021). Textile-based mechanical sensors: A review. *Materials*, 14(20), 6073.

25. J. Fontes, J.S. Wilson (eds.), (2004). Chapter 12: Humidity sensors, in *Sensor Technology Handbook*, Elsevier, Burlington, 271–284.

26. B. Barthod-Malat, C. Cochrane, & F. Boussu, (2021). Development of piezoresistive sensor yarn to monitor local fabric elongation. *Textiles*, 1(2), 170–184.

27. G. Cai, M. Yang, Z. Xu, J. Liu, B. Tang, & X. Wang, (2017). Flexible and wearable strain sensing fabrics. *Chemical Engineering Journal*, 325, 396–403.

28. X. Liu, J. Miao, Q. Fan, W. Zhang, X. Zuo, M. Tian, & L. Qu, (2022). Recent progress on smart fiber and textile based wearable strain sensors: Materials, fabrications and applications. *Advanced Fiber Materials*, 4, 1–29.

29. S. Seyedin, P. Zhang, M. Naebe, S. Qin, J. Chen, X. Wang, & J.M. Razal, (2019). Textile strain sensors: A review of the fabrication technologies, performance evaluation and applications. *Materials Horizons*, 6(2), 219–249.

30. C.R. Merritt, H.T. Nagle, & E. Grant, (2008). Textile-based capacitive sensors for respiration monitoring. *IEEE Sensors Journal*, 9(1), 71–78.

31. K. Du, R. Lin, L. Yin, J.S. Ho, J. Wang, & C.T. Lim, (2022). Electronic textiles for energy, sensing, and communication. *iScience*, 25(5), 104174.

32. I. Wicaksono, C.I. Tucker, T. Sun, C.A. Guerrero, C. Liu, W.M. Woo, & C. Dagdeviren, (2020). A tailored, electronic textile conformable suit for large-scale spatiotemporal physiological sensing in vivo. *NPJ Flexible Electronics*, 4(1), 1–13.

33. R.I. Stefan, J.F. Van Staden, & H.Y. Aboul-Enein, (2004). Biosensor technology, in *Analytical Instrumentation Handbook*, Third Edition. CRC Press, Boca Raton, 707–712.

34. Y.M. Gu, (2021, February). Advances in textile-based flexible temperature sensors. *Journal of Physics: Conference Series*, 1790(1), 012021.

35. A. Sinha, A.K.S. Dhanjai, & G.M. Stojanović, (2022). Textile-based electrochemical sensors and their applications. *Talanta*, 244, 123425.

36. I. Jeerapan, & S. Poorahong, (2020). Flexible and stretchable electrochemical sensing systems: Materials, energy sources, and integrations. *Journal of the Electrochemical Society*, 167(3), 037573.

37. Z. He, G. Zhou, Y. Oh, B.M. Jung, M.K. Um, S.K. Lee, & T.W. Chou, (2022). Ultrafast, highly sensitive, flexible textile-based humidity sensors made of nanocomposite filaments. *Materials Today Nano*, 18, 100214.
38. L. Santos, N. Carbonaro, A. Tognetti, J.L. González, E. De la Fuente, J.C. Fraile, & J. Pérez-Turiel, (2018). Dynamic gesture recognition using a smart glove in hand-assisted laparoscopic surgery. *Technologies*, 6(1), 8.
39. K. Meng, S. Zhao, Y. Zhou, Y. Wu, S. Zhang, Q. He, X. Wang, Z. Zhou, W. Fan, X. Tan, J. Yang, & J. Chen, (2020). A wireless textile-based sensor system for self-powered personalized health care. *Matter*, 2(4), 896–907.
40. Y. Jeong, J. Park, J. Lee, K. Kim, & I. Park, (2020). Ultrathin, biocompatible, and flexible pressure sensor with a wide pressure range and its biomedical application. *ACS Sensors*, 5(2), 481–489.
41. W. Fan, Q. He, K. Meng, X. Tan, Z. Zhou, G. Zhang, & Z.L. Wang, (2020). Machine-knitted washable sensor array textile for precise epidermal physiological signal monitoring. *Science Advances*, 6(11), eaay2840.
42. J. Ferri, C. Perez Fuster, R. Llinares Llopis, J. Moreno, & E. GarciaBreijo, (2018). Integration of a 2D touch sensor with an electroluminescent display by using a screen-printing technology on textile substrate. *Sensors*, 18(10), 3313.
43. I. Kim, J.S. Heo, & M.F. Hossain, (2020). Challenges in design and fabrication of flexible/stretchable carbon-and textile-based wearable sensors for health monitoring: A critical review. *Sensors (Switzerland)*, 20(14), 1–29.
44. Data Bridge, (2021). *Global Wearable Sensors Market - Industry Trends and Forecast to 2029*. https://www.databridgemarketresearch.com/reports/global-wearable-sensors-market
45. C.W. Kan, & Y.L. Lam, (2021). Future trend in wearable electronics in the textile industry. *Applied Sciences*, 11(9), 3914.

17

Color-Based Flexible and Wearable Sensors

Sahar Foroughirad, Behnaz Ranjbar, and Zahra Ranjbar

CONTENTS

17.1 Introduction

Today, wearable sensing technologies that can be attached to the body have attracted attention for making new medical products such as health monitoring and prosthetics, and also creating human-machine interfaces and smart robots. The required strategies for making a smart wearable sensing system (SWSS) mainly depend on their application. However, some key parameters should be kept in mind when developing SWSSs. Their substrate should be flexible and/or stretchable, they should have reversible sensing materials, the results should be monitored easily, and they should have a power supply [1].

Color-changing sensors help extensively in wireless monitoring. That is why they are so important. A lot of color-changing sensors are inspired by nature. For example, Chou et al. [2] made a monitoring wearable tactile sensing system for an e-skin, inspired by a chameleon. They showed that the e-skin can sustain its skin color with no need to apply voltage. Yokota et al. [3] also made a photonic system, which could measure the oxygen level of blood from the ratio of the two pulsatile components when attached to a finger. The color indicator they made directly displayed the data from the body.

As color-changing sensors are mainly nature-inspired, they should be investigated more precisely. In nature, there are two groups of coloration, one is coloration based on pigments and the other is coloration based on changing the structure (structural colors). In the first group, the color change is because of the chemical structure of pigments. This kind of coloration is likely the most popular type of coloration that exists in nature. In the second type, coloration, based on structural change, is made due to the specific microstructure of the surface that is impacted by specific light wavelengths. Although suggesting some specific benefits, this coloration is not so popular.

DOI: 10.1201/9781003299455-17

Structural colors are very stable and long-lasting because they are not influenced by the degradation of pigment photo-chemicals. Structural colors are made of different surface structural properties such as photonic crystals, crystal fibers, surface gratings, and spiral coils. Different mechanisms are utilized on the basis mentioned above to make a specific color. It is vital to know the mechanism precisely to be able to make artificial colors with the same properties [4].

Color-changing sensors (CCSs) can operate with different stimuli. These stimuli are gas and any liquid, temperature, ions, electric and/or magnetic field, and mechanical force. Based on the final application, the structure of the sensing system can be different [5]. In this chapter, we introduce some of the recent research on the fabrication of CCSs such as bulk and colloidal hydrogels, elastomeric hydrogels, thin films, fibers, and core-shell nanoparticles. Finally, the current challenges and future perspectives in this area are discussed.

17.2 Classification of Color-Changing Sensors Based on Their Structure

CCSs are gaining attention on account of their promising applications. The most significant applications of these devices are collecting public and/or personal healthcare information and accurate measurement of specific vital signs such as blood pressure, body temperature, and glucose level [6]. These sensors are usually prepared with the aid of hydrogels, as bulk, colloidal, or elastomeric. They can be prepared as fibers, thin films, and core-shell structures that can be further employed in weaving an intelligent fabric. The following sections are a detailed discussion of each synthesis strategy.

17.2.1 Bulk and Colloidal Hydrogels

Hydrogels can be employed in color-based sensors either by the application of light-absorbing materials such as dyes and pigments or by the use of structural colors. The first category of materials exhibit color by selective absorption of light in UV or visible regions. However, the second classification belongs to the materials that can diffract light by their specific orientation and crystal structure. Both aforementioned categories can be employed in the fabrication of color-changing hydrogels for flexible and wearable CCSs.

Recently, Deng et al. [7] published a novel color-changing flexible contact lens for sensing glucose in human tears. The fluorescent color change from pink to blue, as a result of the increase in glucose amount, can be detected with the aid of a smartphone. The transformation of the obtained image into RGB signals can then be employed for the calculation of the exact glucose amount in the patient's body. The wearable sensor was prepared with 2-hydroxyethyl methacrylate (HEMA) as a hydrogel component and an anthracene-based fluorescent material as glucose-sensitive species. The leakage concern was eliminated by grafting the probe onto the hydrogel network. The fabricated sensor had a fast response time of about 3–5 seconds and a low detection limit of about 9 μM.

Figure 17.1 [7] shows the normalized %RGB values obtained from the real images of the contact lenses at various amounts of glucose with the aid of Photoshop software. As can be seen, by increasing the amount of glucose in artificial tears, the B/R ratio will be increased linearly. This optical behavior was chosen for the calculation of the specific glucose amount. The ratio of blue to red channels was employed to minimize the background

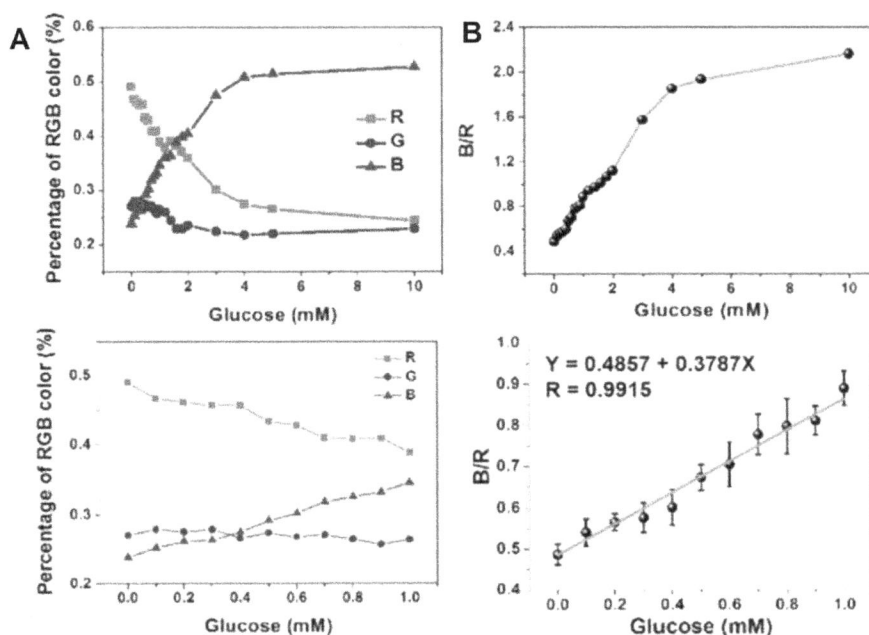

FIGURE 17.1
Glucose detection in real-time with the aid of a smartphone and Photoshop software. (A) Normalized %RGB values of the images at various glucose amounts. (B) B/R of the images at various glucose amounts. Adapted with permission [7]. Copyright (2021), Elsevier.

and environmental interference. Moreover, the authors claimed that the fabricated contact lenses have flexibility similar to the commercial ones, have a transparent structure with no absorption in visible regions, and they also have a beneficial effect of UV absorption, which can absorb UV radiation and thus protect the eyes.

In another study, Qiu et al. [8] synthesized artificial skin with potential application in wearable devices. This novel hydrogel can be stimulated by both mechanical and/or ionic strength stimulants. One of the important challenges faced in the fabrication of conductive sensors is that the color changes instantaneously by mechanical force, however, the force should remain to get the maintained color that is not applicable in real life. The authors employed mechano-chromophores based on spiropyran (SP). The SP structure is changed under mechanical forces and the new structure of merocyanine changes the color from pale yellow to blueish purple. The solvent replacement technique was employed for the fabrication of organohydrogels and ethylene glycol was used for eliminating the drying challenge. The obtained organohydrogel had high mechanical properties that could bear 300% elongation and fracture stress of 0.3 MPa.

In addition, the application of light-absorbing materials such as fluorescent dyes, pigments, and quantum dots, and photonic crystals are widely employed in the synthesis of flexible sensors as they can modify the emitted light by their crystal structure. Their light diffraction ability, and thus their structural color, can be attributed to the change in their refractive indices and/or their lattice spacing represented by Bragg's stack equation [5]. In this regard, Xuan et al. [9] reported the fabrication of magnetic sensors for humidity-sensing applications. In this research, the $Fe_3O_4@SiO_2$ colloids were synthesized by a modified Stöber process and polyethylene glycol-based acrylates were employed as polymeric liquid precursors. The mixture was sandwiched between standard and hydrophobe glass and was oriented with the aid of an external magnet. The photo-polymerization was carried out and the final coating was achieved. The obtained coating represented high humidity sensitivity. The authors claim that in their previous studies, the effect of the refractive index is neglectable while the diffraction change of about 160 nm can be attributed to the change in lattice constant.

In another study in this field, Wang et al. [10] synthesized a novel photonic hydrogel with a sensitive and reversible response to mechanical stimuli. For this purpose, magnetite nanoclusters were encapsulated by carbon, assembled by an external magnetic field, and then polymerized to prepare a 1D magnetic colloidal structure. Different structures were fabricated by varying the magnetic field direction and strength. For touch-sensing investigation, the fabricated colloidal structure was coated on a flexible plastic. The coating exhibited a color-changing ability by mechanical force. Any introduced force including bending, tiny touch, and compression could result in color changing from green to orange. By introducing a pressure of 1.2 N, the diffraction wavelength (λ_d) is shifted from 527 nm to 643 nm, which can be classified in touch-sensing materials (less than 10 KPa).

Reverse opals are a structure in which the solid spheres are replaced by air, while the space between the spheres is filled with some new materials. Reverse opals can be classified as 3D structured photonic crystals. In this area, Wang et al. [11] recently reported a novel shape-memory hydrogel based on reverse opals. In this article, reverse opals were fabricated by using SiO_2 as a sacrificial template. For this purpose, the synthesized silicon dioxide nanoparticles, with a size of 250–290 nm, were assembled on a glass substrate (4 wt% in ethanol), calcined at 600°C (4 hours), and then the polymer precursors including stearyl acrylate (SA) and N-isopropylacrylamide (NIPAM) as monomers were added to the colloidal crystal structure. After polymerization, the film was embedded in HF solution for template removal and fabrication of the reverse opal structure

17.2.2 Elastomeric Hydrogels

Elastomers, due to their ability for deformation and recovery, have attracted the attention of researchers due to their use as (strain) mechano-responsive materials or stress sensors. The other advantage of using elastomers in these classes of materials is their enhanced mechanical properties and absence of solvent, which leads to fewer worries about solvent evaporation. By embedding photonic crystals in elastomers while considering the interfacial compatibility between phases, it is possible to fabricate mechano-chromic materials.

Ito et al. [12] dispersed the photonic particles in the solution of a pre-polymer consisting of an initiator, crosslinker, and ethyl acrylate monomer. They showed when the final crosslinked elastomer is stretched, the color will show a reversible change. In another research work, Shi et al. [13], inspired by a kind of frog, dispersed silica particles with 180 nm diameter in di(ethylene glycol) methyl ether methacrylate monomer, crosslinker, and initiator. Different amounts of diarylethene (1,2-bis(2-methyl-5-phenyl-3-thienyl) perfluorocyclopentene (DAE-TZ), 1,2-bis(3,5-dimethylthiophen-2-yl) perfluorocyclopentene) (DAE-MT), and 1,2-bis(5-methyl-2-phenyl-4-thiazolyl) perfluorocyclopentene (DAE-MP) were mixed in the pre-crosslinked solution in the crosslinking process. The obtained elastomeric structure showed an obvious color change with uniaxial stretching because of the orientation of silica particles combined with background colors.

Another elastomeric structure in this category is the core shell – a rigid core, normally a rigid polymer like polystyrene, with an elastomeric shell [14–17]. Besides mechano-chromic materials, there are other applications of elastomeric-based color-sensing materials that are classified as light- and temperature-responsive materials [14,18], magnetic field-responsive materials [17], and pH-responsive materials [16]. Some researchers showed that by using magnetic nanoparticles like iron oxide and cobalt ferrite particles in elastomers, the magnetic particles could be self-assembled into arrays with the magnetic field, and in special conditions, it could change the color appearance of the structure [17].

Winter et al. [16] revealed additional features for a kind of core-shell structured elastomeric opal film based on polystyrene as a hardcore, crosslinked interlayer (poly(methyl methacrylate-co-allyl methacrylate), and soft poly(ethyl acrylate-co-butyl acrylate-co-(2-hydroxyethyl) methacrylate) as the shell, which can be conveniently introduced by incorporating organic dye molecules that can show a pH-responsive property. It could be interesting for the fabrication of anti-counterfeiting double safety materials or optical sensors.

17.2.3 Thin Films

Thin films are another category of the sensing system. One of their applications is humidity sensing. For humidity sensing, synthetic melanin nanoparticle thin film was prepared and showed a fast and reversible change in color change by changing the humidity level. The mechanism was the swelling of nanoparticles on water uptake in a wet milieu and increasing the thickness of the film [19].

Lim et al. [20] fabricated a polydopamine nanoparticles (PdNP) thin film with perfect solvatochromic performance for the diagnosis of the level of alcoholic solvents. The PdNP thin film exhibited a high refractive index, effective swelling capability, and measurable change in the reflection spectrum by exposure to alcoholic solvents.

For making synthetic structural colors, melanin is a common material. One of the differences between melanin with other materials that are utilized in making structural colors, for example, cellulose and silk, is its disjointed scattering of light to elevate the impregnation of colors [21]. Besides, it is possible to be utilized as the key part of the sensor directly through diffraction or scattering light. Thin film is one of the common morphologies of melanin. One example of this application is the thin film (with a thickness of 50–200 nm) that is made of dopamine and melanin and can be used as a structural color sensor. The reflected color of the dopamine-melanin thin film was shown to be dependent on the film thickness and dopamine concentration [22].

In another study, Zhang et al. [23] polymerized a mixture of isopropyl acrylamide and dopamine and utilized $CuSO_4/H_2O_2$. They made a thin film of this composite by applying it to a silicon wafer. They revealed that the color of the thin film was not dependent on the angle of exposure of light and the colors of the film were dependent on both temperature and film thickness. Chi et al. [24], using a dip coating method, made a colorimetric humidity sensor based on graphene oxide thin film on various substrates including aluminum foil, silicon wafer, and metallized polyethylene terephthalate (mPET). Their results showed that under 10,000 cycles of bending, the bending test of the graphene oxide film on mPET showed perfect spectral stability. It shows the manufactured thin film is suitable for flexible device applications in the detection of environmental condition change and also personal physiological condition monitoring.

Yu et al. [25] introduced a thin film with polyelectrolyte poly-(2-vinyl pyridine) (P2-VP) gel infiltrated in polystyrene (PS) monolayers for use as humidity and counter-ions. P2-VP gel will swell/deswell by changing pH. It makes a drastic variation in the thickness of the film and the shift of the color and reflectance peak. The benefits of this thin film sensor are being reliable, highly reversible, linear, and with a rapid response to changes in pH. Besides, achieving a colorimetric readout makes it favorable for naked-eye monitoring of environmental analysis. Fei et al. [26] synthesized a *Papilio paris* wing covered with a copolymer made of poly (acrylic acid) and acrylamide. Swelling and deswelling of the polymer affect a change in the refractive index of the lamellar interspacing and it leads to a change in reflection wavelength. They also proved their experimental data by finite-difference time-domain simulation.

Men et al. [27], used the two-dimensional nanospheres of a gold array bonded on a hydrogel-type film based on polyacrylic acid. This thin film exhibited a strong diffraction intensity and visual diffraction color because of its large scattering cross-section and periodic structures of the Au nanospheres. The presented study could be extended to introduce new types of visualized sensors. Xue et al. [28] integrated polymethacrylic acid (PMAA) onto the gyroid frame of the *Callophrys Rubi* butterfly wing to make a pH sensor. The most interesting property of the manufactured film was its single-value corresponding relationship between the maximum reflection peak wavelength (λ_{max}) and pH, with a color distinction degree of 18 nm/pH. This result can assure the user of the authenticity and accuracy of the sensor. The structural changes were shown to be completely reversible.

Moirangthem et al. [29] made a cholesteric liquid crystalline polymer containing crown ether moieties to be sensitive to calcium concentration in serum. The sensor exhibited a visible selective response to Ca^{2+}, with a change in color from green to blue. The accuracy of the sensor was in the range of 10^{-4} to 10^{-2} M. From the other advantages of these films, we can mention its price, lack of battery, and ease of use.

17.2.4 Fibers

Fabrication of wearable sensors based on fibers is highly attractive due to the outstanding properties they can provide. High flexibility, low production cost, comfortability, large-scale production potential, and long lifetime are some of the most significant features of the sensors prepared with the aid of fibers [30]. Sheng et al. [31] reported novel color-changeable fibers with a combination of polymer-dispersed liquid crystals as the inner layer and transparent hydrogels as the outer part. This combination resulted in highly applicable fibers with good environmental, mechanical, and chemical stabilities.

For fiber preparation, cholesteric liquid crystals were mixed with silicon dioxide microparticles with a mass ratio of 1:0.03, followed by adding polymethylmethacrylate and dichloromethane for obtaining a polymer-dispersed liquid crystal solution. The mass ratio was set to be 1:15:8 for polymethylmethacrylate, cholesteric liquid crystals, and CH_2Cl_2, respectively. The crosslinking agent, initiator, and other polymerization reagents were mixed to prepare the pre-polymerization mixture; 25 µL tetramethyl ethylenediamine was added and the solution was kept at 25°C. The core fibers were passed slowly (5 cm/s) from the first polymer-dispersed liquid crystal solution followed by the second mixture passing and immediate vacuum drying at 40°C for 2 hours to obtain the final fibers.

The schematic illustration of the cholesteric liquid crystals is represented in Figure 17.2 [31]. It shows that the liquid crystals may display a specific color as they keep their spiral structure under zero electric field. This may be attributed to the good dispersion of the liquid crystals in the polymeric matrix. The unscrewing of the liquid crystals will happen by applying higher voltages in which the liquid crystals will transmit the incident light and the color is obtained from the core fiber.

The reflectance of the liquid crystal-based fibers was plotted from 400 nm to 800 nm while employed at various voltages, represented in Figure 17.3 [31]. The reflectance at 680 nm was observed to have an obvious change from about 46% to 3% while the voltage changed from zero to 26 V, representing the color change from red to black.

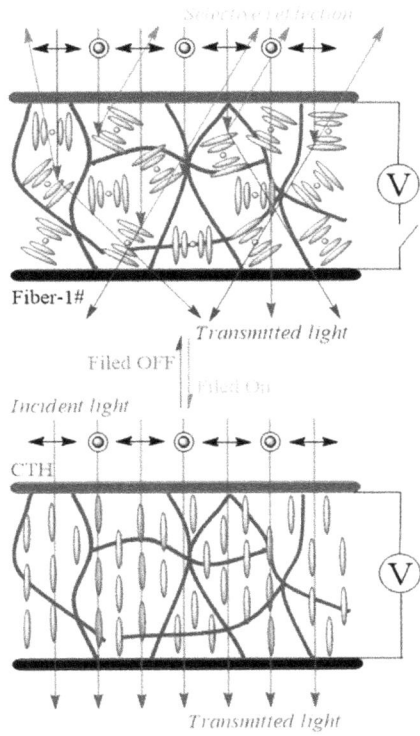

FIGURE 17.2
Schematic illustration of the optical behavior of the liquid crystal-based fibers by removing or applying the voltage at ambient temperature. Adapted with permission [31]. Copyright (2021), Elsevier.

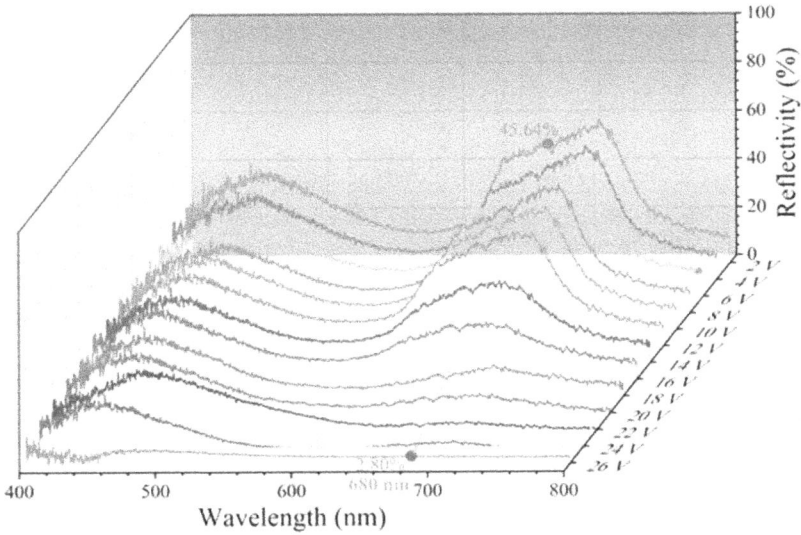

FIGURE 17.3
Visible reflectance spectra of the liquid crystal-based fiber employed under various voltages at 25°C. Adapted with permission [31]. Copyright (2021), Elsevier.

The reported strategy is applicable for the facile preparation of color-changing fabrics with continued coating technology and the fibers can be employed in woven fabrics for further applications. The high stretchability, mechanical strength, and permeability of the fabrics woven by these fibers are promising candidates for intelligent applications. Wang et al. [32] have reported a novel fabrication of electronic wearable sensors with the aid of periodic photonic crystals. Structural colors, as mentioned earlier, are a category of periodic microstructures in which various colors can be obtained by varying the crystals' spacing. These structures are fabricated by inspiration from nature and some animals such as chameleons, fish, and insects. In this paper, a new stretchable fiber-based mechano-chromic sensor was introduced with various color appearances from blue to red.

The photonic structure was prepared by stacking thousands of polydodecylglyceryl itaconate (PDI) bilayers in a crosslinked polyacrylamide hydrogel network. The fiber structure was obtained by injection of the liquid precursors into a tubular template and then the polymerization was carried out. The fabricated polymer was then swelled in deionized water for which the increase in d-spacing of the crystals resulted in the red appearance of the fibers. The effect of crosslinking agents and the template size were investigated on the swelling and optical behavior of the structure.

It was observed that higher crosslinking density will result in a lower degree of swelling and thus a smaller increase in d-spacing will occur. By controlling the crosslinking density, various structural colors could be successfully obtained in visible regions. The H-bonding between the PDI and the polyacrylamide network was responsible for preventing the bilayers from escaping from the gel structure. The final color-changeable electrically responsive sensor was obtained by dipping the swelled silicon fibers into the multiwall carbon nanotube (MWCNT) mixture. The MWCNTs were loaded on the surface of silicon fiber by van der Waals forces and the second layer of silicon layer was applied to obtain the piezo-resistive, electrically conductive fibers. The electrical conductivity was obtained by shrinking the swollen silicon fibers leading to the compact layer of MWCNTs. The optical and electrical responsive sensors were reported to have a fast responding time, about 80 ms, with high strain and application range (up to 200%), which make these fibers a promising candidate for further applications in wearable sensors, intelligent devices, robots, and so on.

17.2.5 Core-Shell Nanoparticles

Core-shell nanocomposites (CSNs) are a kind of nanomaterial with a core-shell structure that consists of an outer layer of one material (shell) and an inner layer of another material (core). The main benefits of using CSNs are their improved and unique properties of a single material that are unavailable or inaccessible from the individual components but are available with some of the properties of the core and shell with distinctive chemical and physical properties. One of the applications of the core-shell structure is using them in wearable sensors. The main reason behind the selection of this structure for sensing devices is that the structure enables them to make miniaturized devices with a reversible and reliable output. These sensors show many interesting characteristics like simple device structure and high pixel density. It is possible to take a high sensitivity from metal-foil-based sensors and traditional semiconductors but they are not good for flexible devices and small or large strains, which can limit their further applications. Therefore, elastomeric materials or conducting polymers need to be combined with conducting materials such as carbon nanotubes, graphene, carbon black, and metal nanoparticles for increasing the sensitivity of wearable pressure sensors.

Today, CSNs have been widely used for making flexible, large deformation, and highly sensitive monitoring piezo-resistive sensors [33]. Ai et al. [34] made a sensitive and reliable flexible pressure sensor by putting a polystyrene ball@reduced graphene oxide core-shell structure (PS ball@rGO) between two thin flexible polydimethylsiloxane sheets (sandwich structure). It was reversible and showed reliable results even after 20000 cycles. Its sensitivity was shown to be 50.9 kPa^{-1} in the range of 3–1000 Pa, and its energy consumption and detection limit were low (~1 μW) and 3 Pa, respectively. The speed of the response was also fast (50 ms). In addition, this highly sensitive and cost-effective sensor can effectively be used in complex forces such as stretching, pressing, vibration, and torsion. Therefore, it can be used for detecting swallowing, blood pulse, articular activities, and skin pulling by attaching the sensor to the body.

Melanin with different morphologies has been widely used to make structural colors, such as core-shell melanin nanoparticles or solid melanin film [21]. Hollowing melanosomes in bird feathers have inspired researchers to synthesize a structural color. They achieved it by making the nanoparticles of melanin in a core-shell structure. The first benefit of core-shell nanoparticles based on melanin is the flexibility to improve surface physics compared with solid melanin nanoparticles. They also enable the producers to have more control in scattering light and refractive index. This can affect the packing structure, self-assembly, and macroscopic properties.

Xiao et al. [35] coated silica on melanin nanoparticles. The prepared nanoparticle was shown to have a core with a refractive index and a shell with a high refractive index. To make these structures into supraball inks of a small size, they utilized a simple reverse emulsion method. By compounding two different sizes of nanoparticles or changing the core size or shell thickness it is possible to modify the tint. Kohri et al. [36] grafted HEMA onto polydopamine nanoparticles to make a structural color. Kohri et al. [37], in another study, synthesized nanoparticles based on core-shell melanin in an ellipsoid shape for creating structural colors. They stretched the spherical particles of PS particles to be ellipsoidal, then coated the PS particles with a layer of melanin to make core-shell particles. They showed that the higher ratio of the core and shell diameters, leads to greater shift in light reflection.

In another study, Kohri et al. [38] made the structure of a core-shell nanoparticle, with melanin as a shell and polystyrene as a core. They showed that a thin melanin layer leads to iridescent structural colors. The colors were shown not to be dependent on an angle with thicker melanin shells. Because the surface of the CSNs with thicker melanin shells was rougher, it disrupts the formation of the crystalline structure. Although melanin can also be used in the shell structure of the CSNs, they were not able to improve the reflecting of assembly because the refractive index of the core was lower while the shell was higher.

17.3 Challenges and Future Perspective

As wearable sensors are in their initial stage of development, there are a lot of challenges in customizing them. For instance, these sensors need to be integrated into acceptable shape surfaces with durability, compatibility, and abrasion resistance [1]. The commercial wearable sensing systems that exist right now are usually accomplished by rigid packaged solid substrates that contain integrated circuits. These rigid packages are not compatible with curved and soft human bodies. It makes the results unreliable because of insufficient skin contact. Besides, these materials need complex sensing systems that can complete a range of functional electronic ingredients, such as equipment operation control, data attainment and processing, self-powered operation, and intelligent data monitoring [39].

The accuracy and practical application of wearable and flexible sensors are highly affected in complex mixtures. The decoupling of the sensor in an environment of multiple stimuli (e.g., pressure, humidity, temperature, and physiological components) is a critical challenge. Choosing a promising substrate, using layer-by-layer deposition techniques, and/or applying a final protective coating are some of the solutions for this issue [40]. By overcoming the aforementioned issues, wearable sensors are finding increased attention these days. Flexible and wearable CCSs are employed for developing fast and accurate, easy-to-fabricate, and stretchable sensing materials.

17.4 Conclusion

In summary, color-changing wearable and flexible sensors are highly attractive and they are employed for their promising applications in various fields. These sensing materials can be used in healthcare monitoring systems, human-machine interfaces, the Internet of Things, intelligent and smart robots, and so on. The color appearance in many of these sensors is prepared by the incorporation of structural colors and hydrogels. These structures can be fabricated in thin films, fibers, or core-shell nanoparticles. The fundamental basics of these structures, fabrications, and final applications have been discussed in detail. The current challenges and future perspectives of the application of color-changing flexible and wearable sensors have been considered to commercialize them. Flexible and wearable sensing materials, based on color-changing systems, have gained significant attention regarding the numerous application potentials they have, which will make them a promising candidate for future human life.

References

1. Z. Lou, L. Wang, and G. Shen, "Recent advances in smart wearable sensing systems," *Adv. Mater. Technol.*, vol. 3, no. 12, p. 1800444, 2018.
2. H.-H. Chou, A. Nguyen, A. Chortos, J. W. F. To, C. Lu, J. Mei, T. Kurosawa, W.-G. Bae, J. B.-H. Tok, and Z. Bao, "A chameleon-inspired stretchable electronic skin with interactive colour changing controlled by tactile sensing," *Nat. Commun.*, vol. 6, no. 1, p. 8011, 2015.
3. T. Yokota, P. Zalar, M. Kaltenbrunner, H. Jinno, N. Matsuhisa, H. Kitanosako, Y. Tachibana, W. Yukita, M. Koizumi, and T. Someya, "Ultraflexible organic photonic skin," *Sci. Adv.*, vol. 2, no. 4, p. e1501856, 2016.
4. H. Inan, M. Poyraz, F. Inci, M. A. Lifson, M. Baday, B. T. Cunningham, and U. Demirci, "Photonic crystals: Emerging biosensors and their promise for point-of-care applications," *Chem. Soc. Rev.*, vol. 46, no. 2, pp. 366–388, 2017.
5. G. Isapour and M. Lattuada, "Bioinspired stimuli-responsive color-changing systems," *Adv. Mater.*, vol. 30, no. 19, pp. 1–36, 2018.
6. X. Li, S. Chen, Y. Peng, and Z. Zheng, "Materials, preparation strategies, and wearable sensor applications of conductive fibers: A review," *Sensors (Basel)*, vol. 22, no. 8, p. 3028, 2022.

7. M. Deng, G. Song, K. Zhong, Z. Wang, X. Xia, and Y. Tian, "Wearable fluorescent contact lenses for monitoring glucose via a smartphone," *Sensors Actuators B. Chem.*, vol. 352, no. P2, p. 131067, 2022.

8. W. Qiu, C. Zhang, G. Chen, H. Zhu, Q. Zhang, and S. Zhu, "Colorimetric ionic organohydrogels mimicking human skin for mechanical stimuli sensing and injury visualization," *ACS Appl. Mater. Interfaces*, vol. 13, no. 22, pp. 26490–26497, 2021.

9. R. Xuan, Q. Wu, Y. Yin, and J. Ge, "Magnetically assembled photonic crystal film for humidity sensing," *J. Mater. Chem.*, vol. 21, no. 11, pp. 3672–3676, 2011.

10. X. Wang, C. Wang, Z. Zhou, and S. Chen, "Robust mechanochromic elastic one-dimensional photonic hydrogels for touch sensing and flexible displays," *Adv. Opt. Mater.*, vol. 2, no. 7, pp. 652–662, 2014.

11. Y. Wang, Z. Zhang, H. Chen, H. Zhang, H. Zhang, and Y. Zhao, "Bio-inspired shape-memory structural color hydrogel film," *Sci. Bull.*, vol. 67, no. 5, pp. 512–519, 2022.

12. T. Ito, C. Katsura, H. Sugimoto, E. Nakanishi, and K. Inomata, "Strain-responsive structural colored elastomers by fixing colloidal crystal assembly," *Langmuir*, vol. 29, no. 45, pp. 13951–13957, 2013.

13. P. Shi, E. Miwa, J. He, M. Sakai, T. Seki, and Y. Takeoka, "Bioinspired color elastomers combining structural, dye, and background colors," *ACS Appl. Mater. Interfaces*, vol. 13, no. 46, pp. 55591–55599, 2021.

14. C. G. Schäfer, M. Gallei, J. T. Zahn, J. Engelhardt, G. P. Hellmann, and M. Rehahn, "Reversible light-, thermo-, and mechano-responsive elastomeric polymer opal films," *Chem. Mater.*, vol. 25, no. 11, pp. 2309–2318, 2013.

15. X. Wang, Y. Li, J. Zheng, X. Li, G. Liu, L. Zhou, W. Zhou, and J. Shao, "Polystyrene@poly(methyl methacrylate-butyl acrylate) core–shell nanoparticles for fabricating multifunctional photonic crystal films as mechanochromic and solvatochromic sensors," *ACS Appl. Nano Mater.*, vol. 5, no. 1, pp. 729–736, 2022.

16. T. Winter, A. Boehm, V. Presser, and M. Gallei, "Dye-loaded mechanochromic and pH-responsive elastomeric opal films," *Macromol. Rapid Commun.*, vol. 42, no. 1, p. 2000557, 2021.

17. W. Hong, Z. Yuan, and X. Chen, "Structural color materials for optical anticounterfeiting," *Small*, vol. 16, no. 16, p. 1907626, 2020.

18. Q. Yin, S. Tu, M. Chen, and L. Wu, "Self-detecting and self-healing reinforce elastomer doped with aggregation-induced emission molecules," *Macromol. Mater. Eng.*, vol. 305, no. 4, p. 2000013, 2020.

19. M. Xiao, Y. Li, J. Zhao, Z. Wang, M. Gao, N. C. Gianneschi, A. Dhinojwala, and M. D. Shawkey, "Stimuli-responsive structurally colored films from bioinspired synthetic melanin nanoparticles," *Chem. Mater.*, vol. 28, no. 15, pp. 5516–5521, 2016.

20. Y.-S. Lim, J. S. Kim, J. H. Choi, J. M. Kim, and T. S. Shim, "Solvatochromic discrimination of alcoholic solvents by structural colors of polydopamine nanoparticle thin films," *Colloid Interface Sci. Commun.*, vol. 48, p. 100624, 2022.

21. M. Xiao, M. D. Shawkey, and A. Dhinojwala, "Bioinspired melanin-based optically active materials," *Adv. Opt. Mater.*, vol. 8, no. 19, p. 2000932, 2020.

22. T.-F. Wu and J.-D. Hong, "Dopamine-melanin nanofilms for biomimetic structural coloration," *Biomacromolecules*, vol. 16, no. 2, pp. 660–666, 2015.

23. C. Zhang, B.-H. Wu, Y. Du, M.-Q. Ma, and Z.-K. Xu, "Mussel-inspired polydopamine coatings for large-scale and angle-independent structural colors," *J. Mater. Chem. C*, vol. 5, no. 16, pp. 3898–3902, 2017.

24. H. Chi, L. Jun, X. Zhou, and F. Wang, "GO film on flexible substrate: An approach to wearable colorimetric humidity sensor," *Dye. Pigment.*, vol. 185, p. 108916, 2021.

25. S. Yu, Z. Han, X. Jiao, D. Chen, and C. Li, "Ultrathin polymer gel-infiltrated monolayer colloidal crystal films for rapid colorimetric chemical sensing," *RSC Adv.*, vol. 6, pp. 66191–66196, 2016.

26. X. Fei, T. Lu, J. Ma, W. Wang, S. Zhu, and D. Zhang, "Bioinspired polymeric photonic crystals for high cycling pH-sensing performance," *ACS Appl. Mater. Interfaces*, vol. 8, no. 40, pp. 27091–27098, 2016.

27. D. Men, F. Zhou, L. Hang, X. Li, G. Duan, W. Cai, and Y. Li, "A functional hydrogel film attached with a 2D Au nanosphere array and its ultrahigh optical diffraction intensity as a visualized sensor," *J. Mater. Chem. C*, vol. 4, no. 11, pp. 2117–2122, 2016.
28. R. Xue, W. Zhang, P. Sun, I. Zada, C. Guo, Q. Liu, J. Gu, H. Su, and D. Zhang, "Angle-independent pH-sensitive composites with natural gyroid structure," *Sci. Rep.*, vol. 7, no. 1, p. 42207, 2017.
29. M. Moirangthem, R. Arts, M. Merkx, and A. P. H. J. Schenning, "An optical sensor based on a photonic polymer film to detect calcium in serum," *Adv. Funct. Mater.*, vol. 26, no. 8, pp. 1154–1160, 2016.
30. Z. Liu, T. Zhu, J. Wang, Z. Zheng, Y. Li, and J. Li, "Functionalized fiber based strain sensors: Pathway to next generation wearable electronics," *Nano-Micro Lett.*, vol. 14, no. 61, pp. 1–39, 2022.
31. M. Sheng, W. Wang, L. Li, L. Zhang, and S. Fu, "All-in-one wearable electronics design Smart electrochromic liquid-crystal-clad fibers without external electrodes," *Colloids Surfaces A Physicochem. Eng. Asp.*, vol. 630, no. June, p. 127535, 2021.
32. Y. Wang, W. Niu, C. Lo, Y. Zhao, X. He, G. Zhang, S. Wu, B. Ju, and S. Zhang, "Interactively full-color changeable electronic fiber sensor with high stretchability and rapid response," *Adv. Funct. Mater.*, vol. 30, no. 19, pp. 1–10, 2020.
33. P. K. Kalambate, Dhanjai, Z. Huang, Y. Li, Y. Shen, M. Xie, Y. Huang, and A. K. Srivastava, "Core@shell nanomaterials based sensing devices: A review," *Trends Anal. Chem.*, vol. 115, pp. 147–161, 2019.
34. Y. Ai, T. H. Hsu, D. C. Wu, L. Lee, J.-H. Chen, Y.-Z. Chen, S.-C. Wu, C. Wu, Z. M. Wang, and Y.-L. Chueh, "An ultrasensitive flexible pressure sensor for multimodal wearable electronic skins based on large-scale polystyrene ball@reduced graphene-oxide core–shell nanoparticles," *J. Mater. Chem. C*, vol. 6, no. 20, pp. 5514–5520, 2018.
35. M. Xiao, Z. Hu, Z. Wang, Y. Li, A. D. Tormo, N. Le Thomas, B. Wang, N. C. Gianneschi, M. D. Shawkey, and A. Dhinojwala, "Bioinspired bright noniridescent photonic melanin supra-balls," *Sci. Adv.*, vol. 3, no. 9, pp. e1701151–e1701151, 2017.
36. M. Kohri, K. Uradokoro, Y. Nannichi, A. Kawamura, T. Taniguchi, and K. Kishikawa, "Hairy polydopamine particles as platforms for photonic and magnetic materials," *Photonics*, vol. 5, no. 4, pp. 1–10, 2018.
37. M. Kohri, Y. Tamai, A. Kawamura, K. Jido, M. Yamamoto, T. Taniguchi, K. Kishikawa, S. Fujii, N. Teramoto, H. Ishii, and D. Nagao, "Ellipsoidal artificial melanin particles as building blocks for biomimetic structural coloration," *Langmuir*, vol. 35, no. 16, pp. 5574–5580, 2019.
38. M. Kohri, K. Yanagimoto, A. Kawamura, K. Hamada, Y. Imai, T. Watanabe, T. Ono, T. Taniguchi, and K. Kishikawa, "Polydopamine-based 3D colloidal photonic materials: Structural color balls and fibers from melanin-like particles with polydopamine shell layers," *ACS Appl. Mater. Interfaces*, vol. 10, no. 9, pp. 7640–7648, 2018.
39. R. C. Webb, Y. Ma, S. Krishnan, Y. Li, S. Yoon, X. Guo, X. Feng, Y. Shi, M. Seidel, N. H. Cho, J. Kurniawan, J. Ahad, N. Sheth, J. Kim, J. G. 6th Taylor, T. Darlington, K. Chang, W. Huang, J. Ayers, A. Gruebele, R. M. Pielak, M. J. Slepian, Y. Huang, A. M. Gorbach and J. A. Rogers, "Epidermal devices for noninvasive, precise, and continuous mapping of macrovascular and microvascular blood flow," *Sci. Adv.*, vol. 1, no. 9, p. e1500701, 2015.
40. M. Segev-bar and H. Haick, "Flexible sensors based on nanoparticles," *ACS Nano*, vol. 7, no. 10, pp. 8366–8378, 2013.

18

Self-Powered Sensors

Sarbaranjan Paria, Haradhan Kolya, Changwoon Nah, and Chun-Won Kang

CONTENTS

18.1 Introduction

We are living in the "World of Sensors". A sensor is a device that provides an output signal concerning a specific input of a physical quantity. Several categories of sensors are found in our homes, offices, cars, and different places to make work easier, such as turning on lights when we come into the room, regulating temperature, perceiving smoke or fire, serving us coffee, opening garage doors as our car arrives, and many more. Historically, in China, the Spring and Autumn period (770–476 BCE) created the time witnessing device called the sundial, and during the Warring States period (475–221 BCE), there was a device called the "Sinan" ("south-pointing ladle") indicating direction, which was made with a magnet [1]. The developed compass unlocked the maritime navigation age and encouraged the advancement in trading internationally. Since then, human civilization has enjoyed outstanding social growth. In the modern era, with the evolution of science and technology and improved industrial insurgency, sensor technology has achieved huge innovations in categories and functions.

Moreover, with fast progress in ultrathin sensors and actuator design, optoelectronic and other electronic devices are anticipated to expand the opportunities for flexible electronics (Figure 18.1a). This emerging area of flexible electronics reflects technology having compatibility with movable parts or arbitrarily curved surfaces with capability in a new application paradigm in the greater area of electronics. Like flow sensors, as shown in Figure 18.1b, other types of sensory networks (SNs) can become principal technological and economic directors for industries globally, with broad applications in structural

DOI: 10.1201/9781003299455-18

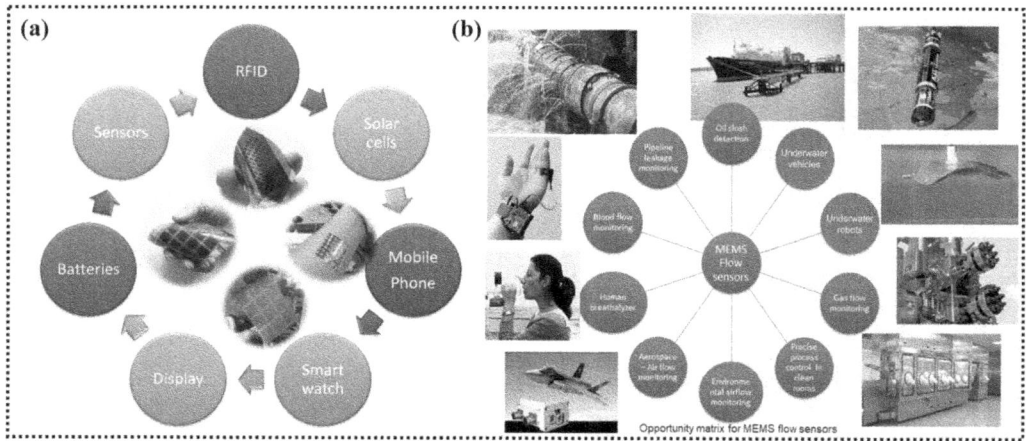

FIGURE 18.1
(a) Development of flexible electronics across a broad range of applications. Adapted with permission [2]. Copyright (2017) Wiley. (b) Diverse application of microelectromechanical system (MEMS) sensors in recent times. Adapted with permission [3]. Copyright (2019), Elsevier.

health, the environment, civil engineering structures/infrastructure, national security, and more. By replacing a definite number of traditional distinct sensors with a vast number of mobile and self-reliant sensors, a statistical investigation of collected signals via the internet over scattered sensors can produce a clear message. An Internet of Things (IoT) that connects daily use objects and systems to a set of data and networks (the internet) is the future for controlling the infrastructure/environment, tracking products, healthcare, medical monitoring, and the smart home. A nanosystem is a combination of nanodevices fulfilling distinct functions with the competence to sense, regulate, connect, and actuate or reply. Sensors are characteristically thought to consist of four parts: the transducer, the recognition element/chemically selective layer, the energy source for the transducer, and the signaling system. The low power consumption of these systems gives possibilities for utilizing the harvested energy from the sensor to power themselves. Power in the micro-watt order is typically required for autonomous, endurable, maintenance-free operations of several sensors like the microelectromechanical system (MEMS), remote and mobile environmental sensors, implantable biosensors, nanorobotics, and even portable/wearable personal gadgets. Nanorobots, for instance, can detect and adjust to the environment, control objects, take actions, and accomplish tedious jobs. Still, the main challenge is to search for an origin of power that can run a nanorobot without too much weight.

Nevertheless, such SNs become almost unrealistic if every sensor needs to be driven by a battery due to the many environmental and health issues. Thus, the new technologies that can collect and transform energy from the environment offer a potential solution for endurable and self-sufficient micro/nanopower sources. However, like the abovementioned applications, self-powered sensors (SPSs) are also needed to monitor oil/gas transportation lines over long distances. In this chapter, we cover the up-to-date advancement of SPSs based on different types of nanogenerators (NGs) in several application areas including physical sensors, biomedical and healthcare devices, wearable devices, human-machine interfaces (HMIs), chemical and environmental monitoring, intelligent traffic, smart cities, robotics, and fiber and fabric sensors, as well as the obstacles and future research guidelines in this field. Moreover, considering the applicability of these sensors,

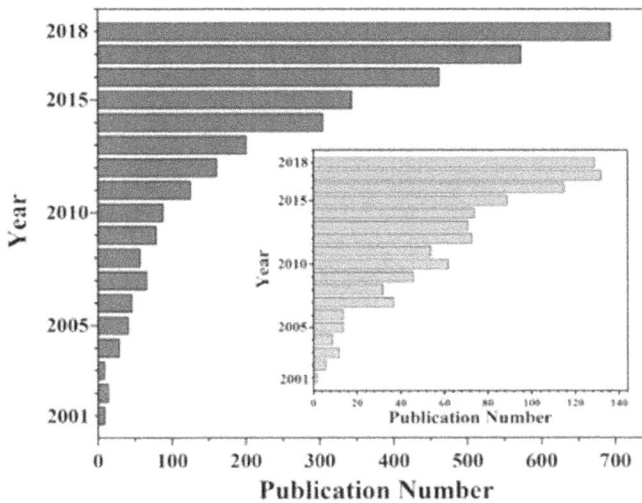

FIGURE 18.2
SPS-based articles published between 2001 and 2018. Adapted with permission [4]. Copyright (2019), Elsevier.

numerous articles regarding this area have been published between 2001 and 2018, as shown in Figure 18.2 [4].

18.2 SPSs and Systems: Origin and Development

The SPS is not a new idea as it was first utilized for nuclear sensors. Lately, those transitioned into chemical and biological sensing. Recently, scientists have extended the term "SPSs" to sensors that can harvest energy from the atmosphere and, thus, do not need any conventional battery or AC power source. The term "SPS" can be explained in two kinds of interpretations. Foremost, it is one type of sensor that inevitably provides an electric signal on mechanical triggering without using any external power. It is worth mentioning that, nowadays, most sensors are passive, that is, in the absence of a power supply, they do not generate any signal. The second type requires the power source for running the sensor to be self-produced. This is possible by using the active and sleeping modes of the sensor. In several cases, sensors do not need to send signals every second; a call at a substantial time interval is sufficient, for example, environmental monitoring. For these cases, the harvested energy during the "sleep" mode of the sensor can power it up once it is in an "active" way, that is, a sustainable operation of the sensors. Thus, the self-powered nanosystem's goal is to harvest energy from the ambient waste energy for powering itself, which makes it capable of working independently, wirelessly, and sustainably.

SPSs comprise different parts capable of sensing, interacting, regulating, and responding. In addition to sensing, processing, and conveying elements, harvesting energy and its storage are essential components of the system. Potential resources are wind, solar, chemical, thermal, mechanical, biomass, and so on. As mentioned earlier, fluid and mechanical energy can exist in many places where sensors can be installed – they are useful for energy harvesting. Recently, hybrid energy harvesters have become a potential sustainable energy harvesting system for powering several electronics and sensing applications. Also, utilizing active sensing technologies to decrease demand is worth realizing in SPSs.

18.3 Different Types of SPSs

Energy harvesting can be defined as capturing a minimal amount of energy from one or more energy sources in the surroundings, collecting those energies, and storing them for subsequent use. Energy scavenging or power harvesting can also be termed energy harvesting. Current advances in wireless and MEMS technology can transform energy harvesting technologies into potential replacements for conventional batteries. Portable electronics and wireless sensors with ultra-low power use traditional batteries as their power supply. However, the battery's lifespan is inadequate and short in comparison with the shelf-life of the devices. Thus, the battery's restoration and recharging become inefficient and sometimes impossible. Therefore, as discussed in the following sections, an enormous amount of research has been paid to energy-scavenging technology as a self-powered source for running easy-to-carry devices or wireless sensor network systems.

18.3.1 Nuclear Energy-Based SPSs

The energy released from nuclear reactions or radioactive decay can be utilized to power a sensor. These types of sensors are feasibly classified into two categories according to their functions in two atmospheres, as illustrated in Figure 18.3. The first form of this sensor has been coined "non-thermoelectric" and largely covers the sensors that generate an electrical current owing to the charged particles produced from radioactive decay. Generally, beta decay causes high-energy electrons, which is an excellent example of this type of sensor. In this sensing device, comprehensive bandgap materials are exploited to construct a diode junction that permits the jumping of high-energy electrons through semiconductors, however, electron transfer in the opposite direction is forbidden. This enables the electrons to go through the minimum resistant pathway, which could be the circuitry sensing device,

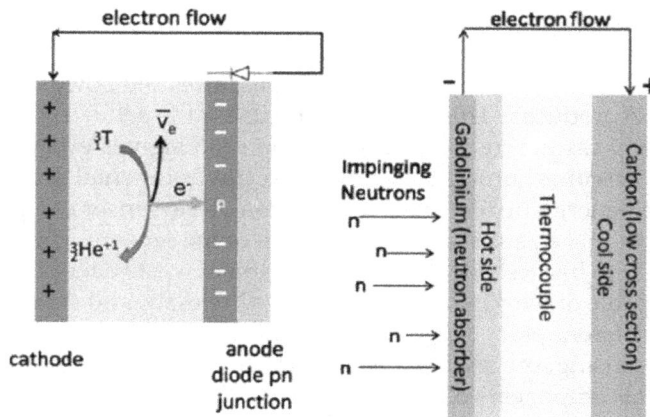

FIGURE 18.3
Principle of working of the nuclear-based SPSs. Adapted with permission [5], Copyright (2011) Springer.

where it can power the electronics for producing signals. Another electrode (near the other one) is placed with the electronics so that, during beta decay, the generated positively charged nuclei capture the electrons effectively. This kind of sensor (non-thermoelectric) could be effective in an environment that is not exposed to such type of radiation where those are most likely to be utilized in signal alarms.

The second kind of nuclear-based SPS essentially includes the thermoelectric (TR) type. These sensors generally produce their electrical energy by using neutrons from radioactive decay. In a distinctive self-powered TR sensor, one terminal of the thermocouple is in touch with a material with a greater cross-section for neutrons. The other terminal of the thermocouple is connected with a material of lower cross-section for neutrons. On exposing neutron radiation to the sensor, the side with the higher cross-sectional material captivates the neutrons and is heated. On the other hand, the side with a lower cross-section does not engross the neutrons and stays at an equal temperature. This developed temperature fluctuation along the thermocouple joint produces an electrical current that can then be utilized to power up the sensors. This sensor is most likely suitable for an environment that is frequently exposed to increased radiation levels, like near a reactor pool. Thus, both kinds of sensors can be operated for a longer time due to their solid-state behavior.

18.3.2 Solar Cell-Based SPSs

A photovoltaic (PV) cell, also known as a solar cell (SC), is an electrical device for directly converting light into electricity through a PV effect consisting of a physical and chemical change. The solar-powered system is well developed and it has extensive use in large-scale systems like satellites and small-scale portable devices. However, due to SCs' limited efficiency, to run a system, the size of solar panels needs to be large for producing higher electricity. In contrast, micro-sized SCs are capable of miniature sensing systems that require less energy. Considering the perception, an SC-based gas sensor is joined in a series to build a self-powered gas sensor (SPGS) at the nanoscale level. As a prototype, Hou et al. developed an integrated system consisting of a humidity sensor (HS) and SC [6]. For the self-powered system, a triple-junction of GaInP/GaAs/Ge was built where a thin Au/AuGeNi (2 nm/5 nm) layer works as the upper electrode. A Schottky diode of Au/ZnO was created on the Au/AuGeNi layer for sensing the humidity. Zinc oxide nanowires (ZnO NWs) were developed on the structure to enhance the sensitivity, where the ratio of surface-to-volume for ZnO NWs was enhanced.

Figure 18.4a illustrates the schematic representation of an integrated system of ZnO NW-based HS and a triple-junction SC. On AM 1.5G (intensity: 100 mW/cm^2) solar irradiating, the fabricated triple-junction SC generated the open-circuit voltage (OCV) of 2.5 V at ambient temperature, which can be exploited to power up the HS. On applying a fixed bias voltage for the Au/ZnO Schottky junction, the change in surface conductivity in water vapor indicates an upsurge in current, which is the plausible reason behind the humid sensing mechanism. At a relative humidity (RH) of 35, 50, 70, and 85%, the current responses are boosted from 75.9, 123.8, 181.2, and 222.6 μA to 141.2, 268.1, 365.7, and 572.5 μA without ZnO NWs and with ZnO NWs, respectively. Moreover, the hysteresis effect illustrated that the current signals could be fully recovered, as shown in Figure 18.4b. On intermittent conditions of 35% and 85% RH, the transient responses exhibited a response time of 31 s with ZnO NWs and 53 s without ZnO NWs. This report provides a general concept about the innovation of integrating gas sensors and SCs. Recently, Tanuma and Sugiyama et al. reported a carbon dioxide sensor with the integration of SCs [7]. The constructed device comprised a heterojunction of p-NiO:Li/n-ZnO structure as an SC unit,

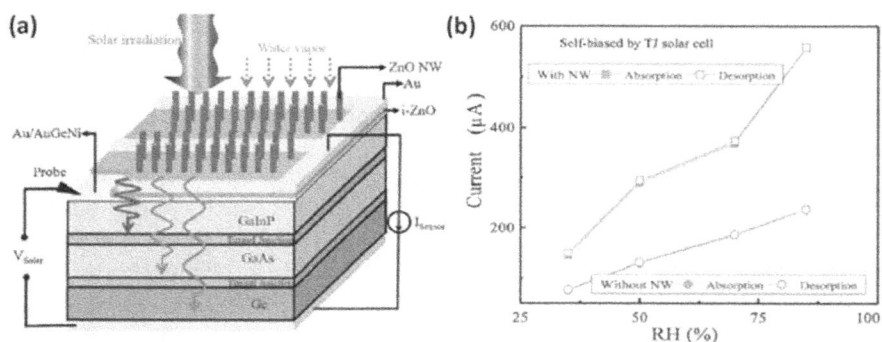

FIGURE 18.4
Integrated triple-junction SCs and HS based on ZnO NWs. (a) Schematic representation of the device assembly and (b) hysteresis behavior of the devices versus different RH. Adapted with permission [6], Copyright (2014) Elsevier.

and for carbon dioxide gas sensing, a thin layer of LaOCl/SnO$_2$ was formed. The SiO$_2$ layer insulated the sensing and SC layers, which are in series connection via Au electrodes. The fabricated devices showed an OCV of 0.16 V and 1.04 nA/cm^2 of short-circuit current (SCC) density under AM 1.5 irradiation. Though the resistance response is observed to be satisfactory, the results are yet to be developed, and some optimizations are expected, such as promoting the SC's performance.

On the other hand, with the progress of MEMS technology, several sensors can be conjoined into one microchip to accomplish several functions. For instance, Juan et al. made up a photodetector (PD) and a self-powered HS by combining crystalline silicon (Si) interdigitated back-contact (IBC) PV cell with tungsten oxide (WO$_3$) and gallium oxide (Ga$_2$O$_3$) thin films in a parallel condition [8]. The IBC cell, under the illumination of AM 1.5G (100 mW/cm^2), generated an OCV, SCC, and conversion efficiency of 0.598 V, 35.58 mA/cm^2, and 10.29%, respectively. Under constant solar illumination, both photodetection and HS work effectively. Considering photodetection, the thin films of WO$_3$ and Ga$_2$O$_3$ have cutoff wavelengths of 370 and 250 nm, correspondingly, which can be helpful in the visible-blind and solar-blind dual-band-based PDs. For the sensing of humidity, with the increase in relative humidity, the obtained current of the WO$_3$ thin film increases monotonously. Thus, it can be said that this method of integration toward PV self-powered gas sensing is based chiefly on the gas sensor (top)/insulator/SC (bottom) layer-by-layer structures. Considering the effect on electricity-generating efficiency, the gas sensor and insulator layers must be thin or transparent. Thus, the irradiated light can pass through them quickly into the SC. It is noteworthy that ample well-characterized PV and gas sensing materials are promisingly combined through several micromachining methods. This foundation should be considered for designing and optimizing integrated structures to achieve added functions and desirable performances of these devices in the future.

18.3.3 Electrochemical Cell-Based SPSs

A fuel cell (FC) generates electrical energy directly from the chemical energy of fuels by using metal catalysts in an electrochemical process. Contrasting with traditional energy sources, during the reaction, FC does not require any combustion; thus, its energy alteration is not restricted by the Carnot cycle. Considering the features of high efficiency, environmental friendliness, and cleanliness, FC remarkably impacts the preference for efficient and clean power-generating procedures in the 21st century. Moreover, after the three power-generating techniques viz atomic energy, fire power, and water power, FCs are the fourth power-generating process, and they have engaged more research interest. Several FCs are expected to be up-and-coming technologies for concurrent electricity generation and pollutant degradation due to their advantages of no subsidiary pollution, comprehensive fuel supply, higher catalytic activity, and the durable physicochemical nature of photocatalysts. With these attractive features, in 2001, I. Willner's group presented an innovative idea of emerging biosensing devices using chemical-to-electrochemical energy conversion in biofuel cells (BFC) [9]. In these BFCs, cathode-immobilized cytochrome c oxidase reduces oxygen with the associated oxidation of cytochrome c, while the anode-immobilized glucose oxidase catalyzes the oxidation of glucose for producing electrons, thus generating an electric corridor to provide OCV for glucose monitoring. These sensing systems do not utilize any external power source, and the generated electrical energy from the cell is utilized for detecting the analyte via its reduction-oxidation reactions, which in turn opened a new concept to construct the self-powered electrochemical sensor systems for sensing different substances like alcohol, fructose, and amino acids.

Figure 18.5 shows the schematic of the different types of sensors and their sensing strategies. In researching affinity-based biorecognition elements, a self-powered immunosensor was developed by Cheng et al. in which the cathode enzyme (bilirubin oxidase) was not immobilized on the cathode of the glucose/oxygen enzymatic BFC; instead, it was involved with an antibody to the carbon nanotubes (CNTs) [10]. On passing the analyte over the cathode, it is changed with a secondary antibody that permits the analyte to attach, followed by CNT. When the analyte exists, the sandwich-type assay restrains the bilirubin oxidase toward the cathode and allows enzymatic and electrocatalytic power generation. A similar idea was followed for self-powered DNA sensors. The concept of hybridization has been exploited to immobilize the enzyme at either the anode or cathode of the enzymatic BFC (EBFC). For example, Wang et al. designed a device for DNA sensing in which a platinum electrode acts as the cathode for catalyzing oxygen reduction.

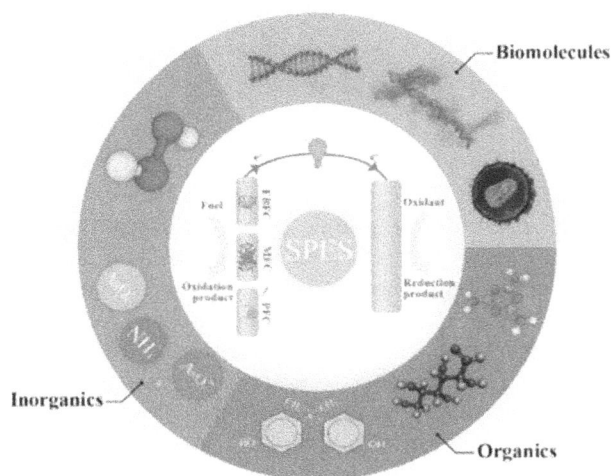

FIGURE 18.5
Schematic diagram of self-powered electrochemical sensors (SPES) detecting various biochemical molecules. PFC: photocatalytic FC; EBGC, enzymatic biofuel cell; MFC, microbial FC. Adapted with permission [4], Copyright (2019) Elsevier.

TABLE 18.1

Application of SPES for Different Types of Analytes [4]

Analyte	Analyte species	FC
Biomolecules	DNA	EBFC
	DNA	MFC
	miRNA-21	EBFC
	p53 protein	EBFC
	CEA	EBFC
	Glycoprotein	EBFC
	CCRF-CEM cell	EBFC
	CCRF-CEM cell; MCF-7 cell	EBFC
Inorganics	H_2O_2	MFC
	H_2O_2	PFC
	HMI toxicity	MFC
	KCN	EBFC
	Arsenite; arsenate	EBFC
Organics	Glucose	EBFC
	Glucose	PFC
	Glucose	EBFC
	LL-cysteine	EBFC
	Glutathione	PFC
	Lactate	EBFC
	Gallic acid	PFC
	Levofloxacin	MFC
	Bisphenol A	PFC
	Melamine	EBFC

The modification of the anode was done with a tiny single strand of DNA that hybridized with the analyte trailed by hybridizing the other analyte DNA with a single strand of DNA containing glucose oxidase and peroxide, which oxidize glucose [11]. The detection limit of the sensor is 6.3 fM. This concept has been expanded toward different materials, as shown in Table 18.1.

18.3.4 Electromagnetic Energy-Based SPSs

In the real world, kinetic energy is one of the most available and abundant energy sources owing to its irregular vibrations in several mechanical systems. Vibrations of Hz to MHz frequencies convey the potential to produce power in the range of a few microwatts to milliwatts per volume energy density. Energy scavenging from high-frequency mechanical motion can be exploited for minimal mass sensing. For example, Asadi et al. established nanoscale SPSs composed of a movable minute permanent magnet placed on a nanoplate and a fixed electromagnetic coil [12]. Utilizing the proposed models, the device's power capacity was studied under different excitations and linear/non-linear modes. Keeping the gap between the magnet and coil of 1.5 nm and at 330 MHz of primary resonance frequency, the device can produce 1.7 and 2.24 mW/cm^3 for linear and non-linear modes. In another work, Wang et al. developed an on-rotor-based electromagnetic energy harvester to wirelessly power a surveillance system on bogie frames [13]. The obtained performance in the 420–820 rpm broad train speeds thoroughly proves the power density enhancement up to 1982 W/m^3. The transformed electricity effectively powers everyday electric utilization and commercial Bluetooth sensors wirelessly. Recently, Li et al. established a self-powered wireless sensor system based on a low-frequency electromagnetic-pendulum energy harvester to monitor water quality [14]. The energy harvester operated under a minor frequency of 1.5 Hz, and the maximum power reached 14.76 mW. With the support of a power management process, a low-power water-quality wireless sensor was implanted to assess the total dissolved solids (TDS) and temperature.

18.3.5 Piezoelectric Nanogenerator-Based SPSs

The word "piezoelectricity" originates from Greek, which means "squeeze or press", and states the characteristics of such materials that produce an electric field with the application of mechanical force. There are several definitions: Briscoe and Dunn described piezoelectricity as the "electric charge that accumulates in response to applied mechanical stress in materials that have non-centrosymmetric crystal structures", while Erturk and Inman described it as "a form of coupling between the mechanical and electrical behaviors of ceramics and crystals belonging to certain classes" [15]. This effect can be categorized into two phenomena: direct and indirect piezoelectric (PE) effect. In 1880, Pierre and Jacques Curie invented the nature of some objects to produce an electrical field when a strain is enforced on those materials, termed the direct PE effect. A year later, considering the principles of thermodynamics, the inverse or converse PE effect was deduced mathematically by Lippmann, which dictates that a PE material will distort when an external electrical force is imposed. Figure 18.6 represents the utilization of PENG-based SPSs in different fields of applications with their different functioning modules. Considering ZnO nanowires as the basic unit, the first device was fabricated with vertically aligned ZnO nanowires [16]. Apart from ZnO NWs, cadmium telluride (CdTe), cadmium sulfide (CdS), gallium nitride (GaN), sodium niobite ($NaNbO_3$), two-dimensional molybdenum disulfide ($2D-MoS_2$), and lead zirconium titanate ($PbZr_xTi_{1-x}O_3$) (PZT) with different morphologies have been utilized for the fabrication of different PENGs [1]. The designed PENGs have been fruitfully applied in energy scavenging. They can also be helpful for sensing, chemical and environmental protection, biomedical and healthcare sensors, robotics, smart cities, fiber, and fabric applications. For example, Lin et al. fabricated packaged transparent flexible PENGs by using a flexible polydimethylsiloxane (PDMS) base to grow ZnO NWs, which were used as an SPS for determining the weight of the vehicle and intensive care of the vehicle's speed [17].

Interestingly, a contactless and magnetic force-driven PENG was fabricated by Cui et al. comprising a ZnO NW with a magnetic capping that remained sandwiched by the two electrodes [18]. Introducing a magnetic lid, the fabricated PENG can harvest electrical energy without contact and function as a magnetic sensor. An SPGS was made with a high humid sensitivity of cadmium (Cd) doped ZnO NW [19]. On exposing 70% RH, a response of 85.7 was found for Cd:ZnO (1:10) NWs, which is much higher than undoped NWs. The inclusion of Cd enhances the number of oxygen vacancies in the NWs, thus, resulting in more adsorption sites on the NWs' surface. Recently, two GaN-based flexible motion sensors and ultraviolet light-emitting diode-based flexible PENGs were fabricated by Lee et al. [20]. Asymmetric polarization across the flexible GaN film plays a vital role in manufacturing these multifunctional sensors. The healthy relationships of the sensors empower the exact measurement of motion. This integration quantifies the magnitude of the strain and identifies the bending direction for understanding multifunctional motion identification. Later, Zhao et al. created a concept of Schottky and Ohmic reversible biosensors for sensitively monitoring neural electric impulses and neurotransmitters [21].

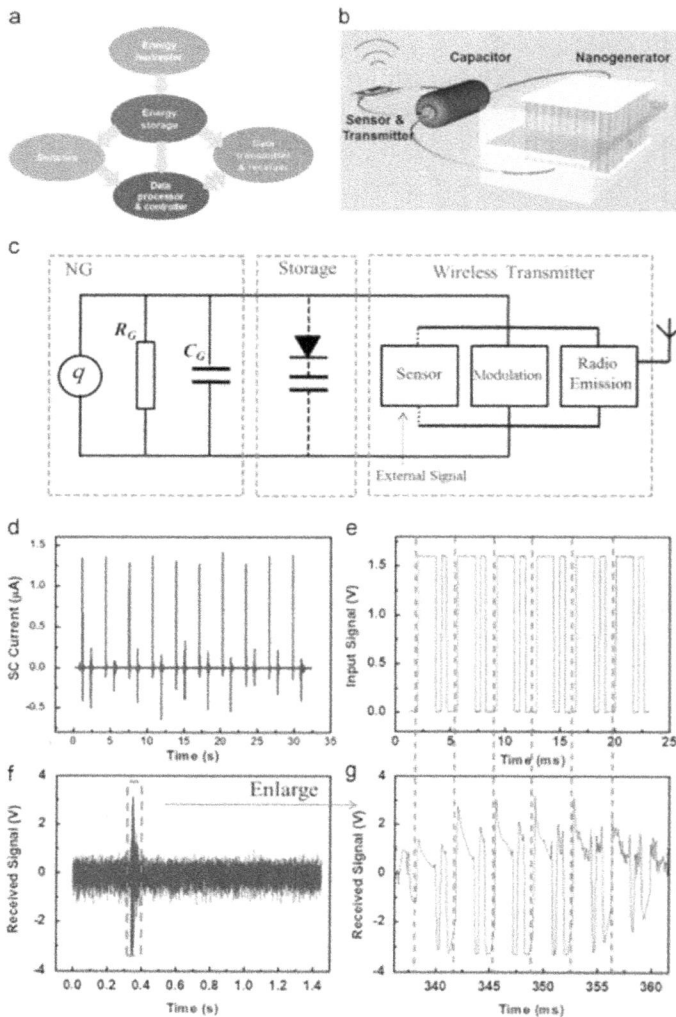

FIGURE 18.6
Wireless data communication for photon detecting via a self-powered system. (a) Schematic representation of the integrated self-powered system; (b) an integrated self-powered prototype system using a piezoelectric nanogenerator (PENG) as the mechanical energy harvester; (c) schematic illustration of the fabricated self-powered system; (d) performance of the designed PENG utilized in the demonstration (e) output voltage pulse used to trigger the LED; (f) recorded signals coming from the headphone plug of the radio; (g) amplification of the pulse in (f). Adapted with permission [16], Copyright (2015) Elsevier.

18.3.6 Triboelectric Nanogenerator-Based SPSs

The triboelectric nanogenerator (TENG), which works via a dual effect of contact electrification and electrostatic induction, was developed by Wang's group to harvest asymmetrical, random, and waste mechanical energy of low frequency and convert the energy into electrical power. Maxwell's displacement current is the governing force for this effective energy-harvesting methodology. On the basis of reported results, TENGs involve four working styles: vertical contact-separation mode, contact sliding mode, single-electrode mode, and free-standing mode, as illustrated in Figure 18.7 [22].

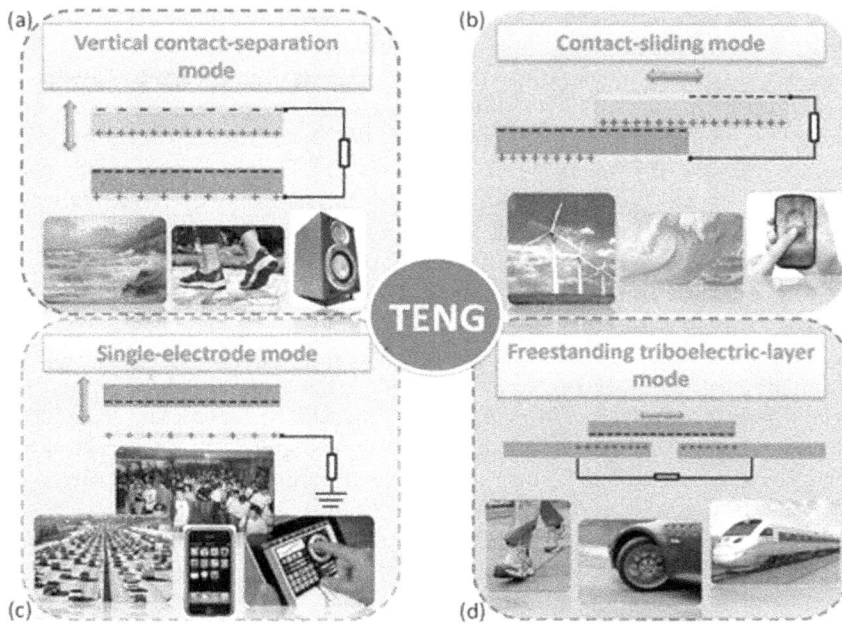

FIGURE 18.7
Fundamental working modes of TENG. Adapted with permission [22], Copyright (2014) The Royal Society of Chemistry.

Since 2012, TENG technology has successfully powered many electronics and developed self-powered viable electronic networks. The following segment includes current advancements of TENGs in numerous applications involving HMI, physical sensors, monitoring chemical and environmental issues, wearable systems, biomedical and healthcare sensors, intelligent traffic, homes and cities, robotics, and fiber-fabric sensors. Depending on triboelectricity for diverse motion types, the sensors can be classified into position, speed, and acceleration. Recent mechanical motion sensors are mostly restricted by their complicated structures, cost ineffectiveness, and problems in fabrication and assembly. Thus, researchers extensively consider TENG-based mechanical motion sensors owing to their simple and flexible systems, high integration levels, and lower costs. A micro-grated triboelectrification-based dynamic motion sensor has been manufactured that displayed a displacement resolution of 173 nm and a 0.02% linearity error [23]. Later, Wu et al. proposed an angle sensor utilizing four coded copper (Cu) foil channels [24]. However, due to the limitations of test motions, the progress of one-dimensional motion sensors is restricted, which enforces the development of multidimensional motion sensors. For instance, Han et al. fabricated a trajectory-tracking sensor composed of planar TENG pixels with a resolution of 250 μm [25]. Again, Chen et al. forged a self-powered motion-tracking device that can detect a velocity range of ±0.1 m/s and 0.02 m/s² of acceleration resolution [26]. Recently, Zhang et al. established acceleration sensors based on the TE phenomenon via utilizing vibration energy to understand self-power and self-sensing, which can be exploited to assess the object's acceleration [27]. The sensor displayed a 0–60 m/s² measurement range with a sensitivity of 0.26 V·s/m² [27]. In recent times, different TENG acceleration sensors have been developed, such as a magnetically levitated TENG-based vibration monitoring sensor and a TE accelerometer that proved that TE sensors could be utilized as easy and combined acceleration sensors.

Fluid sensors are one more essential fragment of physical sensors. The most mechanical energy in the environment is fluid transmission, like wind energy, droplet energy, wave energy, flowing water energy, and gas energy. Thus, domestic and overseas researchers have conducted many works on fluid sensors. For example, Chengkuo Lee and his group developed a self-powered TENG buoy ball for environment monitoring by harvesting water wave energy to construct a self-powered fishing sensor [28]. Recently, Chen et al. projected a real-world bionic jellyfish TENG to harvest energy from water waves and made a self-powered fluctuation sensor for liquid surfaces [29]. The fabricated device can accurately and wirelessly examine the instability of the liquid surface. This research is significant for construction in marine engineering, oceanic reserve development and exploitation, and marine protection.

Sensing of gas is extensively used to monitor the nature and concentration of ambient gases, specifically for recognizing toxic and/or explosive gases. Guo et al. fabricated a self-powered TE sensor that can precisely detect humidity and airflow rate [30]. A TENG was fabricated to supply the power of a chemoreceptive gas sensor composed of ZnO and reduced graphene oxide-based films to perceive the concentration of nitrogen dioxide (NO_2) [31]. The composite-based SPGS showed a better response, selectivity, and sensitivity than ZnO film without reduced graphene oxide. Furthermore, the TENG can be established as an active gas sensor. For instance, Shin et al. developed a functionalized palladium surface for fabricating an H_2 sensor [32]. When the device is in contact with H_2, its generated voltage is proportional directly to the amount of H_2. This phenomenon is due to changing the Pd-coated surface's work function, modifying the TE charging nature. The sensitive and reproducible sensor response was noticed for up to 1% H_2 exposure. With these brief applications, this self-powered sensing technique is also beneficial for self-powered pulse sensors, self-powered endocardial pressure sensors, transdermal drug delivery systems, anti-interference voice recognition, self-powered air filtering systems, Hg^{2+} ion detection, water quality monitoring, tire pressure monitoring systems, and so on [33–40].

18.3.7 Pyroelectric Nanogenerator-Based SPSs

Waste heat in natural environments and industrial processes promotes its candidature as a valuable resource for harvesting thermal energy. The wasted heat can be harvested and utilized by several solid-state energy alteration methods like TR and pyroelectric (PY) phenomena. TR and PY generators (TRG and PYG) have been performed to produce electrical energy from waste heat through Seebeck and PY effects, respectively [41]. TRGs are based on developing temperature differences spatially between two thermocouple terminals composed of two different metals to generate electrical energy converted from heat energy. But such a steady gradient is very intermittent in the environment. Whereas PY effect-based devices have been explored as a substitute for harvesting thermal energy, they do not need spatial temperature fluctuations. However, PYG requires a temporary temperature rise and fall to excite the process of energy conversion. Generally, pyroelectric materials are non-centrosymmetric crystal structures and subcategories of PE materials. Recently, Yang et al. established a PYG as an SPS to detect temperature change [42]. The device comprises a single micro/nanowire of PZT mounted on a thin glass substrate and fused at its two terminals. The whole device was wrapped with PDMS. It was found that the generated voltage is linearly augmented with the rise of the rate of temperature change. The reset and response times of the fabricated device are about 3 and 0.9 s, respectively. The minor sensing limit of the temperature change is 0.4 K at ambient temperature.

In another work, Wang et al. developed a light-triggered PEG comprising a p-n junction (p-Si/n-ZnO) for self-powered near-infrared photo-sensing [43]. A wearable PYG and self-powered breathing sensor have been established by Xue et al., where a polyvinylidene fluoride (PVDF) thin film was integrated into an N95 respirator [44]. An NW-composite-based flexible TRG as a self-powered temperature sensor has been developed by Yang et al., where tellurium (Te)-NW/poly(3-hexyl thiophene) (P3HT) composite acts as the TR material [45]. The fabricated TRG displays a response time and a reset time of 17 and 9 s, respectively, with a sensitivity of 0.15 K in normal atmosphere. Recently, Xie et al. fabricated a flexible TRG made of MoS_2/graphene nanocomposite for a self-powered temperature sensor [46]. The increased electrical conductivity from the added graphene acted as a charge transfer channel in the formulated nanocomposites.

18.3.8 Hybrid Nanogenerator-Based SPSs

Apart from the single harvesting technique, a hybrid cell equipped with different frictional units capable of working as a highly efficient energy harvester and an SPS has been extensively explored. For example, Zhong et al. developed an electromagnetic and TE hybrid structure operated by magnetically coupling to harvest fluid energy and monitor fluid, temperature, and humidity in a smart home/city [47]. In another work, Guo et al. made an all-fiber-based wearable hybrid nanogenerator based on PE and TE phenomena to monitor gestures where silk fibroin and PVDF act as active components [48]. A self-powered PD system has been established considering two harvesting techniques, PV and PE energy harvesting processes, for developing a self-powered PD system [49]. Here, the self-powered PD composed of bismuth ferrite ($BiFeO_3$) film enables fast sensing of 450 nm light illumination. Interestingly, Wang and his coworkers made up a hybrid cell involving TE, PY, and PENG based on sliding mode TENG, and a pyroelectric–piezoelectric nanogenerator (PYPENG) showed an excellent performance not only in energy harvesting but also in sensing temperature and force during friction [50]. The fabricated TENG of 63.5 cm² area generated a power density of 0.15 W/m² by harvesting the typical mechanical energy at 4.41 Hz sliding frequency during sliding motion. The integrated PYPENG can harvest thermal energy from friction-related heat, and the normal force induces mechanical energy. These outstanding presentations validate the hybrid system as a talented energy harvester and an SPS for future applications.

18.4 Conclusions and Outlook

This chapter summarized research development on the road to SPSs based on different nanogenerators. Compared with battery technology with an inadequate lifetime, nanogenerators have been established as green and sustainable power sources by harvesting ambient energy. Because of the randomness of ambient energy, the energy storage system is essential in nanogenerators. Using nanogenerators, the construction of active sensors can diminish energy consumption. Although substantial development of SPSs and nanogenerator systems has been achieved, there are still gaps between experimentation and practical commercial implementation. Here, possible challenges and difficulties for the widespread marketable applications of nanogenerators as power sources and active sensors of SPSs and systems are shown in the following section.

Output performance. The progress of nanogenerators as power suppliers with high output performance is a continual challenge. Thus, looking for advanced active materials and highly effective integration associated with the structural design of the device is essential.

Durability. PENGs and TENGs are generally made of rigid materials. The friction between a solid and liquid substance could be a practical choice to solve this issue. Abrasion will decrease the performance and lifetime of the fabricated device.

Power management. A power management circuit is needed to obtain a sustained and stable power supply. A DC-DC circuit, a steadying voltage circuit, and a defense circuit with a small consumption are essential.

Packaging. Open-type nanogenerators might be attacked by water, dust, and other foreign particles, which would neutralize generated electric charges and decrease performance and mechanical stability. Thus, one should take care of an appropriate packaging design without influencing the operating mode.

Sensitivity. Outstanding sensitivity is a continuous demand when evolving a sensor. The necessities for output performance explained previously are also applicable to this issue. Some innovative constructions can also be employed to augment sensitivity.

Stability. For qualitative analysis, stability is not an important issue for the sensor, but it becomes one of the chief performance characteristics for quantitative analysis.

Multifunction. A multifunctional sensor will decrease the energy and volume needed for SPSs. Consequently, this needs to be taken care of in the future.

Industrial fabrication. Improvement in the structural design, stability, packaging, and sensitivity of SPSs should be made to ensure that production is at an industrial scale. Thus, developing the products for wide or commercial use is essential.

References

1. Z. Wu, T. Cheng and Z. L. Wang, *Sensors*, 2020, 20, 2925.
2. S. T. Han, H. Peng, Q. Sun, S. Venkatesh, K. S. Chung, S. C. Lau, Y. Zhou and V. A. L. Roy, *Advanced Materials*, 2017, 29(33), 1700375.
3. F. Ejeian, S. Azadi, A. Razmjou, Y. Orooji, A. Kottapalli, M. E. Warkiani and M. Asadnia, *Sensors and Actuators A: Physical*, 2019, 295, 483–502.
4. Y. Chen, W. Ji, K. Yan, J. Gao and J. Zhang, *Nano Energy*, 2019, 61, 173–193.
5. R. L. Arechederra and S. D. Minteer, *Analytical and Bioanalytical Chemistry*, 2011, 400, 1605–1611.
6. J.-L. Hou, C.-H. Wu and T.-J. Hsueh, *Sensors and Actuators B: Chemical*, 2014, 197, 137–141.
7. R. Tanuma and M. Sugiyama, *Physica Status Solidi (a)*, 2019, 216, 1800749.
8. Y. Juan, S. Chang, H. Hsueh, T. Chen, S. Huang, Y. Lee, T. Hsueh and C. Wu, *Sensors and Actuators B: Chemical*, 2015, 219, 43–49.
9. E. Katz, A. F. Bückmann and I. Willner, *Journal of the American Chemical Society*, 2001, 123, 10752–10753.
10. J. Cheng, Y. Han, L. Deng and S. Guo, *Analytical Chemistry*, 2014, 86, 11782–11788.
11. Y. Wang, L. Ge, C. Ma, Q. Kong, M. Yan, S. Ge and J. Yu, *Chemistry–A European Journal*, 2014, 20, 12453–12462.
12. E. Asadi, H. Askari, M. B. Khamesee and A. Khajepour, *Measurement*, 2017, 107, 31–40.

13. Z. Wang, W. Wang, F. Gu, C. Wang, Q. Zhang, G. Feng and A. D. Ball, *Energy Conversion and Management*, 2021, 243, 114413.
14. M. Li, Y. Zhang, K. Li, Y. Zhang, K. Xu, X. Liu, S. Zhong and J. Cao, *Energy*, 2022, 251, 123883.
15. C. Covaci and A. Gontean, *Sensors*, 2020, 20, 3512.
16. H. Youfan and Z. L. Wang, *Nano Energy*, 2015, 14, 3–14.
17. L. Lin, Y. Hu, C. Xu, Y. Zhang, R. Zhang, X. Wen and Z. L. Wang, *Nano Energy*, 2013, 2, 75–81.
18. N. Cui, W. Wu, Y. Zhao, S. Bai, L. Meng, Y. Qin and Z. L. Wang, *Nano Letters*, 2012, 12, 3701–3705.
19. B. Yu, Y. Fu, P. Wang, Y. Zhao, L. Xing and X. Xue, *Physical Chemistry Chemical Physics*, 2015, 17, 10856–10860.
20. J. W. Lee, B. U. Ye, Z. L. Wang, J.-L. Lee and J. M. Baik, *Nano Energy*, 2018, 51, 185–191.
21. L. Zhao, H. Li, J. Meng, A. C. Wang, P. Tan, Y. Zou, Z. Yuan, J. Lu, C. Pan and Y. Fan, *Advanced Functional Materials*, 2020, 30, 1907999.
22. Z. L. Wang, *Faraday Discussions*, 2014, 176, 447–458.
23. Y. S. Zhou, G. Zhu, S. Niu, Y. Liu, P. Bai, Q. Jing and Z. L. Wang, *Advanced Materials*, 2014, 26, 1719–1724.
24. Y. Wu, Q. Jing, J. Chen, P. Bai, J. Bai, G. Zhu, Y. Su and Z. L. Wang, *Advanced Functional Materials*, 2015, 25, 2166–2174.
25. C. B. Han, C. Zhang, X. H. Li, L. Zhang, T. Zhou, W. Hu and Z. L. Wang, *Nano Energy*, 2014, 9, 325–333.
26. M. Chen, X. Li, L. Lin, W. Du, X. Han, J. Zhu, C. Pan and Z. L. Wang, *Advanced Functional Materials*, 2014, 24, 5059–5066.
27. Y. Zhang, Y. Fang, J. Li, Q. Zhou, Y. Xiao, K. Zhang, B. Luo, J. Zhou and B. Hu, *ACS Applied Materials & Interfaces*, 2017, 9, 37493–37500.
28. Q. Shi, H. Wang, H. Wu and C. Lee, *Nano Energy*, 2017, 40, 203–213.
29. B. D. Chen, W. Tang, C. He, C. R. Deng, L. J. Yang, L. P. Zhu, J. Chen, J. J. Shao, L. Liu and Z. L. Wang, *Materials Today*, 2018, 21, 88–97.
30. H. Guo, J. Chen, L. Tian, Q. Leng, Y. Xi and C. Hu, *ACS Applied Materials & Interfaces*, 2014, 6, 17184–17189.
31. Q. Shen, X. Xie, M. Peng, N. Sun, H. Shao, H. Zheng, Z. Wen and X. Sun, *Advanced Functional Materials*, 2018, 28, 1703420.
32. S.-H. Shin, Y. H. Kwon, Y.-H. Kim, J.-Y. Jung and J. Nah, *Nanomaterials*, 2016, 6, 186.
33. H. Ouyang, J. Tian, G. Sun, Y. Zou, Z. Liu, H. Li, L. Zhao, B. Shi, Y. Fan and Y. Fan, *Advanced Materials*, 2017, 29, 1703456.
34. Z. Liu, Y. Ma, H. Ouyang, B. Shi, N. Li, D. Jiang, F. Xie, D. Qu, Y. Zou and Y. Huang, *Advanced Functional Materials*, 2019, 29, 1807560.
35. Z. Liu, J. Nie, B. Miao, J. Li, Y. Cui, S. Wang, X. Zhang, G. Zhao, Y. Deng and Y. Wu, *Advanced Materials*, 2019, 31, 1807795.
36. J. Yang, J. Chen, Y. Su, Q. Jing, Z. Li, F. Yi, X. Wen, Z. Wang and Z. L. Wang, *Advanced Materials*, 2015, 27, 1316–1326.
37. S. Chen, C. Gao, W. Tang, H. Zhu, Y. Han, Q. Jiang, T. Li, X. Cao and Z. Wang, *Nano Energy*, 2015, 14, 217–225.
38. Z. H. Lin, G. Zhu, Y. S. Zhou, Y. Yang, P. Bai, J. Chen and Z. L. Wang, *Angewandte Chemie*, 2013, 125, 5169–5173.
39. Y. Bai, L. Xu, C. He, L. Zhu, X. Yang, T. Jiang, J. Nie, W. Zhong and Z. L. Wang, *Nano Energy*, 2019, 66, 104117.
40. H. Zhang, Y. Yang, X. Zhong, Y. Su, Y. Zhou, C. Hu and Z. L. Wang, *ACS Nano*, 2014, 8, 680–689.
41. A. Sultana, M. M. Alam, T. R. Middya and D. Mandal, *Applied Energy*, 2018, 221, 299–307.
42. Y. Yang, Y. Zhou, J. M. Wu and Z. L. Wang, *ACS Nano*, 2012, 6, 8456–8461.
43. X. Wang, Y. Dai, R. Liu, X. He, S. Li and Z. L. Wang, *ACS Nano*, 2017, 11, 8339–8345.
44. H. Xue, Q. Yang, D. Wang, W. Luo, W. Wang, M. Lin, D. Liang and Q. Luo, *Nano Energy*, 2017, 38, 147–154.
45. Y. Yang, Z.-H. Lin, T. Hou, F. Zhang and Z. L. Wang, *Nano Research*, 2012, 5, 888–895.

46. Y. Xie, T.-M. Chou, W. Yang, M. He, Y. Zhao, N. Li and Z.-H. Lin, *Semiconductor Science and Technology*, 2017, 32, 044003.
47. Y. Zhong, H. Zhao, Y. Guo, P. Rui, S. Shi, W. Zhang, Y. Liao, P. Wang and Z. L. Wang, *Advanced Materials Technologies*, 2019, 4, 1900741.
48. Y. Guo, X.-S. Zhang, Y. Wang, W. Gong, Q. Zhang, H. Wang and J. Brugger, *Nano Energy*, 2018, 48, 152–160.
49. J. Qi, N. Ma and Y. Yang, *Advanced Materials Interfaces*, 2018, 5, 1701189.
50. Y. Zi, L. Lin, J. Wang, S. Wang, J. Chen, X. Fan, P. K. Yang, F. Yi and Z. L. Wang, *Advanced Materials*, 2015, 27, 2340–2347.

19

Stretchable and Self-Healing Sensors

Hatice Aylin Karahan Toprakci and Ozan Toprakci

CONTENTS

19.1 Introduction

Recent developments in technology have opened a new area known as wearable electronics to improve comfort, safety, and quality of life. Although the wearability concept was based on the attachment of heavy, bulky electronics on textiles around 30 years ago, today, it is completely different, and state-of-art studies aim to develop organ-type electronics that can easily locate on/into the human body, therefore, most research focuses on minimizing not only the size and weight of the electronics but also the distance between the body and electronics. To develop such wearable electronics, two important factors should be considered. The first factor is the wearability of the system, while the second factor is the reliability of the electronic function. Wearability is directly related to the material selection and 3D design of the products. The wearability function of an electronic is fulfilled by flexible and/or stretchable polymeric surfaces such as films [1], layered structures [2], coatings [3], prints [4], or textiles [5]. The electronic function can be obtained by the polymer itself or additional functional materials with electronic functionality can be incorporated into the polymer. One of the basic functions of wearable electronics is the signal/data extraction that is obtained by sensors. Sensors are systems used for perceiving, monitoring, and interpreting the variables around us. Sensing is performed by the conversion of stimuli into an electrical signal. The stimulus is a significant factor that needs to be controlled discretely and/or monitored dynamically in forms such as mechanical, optical, thermal, chemical, biochemical, and so on. Determination of the stimuli and type, degree,

DOI: 10.1201/9781003299455-19

magnitude, level, and concentration of the stimuli are the basic functions of the sensors. Although sensors can be fabricated by using various materials including metals, semiconductors, and ceramics, for wearable electronics, polymers and polymer composites are important because they are lightweight, flexible, and tunable [6–9].

Polymer-based sensors are still in the development stage and any development in polymer science/technology can be quickly adapted to this area. While initial sensors were designed to monitor one type of stimuli, the recent trend is to focus on the design and fabrication of multifunctional sensors with advanced polymeric systems that have self-healing properties [10]. Self-healing is a biomimetic approach and is adapted to polymers from nature because almost all biological systems can heal themselves to some level. Self-healing is a material property used for the determination of recoverability or damage repair ability that is triggered by various mechanisms after internal and/or external damage. It is required due to sensor aging during usage by the impact of environmental, mechanical (abrasion, friction, force, pressure, strain), electrical, and thermal factors. As the consequences of material aging, structural, dimensional, and mechanical integrity might be degraded; stability and reliability problems occur, which cause sensors to lose their function [7,11–15].

The performance of a self-healing sensor depends on the healing degree of the polymeric system mostly governed by the material type that determines the mechanism of self-healing. Self-healing can be evaluated in two basic categories. One of them is based on the requirement of external stimuli. If self-healing is triggered by the material itself, it is referred to as autonomic self-healing. On the other hand, if external stimuli (light, heat, pressure, magnetic field, electrical field, etc.) are required for self-healing, it is called non-autonomic self-healing. The other classification is based on the requirements of the healing agent. If an additional healing agent is required, it is referred to as extrinsic self-healing. This is generally obtained by the incorporation of encapsulated healing agents into polymeric systems during fabrication. The amount of healing agent and the performance of the system can be controlled by various morphologies including capsules, hollow fibers, and vascular tubes. Under any environmental factor or stimuli, the material is damaged and self-healing material is released from the cracks and fills the damaged area. In this case, self-healing is based on repairing so it is not easy to obtain cyclic healing without modification of the encapsulated system. If self-healing inherently takes place without any additional agents, it is called intrinsic self-healing and this mechanism provides cyclic self-healing after repeated damage. This process might require additional stimuli as a trigger for the reformation of the bonds from chemical and/or physical interactions. Disulfide bonds, boron-based bonds, imine bonds, Diels–Alder chemistry, di-selenide bonds, trans-esterification reactions, and ditelluride bonds can be given as the covalent interactions. Hydrogen bonding, van der Waals interactions, ionic interactions, dipole-dipole interactions, π–π stacking, metal-ligand coordination, and host–guest interactions are the most common noncovalent physical interactions [7,11,12]. Comprehensive information about self-healing mechanisms can be obtained from related literature [7,11].

19.2 Requirements for Stretchable and Self-Healing Sensors

Stretchable self-healing sensors (SSS) are electronic materials that can convert any type of stimuli into a meaningful electrical signal under a high level of deformation. Self-healing polymers are promising materials not only for increasing the performance, durability, and service life of the sensors but also for reducing the pollution caused by electronic waste and maintenance costs [16]. However additional properties are required for stretchable self-healing polymers for their application as sensors that are generally used for organ monitoring [17], motion monitoring [18], human-machine interfaces [19], epidermal electronics, and electronic skins [20]. Also shown in Figure 19.1, mechanical, sensing, self-healing, and other required properties are detailed in the following sections.

19.2.1 Mechanical Properties

Stretchability, coverability, dimensional stability, robustness, and resilience are the most important mechanical properties of SSS.

- *Stretchability.* The deformability of a polymeric sensor can be classified as flexible and stretchable. While flexibles show moderate, limited deformation like bending and twisting, stretchables show large strain deformations. Generally, low-modulus polymers such as elastomers and hydrogels show a high degree of mechanical compliance and reversible stretchability. Since stretchability is directly related to deformation, it is generally expressed as strain (mm/mm, %) [21].

- *Coverability.* SSS are not only used as strain gauges but they are also integrated on various surfaces. Coverability is a measure of conformal integration on curved surfaces. SSS should show good coverability on complex surfaces that have arbitrary shapes with good conformal integration and the ability to maintain function under these circumstances. Coverability is of significance for human-machine-type interfaces [22].

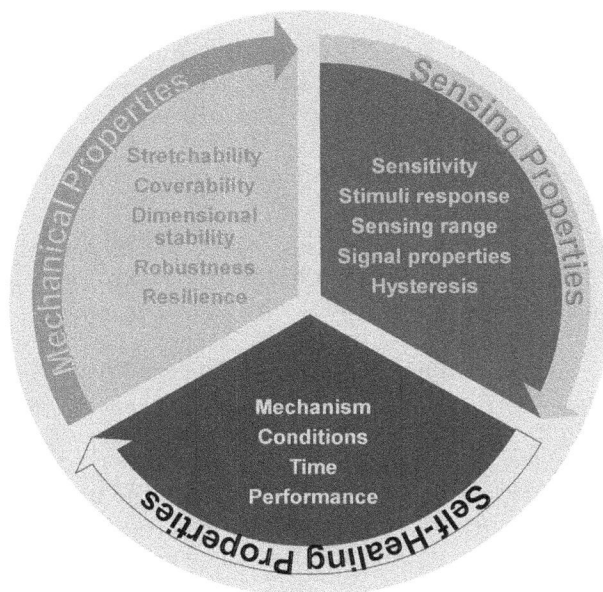

FIGURE 19.1
Basic requirements for stretchable and self-healing sensors.

- *Dimensional stability.* Dimensional stability is a material property that is used as an indication for maintaining size under various stimuli such as force, strain, pressure, heat, and so on. Sensor materials should be compatible with the micro and macro environment. Since stretchable self-healing sensors are exposed to various stimuli for many cycles, cyclic dimensional stability is expressed as the change in surface area% and/or thickness% for hydrogel-based materials shrinkage% due to the removal of water can also be given [5].

- *Robustness.* Robustness is another property that shows the capability of resisting mechanical damage. For stretchable self-healing sensors, the optimum conditions that the system shows robustness should be considered for the determination of its working range and it is significant for the reliability of the data [23].

- *Resilience.* Resilience is a critical property that basically shows the recoverability of the sensing material after cyclic stimuli exposure. Mechanical resilience can be controlled by the determination of the unrecovered strain (mm/mm, %) or stress softening (MPa). Stretchable self-healing sensors function under a high level of strain, so resilience not only affects mechanical properties but also self-healing and sensing [24].

19.2.2 Sensing Properties

Stimuli-specific response, signal quality and stability, sensitivity, sensing range, response time, delay, cyclic behavior, and hysteresis are the most important properties used for the determination of performance.

- *Stimuli-specific response.* Stretchable self-healing sensor materials should be sensitive to stimuli and generate reliable and reproducible electrical signals under various levels of the stimuli without influencing the measurand and being insensitive to other stimuli that can ruin the required signal [25].

- *Sensing range.* The sensing range is the minimum and maximum values of the stimuli where meaningful, accurate signals are obtained. The range can be given according to stimuli such as force (N) [26], strain (mm/mm, %) [27], stress (Pa) [13], temperature (°C) [15], concentration (ppm, %, wt%) [25], and so on. Since SSS are highly deformable structures with self-healing ability, they are promising materials for a wide sensing range.

- *Signal properties.* Signal quality, linearity, response time, delay, and so on are the most important signal properties for SSS. Electrical signal generation is the basic function of a sensor. However, the quality of the signal in terms of strength and noise is of vital importance for performance. The noisy signal might require additional electronics for obtaining reliable data. The materials used for SSS should give meaningful data with the least noise. Linearity is also important for monitoring the stimuli. In the application range, a linear response is very useful in terms of interpreting the data and calibration of the sensor. Response time is the period for obtaining the electrical signal and its magnitude and it might be related to the material itself, conditions, or the electronics. Response time should be as low as possible with the least delay. Other criteria can be given as low-potential requirements for electrical signal generation [20,28–30].

- *Sensitivity.* Sensitivity is the most important criterion that is used for the determination of sensor performance. Although formulation might change depending on the mechanism of sensing, generally, it is a ratio between output and input change. Since it defines the optimum sensing region, it also defines the required minimum input for obtaining meaningful data [14,25,26,29–34].

- *Hysteresis*. Hysteresis is a measure of deviation from the initial and/or previous state and can be defined as electrical resilience. It is also related to the precision, reproducibility, and accuracy of the signal. Since sensors are exposed to cyclic stimuli, hysteresis is directly related to the stability of the signal. In the case of high stability, highly reversible data is obtained. Structural, self-healing, mechanical properties of the sensor, environmental conditions, and some electronic interferences might be the reasons for hysteresis [2].

19.2.3 Self-Healing Properties

Samples are generally partially [35] or completely cut [36], notched [35], scratched [37], or punctured [38] before self-healing analysis. The most important properties considered for self-healing performance are given in the following section.

- *Self-healing mechanism*. The mechanism of self-healing can be counted as the baseline of material performance. The process can be autonomic, non-autonomic, intrinsic, or extrinsic. To develop stable SSS, an effective and repeatable self-healing mechanism should be achieved [7,11,12].
- *Self-healing conditions*. Depending on the mechanism of self-healing, external stimuli might be required such as temperature [36] and water vapor [4]. While some materials can self-heal under room conditions [3], some require higher temperatures [36].
- *Self-healing time/speed*. Self-healing time is the period for recovery of the polymer. While for some materials this value is very low such as milliseconds [6] to minutes [36], for some systems, it can be as long as 48 hours [35]. In the case of multiple molecular interactions, time decreases, so the mechanism of self-healing directly affects the required time. As reported in the literature, this is not only related to polymer type but also filler type and concentration [26]. Healing time is cross-checked by mechanical and electrical properties and their self-healing time might be different [28].
- *Self-healing cycles*. The number of cut-healing cycles is an important indicator of SSS performance. Based on the literature generally, one to ten cycles are performed [14]. Not only the number of cycles but also the degree of self-healing is important.
- *Self-healing of physical and morphological appearance*. The basic evaluation of self-healing is controlling the sample integrity from its physical appearance by the naked eye. There can be no, partial, or complete healing. Even if the sample looks like completely recovered, this may not reflect its performance. After separated pieces are self-healed, a sample is generally analyzed by an optical microscope or scanning electron microscope to investigate the morphology [35,36].
- *Mechanical self-healing*. Tensile strength, tensile strain, elastic modulus, coverability, dimensional stability, robustness, stress softening, and resilience can be evaluated after self-healing, and performance% is calculated based on the virgin sample. These properties can be analyzed as a function of (cut-healing) cycle number [39].

- *Electrical self-healing.* Electrical property is determined by the sensing mechanism. While piezoresistive SSS, electrical, or ionic conductivity/resistance [35] is evaluated, for capacitive and piezoelectric sensors, capacitance [14] and voltage [26] are evaluated, respectively, after self-healing as a function of time, cycle number and performance% are calculated according to the virgin sample. A higher percentage indicates better self-healing performance.
- *Self-healing of the sensing properties.* Basic sensing properties mentioned previously, such as sensing range, signal properties, sensitivity, and hysteresis, should be determined after self-healing cycles, and if possible, performance% should be calculated.

19.2.4 Other Properties

In addition to mechanical, sensing, and self-healing properties, SSS fabrication should be done. The system should be easily adaptable to various surfaces and interfaces. Additional adhesiveness is required for applications such as e-skin [6]. Signal processing systems should be affordable. Transparency [40], biocompatibility [3], biodegradability [41], anti-drying [42], anti-freezing [20], and recyclability [6] of the system might be required for certain applications.

19.3 Classification of Stretchable Self-Healing Sensors Based on the Mechanism of Sensing

Stretchable self-healing sensors can be classified based on material type, application, and healing or sensing mechanism. Since the components of the sensing system are chosen based on the sensing mechanism, here SSS are classified as piezoresistive, capacitive, and piezoelectric sensors. Applications of the sensors are given under each mechanism.

19.3.1 Piezoresistive Stretchable Self-Healing Sensors

Piezoresistive stretchable self-healing sensors (PSSS) can be used for strain, pressure, gas, humidity, and temperature monitoring.

- *Strain sensors.* PSSS can be used for monitoring any action that is based on strain deformation including bending, motion, stress, and force monitoring as given in Figure 19.2 [27]. The sensitivity of a PSSS is determined by GF. $GF = (\Delta R/R_0)/\varepsilon = (R-R_0/R_0)/(l-l_0/l_0)$ or $(\Delta S/S_0)/\varepsilon = (S-S_0/S_0)/(l-l_0/l_0)$ where R is the electrical/ionic resistance, R_0 is the initial resistance, S is the electrical/ionic conductance, S_0 is the initial conductance, l is the length, and l_0 is an initial length of the sensor. Sensing range, minimum and maximum detectable strain level, hysteresis, recoverability, aging, and self-healing properties should be considered [4,28,36,39].

One approach is the addition of electrically conductive fillers to obtain piezoresistance. Liu et al. fabricated a self-healing strain sensor with poly(ε-caprolactone)-based microspheres (mPCL) [36]. Silver nanowires (AgNWs) were used for the conductive network formation and graphene oxide (GO) was used for stable conductivity; m-PCL/GO/AgNWs (PGA)

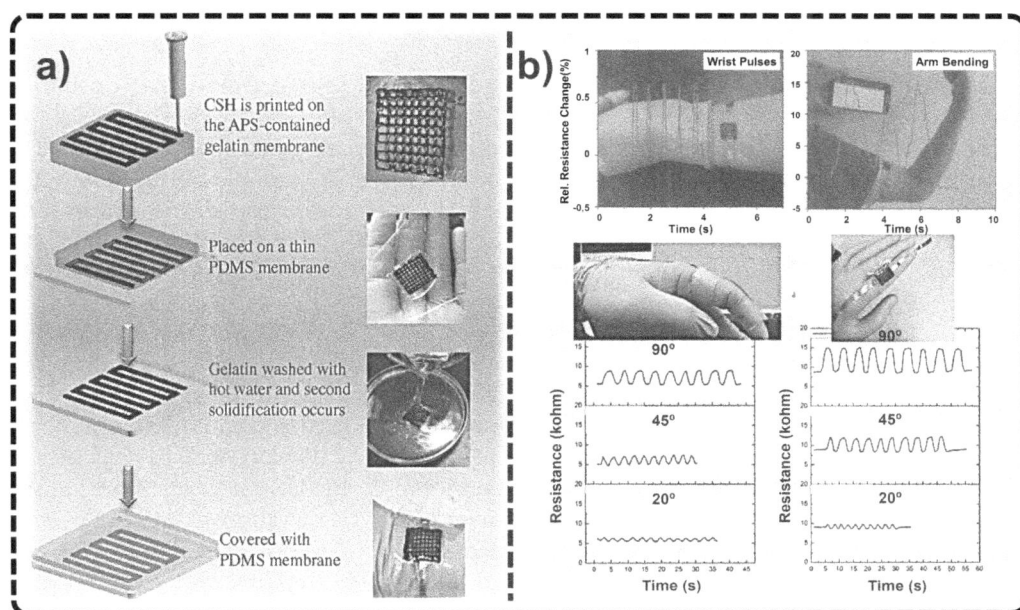

FIGURE 19.2
(a) Preparation of 3D-printed sensor. (b) A real-time body motion monitoring system using a smartphone and 3D-printed wearable and flexible sensors attached on the index finger, biceps, and wrist, and corresponding resistance variation upon repeated bending and relaxing motion from 0° to 20°, 45°, and 90°. Adapted with permission [27]. Copyright (2021), John Wiley and Sons.

films were prepared, and liquid poly(dimethylsiloxane) (PDMS) was poured on them. The conductive layer was scratched, complete self-healing was obtained after 3 min at 80°C, and the mechanism was reported to be based on melting, diffusion, and interaction between polymer and fillers. The conductivity was 0.45 S/cm and the sample was analyzed for its bending and strain-sensing performance before and after self-healing. While conductivity showed a decrease as a function of the healing cycle; sensitivity under bending and strain increased and remained almost constant after the second cycle. The sensitivity of the material under different bending angles (0–180°) was 0.26 rad^{-1}. The sample showed positive piezoresistance under 10, 20, and 30% strain loading and was used for index finger, elbow, and neck movement monitoring [36].

In addition to films, unique product design based on sacrificial bonds was also reported. Wu et al. developed self-healing carbon nanotubes (CNT) filled conductive chitosan (CS) nanocomposites [4]. The fibrous morphology with various sacrificial bonds was fabricated by the solvent-cast 3D printing method. The healing of mechanical properties by water vapor exposure for 10 s was reported to be enough. However, a CS/CNT strain sensor was prepared by the deposition of a coiling CS/CNT on a flexible PDMS film, and GF was found around 4 under 10% strain for 1000 cycles. So, the high level of stretchability of the system is questionable [4].

One of the most comprehensive studies was carried out by Guo et al. [39]. They used N-acryloylmorpholine (ACMO) and carboxyl multi-walled carbon nanotubes (c-CNT) as the monomer and conductive filler, respectively, for the preparation of the slurry with a dispersing agent (BYK). For the fabrication of stretchable strain sensors, c-CNT/BYK/ACMO (CBA) was printed by digital light processing-based technique and polymerized under UV exposure. The mechanism of self-healing was reported to be based on the formation of hydrogen bonds between water molecules, c-CNT, and poly(ACMO). To investigate the self-healing performance, ten cut-healing cycles were repeated. Mechanical properties were reported to decrease after self-healing. While strain at break and elastic modulus were 1209.2%, and 0.115 MPa, respectively, they dropped to 246.2% and 0.095 MPa, respectively, after the tenth healing cycle because of unhealable bonds. The virgin sample had GF of 2 under 100% strain. The level of strain was found significant in terms of the stability of the data. Samples were mounted on the index finger before and after self-healing. Although the sensing pattern was similar even after the second healing, the sensitivity increased [39].

In addition to incorporating conductive fillers into self-healing polymers, conductive polymer incorporation into self-healing polymeric networks is one of the recent trends for PSSS. Wang et al. developed highly stretchable composites with PAni and polyacrylic acid (PAA) [28]. While PAni was used for conductivity; the function of PAA was to obtain a soft, stretchable network. Sample mechanically and electrically self-healed ~99% after 24 h at 25°C, under 60% RH. However, electrical conductivity healed faster than mechanical properties and that was attributed to a quick reformation of a conductive network because of electrostatic interactions. The sensing performance was determined before and after self-healing and GFs were determined between 0 and 100% and 100 and 400% strain as 11.6/4.7 and 10.5/4.8, respectively. Samples showed stable responses for finger and knee movement monitoring. However, no data was given for wearables after self-healing [28].

- *Pressure sensors.* PSSS can be used for monitoring pressure, applied force, and compression% because of their highly deformable structure. They are promising materials for touch sensors, e-skins, pressure mapping, and so on. The sensitivity can be determined by GF as $(\Delta R/R_0)/\varepsilon$, $(\Delta S/S_0)/\varepsilon$, or $\Delta I/\Delta P$ where ΔI is the current change and ΔP is the pressure change. Sensing range, minimum and maximum detectable compression ratio/pressure/force, hysteresis, recoverability, aging and self-healing mechanisms, conditions, time, and performance should be considered [1,27,28].

Tee et al. developed one of the first self-healable stretchable pressure sensors from supramolecular organic-inorganic material by using Empol 1016 (diacids, triacids mixture), diethylenetriamine, nickel particles, and urea [1]. Nickel was varied between 0 and 75 v%, and percolation concentration was determined as 15 v%. While 15 s was reported to be enough for 98% electrical healing, higher nickel concentrations led to lower efficiency because of lower interaction through the self-healing network. Time and temperature were also important for mechanical self-healing. At 50°C, after 5 min, the sample showed almost the same mechanical properties as the virgin sample. Hydrogen bonding was given as the basic factor for the self-healing mechanism. The system showed negative piezoresistance between 0.05 and 0.4 MPa because of a decrease in the filler-filler distance [1].

Darabi et al. prepared injectable conductive highly stretchable self-healing (CSH) hydrogels by using polypyrrole-grafted chitosan (Ch-PPy), acrylic acid (AA) monomers, and ferric irons to form a PAA-based double network. To investigate the composition effect on morphological, mechanical, electrical, and self-healing properties, various amounts of Ch-PPy and crosslinkers were used to obtain a different crosslinking degree. While a higher amount of Ch-PPy led to an increase in conductivity, higher crosslinking concentrations led to slower self-healing, lower strain at break, and less porous morphology. Two minutes was found enough for a complete recovery because of the hydrogen and ionic

Ch-PPy-PAA interactions. Electrical self-healing was calculated as 90 and 96 % after 30 and 60 s, respectively. Negative piezoresistance was observed between 0 and 500 Pa. The sample was dynamically cycled between up to 25 and 50% compression and resulted in stable negative piezoresistance. The CSH was also 3D printed and used for muscle, pulse, and respiration monitoring. Although composites were sensitive to shear deformation that occurred during 3D printing, good self-healing performance led to quick recovery [27].

In another study, Wang et al. developed pressure sensors from PAni, PAA, and phytic acid (PA) and investigated the effects of sensor geometry with flat and patterned samples [28]. The patterned sample showed higher sensitivity between 0 and 300 Pa before and after self-healing. The sensitivity of the sensor was determined as 37.6 kPa^{-1}, 7.1 kPa^{-1}, and 1.9 kPa^{-1} for different levels of pressure as 0–0.8, 0.8-4.5, and P > 5 kPa, respectively, with a low response time of 50 ms. In addition, these samples showed almost similar sensing performance even after eight months. Inhale-exhale cycles and speaking cycles of a person were successfully monitored. The sample showed different sensitivity to different calligraphy brushes. Higher sensitivity was observed for the brush with the highest elastic modulus. Furthermore, as given in Figure 19.3, for pressure mapping, a pixel array was prepared, and four different chess pieces were successfully monitored. The system was also used for the determination of finger position [28].

- *Gas sensors.* Gas sensors are a subclass of chemical sensors, and they can monitor the type and concentration of gas. The gas sensor's function is based on chemical interaction and the sensor material should be sensitive/specific to the gas to be characterized. Gas sensors are used for various applications including health, environmental, and indoor monitoring. The sensitivity of the gas sensors can be determined as a function of gas concentration or at a constant concentration as a function of time. The slope of the $(\Delta R/R_0)/\Delta c_g$, $(\Delta S/S_0)/\Delta c_g$, $(\Delta R/R_0)/\Delta t$, $(\Delta S/S_0)/\Delta t$, can be given as the sensitivity where Δc_g is changing gas concentration and Δt is changing in time. One important feature of a gas sensor is the limit of detection (LOD) value that determines the lowest detectable gas concentration under given

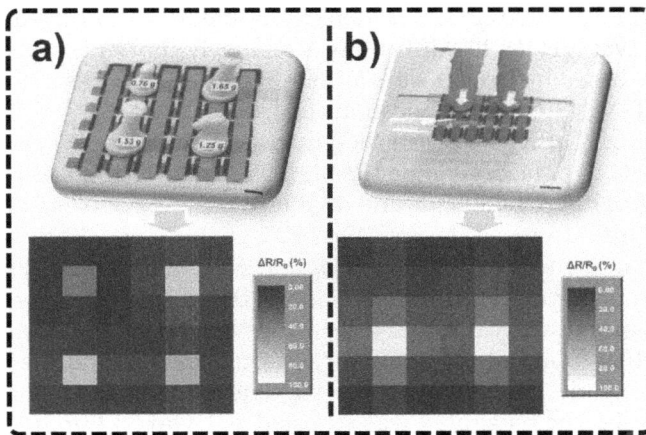

FIGURE 19.3
(a) A photograph of four chess pieces with different weights distributed on the surface of an integrated 6 × 6 pressure sensor array and a corresponding signal map showing the pressure distribution. (b) A photograph of a bent pressure sensor pressed by two fingers and a corresponding signal map showing the pressure distribution. Adapted with permission [28]. Copyright (2021), John Wiley and Sons.

conditions. The lower the LOD value shows, the higher interaction between sensor and gas. For all gas sensors, the optimum sensing region should be determined based on the minimum and maximum detectable gas concentration. However, sensing behavior for SSS should be determined under strain or any kind of deformation including bending, twisting, or compression by considering self-healing. Threshold limit values (TLV) might be required for actual gas monitoring applications. TLV is a value that shows the maximum gas concentration that a person can be exposed to without any harm. For electro-mechanical sensors, applied strain or pressure have a direct/quick effect on deformation, and signal generation is quick. However, absorption of gas and change in resistance might take a longer time for gas sensors. Similarly, recoverability might require more time. The design of the gas sensors with a higher ratio of gas selective functional groups can accelerate the polymer-gas interaction and lower the response time. The sensing mechanism is generally based on the change in conductive network structure caused by gas absorption. Stretchable self-healing gas sensors might be sensitive to humidity or temperature, which can affect the performance negatively/positively. SSS gas sensors with low power consumption and tunable and predictable sensing performance under different levels of strain at room temperature with transparency are some of the recent trends [25,33].

Since NH_3 and NO_2 are counted as harmful gases to human health, in most of the studies about SSS, their detectability was investigated. Duy et al. developed gas sensors from stretchable graphene-based hydrogels functionalized with a PU-diol oligomer [25]. The virgin sample was found sensitive to both gases but with higher sensitivity to NO_2 (3.5 ppm) compared with NH_3 (25 ppm). The gas sensing performance at 25°C (dry air) under NO_2 and NH_3 (25 ppm) exposure was analyzed for virgin, under strain (0–30%). Samples were exposed to NO_2 and NH_3 at various concentrations at 0.5–3.5 and 2–20 ppm, respectively. The LOD value under 0 and 30% strain was determined as 0.8 and 2.46 ppm, respectively, which was caused by the change in the network structure. The gas sensing performance after healing showed a similar trend as the virgin sample, however, the noise level increased from 0.013 to 0.021%. The detection time was around 4 min for both gases. The recovery time was NH_3 (< 10 min) and was relatively shorter compared with NO_2 (> 1 h). These graphene-based hydrogels were found promising in terms of their TLV for actual NH_3 monitoring (TLV ≤ 25 ppm) applications [25].

Wu et al. synthesized ethylene glycol (EG)-modified double network organo-hydrogels from carrageenan (CA)/polyacrylamide (PAM) [33]. The electrical self-healing was found to be good because of hydrogen bonds and double helices morphology of CA. The conductivity value of the film decreased from 3.43 to 1.72 μS, for 0 and 100% stretched samples, respectively. The sample was exposed to NH_3 and NO_2 under 0 and 100% strain at various concentrations under dynamic on-off mode to determine the response and recovery time. The conductance directly decreased regardless of the gas concentration, which was attributed to the dissolution of the analyte gas in the solvent and the prevention of ion transport in the network. The LOD values were determined as 91.6 and 3.5 ppb for NH_3 and NO_2, respectively. The sensitivity was reported as 1.4 and 8.4 ppm^{-1} for NH_3 and NO_2, respectively. Tunable gas sensing was obtained by changing the strain% of the sample. The sample also showed good NO_2 sensing performance after self-healing [33].

- *Humidity sensors.* PSSS can be used to determine the humidity. The humidity sensing capacity of wearable electronics has increasing importance for many aspects such as monitoring the moisture level of the environment, skin, or any metabolic data related to human health. The sensitivity of the sensor can be measured as a function of humidity or at constant humidity as a function of time. The slope of the $(\Delta R/R_0)/\Delta H$ or $(\Delta R/R_0)/\Delta t$ can be given as the sensitivity where ΔH is a change in humidity and Δt is a change in time. Since relative humidity (RH%) is directly related to temperature, the optimum temperature should be determined for the sensing range by considering minimum and maximum detectable RH%. Humidity sensing should be investigated under strain before and after self-healing. Since humidity is the vapor phase of water, the sensing mechanism is based on water absorption. Because of this process, network structure and conductivity change. The performance of a humidity sensor depends on polymer-humidity interaction, absorption speed, and required time for equilibrium during and after exposure. Since this process is relatively slow, response and recovery time can be longer. Hydrogels show superior properties for stretchable self-healing electronic systems. However, time-depended evaporation of water negatively affects most of the properties including mechanical, electrical, and self-healing. In this case, not only the so-called properties but also the homogeneity and repeatability of the system become questionable. Since the system is sensitive to humidity, the liquid phase captured in the network should be in balance. To minimize this risk, increase the stability and sensitivity to humidity, and obtain anti-drying sensors, some diols including glycerol (Gly) and ethylene glycol (EG) can be used [33].

Wu et al. developed self-healable stretchable CS/CNT fibers with sacrificial bonds prepared by 3D printing [4]. The self-healing mechanism was reported to be based on increased chain movement and electrostatic interactions because the swelling stemmed from water vapor. As expected, swelling led to a change in the matrix structure and conductive network. The humidity sensing performance was determined under various reversible RH conditions. The system showed positive piezoresistance from 35 to 85% RH – probably destruction of a conductive network because of matrix swelling. However, no information was given about humidity sensing under strain [4].

To develop anti-dry stretchable self-healing sensors, Gly and EG were used for ionic CA/PAM double network (DN) organohydrogels. DN, EG-DN, and Gly-DN were evaluated in terms of mechanical and humidity sensing performance. Ionic organohydrogel-based EG-DN and Gly-DN sensors with a high stretchability of around 1225% showed high sensitivity to humidity monitoring from 4–90% RH. However, DN-Gly showed the best performance of more than 500% with a relatively fast response/recovery time (0.27/0.30 s) that was attributed to the hygroscopic character of Gly. The most important contribution of this sensor was its good performance under strain, twisting, and bending, however, mechanical and sensing properties were not investigated after self-healing [33].

- *Temperature sensors.* A temperature sensor is a device that can convert temperature change into an electrical signal, and the change in resistance in a certain period reflects the sensitivity that is also expressed as the temperature coefficient of resistance (TCR, $°C^{-1}$). The slope of the $(\Delta R/R_0)/\Delta t$ can also be used for sensitivity. Response time and recovery time are critical for applications that require a quick response. Temperature sensors are used for various applications such as body and ambiance monitoring. Since stretchable self-healing materials are thermally degradable polymers and show dimensional change under high temperatures, minimum and maximum detectable temperatures should not exceed the thermally stable region. The minimum detectable temperature range is a value that defines the minimum temperature change that can be monitored and it is also an indication of sensitivity. In addition to these, for SSS, temperature monitoring should be determined under deformation before and after self-healing. One important problem regarding ionic conductive sensors is the evaporation of liquids during temperature measurement, which can destroy the structure and prevent the sensing ability of the system. To lower the rate of evaporation and keep the liquid phase in equilibrium, anti-freeze and anti-dryness can be obtained by using binary solvents such as Gly/water. In addition to electrical and/or ionic conductivity, the thermal conductivity of the material is also important in terms of heat dissipation and homogeneous temperature monitoring, especially for mapping applications. The temperature-discrimination capacity of the sensor is also important for stepwise temperature analysis to determine the sensitivity. Since temperature sensors can be operated in contact and non-contact modes, distance-depended sensing performance is also important. To determine that, the sample is placed at a known distance and the measured value and exact value are compared [15,25].

Ge et al. synthesized PAA/PAni binary networked-hydrogel (PPBN-hydrogel), and to mimic the muscle structure, PAni nanofibers (PAni NFs) were used [15]. Reversible hydrogen bonding, metal-coordination, and electrostatic interactions were found effective for stretchability and high healing efficiency (90.8%). Temperature sensing was analyzed in a broad range from 40 to 110°C. The TCR value was calculated as (-0.0164°C^{-1}) with good repeatability. This performance was attributed not only to inherent conductivity but also to the network structure of PAni NFs. The temperature sensor showed good cyclic performance between 50 and 90°C in terms of temperature-discrimination capacity and good instantaneous electrical response between 30 and 80°C. In addition to these, the PPBN-hydrogel sensor showed good distance-depended sensing ability between 3 and 15 mm. By increasing the distance, a larger resistance readout was obtained. The distance-based sensitivity was determined as -0.022°C mm^{-1}. To specify the instantaneous sensitivity, a sample was tested at 25.4, 35.1, and 45.1 °C continuously and showed stability for 50 s. To monitor the body temperature performance, a sensor was placed on the forehead of a person. From 36.4 to 39.1°C and from 39.9 to 36°C a noticeable electrical signal was monitored for 50 s and the sensor was reported to be applicable for fever monitoring. Although the study is of significance in terms of the characterization of temperature sensors, no data was presented for sensing performance after self-healing with or without strain [15].

Duy et al. prepared SSS from graphene-based hydrogels functionalized with polyurethane diol (PU-diol) oligomer [25]. Hydrogels were coated on a PDMS substrate and electrical self-healing performance was investigated. The performance was around 90, 89, and 50% for the first, second, and third healing, respectively. The temperature sensing behavior was analyzed between 30 and 100°C and the sample was tested under 30% strain between 13 and 100°C. Although the sample showed similar behavior, sensitivity was lower compared with the unstretched sample. Virgin, self-healed, and stretched samples were compared in terms of the minimum sensible temperature range. The virgin sample showed good performance, even at $\Delta T = 0.2$°C. The sample showed 1.5, 2.3, 4.0, 6.7, and 12.5% differences for the values of $\Delta T = 0.5, 1.0, 2.5, 5.0,$ and 10°C, respectively. The sample under 30% strain showed similar behavior with a shift in sensitivity and sensible temperature range ($\Delta T = 0.5$°C). The self-healed sample also showed a similar trend as the virgin sample with a minimum ΔT value of 0.2°C; 1.5, 2.5, 3.9, 8, and 13.3 % difference was observed for $\Delta T = 0.5, 1.0, 2.5, 5.0,$ and 10°C, respectively [25].

19.3.2 Capacitive Stretchable Self-Healing Sensors

Capacitive stretchable self-healing sensors (CSSS) are capacitors that consist of a dielectric material sandwiched between electrodes. Traditional capacitive sensors have some problems because of their planar, hard-brittle configuration, however, stretchable capacitive sensors have many advantages because of their elastic nature. For CSSS, self-healing material can be used as the dielectric [14,29,43] electrodes can be prepared by the addition of conductive filler into the self-healing material [14], or conductive materials [30] can be coated [14] or attached [31] on the surface. There are limited studies in the literature about CSSS and most of them are focused on strain [29,30] and pressure [14,31] sensors. Capacitance (C) is the generated electrical signal for CSSS with the effect of change in stimuli. It defines the stored electrical charge under a given voltage and is expressed as $C = \varepsilon_r\varepsilon_0 A/d$ where ε_r is the permittivity of the free space, ε_0 is the dielectric constant of the material, A is the surface area, and d is the distance between electrodes. In the case of constant ε_r, C can be changed by a change in ε_0, A, and d [43]. Since dimensional change directly affects the capacitance, the dimensional recoverability of the material is of significance in terms of obtaining reliable results [14,29–31,43]. Although ε_0 is a bulk property of a material, this value might change as a consequence of the evaporation of water for hydrogel-based self-healing materials, so the material might need extra encapsulation [29]. Measurement conditions such as applied frequency [29] or characterization environment (i.e., under air or in water) [32] can also affect the outcomes [31]. In addition to strain and pressure monitoring, CSSS can be used for touch [31], muscle activity [31], and human movement monitoring including finger bending [29,31,32], elbow monitoring [30], or wrist monitoring [29]. Although CSSS can have a single function such as strain or pressure monitoring, dual-function CSSS were also reported in the literature [30].

- *Strain sensors*. CSSS can be used for monitoring strain deformation. The sensitivity of a sensor is determined by GF as $GF = (\Delta C/C_0)/\varepsilon$, or normalized capacitance (C/C_0), where ΔC is the capacitance change $(C–C_0)$, and C_0 is the initial capacitance. Sensing range, minimum and maximum detectable strain/capacitance level, linear region, hysteresis, recoverability, anti-stress relaxation degree, aging, self-healing, and dielectric constant of the material should be considered. Under strain, loading distance between electrodes decreases and this leads to an increase in the C value. During unloading, the d value increases and the C value decreases [29,30].

Bin Yin et al. synthesized polyurethane (PU) with donor (D) and acceptor (A) groups [30]. Naphthalene ring (D) and imide group (A) were incorporated into the PU main chain alternately to obtain a DA-PU structure. In this way, not only an intra- but also an inter-chain interacted network was obtained with 1900% strain at break. The self-healing speed was determined as 1.0–6.15 µm/min for 60 and 80°C, respectively, and healing time decreased from 400 min to 65 min. DA-PU recovered the mechanical properties after the self-healing process. As given in Figure 19.4, a five-layered capacitive sensor was prepared. In Figure 19.4a, the relative capacitance was also given as a function of strain (0–100%) for both virgin and self-healed DA-PU, and almost the same response was obtained with a GF of 0.97. The capacitance showed a linear increasing trend that was attributed to a decrease in the distance between electrodes because of strain loading. To test the capacitive sensing performance for real case monitoring, it was located on the elbow of a badminton player. As given in Figure 19.4b, two movements were determined as standard and wrong serve. The capacitance change was found remarkably different for the wrong service and this sensor was found useful in terms of movement monitoring [30].

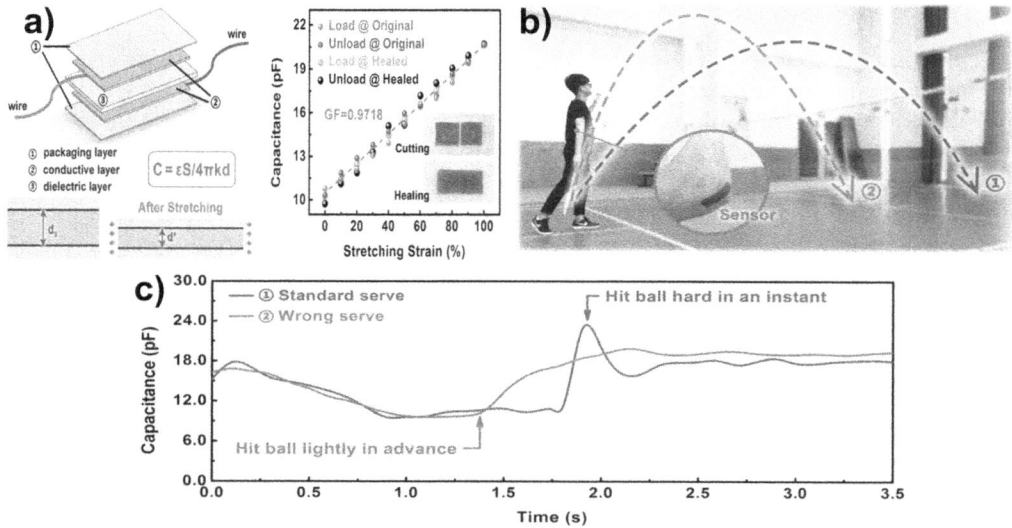

FIGURE 19.4
(a) Schematic representation of a capacitive sensor and corresponding capacitance change upon elongational deformation of original and self-healed sensors; (b) demonstration of standard and wrong serves in badminton; and (c) corresponding real-time capacitance responses. Adapted with permission [30]. Copyright (2021), John Wiley and Sons.

Rao et al. fabricated a capacitive strain sensor by using polydiacetylene (PDA)-based hydrogel [29]. The self-healing behavior of PDA-PAA-Cr^{3+} was reported to be good. Hydrogen bonding and electrostatic interactions were given as the basic mechanism for self-healing. The capacitive sensing was determined as a function of strain between 20 and 500%. Stable stepwise cyclic characterization was also performed between 50 and 400%. The GF for 100 and 500% were determined as 3 and 161, respectively, under 20 Hz. The frequency was also found to affect the GF. While the sample showed a linear response between 20 and 130%, after that region, a non-linear response was obtained because of the changes in the network structure such as the breakdown of the physical interactions between PDA and PAA. The sensor was also used for index finger and wrist motion monitoring. Although a high GF was obtained, no data was given about the sensing performance of self-healed samples [29].

- *Pressure sensors.* Pressure sensors are one of the common applications of capacitive sensors. Stretchable self-healing polymers with dielectric properties are promising materials for capacitive sensors and are used for pressure [31], force [32] monitoring, or pressure mapping [14]. The sensitivity can be determined by GF as $(\Delta C/C_0)/\varepsilon$, or pressure sensitivity PS $= (\Delta C/C_0)/\Delta P$ or normalized capacitance ΔC, where $\Delta P = P–P_0$, P is the pressure and P_0 is the initial pressure. Under pressure loading, generally surface area of the stretchable material increases and distance between electrodes decreases and C increases. The unloading process shows the reverse behavior. Since the dielectric constant of the material is affected by water content, evaporation of water might lead to a change in the dielectric constant and that results in a change in capacitance and sensitivity of the sensor. Sensing range, frequency, minimum and maximum detectable capacitance level/

compression ratio/pressure/force, linear region, hysteresis, recoverability, aging, and self-healing mechanisms should be considered [14,31,43].

Lei et al. synthesized supramolecular mineral hydrogel by using amorphous calcium carbonate (ACC) nanoparticles, PAA, and alginate [31]. The sample showed a high tensile strain of ~1000% with good healing performance. Recovered capacitance was higher because of unrecovered dimensional changes. For the capacitive sensors, a polyethylene (PE) film was used as a dielectric layer between two layers of ACC/PAA/alginate hydrogel to which two electrodes were attached. To hinder the removal of water, the capacitive sensor was encapsulated by PE film. The sensor was exposed to an increasing level of compression from 0–1 kPa and PS was calculated as 0.17 kPa^{-1}. The applied pressure during finger touch and rain was successfully monitored by the sensor. One important contribution of this work was the recycling process. To observe the sensing behavior, the sample was recycled ten times by drying and swelling and showed almost full recovery for touch sensing [31].

Zhang et al. developed a sensor by using a curcumin polymer block (P-Cur) and Eu(III) [14]. P-Cur-Eu showed high tensile strength and strain of 1.8 MPa and 900%, respectively. Time was reported to be significant for room temperature healing and 48 h was reported to be enough for 98% recovery that was caused by a high level of interactions including metal-ligand coordination and hydrogen bonds. To fabricate capacitive sensors, silver was incorporated into P-Cur and used as the conductive electrode. P-Cur-Eu was used as the dielectric material and top and bottom surfaces were coated with P-Cur-Silver. It was shown that the system was touch-sensitive before and after healing. This study is of significance in terms of using a conductive self-healing electrode layer [14].

19.3.3 Piezoelectric Stretchable Self-Healing Sensors

Piezoelectric sensors are dielectric materials with the ability to transduce mechanical stimuli into an electrical signal in the form of voltage generation, and conversely, under an electrical field, they show deformation. Conventional piezoelectric materials, piezoelectric ceramics (PZT), function based on the internal polarization that results in a charge formation. PZT are rigid materials, and they are not feasible for wearables. Polymers can be classified as piezoelectric polymers (PP) and piezoelectric polymer composites (PPC). PP have inherent piezoelectricity, and its response is based on the orientation of the molecular dipoles. PPC are fabricated by the addition of PZT into the polymeric matrix and they combine the advantages of polymers and PZT [8,12,43–45]. However, these materials have some drawbacks such as a low limited level of deformability. Piezoelectric stretchable self-healing sensors (PeSSS) are promising materials for wearable electronics and artificial skin [13,26]. Since they are highly deformable, they can quickly generate voltage even under low-stress levels with high sensitivity, low power requirement, and low cost [8,12]. Leakage current and short-circuit formation should be controlled and prevented by some precautions such as anti-short-circuit treatment or packaging [26,46]. PeSSS generally consist of a dielectric layer that is responsible for electrical charge generation and two conductive electrode layers that consist of conductive polymer composites (Figure 19.5) [13] or hydrogels [46]. Therefore, the optimum frequency should be determined for PeSSS. In the case of partial recovery, the voltage signal might be noisy at high-frequency values [13].

- *Strain sensors*. PeSSS can be used for strain monitoring. The sensitivity can be determined by comparison of the generated voltage values or by considering the ratio between voltage change and strain GF = $(\Delta V/V_0)/\varepsilon$, or normalized voltage ΔV

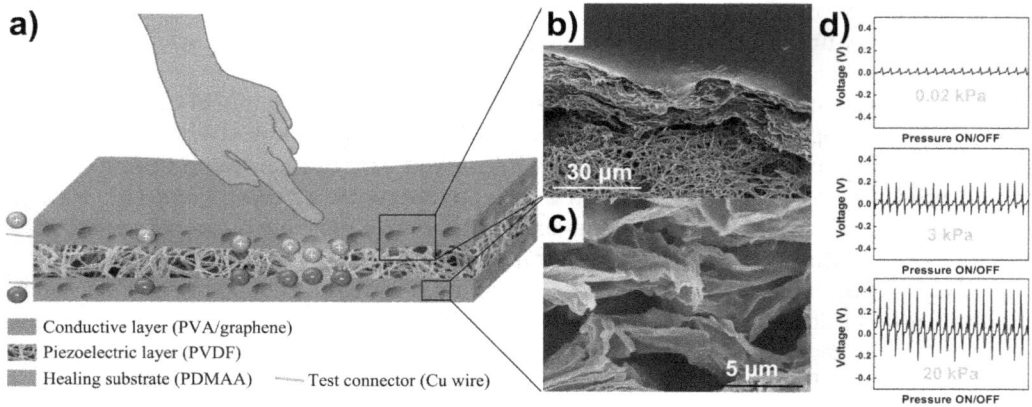

FIGURE 19.5

(a) Schematic representation of piezoelectric stretchable and self-healing sensor. (b and c) SEM images of a cross-sectional view of the sensor with low and high magnifications. (d) Pressure response curves of PeSSS under different pressures. Adapted with permission [13]. Copyright (2013), Springer Nature.

(V/V_0) or voltage change $V–V_0$ where V is the voltage and V_0 is the initial voltage. Sensing range, minimum and maximum detectable strain/voltage level, linear region, hysteresis, recoverability, aging, self-healing, and dielectric constant of the material should be considered [34].

Polymer-based systems might suffer from low voltage generation. Some molecular level modifications to improve piezoelectric coefficient by incorporation of flexible chain segments into the polymer is a way of increasing this. Wang et al. developed a piezoresistive material from a lactate-based piezoelectric elastomer (LBPE) [34]. A poly(butanediol/lactate/sebacate/itaconate)-based LBPE was prepared, and the sample showed high strain at break around 439% with low elastic modulus and good self-healing in 15 min; 77.38 and 44.98% were determined as recovery values for electrical and mechanical properties, respectively, after five healing cycles. Piezoelectric performance of the material was determined by performing a free-fall dropping test. The distance between the dropping point and the material varied between 5 and 100 cm and maximum voltage was measured between 1 and 4 V with a low response time of 1.39×10^{-3} s. This increase in the output voltage was attributed to increased lattice distortions and polarization. In addition to strain sensors, materials can also be used as a generator because of their good open-circuit voltage densities (0.07–0.24 V cm^{-2}) and short-circuit current densities (4.25–19.75 μA cm^{-2}) [34].

- *Pressure sensors.* Pressure sensors are the most common application of piezoelectric sensors. Dielectric stretchable self-healing polymers with piezoelectric properties can be used for pressure monitoring [13] and force monitoring [26]. The sensitivity can be determined by comparison of the generated voltage values or GF as $(\Delta V/V_0)/\varepsilon$, or pressure sensitivity PS = $(\Delta V/V_0)/\Delta P$ or normalized voltage or voltage change. Sensing range, minimum and maximum detectable force/pressure/voltage level, linear region, hysteresis, recoverability, aging and self-healing properties, and dielectric constant of the material should be considered.

One of the first studies about SSS was a piezoelectric pressure sensor. In this study, Hou et al. developed a three-layered sensor with PVDF nanofibers as the piezoelectric layer that was covered by stretchable, self-healing, conductive poly(N,N-dimethylacrylamide)-PVA/rGO (PDMAA-PVA/rGO)-based top and bottom electrodes [13]. The sample showed good sensitivity to pressure levels from 0.02 kPa to higher levels of 3 and 20 kPa with maximum voltage values of 0.05, 0.2, and 0.4 V, respectively. Electrical, mechanical, and sensing behavior was found to have a good recovery after self-healing. The sensor was reported to generate a low voltage (1 < V) and stable voltage generation was obtained at lower frequencies (0.2 to 1 Hz) [13].

In addition to inherent piezoelectric polymers, PZT-filled polymer composites can also be used for PeSSS. Yang et al. developed a three-layered self-healing polydimethylsiloxane (H-PDMS) based sensor [26]. PZT particles were added at different ratios from 30–80 wt%. The electrodes were prepared by conductive AgNWs/H-PDMS mixture. The piezoelectric layer was sandwiched between the top and bottom electrodes. Mechanical properties of H-PDMS and 70/30 PZT/H-PDMS were determined before and after self-healing. The virgin sample showed tensile stress/strain values of 32 kPa and 1100%, respectively, and these values were almost kept similar after 12 hours of self-healing. However, PZT addition led to a decrease in stress/strain values after self-healing. The piezoelectric properties were analyzed as a function of the PZT ratio. The maximum voltage was generated for the 70/30 PZT/H-PDMS-based sensor with a value around 3 V. Same sample was exposed to a cyclic load test under 1–30 N and voltage increased linearly parallel with the force between 1.5 and 3 V. To monitor force based on the position, a sensor array was prepared and located on a human hand. Both position and force were successfully monitored even after self-healing [26].

19.4 Design of Stretchable and Self-healing Sensors

The SSS design process starts with the definition of the problem followed by the idea generation. At this point, the idea generally covers the basic strategy to be followed during the design and production processes and the determination of the application(s) and function(s) of the sensor. Strain, bending, pressure, temperature, gas, and humidity sensors can be given as examples for SSS applications. The next step is the decision of the sensing mechanism that directly affects the size, performance, number of additional equipment, and cost of the system. For SSS, piezoresistive, capacitive, piezoelectric, chemical, or inductive sensing mechanisms can be used. After that, requirements, restrictions, limits, and standards should be known. The requirements should be considered based on the application. While stretchable, self-healing strain sensors require a high level of deformation and quick recovery, gas sensors require specific chemical interaction between gas and material. More specifically for strain sensors, cyclic stability and gas sensors, LOD, and TLV values should be considered for commercial applications. In addition to these, sensing performance should be considered based on the application and sensing mechanism. Based on the mechanism, and expected minimum/maximum outcomes, sensitivity should be evaluated such as GF, normalized capacitance, and voltage difference. The next step is the determination of the self-healing mechanism and material selection. The sensing performance is mostly governed by the type of material. As given previously, natural or synthetic polymers [4], conductive composites [36], ICPs [47], hydrogels [10], organogels [20], and single or dual network gels [6] can be used for the fabrication of SSS. Since the healing mechanism is directly related to the material, these two titles are evaluated together. Fabrication and the form of the sensor is the other important step. In this case, not only the application but also the mechanism and material type have to be considered. The form of the sensor should be compatible with the application and sensing mechanism. While capacitive SSS are three-layered systems with electrodes [26]; piezoresistive sensors can be in single or double-layered films [3]. Films [48], nanofibers [49], fibers [5], meshed structures [40], and other 3D geometries can be fabricated by solution-based processes such as solution casting [40], 3D printing [39], spraying [50], spin-coating [26], or combinations of these. As known, based on the mechanism of sensing, some electronics are required such as a current source, voltage source, LCR-meter, and amplifier for signal extraction and processing. Calculation of the cost is the final step of the design, and it is important for understanding the feasibility of the sensor for scale-up. Sustainability is one of the increasing trends in sensor science and technology. SSS can be evaluated in the sustainable electronics category for providing longer life and better performance. This contribution can be doubled by biodegradability and recycling functions. For the design of SSS, all the steps given earlier, from material selection to fabrication, should be considered from the sustainability perspective because it determines not only environmental and economic impacts but also the social impact of the system.

19.5 Conclusions and Future Perspectives

Self-healing is a material property used for the determination of recoverability or repair ability after damage. Sensors are systems for perceiving, monitoring, and interpreting the variables by conversion of stimuli into an electrical signal. Like all materials, sensors age during usage with the impact of external factors. For increasing performance, life, and decreasing maintenance costs of the sensors, intrinsically self-healing polymers are promising materials. Herein, pioneer and recent prominent studies were highlighted based on the mechanism and applications of stretchable, self-healing sensors. Requirements for SSS were summarized based on their mechanical, electrical, and self-healing properties. Three common mechanisms – piezoresistive, capacitive, and piezoelectric sensors – were discussed based on their current applications, challenges, and performance. While stretchable, self-healing piezoresistive sensors are used for strain, bending, pressure, gas, humidity, and temperature monitoring, capacitive and piezoelectric sensors are generally used for strain, bending, force, and pressure monitoring. As summarized previously, in the last ten years, remarkable studies were carried out and significant achievements were obtained in terms of wearable electronics. This shows self-healing technology will have a place in the future of electronics. However, for commercial applications, restrictions, standards, proper batch processing methods, cost, biodegradability, recyclability, and sustainability should be considered for the design of SSS. To lower the cost, obtain any size, shape, and geometry, and fabricate commercial SSS, batch processing methods used for in-mold electronic products should be modified.

References

1. B.C.K. Tee, C. Wang, R. Allen, Z. Bao, An electrically and mechanically self-healing composite with pressure- and flexion-sensitive properties for electronic skin applications, *Nat. Nanotechnol.* 7 (2012) 825–832.

2. Y.J. Liu, W.T. Cao, M.G. Ma, P. Wan, Ultrasensitive wearable soft strain sensors of conductive, self-healing, and elastic hydrogels with synergistic "soft and hard" hybrid networks, *ACS Appl. Mater. Interfaces.* 9 (2017) 25559–25570.

3. L. Han, M. Liu, B. Yan, Y.S. Li, J. Lan, L. Shi, R. Ran, Polydopamine/polystyrene nanocomposite double-layer strain sensor hydrogel with mechanical, self-healing, adhesive and conductive properties, *Mater. Sci. Eng. C.* 109 (2020) 110567.

4. Q. Wu, S. Zou, F.P. Gosselin, D. Therriault, M.C. Heuzey, 3D printing of a self-healing nanocomposite for stretchable sensors, *J. Mater. Chem. C.* 6 (2018) 12180–12186.

5. L. Shuai, Z.H. Guo, P. Zhang, J. Wan, X. Pu, Z.L. Wang, Stretchable, self-healing, conductive hydrogel fibers for strain sensing and triboelectric energy-harvesting smart textiles, *Nano Energy.* 78 (2020) 105389.

6. W. Peng, L. Han, H. Huang, X. Xuan, G. Pan, L. Wan, T. Lu, M. Xu, L. Pan, A direction-aware and ultrafast self-healing dual network hydrogel for a flexible electronic skin strain sensor, *J. Mater. Chem. A.* 8 (2020) 26109–26118.

7. Y. Zhao, A. Kim, G. Wan, B.C.K. Tee, Design and applications of stretchable and self-healable conductors for soft electronics, *Nano Converg.* 6 (2019) 25.

8. Q. Zhang, L. Liu, C. Pan, D. Li, Review of recent achievements in self-healing conductive materials and their applications, *J. Mater. Sci.* 53 (2018) 27–46.

9. Y. Gai, H. Li, Z. Li, Self-healing functional electronic devices, *Small.* 17 (2021) 2101383.

10. J. Wu, Z. Wu, Y. Wei, H. Ding, W. Huang, X. Gui, W. Shi, Y. Shen, K. Tao, X. Xie, Ultrasensitive and stretchable temperature sensors based on thermally stable and self-healing organohydrogels, *ACS Appl. Mater. Interfaces.* 12 (2020) 19069–19079.

11. S. Utrera-Barrios, R. Verdejo, M.A. López-Manchado, M. Hernández Santana, Evolution of self-healing elastomers, from extrinsic to combined intrinsic mechanisms: A review, *Mater. Horizons.* 7 (2020) 2882–2902.

12. M. Khatib, O. Zohar, H. Haick, Self-Healing Soft Sensors: From material design to implementation, *Adv. Mater.* 33 (2021) 2004190.

13. C. Hou, T. Huang, H. Wang, H. Yu, Q. Zhang, Y. Li, A strong and stretchable self-healing film with self-activated pressure sensitivity for potential artificial skin applications, *Sci. Rep.* 3 (2013) 3138.

14. Q. Zhang, S. Niu, L. Wang, J. Lopez, S. Chen, Y. Cai, R. Du, Y. Liu, J.-C. Lai, L. Liu, C.-H. Li, X. Yan, C. Liu, J.B.-H. Tok, X. Jia, Z. Bao, An elastic autonomous self-healing capacitive sensor based on a dynamic dual crosslinked chemical system, *Adv. Mater.* 30 (2018) 1801435.

15. G. Ge, Y. Lu, X. Qu, W. Zhao, Y. Ren, W. Wang, Q. Wang, W. Huang, X. Dong, Muscle-inspired self-healing hydrogels for strain and temperature sensor, *ACS Nano.* 14 (2020) 218–228.

16. Y.J. Tan, G.J. Susanto, H.P. Anwar Ali, B.C.K. Tee, Progress and roadmap for intelligent self-healing materials in autonomous robotics, *Adv. Mater.* 33 (2021) 2002800.

17. X. Pei, H. Zhang, Y. Zhou, L. Zhou, J. Fu, Stretchable, self-healing and tissue-adhesive zwitterionic hydrogels as strain sensors for wireless monitoring of organ motions, *Mater. Horizons.* 7 (2020) 1872–1882.

18. J. Xu, G. Wang, Y. Wu, X. Ren, G. Gao, Ultrastretchable wearable strain and pressure sensors based on adhesive, tough, and self-healing hydrogels for human motion monitoring, *ACS Appl. Mater. Interfaces.* 11 (2019) 25613–25623.

19. C.Z. Hang, X.F. Zhao, S.Y. Xi, Y.H. Shang, K.P. Yuan, F. Yang, Q.G. Wang, J.C. Wang, D.W. Zhang, H.L. Lu, Highly stretchable and self-healing strain sensors for motion detection in wireless human-machine interface, *Nano Energy.* 76 (2020) 105064.

20. H. Zhang, W. Niu, S. Zhang, Extremely stretchable and self-healable electrical skin with mechanical adaptability, an ultrawide linear response range, and excellent temperature tolerance, *ACS Appl. Mater. Interfaces.* 11 (2019) 24639–24647.

21. G. Cai, J. Wang, K. Qian, J. Chen, S. Li, P.S. Lee, Extremely stretchable strain sensors based on conductive self-healing dynamic cross-links hydrogels for human-motion detection, *Adv. Sci.* 4 (2017) 1600190.

22. T. Wu, B. Chen, A mechanically and electrically self-healing graphite composite dough for stencil-printable stretchable conductors, *J. Mater. Chem. C.* 4 (2016) 4150–4154.

23. S. Xia, S. Song, G. Gao, Robust and flexible strain sensors based on dual physically cross-linked double network hydrogels for monitoring human-motion, *Chem. Eng. J.* 354 (2018) 817–824.

24. Y. Hu, N. Liu, K. Chen, M. Liu, F. Wang, P. Liu, Y. Zhang, T. Zhang, X. Xiao, Resilient and self-healing hyaluronic acid/chitosan hydrogel with ion conductivity, low water loss, and freeze-tolerance for flexible and wearable strain sensor, *Front. Bioeng. Biotechnol.* 10 (2022) 837750.

25. L.T. Duy, H. Seo, Eco-friendly, self-healing, and stretchable graphene hydrogels functionalized with diol oligomer for wearable sensing applications, *Sensors Actuators, B Chem.* 321 (2020) 128507.

26. M. Yang, J. Liu, D. Liu, J. Jiao, N. Cui, S. Liu, Q. Xu, L. Gu, Y. Qin, A fully self-healing piezo-electric nanogenerator for self-powered pressure sensing electronic skin, *Research.* 2021 (2021) 9793458.

27. M.A. Darabi, A. Khosrozadeh, R. Mbeleck, Y. Liu, Q. Chang, J. Jiang, J. Cai, Q. Wang, G. Luo, M. Xing, Skin-inspired multifunctional autonomic-intrinsic conductive self-healing hydrogels with pressure sensitivity, stretchability, and 3D printability, *Adv. Mater.* 29 (2017) 1700533.

28. T. Wang, Y. Zhang, Q. Liu, W. Cheng, X. Wang, L. Pan, B. Xu, H. Xu, A self-healable, highly stretchable, and solution processable conductive polymer composite for ultrasensitive strain and pressure sensing, *Adv. Funct. Mater.* 28 (2018) 1705551.

29. V.K. Rao, N. Shauloff, X. Sui, H.D. Wagner, R. Jelinek, Polydiacetylene hydrogel self-healing capacitive strain sensor, *J. Mater. Chem. C.* 8 (2020) 6034–6041.

30. W. Bin Ying, G. Wang, Z. Kong, C.K. Yao, Y. Wang, H. Hu, F. Li, C. Chen, Y. Tian, J. Zhang, R. Zhang, J. Zhu, A biologically muscle-inspired polyurethane with super-tough, thermal reparable and self-healing capabilities for stretchable electronics, *Adv. Funct. Mater.* 31 (2021) 2009869.

31. Z. Lei, Q. Wang, S. Sun, W. Zhu, P. Wu, A bioinspired mineral hydrogel as a self-healable, mechanically adaptable ionic skin for highly sensitive pressure sensing, *Adv. Mater.* 29 (2017) 1700321.

32. B. Xue, H. Sheng, Y. Li, L. Li, W. Di, Z. Xu, L. Ma, X. Wang, H. Jiang, M. Qin, Z. Yan, Q. Jiang, J.-M. Liu, W. Wang, Y. Cao, Stretchable and self-healable hydrogel artificial skin, *Natl. Sci. Rev.* 9 (2021) nwab147.

33. J. Wu, Z. Wu, W. Huang, X. Yang, Y. Liang, K. Tao, B.R. Yang, W. Shi, X. Xie, Stretchable, stable, and room-temperature gas sensors based on self-healing and transparent organohydrogels, *ACS Appl. Mater. Interfaces.* 12 (2020) 52070–52081.

34. X. Wang, Q. Liu, X. Hu, M. You, Q. Zhang, K. Hu, Q. Zhang, Y. Xiang, Highly stretchable lactate-based piezoelectric elastomer with high current density and fast self-healing behaviors, *Nano Energy.* 97 (2022) 107176.

35. X. Yan, Z. Liu, Q. Zhang, J. Lopez, H. Wang, H.C. Wu, S. Niu, H. Yan, S. Wang, T. Lei, J. Li, D. Qi, P. Huang, J. Huang, Y. Zhang, Y. Wang, G. Li, J.B.H. Tok, X. Chen, Z. Bao, Quadruple H-Bonding cross-linked supramolecular polymeric materials as substrates for stretchable, antitearing, and self-healable thin film electrodes, *J. Am. Chem. Soc.* 140 (2018) 5280–5289.

36. S. Liu, Y. Lin, Y. Wei, S. Chen, J. Zhu, L. Liu, A high performance self-healing strain sensor with synergetic networks of poly(es for stretchable, antitearing, anene and silver nanowires, *Compos. Sci. Technol.* 146 (2017) 110–118.

37. X. Wu, Z. Li, H. Wang, J. Huang, J. Wang, S. Yang, Stretchable and self-healable electrical sensors with fingertip-like perception capability for surface texture discerning and biosignal monitoring, *J. Mater. Chem. C.* 7 (2019) 9008–9017.

38. Y. Ma, K. Liu, L. Lao, X. Li, Z. Zhang, S. Lu, Y. Li, Z. Li, A stretchable, self-healing, okra poly-saccharide-based hydrogel for fast-response and ultra-sensitive strain sensors, *Int. J. Biol. Macromol.* 205 (2022) 491–499.

39. B. Guo, X. Ji, X. Chen, G. Li, Y. Lu, J. Bai, A highly stretchable and intrinsically self-healing strain sensor produced by 3D printing, *Virtual Phys. Prototyp.* 15 (2020) 520–531.

40. L. Chen, M. Guo, Highly transparent, stretchable, and conductive supramolecular ionogels integrated with three-dimensional printable, adhesive, healable, and recyclable character, *ACS Appl. Mater. Interfaces.* 13 (2021) 25365–25373.

41. Y. Jiao, K. Lu, Y. Lu, Y. Yue, X. Xu, H. Xiao, J. Li, J. Han, Highly viscoelastic, stretchable, conductive, and self-healing strain sensors based on cellulose nanofiber-reinforced polyacrylic acid hydrogel, *Cellulose.* 28 (2021) 4295–4311.

42. J. Wu, Z. Wu, H. Xu, Q. Wu, C. Liu, B.R. Yang, X. Gui, X. Xie, K. Tao, Y. Shen, J. Miao, L.K. Norford, An intrinsically stretchable humidity sensor based on anti-drying, self-healing and transparent organohydrogels, *Mater. Horizons.* 6 (2019) 595–603.

43. Z. Ma, Y. Zhang, K. Zhang, H. Deng, Q. Fu, Recent progress in flexible capacitive sensors: Structures and properties, *Nano Mater. Sci.* (2022) in press.

44. K.S. Ramadan, D. Sameoto, S. Evoy, A review of piezoelectric polymers as functional materials for electromechanical transducers, *Smart Mater. Struct.* 23 (2014) 33001.

45. M.T. Chorsi, E.J. Curry, H.T. Chorsi, R. Das, J. Baroody, P.K. Purohit, H. Ilies, T.D. Nguyen, Piezoelectric biomaterials for sensors and actuators, *Adv. Mater.* 31 (2019) 1802084.

46. Y. Zhu, F. Sun, C. Jia, T. Zhao, Y. Mao, a stretchable and self-healing hybrid nano-generator for human motion monitoring, *Nanomaterials.* 12 (2021) 104.

47. J. Chen, Q. Peng, T. Thundat, H. Zeng, Stretchable, injectable, and self-healing conductive hydrogel enabled by multiple hydrogen bonding toward wearable electronics, *Chem. Mater.* 31 (2019) 4553–4563.

48. W.C. Liu, C.H. Chung, J.L. Hong, Highly stretchable, self-healable elastomers from hydrogen-bonded interpolymer complex (HIPC) and their use as sensitive, stable electric skin, *ACS Omega.* 3 (2018) 11368–11382.

49. C. Zheng, Y. Yue, L. Gan, X. Xu, C. Mei, J. Han, Highly stretchable and self-healing strain sensors based on nanocellulose-supported graphene dispersed in electro-conductive hydrogels, *Nanomaterials.* 9 (2019) 937.

50. M. Hao, L. Li, S. Wang, F. Sun, Y. Bai, Z. Cao, C. Qu, T. Zhang, Stretchable, self-healing, transient macromolecular elastomeric gel for wearable electronics, *Microsystems Nanoeng.* 5 (2019) 1–10.

20

Flexible and Wearable Sensors for Biomedical Applications

Hamide Ehtesabi

CONTENTS

20.1 Introduction

A sensor is a device that detects physical or chemical phenomena. After the rise of silicon-based semiconductors in the 1950s, and the triggering of the information technology cascades, silicon-based electronic sensors emerged. A brilliant application of these sensors was in medical and healthcare technology [1]. Thereafter, wireless interactions and internet connections were added to medical sensors to improve their abilities [2]. Today, the pathway of sensor development has arrived at "flexibility". Flexible and wearable sensors can provide continuous monitoring of human health and promote healthcare system

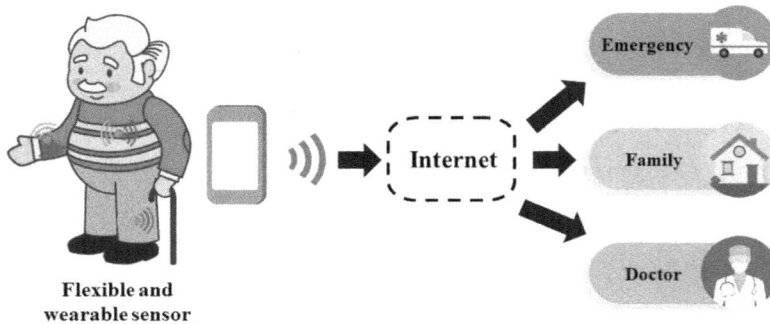

FIGURE 20.1
Schematic representation of the use of flexible and wearable sensors.

services. Besides the fabrication of soft silicone elastomers such as polydimethylsiloxane (PDMS), other flexible polymers including polyurethane (PU), polyethylene terephthalate (PET), and polycarbonate (PC) have been used in wearables [3]. Flexible sensors are made of materials that can be easily changed in size without any functional disruption. These new kinds of sensors have received a lot of attention for their applications in wearable electronics and intelligent systems. These sensors can be installed on the human body for continuous, real-time, and remote monitoring of the physicochemical properties of patients, such as the level of blood glucose, wound infection, heart rate, temperature, and respiration, as well as an electrocardiogram (ECG) (Figure 20.1).

The remote and real-time monitoring of patients and their administration processes significantly improves the efficiency of the treatment and prevents the wasting of vital time from emerging abnormal symptoms in patients to detection by clinical staff. Undoubtedly, flexible sensors will be the key members of artificial electronic skin, smart wound dressings, telemedicine, and in-home healthcare systems in the near future, which can revolutionize the human way of life. To achieve this perspective, flexible sensors should provide some other basic characteristics, including high selectivity, sensitivity, biocompatibility, low energy consumption, durability, and appropriate stability [4–6]. In this chapter, fundamental materials for the fabrication of flexible sensors are introduced. Then, the role of smart communication electronics in the structure of new advanced wearable sensors is described. The types and mechanisms of signal sensing in flexible sensors are discussed in the following section. Moreover, the novel advances in the application of flexible and wearable sensors, such as body monitoring, artificial electronic skins, and drug delivery, are briefly explained. At the end of the chapter, we summarize the future challenges and opportunities for the application and development of flexible sensors.

20.2 Flexible Materials for Wearable Sensors

In the beginning, the most common duty of wearable sensors was to monitor the physical properties of the human body, including heart rate, respiration rate, body motion, muscle activity, and temperature. The wide application of these sensors as wearable tools is restricted by their rigid and low-breathability nature [7]. Because of the difficulty of continuous access to biological fluid, the rigidity of conventional sensors has also created a major obstacle to the monitoring of biochemical signals such as blood glucose and pH [8]. New smart advanced sensors consist of sensing tools, electronics, communication circuits, remote data transmission systems (wireless connection), biocompatible bandages, and energy sources. In addition to sensitivity, biocompatibility, robustness, and air/moisture permeability, it is obvious that new advanced wearable sensors should have the desired flexibility to be applied in medical applications. However, adding the property of flexibility to the structure of such a complicated sensing system requires expensive flexible materials and a high-tech fabrication process that significantly increases the cost of the final product, especially for in-home monitoring and nursing point diagnosis [9].

Consequently, most research is concentrated on the development and achievement of new flexible materials for the fabrication of economic and efficient medical sensors. Cellulose-containing materials are natural-based polymers that are widely used in smart wearable devices due to their flexibility, biocompatibility, and biodegradability. Synthetic and chemical flexible polymers and nanomaterials are also applied to the fabrication of high-performance sensors. They are more economical, efficient, and capable of mass production. However, their low biocompatibility and degradability remain great challenges for biomedical applications. In recent years, conductive nanomaterials such as carbon nanotubes (CNTs) and graphene oxides (GOs) with remarkable mechanical and electrical properties have been combined with flexible elastomers for the fabrication of wearable sensors. Elastomers, or elastic polymers, are made of polymer chains with an appropriate number of crosslink bonds that are able to undergo a large size change in the presence of external forces and return to their primary shapes just after removing the imposed force [8]. Hydrogels are a new candidate for application in wearable sensors that simultaneously possess biocompatibility, degradability, a low fabrication cost, feasibility, and potential for mass production properties [10]. However, there are obstacles for hydrogels in wearable sensing applications such as low toughness and elasticity [11]. According to remarkable progress in the application of conductive hydrogels in flexible sensors, it seems that the development of hydrogels with the desired mechanical properties and structural resilience can open a new gateway toward the real application of scalable, flexible sensors in the monitoring of physiological signals [10].

20.2.1 Application of Flexible Materials in Wearable Sensors

Flexible materials are applied in two main portions of the wearable sensor's structure: one, the sensor substrates, and two, as functional compartments. All the electronic compartments of wearable sensors need to be placed on a flexible substrate that is usually made from fabrics and polymers. The substrate is in direct contact with the skin on one side, and on the other side, it interacts with sensor electronics and energy sources. The sensor substrate possesses flexible, biocompatible, biodegradable, durable, and appropriate mechanical properties to provide high efficiency and comfortability for wearable sensors. In addition to the sensor substrate, flexible materials can also participate in the structure of functional compartments in wearable sensors, including signal receptors/detectors, data processors, data transmitters, energy sources, and wireless connectors (Figure 20.2). For chemical sensors such as pH and glucose sensors, the signal receptors are in direct and continuous contact with body fluids, so their safety, biocompatibility, and durability must be proved before applying to flexible sensor structures [8].

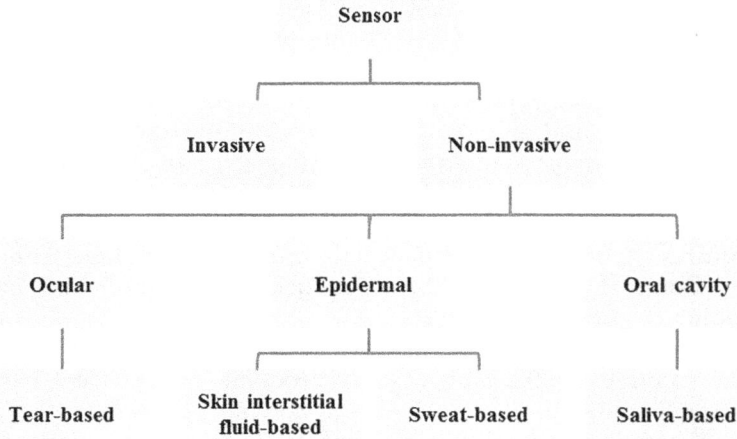

FIGURE 20.2
Some flexible materials for fabricating wearable sensors.

20.3 Sensor Networks for Wearable Flexible Sensors

For a long time, hospitalization has increased, imposing a heavy financial burden on healthcare systems all over the world. Unnecessary referring of patients to the clinical center increases the probability of infection and prolongs their treatment process. With the evolution of electronics and information technology and consequently the emergence of wireless communication, wearable sensors have taken an important step toward real application in the remote monitoring of patients. In this new system, wearable flexible sensors continuously monitor the physiological and chemical properties of the patient and, after processing the receiving monitoring information from the sensor's data processor, analyzed data is remotely sent to the clinical operator and appropriate instructions are sent back to the patient [2]. The schematic network communication in smart wearable flexible sensors for biomedical applications is shown in Figure 20.3.

20.4 Types of Sensing Using Wearable Flexible Sensors

Based on the intrinsic nature of human properties that can be monitored by flexible sensors, two types of physical and chemical sensing procedures have been developed that are briefly explained in the following sections.

20.4.1 Physical Sensing

The detection mechanism of flexible physical sensors is based on the changes in electrical parameters, including capacitance, resistance, piezoelectricity, and triboelectricity [12]. To perform sensing duties, these kinds of sensors are commonly attached to the skin, wrist,

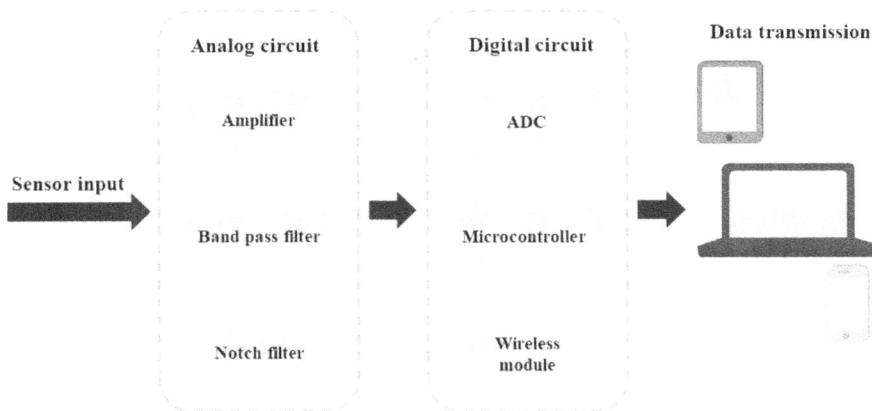

FIGURE 20.3
A schematic diagram of the transmission of data from the sensor to the monitoring unit.

ankle, neck, chest, thigh, and forehead [13]. Physical sensing has been the earliest usage of wearable sensors because of their simplicity and ease of installation on the patient's body. However, its basic problems, including flexibility, heaviness, large size, energy sources, and durability, have still not entirely been overcome. Physical sensors usually monitor the physical activities of humans, such as body motion, voice, and facial expressions [14]. They can also precisely measure the vital physiological symptoms of the body, including heart rate [15], blood pressure [16], temperature [17], calories burned [18], electromyography (EMG) [19], electroencephalography (EEG) [20], and electrocardiography (ECG) [21].

20.4.2 Chemical Sensing

For the monitoring of chemical properties of the human body, chemical sensors realize the composition and concentration of targets via optical and electrochemical sensing mechanisms [22–24]. Chemical monitoring of the human body has challenges. The most common method for chemical sensing is blood sampling, which is invasive, painful, and requires professional staff. In addition, these problems will worsen if continuous and real-time monitoring of patients is required. On the other hand, conventional analytical apparatuses currently existing in medical centers are expensive, time-consuming, and require trained personnel. In addition to overcoming all the mentioned obstacles, the newly achieved flexible wearable chemical sensors are able to use any fluids of the body as sensing samples, such as interstitial fluid, saliva, tears, and even body sweat, without the need for clinical staff [12].

20.4.2.1 *Invasive and Non-Invasive Sensors*

Invasive or intrusive sensors access body fluids via injecting tools. This method of sampling may be used for temporary sensing of chemical analytes, but with continuous monitoring, the story is changed because blood contamination is probable and long injections may bother people. Non-invasive flexible sensors do not need body blood and, consequently, any injections. Instead, they measure or detect their targets within other body fluids such as saliva, sweat, and tears (Figure 20.4). Non-invasive sensing procedures are widely used for monitoring glucose in diabetic patients, oxygen saturation in high-risk patients, cholesterol tracking in cardiac patients, and drug efficacy control [25].

```
┌─────────────────────────┐                    ┌─────────────────────────┐
│  Flexible material as the│                    │  Flexible material as the│
│  substrate of sensor     │                    │  signal conversion part  │
└─────────────────────────┘                    └─────────────────────────┘
```

Flexible material as the substrate of sensor		Flexible material as the signal conversion part
Silk fabric		PDMS
Polyimide	Flexible materials	SEBS substrate
Spandex substrate		Nanofiber electrode
Microbial nanocellulose		Free standing electrochemical sensing system (FESS)
Paper		Silk fabric derived carbon textile

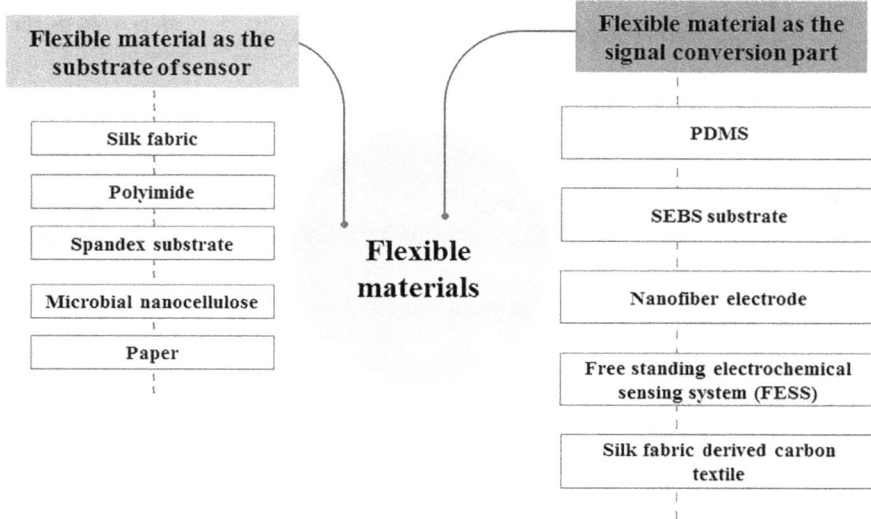

FIGURE 20.4
Classification of flexible and wearable sensors.

20.4.3 Mechanisms of Sensing

The physiological activities of the human body produce mechanical stimulations (force, vibration, motion, strain, and pressure) and changes in electrolyte or metabolite concentrations. These activities are translated into electrical or optical parameters by related sensors. This conversion of physiological activities to reportable parameters works based on particular and precise mechanisms, including resistance, capacitance, and piezoelectricity. The fabrication of flexible parts of such a complicated sensing system without any disruption in functionality and efficiency is one of the greatest milestones in the field of wearable sensor technology [26].

20.4.4 Fabrication Methods

Wearable sensors are fabricated through a wide variety of methods and techniques, which is not possible to explain in this chapter. Among these methods, pattern transferring is the most commonly used procedure that benefits from lithography, micro-scale modeling, handwriting, and also several printing techniques such as screen printing, inkjet printing, and 3D printing [2, 26].

20.5 Biomedical Applications of Wearable Flexible Sensors

With the continuous increase in the number of patients and their related binational burden, the concerns of healthcare systems about managing this issue are rising. As one of the strongest suggestions, wearable flexible sensors can provide remote and real-time monitoring of the patient's condition. In this section, different biomedical applications of wearable sensors are briefly described and explained (Figure 20.5).

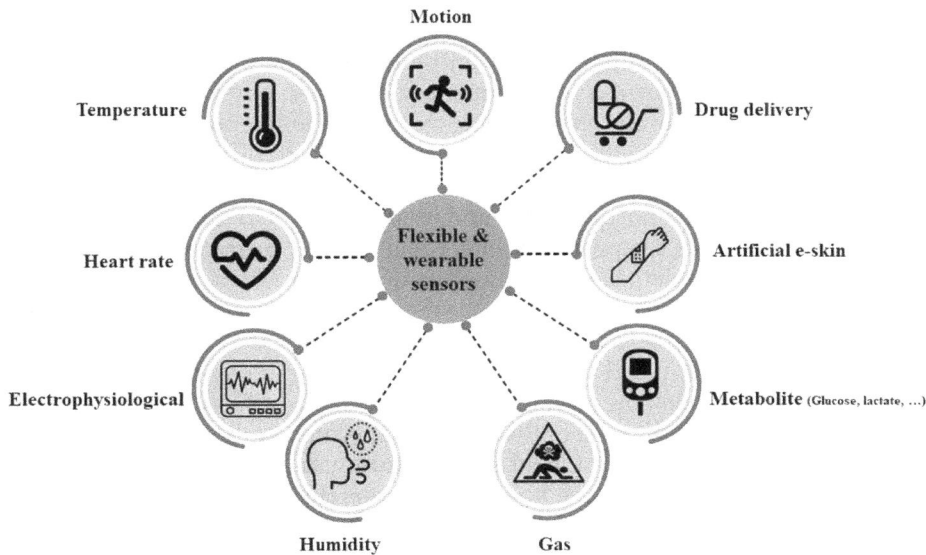

FIGURE 20.5
Different applications of flexible and wearable sensors.

20.5.1 Motion Sensors

The monitoring of physical activity by flexible and wearable sensors is the first step to an understanding of a person's general condition. As a pioneer in the field of biomedical sensing, there are many examples of practicable reports on flexible physical sensors [27]. To obtain a feasible sensor for continuous monitoring of human activities, miniaturization of the sensor compartment should be noted. Recently, the remarkable features of nanomaterials, including ultra-small size, high activity, high conductivity, and their related machines, have been widely used by researchers to fabricate efficient and miniaturized sensors. In a study, Janus nanosheets (JNs) were embedded into a three-dimensional network structure of gum/polyvinyl alcohol (PVA). In addition to their appropriate mechanical strength (1.0 MPa), JN hydrogels also show self-healing efficiency (93.1%) in the presence of reversible interaction. The synthesized resistive-type sensor based on JNs shows high sensitivity (gauge factor = 12.5) and acceptable sensing performance (100% strain, 600 cycles). Various body motions, including bending (wrist, elbow, finger), running, cough vibrations, and pulse rates, can be realized by introducing a physical sensor [28].

20.5.2 Vital Sign Sensors

The common vital symptoms of the human body are blood pressure, temperature, electrophysiological, and heart rate. The new advanced wearable sensors benefit from all flexible materials, miniaturized electronics, wireless data communication, and smartphone applications for remote and real-time monitoring of vital symptoms without any need to refer the patient to clinical centers. These smart flexible systems are applied in the form of smart bracelets, watches, and glasses, which significantly improve doctor-patient relationships and the technology of remote monitoring.

20.5.2.1 Temperature Sensors

The temperature is one of the first, fastest, and most accurate indicators of body state in medical science and has a key role in the primary detection of fever, metabolic functionality, depression, and insomnia. The simplest method of temperature monitoring is skin measurement, which is done by traditional wearable temperature sensors [29]. The preferable application of flexible and wearable temperature sensors is the monitoring of neonates and infants because they are not expensive and easily attached to the axilla or head [12]. An advanced flexible wearable temperature sensor has recently been developed for the early detection of temperature-related ailments such as COVID-19. Multi-wall carbon nanotubes are selected for the fabrication of the proposed sensor via screen-printing and drop-casting procedures. Notably, covering the sensor surface with a sheet of PET enhanced its durability and stability against physical movements and other harsh environmental conditions. To evaluate the produced flexible sensor, it is placed in different locations of the body, and the results are compared with those of the standard thermocouple and IR sensor. Based on recorded results, there are no significant temperature variations between different body locations. Additionally, the response time of this wearable sensor is approximately 13 s with a recovery time of ~38s and a sensitivity of -0.0685% $°C^{-1}$ [30].

20.5.2.2 Heart Rate Sensors

Heart rate is another important monitoring target that is widely set up for hospital patients due to the proven symptoms of cardiovascular disease and other heart-related disorders. The high current consumption, bulk dimensions, rigid sensing tools, and fixed position of common heart monitoring systems in existing clinical centers restrict their feasible and real-time applications [31]. In contrast, flexible wearable sensors with the ability to perform real-time, simple, portable, and accurate monitoring of heart rate are commercially available. The flexible heart rate sensor has been designed based on light variations or photoplethysmography (PPG) to sense the cardiovascular pulse. Basically, light photons are absorbed, scattered, or reflected after collision with biological tissues. Amazingly, human blood absorbs more light than other tissues. Therefore, if the amount of blood decreases, the amount of absorbed light decreases too, and consequently, the intensity of reflected light from the tissue increases. This phenomenon is used for the measurement of the heart rate by light-reflective sensors. Flexible wearable heart rate sensors based on the PPG reflectance mode can easily be installed on the patient's clothes without any annoying attachment to the skin. In more updated systems, integration of these sensors with wireless and smartphone technology results in remote, real-time, flexible, and wearable sensors for measuring the human heart rate [32].

20.5.2.3 Electrophysiological Sensors

The human body continuously responds to external and internal motivations. These physiological responses generate different electrical pulses that are evoked from neurons and their related electrical signaling pathways. Such electrical pulses are powerful enough to be detected from the surface of the body. According to the targeted organ, there are different electrophysiological sensors, including electrocardiogram (ECG, or EKG) for the heart's electrical activity, electroencephalogram (EEG) for brain monitoring, and electromyogram (EMG) related to muscular activity [33]. Despite the undeniable role of electrophysiological sensors in monitoring and medical diagnosis, there is no tangible example of a commercialized wearable platform for these sensors. Based on a recent report, flexible and wearable electrodes were developed by using the printing-transfer method for the measurement of electrophysiological signals (ECG) from human skin. The resultant outcomes from reported flexible ECG sensors are comparable with those recorded by conventional metal clips (SNR > 25 dB). The experience of this study can be used to develop advanced commercialized prototypes of wearable electrophysiological sensors for remote and real-time monitoring of vital physiological symptoms in patients [34].

20.5.2.4 Humidity Sensors

Most acute medical disorders such as lung cancer, acidosis, hypoxia, metabolic alkalosis, and cardiopulmonary arrest can affect the rate and manner of patient respiration. Therefore, real-time and continuous monitoring of respiration is a necessary duty of clinical centers for the permanent evaluation of the patient's health [35]. Similar to other hospital monitoring systems, spirometers and air flow meters are not suitable for real-time and portable monitoring [36, 37]. There are existing flexible respiration sensors with a polymeric nature. However, polymer films within the structure of these sensors have low hygroscopicity and breathability, which can significantly decrease their sensitivity and comfortability. The novel flexible respiration sensor was fabricated by the innovative use of non-woven fabric (NWF) coated with GO. Additionally, bovine serum albumin was used as a mediator to facilitate the attachment of GO sheets to the NWF. The fabricated sensor shows appropriate feasibility under various breathing conditions, including fast breathing, deep breathing, nose or mouth breathing, and simple spoken words. A study by Wang et al. proved the remarkable role of nanomaterials and their derivations in the improvement and development of efficient wearable respiration sensors with excellent advantages such as skin-friendly, portability, breathability, sensitivity, and flexibility [38].

20.5.2.5 Gas Sensors

Continuous monitoring of toxic gases such as ammonia (NH_3) is another application of flexible wearable sensors. The hazardous effects of toxic gases are pulmonary fibrosis, chronic cough, asthma, and mucosal irritation [39]. The sensors of NH_3 must be sensitive enough to its low concentrations (50 ppm for 8 h d^{-1}). A flexible sensor was fabricated from polycrystalline $SrGe_4O_9$ nanotubes (NTs) by using an electrospinning method for monitoring NH_3 at room temperature. To improve the sensing response of the synthesized sensor from 2.49 to 7.08 for 100 ppm NH_3, the $SrGe_4O_9$ NTs were decorated with platinum nanoparticles. By using nanoparticles and nanotubes, the fabricated sensor exhibits appropriate flexibility, mechanical stability, and sensing stability [40].

20.5.2.6 Metabolites Sensors

All biological processes of living organisms, including the human body, take place in a permanent pH, chemical, and ion balance. Any variation in metabolites can disrupt the natural function of the human body. In other words, by monitoring metabolites, the health condition of the body can be checked. However, the monitoring methods and apparatuses available in medical centers are slow, expensive, invasive, and painful. Lots of portable and miniaturized sensors for non-invasive real-time monitoring of free ions, biomarkers, biochemical metabolites, and electrolytes by using body exertions have been well developed in recent years. Due to its non-invasive nature and ease of sampling, sweat is the most common biofluid method of sensing. In addition, the electrolytes and metabolites existing in sweat are significantly related to the blood contents. The usually targeted contents of sweat for detection by wearable sensors are free ions (Ca^{2+}, Na^+, K^+, Cl^-), heavy metals, glucose (for diagnosing and treating diabetes), and lactic acid (evaluation of tissue oxygenation and pressure ischemia in the elderly) [26].

20.5.2.7 Other Biomedical Applications

Artificial electronic skins. The development of human adaptive electronic skins (e-skins) is another application of wearable sensors. The e-skins are able to mimic subtle spatiotemporal sensing such as temperature, vibration, shear, pain, strain, and pressure, and the functionality of human skin, such as transducing sensing data in the form of electronic signals [41]. Therefore, e-skins will have potential applications in continuous and real-time health monitoring, rehabilitation devices, and prosthetic limbs in the near future. The most crucial requirement of e-skins for these applications is that they have a stretchable and flexible structure [42]. In an innovative study, the interlocked microstructures of human epidermal ridges were modeled and inspired by the fabrication of piezoresistive interlocked microdome arrays. The resulting wearable and flexible e-skins arrays are efficient and sensitive to a wide variety of mechanical forces such as bending, twisting, shearing, and stretch forces. The exceptional geometry of interlocked arrays can easily recognize different kinds of mechanical stimuli because each imposed mechanical force with particular directions causes different types of deformation in designed arrays and finally generates different output patterns from the sensor. The final prepared wearable electrical skins can conveniently be attached to the various parts of the body and successfully detect the types, intensity, and direction of mechanical forces, including airflows and vibrations [43].

Therapeutic and drug-delivery platforms. The most advanced progress in the field of flexible biomedical sensors is multi-functional wearable sensors with a collection of competencies ranging from flexibility and smart wireless connections to therapeutics and drug delivery. In addition to remote and real-time health monitoring of patients, remote and smart controlled drug delivery to specific organs and tracking of the effects of drugs on the treatment process are available with these new systems. Based on a new report, wearable drug-delivery devices were designed and successfully fabricated from stretchable elastomer and microgel depots. In order to deliver drugs, the nanoparticles loaded with drugs are embedded in the proposed device. Drug release is triggered after compression caused by applying external tensile strain to the elastomer film. Hence, natural daily motions or intentional administration impose the release of drugs from the fabricated wearable device [44].

20.6 Conclusion

In conclusion, new procedures of healthcare systems all over the world are moving toward smart and remote treatment of patients. The design and development of sensitive, selective, robust, durable, and economical sensors for detecting and measuring biomedical signals have been the first achievements of this procedure. Next, miniaturization and improvement of energy consumption in biomedical sensors are being studied by scientists. Despite all the remarkable studies in the fabrication of wearable biomedical sensors at laboratory levels, there were some obstacles in the way of their practical application. In addition to all the engineering and functional properties, an efficient medical sensor must provide welfare and comfort during treatment time and not create extra pain for the patients. Hence, the practical application of wearable sensors relies on lowering the implantable device volume and improving device flexibility, biocompatibility, and durability. The new advanced wearable flexible sensors can monitor a wide range of important properties of the human body, including heart rate, pressure, temperature, electrophysiological pulses, humidity, gases, metabolites, ions, and heavy metals. Furthermore, wearable sensors are the key components of artificial skins and various drug-delivery systems. As a result, wearable sensors have passed a lot of existing challenges, but there is a long way toward their practical utility in biomedical applications. Accordingly, the next generation of research in the field of wearable sensors is focused on smartphone-based biomedical sensors, artificial skins, smart drug delivery, and batteryless flexible sensors.

References

1. Y. Gu, T. Zhang, H. Chen, F. Wang, Y. Pu, C. Gao, S. Li, Mini review on flexible and wearable electronics for monitoring human health information, *Nanoscale Res. Lett.* 14 (2019) 1–15.
2. A. Nag, S.C. Mukhopadhyay, J. Kosel, Wearable flexible sensors: A review, *IEEE Sens. J.* 17 (2017) 3949–3960.
3. K. Baeg, J. Lee, Flexible electronic systems on plastic substrates and textiles for smart wearable technologies, *Adv. Mater. Technol.* 5 (2020) 2000071.
4. M. Jian, C. Wang, Q. Wang, H. Wang, K. Xia, Z. Yin, M. Zhang, X. Liang, Y. Zhang, Advanced carbon materials for flexible and wearable sensors, *Sci. China Mater.* 60 (2017) 1026–1062.
5. P. Wang, M. Hu, H. Wang, Z. Chen, Y. Feng, J. Wang, W. Ling, Y. Huang, The evolution of flexible electronics: From nature, beyond nature, and to nature, *Adv. Sci.* 7 (2020) 2001116.
6. H. Ehtesabi, S.O. Kalji, L. Movsesian, Smartphone-based wound dressings: A mini-review. *Heliyon* 8 (2022) e09876.
7. C. Ning, K. Dong, R. Cheng, J. Yi, C. Ye, X. Peng, F. Sheng, Y. Jiang, Z.L. Wang, Flexible and stretchable fiber-shaped triboelectric nanogenerators for biomechanical monitoring and human-interactive sensing, *Adv. Funct. Mater.* 31 (2021) 2006679.
8. H. Liu, L. Wang, G. Lin, Y. Feng, Recent progress in the fabrication of flexible materials for wearable sensor, *Biomater. Sci.* 10 (2022) 614–632.
9. A. Pal, V.G. Nadiger, D. Goswami, R.V. Martinez, Conformal, waterproof electronic decals for wireless monitoring of sweat and vaginal pH at the point-of-care, *Biosens. Bioelectron.* 160 (2020) 112206.
10. C. Cui, Q. Fu, L. Meng, S. Hao, R. Dai, J. Yang, Recent progress in natural biopolymers conductive hydrogels for flexible wearable sensors and energy devices: Materials, structures, and performance, *ACS Appl. Bio. Mater.* 4 (2021) 85–121.
11. R. Liu, H. Wang, W. Lu, L. Cui, S. Wang, Y. Wang, Q. Chen, Y. Guan, Y. Zhang, Highly tough, stretchable and resilient hydrogels strengthened with molecular springs and their application as a wearable, flexible sensor, *Chem. Eng.* 415 (2021) 128839.
12. Y. Wang, B. Yang, Z. Hua, J. Zhang, P. Guo, D. Hao, Y. Gao, J. Huang, Recent advancements in flexible and wearable sensors for biomedical and healthcare applications, *J. Phys. D: Appl. Phys.* 55 (2021) 134001.
13. Y. Fan, H. Tu, H. Zhao, F. Wei, Y. Yang, T. Ren, A wearable contact lens sensor for noninvasive in-situ monitoring of intraocular pressure, *Nanotechnology* 32 (2020) 095106.

14. Y. Tu, L. Liu, M. Li, P. Chen, Y. Mao, A review of human motion monitoring methods using wearable sensors, *Int. J. Online Eng.* 14 (2018) 168–179.

15. M. Wójcikowski, B. Pankiewicz, Photoplethysmographic time-domain heart rate measurement algorithm for resource-constrained wearable devices and its implementation, *Sensors* 20 (2020) 1783.

16. J. Fortin, D.E. Rogge, C. Fellner, D. Flotzinger, J. Grond, K. Lerche, B. Saugel, A novel art of continuous noninvasive blood pressure measurement, *Nat. Commun.* 12 (2021) 1–14.

17. Y. Su, C. Ma, J. Chen, H. Wu, H. Luo, Y. Peng, Z. Luo, L. Li, Y. Tan, O.M. Omisore, Z. Zhu, Printable, highly sensitive flexible temperature sensors for human body temperature monitoring: A review, *Nanoscale Res. Lett.* 15 (2020) 1–34.

18. F. Fotouhi-Ghazvini, S. Abbaspour, Wearable wireless sensors for measuring calorie consumption, *J. Med. Signals Sens.* 10 (2020) 19.

19. X. Li, Y. Chai, Design and applications of graphene-based flexible and wearable physical sensing devices, *2D Materials* 8 (2020) 022001.

20. K. Gao, G. Shen, N. Zhao, C. Jiang, B. Yang, J. Liu, Wearable multifunction sensor for the detection of forehead EEG signal and sweat rate on skin simultaneously, *IEEE Sens. J.* 20 (2020) 10393–10404.

21. Y. Qiao, X. Li, J. Jian, Q. Wu, Y. Wei, H. Shuai, T. Hirtz, Y. Zhi, G. Deng, Y. Wang, G. Gou, J. Xu, T. Cui, H. Tian, Y. Yang, T. Ren, Substrate-free multilayer graphene electronic skin for intelligent diagnosis, *ACS Appl. Mater. Interfaces* 12 (2020) 49945–49956.

22. Y. Wang, J. Huang, Recent advancements in flexible humidity sensors, *J. Semicond.* 41 (2020) 040401.

23. J. Guo, C. Yang, Q. Daia, L. Kong, Soft and stretchable polymeric optical waveguide-based sensors for wearable and biomedical applications, *Sensors* 19 (2019) 3771.

24. P.C. Ferreira, V.N. Ataide, C.L.S. Chagas, L. Angnes, W.K.T. Coltro, T.R.L.C. Paixão, W.R. de Araujo, Wearable electrochemical sensors for forensic and clinical applications, *TrAC - Trends Anal. Chem.* 119 (2019) 115622.

25. T. Islam, S.C. Mukhopadhayay, Wearable sensors for physiological parameters measurement: Physics, characteristics, design and applications, in *Wearable Sensors: Applications, Design and Implementation*, IOP Publishing Ltd (2017) 1–31.

26. Y. Liu, H. Wang, W. Zhao, M. Zhang, H. Qin, Y. Xie, Stretchable sensors for wearable health monitoring: Sensing mechanisms, materials, fabrication strategies and features, *Sensors* 18 (2018) 645.

27. A. Derungs, O. Amft, Estimating wearable motion sensor performance from personal biomechanical models and sensor data synthesis, *Sci. Rep.* 10 (2020) 11450.

28. N. Zhang, G. Zhao, F. Gao, Y. Wang, W. Wang, L. Bai, H. Chen, H. Yang, L. Yang, Wearable flexible sensors for human motion detection with self-healing, tough guar gum-hydrogels of GO-P4VPBA/PDA janus nanosheets, *ACS Appl. Polym. Mater.* 4 (2022) 3394–3407.

29. H. Ota, M. Chao, Y. Gao, E. Wu, L.C. Tai, K. Chen, Y. Matsuoka, K. Iwai, H.M. Fahad, W. Gao, H.Y.Y. Nyein, 3d printed "earable" smart devices for real-time detection of core body temperature, *ACS Sens.* 2 (2017) 990–997.

30. K. Thiyagarajan, G.K. Rajini, D. Maji, Cost-effective, disposable, flexible and printable MWCNT-based wearable sensor for human body temperature monitoring, *IEEE Sens. J.* 22(2021) 16756–16763.

31. S.O. Yun, J.H. Lee, J. Lee, C.Y. Kim, A flexible wireless sensor patch for real-time monitoring of heart rate and body temperature, *IEICE Trans. Inf. Syst.* 102 (2019) 1115–1118.

32. A.R. Maria, S. Pasca, R. Strungaru, Heart rate monitoring by using non-invasive wearable sensor, in *2017 E-Health and Bioengineering Conference (EHB)*, IEEE (2017).

33. H. Prance, Sensor developments for electrophysiological monitoring in healthcare, in *Applied Biomedical Engineering*, Intechopen (2011) 265–286.

34. N. Luo, J. Ding, N. Zhao, B.H.K. Leung, C.C.Y. Poon, Mobile health: Design of flexible and stretchable electrophysiological sensors for wearable healthcare systems, in *2014 11th International Conference on Wearable and Implantable Body Sensor Networks*. IEEE (2014).

35. M. Elliott, A. Coventry, Critical care: The eight vital signs of patient monitoring, *Br. J. Nurs.* 21 (2012) 621–625.

36. K.R. Jat, Spirometry in children, *Prim. Care Respir. J.* 22 (2013) 221–229.

37. G.X. Ayala, C. Gillette, D. Williams, S. Davis, K.B. Yeatts, D.M. Carpenter, B. Sleath, A prospective examination of asthma symptom monitoring: Provider, caregiver and pediatric patient influences on peak flow meter use, *J. Asthma* 51 (2014) 84–90.

38. Y. Wang, L. Zhang, Z. Zhang, P. Sun, H. Chen, High-sensitivity wearable and flexible humidity sensor based on graphene oxide/non-woven fabric for respiration monitoring, *Langmuir* 36 (2020) 9443–9448.

39. W. Liu, Y. Liu, J. Do, J. Li, Highly sensitive room temperature ammonia gas sensor based on Ir-doped Pt porous ceramic electrodes, *Appl. Surf. Sci.* 390 (2016) 929–935.

40. T. Huang, Z. Lou, S. Chen, R. Li, K. Jiang, D. Chen, G. Shen, Fabrication of rigid and flexible SrGe4O9 nanotube-based sensors for room-temperature ammonia detection, *Nano Res.* 11 (2018) 431–439.

41. J.C. Yeo, C.T. Lim, Emerging flexible and wearable physical sensing platforms for healthcare and biomedical applications, *Microsyst. Nanoeng.* 2 (2016) 1–19.

42. R.S. Dahiya, G. Metta, M. Valle, G. Sandini, Tactile sensing—from humans to humanoids, *IEEE Trans. Robot.* 26 (2009) 1–20.

43. J. Park, Y. Lee, J. Hong, Y. Lee, M. Ha, Y. Jung, H. Lim, S.Y. Kim, H. Ko, Tactile-direction-sensitive and stretchable electronic skins based on human-skin-inspired interlocked microstructures, *ACS Nano* 8 (2014) 12020–12029.

44. J. Di, S. Yao, Y. Ye, Z. Cui, J. Yu, T.K. Ghosh, Y. Zhu, Z. Gu, Stretch-triggered drug delivery from wearable elastomer films containing therapeutic depots, *ACS Nano* 9 (2015) 9407–9415.

21

The Role of Additive Manufacturing in Flexible and Wearable Sensors

Mansour Mahmoudpour, Zahra Karimzadeh, Abolghasem Jouyban, and Jafar Soleymani

CONTENTS

21.1 Introduction

Human-integrated tools contain implantable and wearable probes for interacting with the body or monitoring vital signals [1]. These devices need to be attached to human organs or skin to perform functions like sensing and energy harvesting. Widespread acceptance of these technologies depends on entirely novel construction methods with critical features such as ultra-thinness, superior flexibility, and stretchability, as well as being lightweight with a rapid reaction in designing more well-organized wearable substrates [2, 3]. These health-monitoring nanoprobes require physiological evidence achieved from wristbands, pads, wearable straps, or skin-conformal tattoos [4]. In this context, with the exploration of functional soft materials, significant efforts were made to realize the desired interaction between biological systems and multifunctional wearable tools [5, 6]. In light of the surprising properties of soft materials like flexibility, stretchability, deformability, and ultra-thinness, flat devices are made and transferred to the analyte at the biological surface to produce a cohesive interface [7, 8]. Until now, the detecting interfaces have been designed with many assembly methods such as lamination, filling, planar printing, coating, micro-channel molding, and lithography [9]. Nevertheless, the adoption of these technologies in the construction of sensitive, wearable, and flexible (bio)sensors has been restricted because of their limited extensibility, high cost in mass production, material wastage, poor durability, and lack of manufacturing scalability [10]. Alternatively, the identifying geometries could be directly generated on the analyte interfaces without the prior need for microstructure facilities by progressing multifunctional three-dimensional (3D) printing technologies. Owing to the developments in device design and multifunction printing capabilities, these printing approaches can open a research space for applications such as non-invasive sweat analysis, wearable (bio)sensors, fully 3D-printed quantum dot-based light-emitting diodes, electromagnetic constructions, and regenerative pathways [11].

Additive manufacturing (AM) processes can be performed as ideal choices for addressing the earlier restrictions of the current microfabrication technologies in producing wearable/flexible printed devices with varied functionalities, for example, actuation, sensing, energy harvesting/storage, and computation, leveraging benefits like scalability, low processing temperature, low cost, versatile ink formulation, and maskless processing [12]. AM constructs a 3D building by constantly adding layer-by-layer materials with the guidance of a computer-aided design (CAD) model. Numerous kinds of materials, comprising biomaterials, metals, polymers, ceramics, and composite materials, have been developed in different types of AM techniques, comprising selective laser melting (SLM), binder jetting (BJ), laminated object manufacturing (LOM), fused deposition modeling (FDM), selective laser sintering (SLS), and stereolithography (SLA). More and more research studies on AM of structural materials have been published in a varied range of journals in recent years as shown in Figure 21.1. In 2020, a special paper in *Nature* proposed that scholars are developing methods for faster, larger, and more innovative printing [13]. Recently, there have been numerous inspiring works on AM for designing ultrafast 3D printing of multiscale constructions and 3D bioprinting of organs or tissues [14, 15]. An extensive range of materials established for AM methods has been advanced, including polymers, metals, and

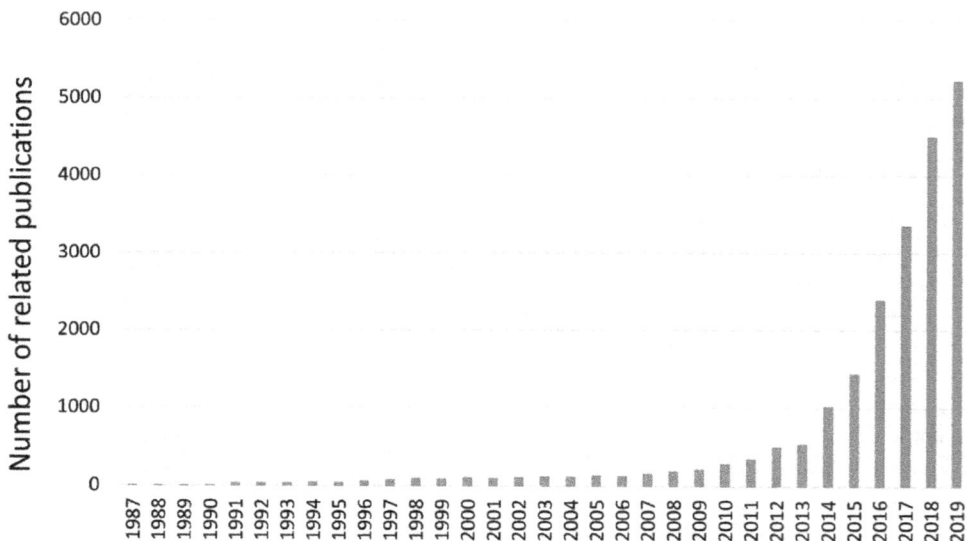

FIGURE 21.1
Representation of statistical data of articles on the "additive manufacturing" topic. Adapted with permission [13]. Copyright (2020) Elsevier.

multi-material systems. Combined or multi-process 3D printing also has the potential for enhancing the efficiency of material structures.

In this chapter, we have tried to provide current advances and progress in 3D AM of different materials, focusing on their potential and capability for constructing wearable human-integrated devices. Important subjects for the AM of structural materials, comprising cost-efficiency, mechanical robustness, geometric complexity, printability, and printing speed/scalability/resolution are discussed, and the recent limitations and future directions for designing 3D printable materials are described.

21.2 Ink-Based Additive Manufacturing Methods

AM techniques can produce predesigned patterns layer by layer to form the final parts. There are various classes of AM methods, containing ink-based, filament-based, and powder-based techniques. Ink-based AM approaches have attracted growing attention in the field of stretchable/flexible electronics, owing to their ability to print/deposit solution-processed materials into designer complex forms in a spatially controlled manner [16]. Ink-based AM methods like aerosol jet printing and inkjet printing, as well as spray coating, are feasible for printing low-viscosity ink materials. These printing approaches split functional ink into micro-droplets and deposit the droplets onto platforms. In the case of aerosol printing, materials with higher viscosity can be simply printed as the ultrasonic energy can split functional inks into micro-droplets. Other strategies, for example, embedded 3D printing and direct ink writing (DIW), are also desirable to print materials with high viscosity in which the functional materials are extruded via a printhead nozzle throughout the printing procedure [17]. For embedded 3D printing, the printhead is immersed into the elastomeric polymer matrix. To form the patterns considered in the DIW method, materials from the nozzle are directly deposited on the scaffold. In comparison to DIW, other AM methods like stereolithography printing allow better surface quality and more-enhanced printing resolutions. Hydrogels, resins, and photo-curable polydimethylsiloxane (PDMS) as usual materials can be employed for stereolithography printing. Nevertheless, stereolithography printing has restricted the utilization of integrated wearable tools, particularly for multi-material printing. This limitation can arise from the inability of the stereolithography procedure to print numerous functional materials simultaneously [18]. These ink-based AM methods possess exceptional uses as a result of their diverse mechanisms and manufacturing demands. For instance, DIW could allow microscale stereo-structure printing and inkjet printing could prefer thin-film pattern production. Other ink-based creating methods like screen printing, which are extremely adaptable with roll-to-roll strategies for mass production, are applied to construct flexible and wearable electronics.

21.3 Structural Materials for Additive Manufacturing

The application of appropriate materials for obtaining the desired AM is prominent while selecting 3D printing technology. Several materials containing polymers, metals, ceramics, bio-ink, biofluids plastics, resins, and composites can be employed in the fabrication of the

3D model [19]. Moreover, the engagement of more compatible and advanced materials has received substantial attention in a typical 3D model using various energy sources such as heat or light to render the structure directly from the raw material. Although this can be reached by the advent of developing 3D printing technologies, exploration of advanced functional materials is application/design specific. With recent advances in 3D printing, the amelioration of AM technologies can be categorized into four subgroups: (i) employment of advanced functional materials; (ii) speed enhancement in printing; (iii) development of application-specific printing technology regarding smart materials; (iv) resolution enhancement. For printed flexible and wearable devices, the biocompatibility, sensitization, irritation, and cytotoxicity of functional materials are also essential [20]. Matrix materials and conductive materials are the main applied materials for printed wearable electronics, which usually have direct interfaces/contacts with human skin. Therefore, seriously considering the biocompatibility of these wearable materials is dominant. At this point, most solution-processable matrix materials with high biocompatibility such as poly(styrene-co-ethylene butylene-co-styrene) (SEBS), poly(styrene-butadiene-styrene) (SBS), and PDMS have been employed to manufacture different types of stretchable skin-wearable electronics [21]. Conductive fillers and biocompatible polymer-based matrices are typical materials used for formulating electrically conductive electrodes [22].

21.3.1 Encapsulation Materials and Matrix

In digital model-based AM technology, materials with low bending stiffness, a rigid nature, bulky interfaces, and unsatisfactory mechanical confrontations with skin are not feasible. They induce discomfort in long-standing wearability. Therefore, softer materials with lower elastic modulus and enhanced stretchability offer an excellent future for making skin-compatible flexible devices. Polymers as the most commonly used encapsulation materials and matrices are of crucial importance to AM in the fabrication of flexible sensors. Currently, the majority of polymeric materials are classified as thermoplastic polymers, elastomers, and thermosetting polymers. Besides the printing materials to produce substances by layered printing, there are various production approaches for composites of polymer, containing 3D drawing, stereolithography, inkjet 3D printing, selective laser sintering, and deposition molding [23]. Therefore, to extend the influence of AM in the construction of end-product components, the development and introduction of advanced polymers are significant for AM considering their biocompatibility.

One of the commonly used 3D printed materials in AM is commodity thermoplastics and their composites. Despite the accessible properties of thermoplastic products, the lack of functionality and strength in pure form is a common issue in AM. Accordingly, different types of processing techniques can alter the mechanical stability of pure thermoplastic parts. Efforts have focused on the fused deposition modeling process to enhance the modulus of elasticity and tensile strength of the 3D printed samples [24]. For example, Eutionnat-Diffo et al. [25] compared the tensile deformation of conductive and non-conductive polylactic acid (PLA) filaments deposited on polyethylene terephthalate (PET) fabrics along with studying fabrics features and the temperature property of the printing platform. After statistical and theoretical optimization, the non-conductive PLA printing guide showed improved durability after washing or adding conductive filler, however, fracture stress of the woven fabric was affected by the washing process.

Generally, silicone elastomer, specifically PDMS, is an alternative polymer for fabricating wearable sensors due to its stretchability, excellent thermal stability, high mechanical flexibility, and biocompatibility, compared with other elastomers. Furthermore, PDMS exhibits well-suiting properties for the skin and provides continuous monitoring. Along with the aforementioned properties, not only do they form the architecture of a flexible 3D system but there is no need for adhesive to hold the sensing portion at a fixed place [26]. As a typical example, He et al. [27] used a direct ink writing 3D printer for printing PDMS ink and making distinct wetted surfaces by patterned ordered porous constructions with different geometric parameters. Under optimum conditions, the sample PDMS film with superhydrophobic properties exhibited excellent thermal aging durability.

In thermosetting plastics, the melted polymer by initial hot processing can undergo chemical reactions to create crosslinked structures. During this irreversible process, the thermosetting plastics harden after curing and can no longer soften when reheated. These materials mainly contain polymethylmethacrylate (PMMA), organosilicon, unsaturated polyester, amino plastics, epoxy, and phenolic plastics [28]. Through 3D printing, Shim et al. [29] printed PMMA in different printing directions (0°, 45°, and 90°) and studied their effect on printing performance. The results prove that the printing direction had an exciting impact on the surface morphology, bending strength, print resolution, and microbial reaction. Furthermore, Lin et al. [30] fabricated ten diverse photopolymerization resins using an ultraviolet 3D printer. The surface hardness values, the bending moduli, and bending strengths of the printed samples were obtained in an acceptable range as well as for clinical resin materials. These results could offer a hopeful perspective for 3D printing and other probable applications.

Despite the variety of available polymer materials produced by AM, pure polymer products suffer from a lack of strength and functionality. Currently, the use of polymer composites has been significantly developed in the case of wearable sensors to overcome these shortcomings [31]. In pursuit of this aim, nanocomposites, fiber-reinforced polymer composites, and particle-reinforced polymer composites have become common today for the development of wearable and flexible sensors with high functional and mechanical properties. Some examples and classifications of reinforcement approaches of polymer composites are given in Table 21.1. In comparison with other polymer composites, particle-reinforced materials are extensively combined with a polymer matrix to improve mechanical and electrical properties due to the low material costs and ease of mixing, encapsulating, attaching, or incorporating particles. These particles as nanofillers increase the electrical conductivity of elastomers even under mechanical stimulus as they

TABLE 21.1

Some Examples and Classifications of Reinforcement Approaches of Polymer Composites

Reinforcement approaches	Properties
Particle-reinforced polymer materials (e.g., $BaTiO_3$, iron, glass bead)	i) Improved electrical conductivity ii) Adjustable resonance frequency iii) Enhanced dielectric permittivity iv) Reduced coefficient of thermal expansion v) Improved storage modulus vi) Reduced elongation at break vii) Improved tensile/compressive modulus
Fiber-reinforced polymer materials (e.g., short carbon fiber or continuous carbon fiber)	Improved compressive modulus and tensile strength property
Polymer nanocomposites (e.g., silver, graphene, graphene oxide (GO))	i) Improved electrical conductivity ii) Enhanced thermal stability iii) Improved tensile modulus, elongation, and strength

can reorganize the conductive networks within the polymer. To accentuate this progress, Nikzad et al. [32] obtained a 3D-printed metal/ acrylonitrile butadiene styrene polymer composite using FDM technology. They showed enhanced stiffness, tensile, thermal, and mechanical properties with the addition of metal fillers. Additionally, Compton et al. [33] prepared an oriented-fiber-filled epoxy resin direct writing technology. The 3D-printed honeycomb structure exhibited high strength, toughness, and performance with the addition of nanomaterials to polymer materials. Another research group [34] obtained graphene-PDMS with different sizes and shapes by 3D printing. The printed constructions can simply be relaxed and compressed.

21.3.2 Conductive Materials

To make conductive patterns, conductive materials can be directly applied on the surface by formulating into printable inks or forming a 3D network by mixing with matrix materials [35]. Numerous kinds of conductive fillers are generally used to form conductive printable materials including metallic materials, liquid metals, conjugated polymers, and carbon-based nanomaterials.

21.3.2.1 Metallic Materials

Metals are promising conductive fillers in structural applications that have been extensively developed by different manufacturing procedures. The current applications of metals can be extended by AM as well as reducing the number of structural components, reducing the production and processing timescale, and forming complex structures. The geometrical properties of metal-based nanomaterials, such as shape and aspect ratio, along with their excellent electrical conductivity, play prominent roles in printing flexible devices with high electrical and mechanical characteristics. Since the features of printing materials are associated with the final properties and microstructures of printed parts, several advanced metallic materials with respectable structural properties have been designed to apply to AM. However, obtaining parts with favored properties or structures is difficult using typical alloys. Thus, taking inspiration from other viewpoints, such as diverse manufacturing methods (e.g., solution methods) is popular, along with the development of novel metallic materials for AM. Although metals show numerous structural applications, they have the potential for functional applications because of the advantages of chemical and physical properties [36]. In the context of translating advanced 3D structures with improved conductivity, materials with higher aspect ratios and low concentrations of metallic fillers have been considered to have potential for printing processes. In wearable devices, a low concentration of fillers is preferred to enhance the optical transparency of the printed electrodes [37]. For example, Kim et al. [38] prepared a stretchable 3D-printed silver-Ecoflex on a thin hydrogel matrix, which shows excellent conductivity. To improve printability and modify the rheology of ink, methyl isobutyl ketone was added. Furthermore, its high stretchability was obtained from the hydrogel materials serving as an energy dissipation matrix. Compared with high-aspect-ratio metal nanowires (NWs), silver and copper NWs are broadly used fillers in conductive inks, owing to their feasibility for scale-up synthesis and high conductivity-mechanical properties. Liquid metals (e.g., EGaIn) have also attracted these solid-state materials due to high surface tensions and the possible occurrence of an oxide layer on the surface [39]. For instance, Wang et al. developed a conductor based on silver flakes and conductive fillers that had a stretchability of 1000% and an initial conductivity of 8331 S/cm with promising cycling durability [40].

21.3.2.2 Carbon-Based Materials

Despite the common use of metals in the 3D printing process for different sensing applications, metallic electrodes are expensive and not biocompatible in some cases. At this state, carbon-based conductive materials as filament sources have been in the spotlight for developing inexpensive biosensing platforms. Carbon-based nanomaterials similar to metallic nanomaterials can also occur in 0D (carbon black), 1D (carbon nanotubes (CNTs)), and 2D (graphene) shapes [41, 42]. Compared with carbon blacks, graphene-based materials show better mechanical and electrical performances. Secor et al. [43] formulated printable inks with ethyl cellulose and graphene and printed the patterns on a stretchy polyimide substrate with high conductivity. An alternative effort has used few-layer graphene as ink for inkjet printing [44]. To formulate high-concentration and stable graphene inks, polymer stabilization and solvent exchange methods were used. The rheological property of these methods displayed compatibility for inkjet printing. Moreover, suitable results were obtained for transmittance and sheet resistance of the printed graphene films. CNTs as conducting and semiconducting fillers are mainly used for forming conductive inks through chemical vapor deposition, laser ablation, and electric arc discharge methods. The superior mechanical properties and lower costs of these carbon-based fillers compared with metal ones make them favorable to use in stretchable/flexible conductors. There is low dispersibility of CNT materials and graphene is the primary matter in a variety of solvents. To overcome this issue, surfactants (e.g., cetryltrimethyl ammonium bromide (CTAB), sodium taurodeoxycholate (STC), sodium cholate (SC)) could be added. Moreover, graphene oxides (GO) could also assist the dispersion of CNTs in an aqueous solution owing to their amphiphilic nature. The advantage of GO addition over surfactants is eliminating the surfactant removal step after film deposition. At present, the combination of conductive carbon materials with metal materials producing hybrid materials leads to significant improvements in co-optimized electrical and mechanical high performance. The stiffness contrast between the hard and soft phases can greatly increase the fracture toughness and disperse energy. Therefore, the development of various conductive composite filaments in wearable electronics and biosensors is of importance.

21.3.2.3 Conjugated Polymers

Conjugated polymers have been directed toward the development of conductors benefitting from their mechanical flexibility and excellent charge transport characteristics. Inherently conductive polymers, namely poly(3,4-ethylene dioxythiophene): polystyrene sulfonate (PEDOT: PSS) and polyaniline (PANI) have aroused much scientific interest in formulating printable inks owing to their chemical stability and solubility in common solvents in composites. PANI as a conducting element in polymer matrix possesses high structural and chemical resistance in harsh conditions without experiencing degradation. Composites or blend formations using various nanomaterials or polymers can effectively improve the mechanical properties of PANI. The electrical properties of conductors-based PANI are significantly influenced by the PANI form (such as emeraldine salt, emeraldine-base, pernigraniline, and leucoemeraldine), level of doping, molecular orientation, polymer chain length, and conjugation. In the context of PEDOT: PSS, conductivity and surface tension can be enhanced by blending with elastomer or modification with additives (e.g., volatile, non-ionic surfactant, Triton X-100, or H_2SO_4 solution) [45]. Endowed with these properties, Yuk et al. fabricated efficient hydrogel networks based on interconnected

PEDOT: PSS nanofibrils [46].]. Aqueous PEDOT: PSS was mixed with the volatile additive dimethyl sulfoxide (DMSO) through controlled rehydration and dry annealing. The formulated hydrogel exhibited excellent mechanical and electrical conductivity, which is highly desired in sensing and bioelectronic applications. They also formulated PEDOT: PSS-based printable ink for the construction of high-performance conducting polymers to measure the in vivo bioelectronic signal [46]. PDMS was used as the substrate and insulation and a conductive polymer was employed as the electronic circuit to fabricate a neural probe. For obtaining high-viscose ink, the PEDOT: PSS was cryogenically frozen before being redispersed in DMSO solutions following a lyophilization process. The formulated polymer ink enables the 3D printing features with good reproducibility, high electrical conductivity, a resolution of over 30 μm, and an excellent aspect ratio. The results revealed that the PEDOT: PSS hydrogel was highly conductive despite the dipping and swelling of dried 3D-printed conducting polymers in the phosphate buffer solution.

21.4 Application of Additive Manufacturing in Flexible and Wearable Sensors

In recent years, flexible electronic interfaces have attracted tremendous attention. Several published works described various wearable/flexible devices like flexible electrodes, strain sensors, nanogenerators, batteries, and capacitors. Owing to the high performance, and material-saving properties of AM, 3D printing makes it feasible for different functional nanomaterials to be used in flexible electronic fields. For mechanical materials, 3D printing could either ensure the material's functionality or increase the robustness of the printed material. Numerous novel types of printable functional nanomaterials pave the way for flexible tools to find utilizations in wearable electronics. In addition, considerable efforts have been put into 3D printing constructions to expand the applications in wearable electronics. Recently, various studies have concentrated on the wearable electronics manufactured via AM, including flexible substrates for electrocardiogram signal determinations, prosthetic skin incorporated with multimodal strain probes, wearable scaffolds for electroencephalogram signal quantification, wearable electrodes for electromyography signal measurements, 3D-printed soft electrodes with conducting polymer, ultra-flexible 3D-triboelectric nanogenerators, and wearable electrodes for electrodermal activity. Therefore, this section summarizes the latest developments in 3D-printing materials in flexible/wearable electronics.

21.4.1 Strain Sensors

Among 3D printing approaches, DIW provides facilities for the manufacture of stretchable devices. Zhao's group designed stable flexible strain sensors based on conductive silicon rubber (CSR) filled with silver-coated-glass fiber (AGF) [47]. The DIW method has been employed to surround the detecting substrates, for example, AGF, in the PDMS matrix as shown in Figure 21.2. The printed arrangement has been evaluated in various strain rates, which results in durability of up to 600 cycles and a stable gauge factor of 8–10. The printed flexible strain probe indicated a promising application in wearable electronics and soft robotics. Guo and colleagues introduced a stretchable tool using silver nanoparticle

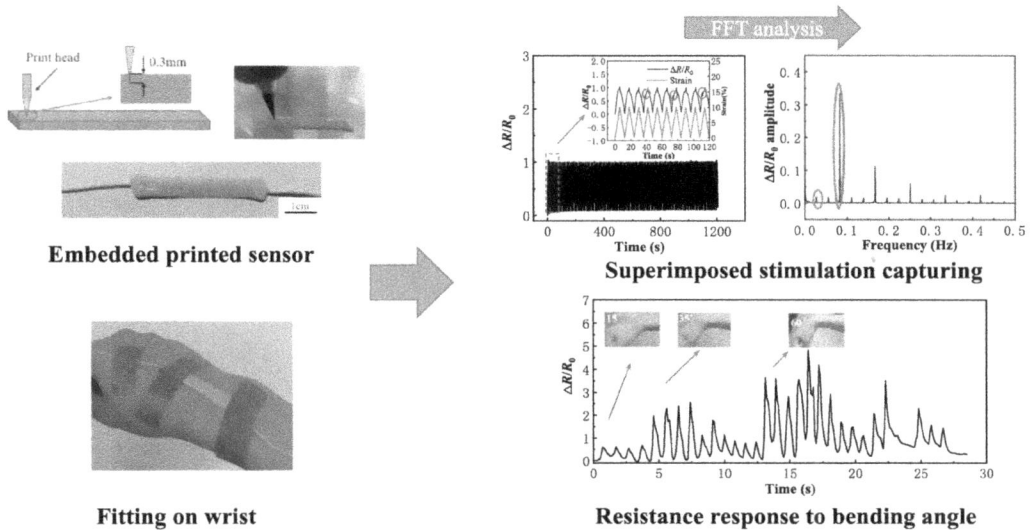

FIGURE 21.2
Fabrication procedure of flexible strain sensor using silicon rubber (CSR) filled with carbon fiber (CF) and silver-coated-glass fiber (AGF), which was embedded into a polydimethylsiloxane (PDMS) matrix based on 3D printing. Adapted with permission [47]. Copyright (2020) Elsevier.

conductive inks [48]. A multiscale, multi-material, and multifunctional 3D printing tactic has been adopted to print tactile sensors. The tactile probes are created via a spiral sensor part, a base layer, an isolating layer, two electrodes, and a supporting constituent. The multilayer sensor was established to be capable of sensing and discriminating human movements, comprising finger motions and pulse monitoring. The SEM outcomes indicated that multiscale buildings were presented to enhance the sensitivity of the tactile probe. In another study, Le's group developed a 3D printing strain sensor by constructing a dipole antenna assembly on a 3D-printed Ninjaflex filament platform. An FDM 3D printer is engaged to print the Ninjaflex filament substrate, and to produce a fully functional antenna, electrically conductive adhesive is applied to fill the substrate cavities [49]. Votzke's group established 3D printing of strain sensors based on liquid metal paste. In this study, gallium-based liquid metal alloys are used as the sensor substrates, encapsulating most of them in silicon sheets to perform leak-proof operations. The designed probe has almost near-zero hysteresis under 375 cycles at 200% strain. The practical utilization of the suggested strain sensor and related outcomes prepared the ground for the detection of human joint motion [50].

The 3D printing capability of silicone elastomers such as long curing time and low viscosity was solved by the Zhou group, which utilized nano-silica particles, by improving curing time and rheology [51]. The modified elastomer can simply be printed using the DIW technique to make a high-stretch elastomeric cavity. Furthermore, a strain sensor can be made by filling the elastomer cavities with conductive ink grease. The accuracy and speed of the DIW technique have been boosted by engaging a theoretical model. The printing procedure as well as the corresponding signal of different hand movements are shown in Figure 21.3. The designed platform can be applied for sensing human motions and can also be employed in soft robotics.

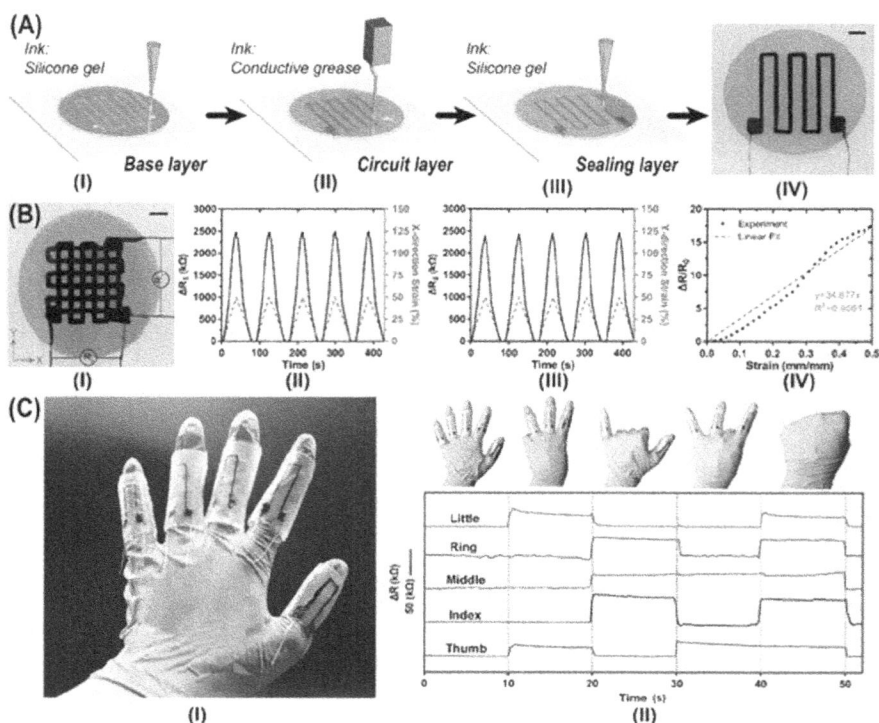

FIGURE 21.3
(A) The fabrication process of the strain sensor. (B) Optical performance and photograph of a printed biaxial flexible electronic. (C) Optical photograph and its electrical resistance change as a function of time for strain sensors within the glove. Adapted with permission [51]. Copyright (2019) American Chemical Society.

21.4.2 Nanogenerators

Nanogenerators of electrical energy can be fabricated by an effective material usage AM in a short period. Triboelectric nanogenerators (TENGs) as developing devices have revealed a considerable improvement over the past few years. Regarding the coupling effects of electrostatic induction and triboelectrification, TENGs can change mechanical motions into electricity [52]. Chen et al. [53] explored a 3D ultra-flexible TENG hybrid that was composed of ionic hydrogel and composite resin elements for designing wearable sensors. In an alternative effort, they fabricated a single integrated TENG made of CNTs and poly glycerol sebacate as the two electrification components [54]. The conductive CNTs acted as not only electrodes but also as a flexible matrix. It has been revealed that enhanced power output could be provided by introducing nano- and microscale topographies regarding increased surface area rather than that of traditionally molded counterparts. The hierarchical porous structures such as aerogels with high specific surface area and ultralow density are promising materials. At this state, Qian et al. [55] 3D-printed TENGs based on cellulose nanofiber (CNF) aerogels with hierarchical nano/micro-3D structure (AP-TENGs). Benefitting from the 3D nano/micro hierarchically patterned structure reached after 3D printing and subsequent drying, the effectivity, mechanical resilience, and surface roughness of the device were enhanced. This enhancement contributes to triboelectric response improvement with higher voltage output (Figure 21.4).

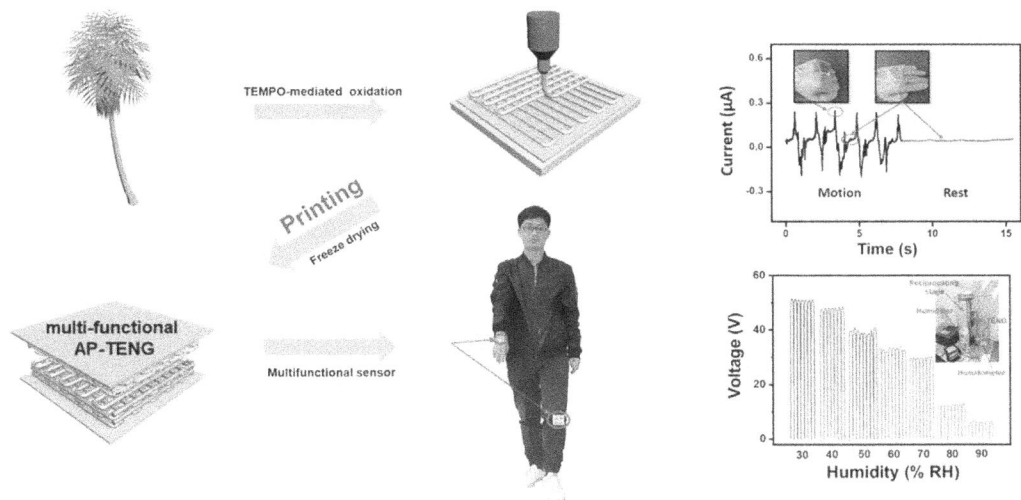

FIGURE 21.4
The whole 3D printing process of AP-TENG and the electrical and mechanical properties. Adapted with permission [55]. Copyright (2019) Elsevier.

Furthermore, Ma et al. [56] utilized an rGO-MnO$_2$-based supercapacitor and TENG in a self-powered system that could convert the mechanical signals into electrical power and additionally store such scavenged power. The measured transfer charge, short-circuit current, and open-circuit voltage of TENG show that rectifying the direct electrical outputs from the TENG is necessary. Moreover, the highest instantaneous power output was achieved when the external load matched the internal impendence of TENG. TENG can successfully harvest the mechanical inputs from finger pressing and convert mechanical energy to electrical energy. The holistic incorporation of storage and energy harvesting units promises the employment of self-powered wearable devices with superior aptitude.

21.4.3 Flexible Electrodes

To monitor physiological indicators, electrodes are widely used. Flexible electrodes can continuously record these indicators in a non-invasive way. To fabricate flexible electrodes, 3D printing technology affords a new technique by introducing excellent multilevel structures. Ho et al. [57] reported a novel construction procedure for 3D microlevel conductive materials. An interconnected sugar scaffold was prepared through 3D powder bed printing. For producing flexible porous electrodes, conductive silicone elastomers were integrated into the scaffold with subsequent elimination of the scaffold. The data of the body outlines with a high resolution used in the 3D printing endow the flexible electrode to be close-fitting, precise, and sensitive. These designed electrodes record the physiological signals of the body successfully. Three-dimensional printing structures adopt a vast number of advantages, such as multilevel fabrication capabilities, material saving, and precise control compared with conventional manufacturing technology. Current studies illustrate the robustness of 3D-printed construction materials in wearable devices. For example, printable elastomers with special constructions are further sensitive to faint distortion, which is respectable for various kinds of sensors. Furthermore, the precise control of the 3D structure with a state-of-art 3D scanning procedure is significant for improving the precision and sensitivity of flexible electrodes. On account of the electrical properties of printing ink used in wearable devices, the performances of 3D-printed materials are almost the same as conventional electronics. Accordingly, improvement of electrical performance along with optimizing printable characteristics should be further explored. Despite recent challenges, enormous efforts have been made toward the development of 3D-printed wearable electronics, such as the introduction of functional and soft structures designed for wearable electronics as a result of recent technological progress and material innovations.

21.4.4 Soft Sensors, Actuators, and Robotics

AM can offer a hopeful perspective to be used in soft sensor, actuator, and robotic applications. Three-dimensional-printed sensors detecting and responding to environmental stimuli can be classified into two main types. First, 3D-printed sensors are fabricated from materials that respond to external stimuli and generate signals directly. Second, 3D-printed sensors as the matrix to support responsive substances (e.g., living cells, biomolecules, and nanoparticles). For instance, Ze et al. [58] designed novel shape memory polymer composites that demonstrated adjustable mechanical properties under the actuation of magnetic fields. The created grippers were confirmed to be efficient actuators, demonstrating great potential for extensive applications. In addition, 3D or 4D printed soft robotics have caught the spotlight owing to their diversity and flexibility. In this regard, Wehner et al. [59] fabricated a completely soft autonomous robot using an integrated fabrication strategy and design. Here, soft lithography and 3D printing were employed to design diverse parts of the robot with integrated capability.

21.5 Conclusion and Future Perspectives

Efforts were made to describe the various characteristics of advanced 3D printing methods developed in the emerging next generation of wearable tools. The feasibility of 3D printing technology is obvious in its various features, which include rapid and on-demand construction abilities, facile operation, and accurate and controlled multifunctional material deposition into customized 3D-printed buildings. In light of these benefits, this developed printing technology can be used to build patient-specific geometry to design new wearable substrates, either directly on the skin or other analyte substrates using functional soft materials. In this regard, 3D printing methods, particularly material extrusion, material jetting, and photopolymerization, have been employed in fabricating a wide range of wearable nanoprobes. It can be noted that the choice of an appropriate material regarding the specific 3D printing method is challenging in the manufacturing procedure to obtain an efficient and fully integrated wearable device. For the progress of wearable devices, further 3D printing technologies can be developed to incorporate new printable multi-materials having excellent conductivity, self-healing capability, recyclability, flexibility, and stretchability in designing 3D architectures. In such a background, advances in material science are possible to lead to rapid developments in properties like stability and multi-functionality of the desired printed wearable devices.

Acknowledgments

The authors would like to acknowledge financial support from the Pharmaceutical Analysis Research Center, Tabriz University of Medical Science (Tabriz, Iran) (registration no: 68444).

References

1. Z. Karimzadeh, M. Mahmoudpour, E. Rahimpour, A. Jouyban, Nanomaterial based PVA nanocomposite hydrogels for biomedical sensing: Advances toward designing the ideal flexible/wearable nanoprobes, *Advances in Colloid and Interface Science* 305 (2022) 102705.
2. M. Dervisevic, M. Alba, B. Prieto-Simon, N.H. Voelcker, Skin in the diagnostics game: Wearable biosensor nano-and microsystems for medical diagnostics, *Nano Today* 30 (2020) 100828.
3. M. Mahmoudpour, A. Saadati, M. Hasanzadeh, H. Kholafazad-Kordasht, A stretchable glove sensor toward rapid monitoring of trifluralin: A new platform for the on-site recognition of herbicides based on wearable flexible sensor technology using lab-on-glove, *Journal of Molecular Recognition* 34(10) (2021) e2923.
4. Z. Zhu, S.Z. Guo, T. Hirdler, C. Eide, X. Fan, J. Tolar, M.C. McAlpine, 3D printed functional and biological materials on moving freeform surfaces, *Advanced Materials* 30(23) (2018) 1707495.
5. P. Wei, H. Leng, Q. Chen, R.C. Advincula, E.B. Pentzer, Reprocessable 3D-printed conductive elastomeric composite foams for strain and gas sensing, *ACS Applied Polymer Materials* 1(4) (2019) 885–892.
6. Z. Karimzadeh, H. Namazi, Nontoxic double-network polymeric hybrid aerogel functionalized with reduced graphene oxide: Preparation, characterization, and evaluation as drug delivery agent, *Journal of Polymer Research* 29(2) (2022) 37.
7. Z. Karimzadeh, M. Mahmoudpour, M.d.l. Guardia, J. Ezzati Nazhad Dolatabadi, A. Jouyban, Aptamer-functionalized metal organic frameworks as an emerging nanoprobe in the food safety field: Promising development opportunities and translational challenges, *TrAC Trends in Analytical Chemistry* 152 (2022) 116622.
8. M. Mahmoudpour, A. Jouyban, J. Soleymani, M. Rahimi, Rational design of smart nano-platforms based on antifouling-nanomaterials toward multifunctional bioanalysis, *Advances in Colloid and Interface Science* 302 (2022) 102637.
9. D.J. Lipomi, M. Vosgueritchian, B.C. Tee, S.L. Hellstrom, J.A. Lee, C.H. Fox, Z. Bao, Skin-like pressure and strain sensors based on transparent elastic films of carbon nanotubes, *Nature Nanotechnology* 6(12) (2011) 788–792.
10. S. Abdollahi, E.J. Markvicka, C. Majidi, A.W. Feinberg, 3D printing silicone elastomer for patient-specific wearable pulse oximeter, *Advanced Healthcare Materials* 9(15) (2020) 1901735.
11. K. Nagamine, A. Nomura, Y. Ichimura, R. Izawa, S. Sasaki, H. Furusawa, H. Matsui, S. Tokito, Printed organic transistor-based biosensors for non-invasive sweat analysis, *Analytical Sciences* 36 (2020) 19R007.
12. S.K. Saha, D. Wang, V.H. Nguyen, Y. Chang, J.S. Oakdale, S.-C. Chen, Scalable submicrometer additive manufacturing, *Science* 366(6461) (2019) 105–109.
13. G. Liu, X. Zhang, X. Chen, Y. He, L. Cheng, M. Huo, J. Yin, F. Hao, S. Chen, P. Wang, S. Yi, L. Wan, Z. Mao, Z. Chen, X. Wang, Z. Cao, J. Lu, Additive manufacturing of structural materials, *Materials Science and Engineering: R: Reports* 145 (2021) 100596.
14. D.A. Walker, J.L. Hedrick, C.A. Mirkin, Rapid, large-volume, thermally controlled 3D printing using a mobile liquid interface, *Science* 366(6463) (2019) 360–364.
15. A. Lee, A. Hudson, D. Shiwarski, J. Tashman, T. Hinton, S. Yerneni, J. Bliley, P. Campbell, A. Feinberg, 3D bioprinting of collagen to rebuild components of the human heart, *Science* 365(6452) (2019) 482–487.

16. Y.-Z. Zhang, Y. Wang, T. Cheng, L.-Q. Yao, X. Li, W.-Y. Lai, W. Huang, Printed supercapacitors: Materials, printing and applications, *Chemical Society Reviews* 48(12) (2019) 3229–3264.
17. Y. Kim, H. Yuk, R. Zhao, S.A. Chester, X. Zhao, Printing ferromagnetic domains for untethered fast-transforming soft materials, *Nature* 558(7709) (2018) 274–279.
18. T.R. Ray, J. Choi, A.J. Bandodkar, S. Krishnan, P. Gutruf, L. Tian, R. Ghaffari, J.A. Rogers, Bio-integrated wearable systems: A comprehensive review, *Chemical Reviews* 119(8) (2019) 5461–5533.
19. N. Ahmad, P. Gopinath, A. Vinogradov, 3D printing in medicine: Current challenges and potential applications, in *3D Printing Technology in Nanomedicine*, Elsevier, 2019, pp. 1–22.
20. M. Bernard, E. Jubeli, M.D. Pungente, N. Yagoubi, Biocompatibility of polymer-based biomaterials and medical devices–regulations, in vitro screening and risk-management, *Biomaterials Science* 6(8) (2018) 2025–2053.
21. S.P. Sreenilayam, I.U. Ahad, V. Nicolosi, V.A. Garzon, D. Brabazon, Advanced materials of printed wearables for physiological parameter monitoring, *Materials Today* 32 (2020) 147–177.
22. N. Matsuhisa, D. Inoue, P. Zalar, H. Jin, Y. Matsuba, A. Itoh, T. Yokota, D. Hashizume, T. Someya, Printable elastic conductors by in situ formation of silver nanoparticles from silver flakes, *Nature Materials* 16(8) (2017) 834–840.
23. X. Wang, M. Jiang, Z. Zhou, J. Gou, D. Hui, 3D printing of polymer matrix composites: A review and prospective, *Composites Part B: Engineering* 110 (2017) 442–458.
24. J. Yang, Q. Chen, F. Chen, Q. Zhang, K. Wang, Q. Fu, Realizing the full nanofiller enhancement in melt-spun fibers of poly (vinylidene fluoride)/carbon nanotube composites, *Nanotechnology* 22(35) (2011) 355707.
25. P.A. Eutionnat-Diffo, Y. Chen, J. Guan, A. Cayla, C. Campagne, X. Zeng, V. Nierstrasz, Stress, strain and deformation of poly-lactic acid filament deposited onto polyethylene terephthalate woven fabric through 3D printing process, *Scientific Reports* 9(1) (2019) 1–18.
26. M. Abshirini, M. Charara, P. Marashizadeh, M.C. Saha, M.C. Altan, Y. Liu, Functional nanocomposites for 3D printing of stretchable and wearable sensors, *Applied Nanoscience* 9(8) (2019) 2071–2083.
27. Z. He, Y. Chen, J. Yang, C. Tang, J. Lv, Y. Liu, J. Mei, W.-m. Lau, D. Hui, Fabrication of Polydimethylsiloxane films with special surface wettability by 3D printing, *Composites Part B: Engineering* 129 (2017) 58–65.
28. A. Bîrcă, O. Gherasim, V. Grumezescu, A.M. Grumezescu, Introduction in thermoplastic and thermosetting polymers, in *Materials for Biomedical Engineering*, Elsevier, 2019, pp. 1–28.
29. J.S. Shim, J.-E. Kim, S.H. Jeong, Y.J. Choi, J.J. Ryu, Printing accuracy, mechanical properties, surface characteristics, and microbial adhesion of 3D-printed resins with various printing orientations, *The Journal of Prosthetic Dentistry* 124(4) (2020) 468–475.
30. C.-H. Lin, Y.-M. Lin, Y.-L. Lai, S.-Y. Lee, Mechanical properties, accuracy, and cytotoxicity of UV-polymerized 3D printing resins composed of Bis-EMA, UDMA, and TEGDMA, *The Journal of Prosthetic Dentistry* 123(2) (2020) 349–354.
31. Y. Liu, W. Zhang, F. Zhang, J. Leng, S. Pei, L. Wang, X. Jia, C. Cotton, B. Sun, T.-W. Chou, Microstructural design for enhanced shape memory behavior of 4D printed composites based on carbon nanotube/polylactic acid filament, *Composites Science and Technology* 181 (2019) 107692.
32. M. Nikzad, S.H. Masood, I. Sbarski, Thermo-mechanical properties of a highly filled polymeric composites for fused deposition modeling, *Materials & Design* 32(6) (2011) 3448–3456.
33. B.G. Compton, J.A. Lewis, 3D-printing of lightweight cellular composites, *Advanced Materials* 26(34) (2014) 5930–5935.
34. K. Huang, S. Dong, J. Yang, J. Yan, Y. Xue, X. You, J. Hu, L. Gao, X. Zhang, Y. Ding, Three-dimensional printing of a tunable graphene-based elastomer for strain sensors with ultrahigh sensitivity, *Carbon* 143 (2019) 63–72.
35. S. Choi, S.I. Han, D. Kim, T. Hyeon, D.-H. Kim, High-performance stretchable conductive nanocomposites: Materials, processes, and device applications, *Chemical Society Reviews* 48(6) (2019) 1566–1595.

36. K. Chen, R. Huang, Y. Li, S. Lin, W. Zhu, N. Tamura, J. Li, Z.W. Shan, E. Ma, Rafting-enabled recovery avoids recrystallization in 3D-printing-repaired single-crystal superalloys, *Advanced Materials* 32(12) (2020) 1907164.
37. C. Tan, M.Z.M. Nasir, A. Ambrosi, M. Pumera, 3D printed electrodes for detection of nitroaromatic explosives and nerve agents, *Analytical Chemistry* 89(17) (2017) 8995–9001.
38. S.H. Kim, S. Jung, I.S. Yoon, C. Lee, Y. Oh, J.M. Hong, Ultrastretchable conductor fabricated on skin-like hydrogel–elastomer hybrid substrates for skin electronics, *Advanced Materials* 30(26) (2018) 1800109.
39. A.R. Jacob, D.P. Parekh, M.D. Dickey, L.C. Hsiao, Interfacial rheology of gallium-based liquid metals, *Langmuir* 35(36) (2019) 11774–11783.
40. J. Wang, G. Cai, S. Li, D. Gao, J. Xiong, P.S. Lee, Printable superelastic conductors with extreme stretchability and robust cycling endurance enabled by liquid-metal particles, *Advanced Materials* 30(16) (2018) 1706157.
41. Z. Li, L. Wang, Y. Li, Y. Feng, W. Feng, Carbon-based functional nanomaterials: Preparation, properties and applications, *Composites Science and Technology* 179 (2019) 10–40.
42. M. Mahmoudpour, J.E.-N. Dolatabadi, M. Hasanzadeh, J. Soleymani, Carbon-based aerogels for biomedical sensing: Advances toward designing the ideal sensor, *Advances in Colloid and Interface Science* 298 (2021) 102550.
43. E.B. Secor, P.L. Prabhumirashi, K. Puntambekar, M.L. Geier, M.C. Hersam, Inkjet printing of high conductivity, flexible graphene patterns, *The Journal of Physical Chemistry Letters* 4(8) (2013) 1347–1351.
44. J. Li, F. Ye, S. Vaziri, M. Muhammed, M.C. Lemme, M. Östling, Efficient inkjet printing of graphene, *Advanced Materials* 25(29) (2013) 3985–3992.
45. J. Guo, Y. Yu, H. Wang, H. Zhang, X. Zhang, Y. Zhao, Conductive polymer hydrogel microfibers from multiflow microfluidics, *Small* 15(15) (2019) 1805162.
46. H. Yuk, B. Lu, S. Lin, K. Qu, J. Xu, J. Luo, X. Zhao, 3D printing of conducting polymers, *Nature Communications* 11(1) (2020) 1–8.
47. C. Zhao, Z. Xia, X. Wang, J. Nie, P. Huang, S. Zhao, 3D-printed highly stable flexible strain sensor based on silver-coated-glass fiber-filled conductive silicon rubber, *Materials & Design* 193 (2020) 108788.
48. S.Z. Guo, K. Qiu, F. Meng, S.H. Park, M.C. McAlpine, 3D printed stretchable tactile sensors, *Advanced Materials* 29(27) (2017) 1701218.
49. T. Le, B. Song, Q. Liu, R.A. Bahr, S. Moscato, C.-P. Wong, M.M. Tentzeris, A novel strain sensor based on 3D printing technology and 3D antenna design, in *2015 IEEE 65th Electronic Components and Technology Conference (ECTC)*, IEEE, 2015, pp. 981–986.
50. C. Votzke, U. Daalkhaijav, Y. Mengüç, M.L. Johnston, Highly-stretchable biomechanical strain sensor using printed liquid metal paste, in *2018 IEEE Biomedical Circuits and Systems Conference (BioCAS)*, IEEE, 2018, pp. 1–4.
51. L.-y. Zhou, Q. Gao, J.-z. Fu, Q.-y. Chen, J.-p. Zhu, Y. Sun, Y. He, Multimaterial 3D printing of highly stretchable silicone elastomers, *ACS Applied Materials & Interfaces* 11(26) (2019) 23573–23583.
52. C. Wu, H. Tetik, J. Cheng, W. Ding, H. Guo, X. Tao, N. Zhou, Y. Zi, Z. Wu, H. Wu, Electrohydrodynamic jet printing driven by a triboelectric nanogenerator, *Advanced Functional Materials* 29(22) (2019) 1901102.
53. B. Chen, W. Tang, T. Jiang, L. Zhu, X. Chen, C. He, L. Xu, H. Guo, P. Lin, D. Li, Three-dimensional ultraflexible triboelectric nanogenerator made by 3D printing, *Nano Energy* 45 (2018) 380–389.
54. S. Chen, T. Huang, H. Zuo, S. Qian, Y. Guo, L. Sun, D. Lei, Q. Wu, B. Zhu, C. He, A single integrated 3D-printing process customizes elastic and sustainable triboelectric nanogenerators for wearable electronics, *Advanced Functional Materials* 28(46) (2018) 1805108.
55. C. Qian, L. Li, M. Gao, H. Yang, Z. Cai, B. Chen, Z. Xiang, Z. Zhang, Y. Song, All-printed 3D hierarchically structured cellulose aerogel based triboelectric nanogenerator for multi-functional sensors, *Nano Energy* 63 (2019) 103885.

56. C. Ma, R. Wang, H. Tetik, S. Gao, M. Wu, Z. Tang, D. Lin, D. Ding, W. Wu, Hybrid nanomanufacturing of mixed-dimensional manganese oxide/graphene aerogel macroporous hierarchy for ultralight efficient supercapacitor electrodes in self-powered ubiquitous nanosystems, *Nano Energy* 66 (2019) 104124.
57. D.H. Ho, P. Hong, J.T. Han, S.Y. Kim, S.J. Kwon, J.H. Cho, 3D-printed sugar scaffold for high-precision and highly sensitive active and passive wearable sensors, *Advanced Science* 7(1) (2020) 1902521.
58. Q. Ze, X. Kuang, S. Wu, J. Wong, S.M. Montgomery, R. Zhang, J.M. Kovitz, F. Yang, H.J. Qi, R. Zhao, Magnetic shape memory polymers with integrated multifunctional shape manipulation, *Advanced Materials* 32(4) (2020) 1906657.
59. M. Wehner, R.L. Truby, D.J. Fitzgerald, B. Mosadegh, G.M. Whitesides, J.A. Lewis, R.J. Wood, An integrated design and fabrication strategy for entirely soft, autonomous robots, *Nature* 536(7617) (2016) 451–455.

22

Current Challenges and Perspectives of Flexible and Wearable Sensors

Shubham Mishra, Tania K. Naqvi, and Prabhat K. Dwivedi

CONTENTS

22.1 Introduction

Wearable devices are next-generation sensing tools, generally in support of smartphone applications, which are primarily focused on sensing, gathering, and processing untapped data of the human body for health, behavior, and fitness monitoring. From very simple adhesive bandages to high-end sensors for physiological monitoring wearable devices, the development of wearable technology is depicted in Figure 22.1. Although wearable device development has had a long journey, it is still in the nascent stage of internet-enabled wearable technology. Recently, the technology domain has focused on the proliferation of internet and communication technology (ICT)-enabled interconnected small sensing devices termed Internet of Things (IoT) devices. Low-cost, lightweight, compact IoT devices have penetrated various domains of human life to obtain real-time analysis of conditions. This chapter is mainly focused on the non-conventional emerging aspects of wearable technology. Starting with development and state-of-the-art wearable devices, we move toward non-conventional wearable technology after a brief section on conventional applications

DOI: 10.1201/9781003299455-22

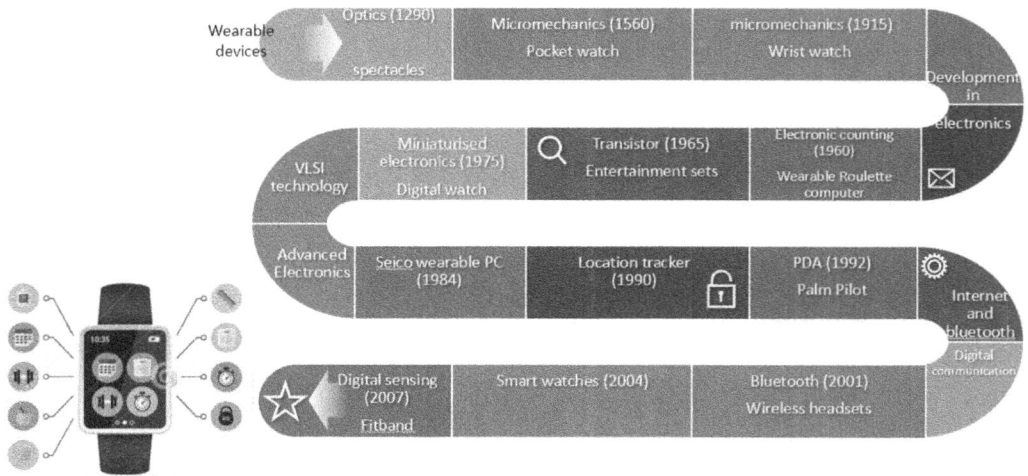

FIGURE 22.1
Different phases in the development of wearable devices.

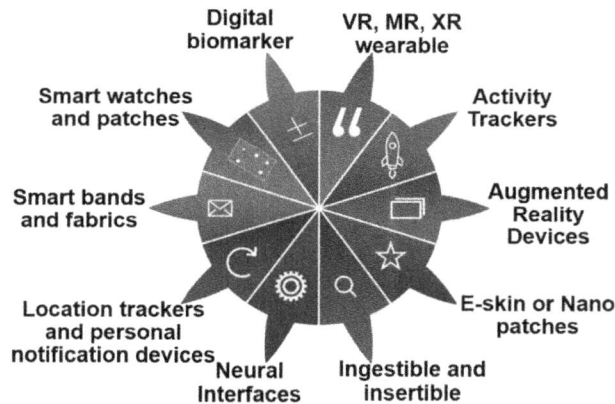

FIGURE 22.2
Classification of wearable devices based on their functionality and applications.

and technologies. Non-conventional wearable technology is studied from various aspects of sensing technologies such as materials, applications, sensing modules and domains, fabrication techniques, and integration methods. Various emerging technologies and possibilities are discussed from each point of view. Further, a future direction of the development of wearable technology from different technical aspects and emerging consumer demands and interest in wearable products is discussed. Finally, an overview of challenges in such developments, technical and ethical concerns are discussed.

22.1.1 Classification of Wearable Devices

Wearable devices range from a variety of applications from healthcare monitoring to augmented reality-based human interaction. Wearable devices have been developed to a stage where the real world can be successfully merged with the digital world. A holistic overview of the different applications of wearable devices is depicted in Figure 22.2. Progress in different technologies such as nanofabrication, imaging, and spectroscopy have opened a wide possibility of device development and fabrication, however, digital communication and internet-enabled technologies have impacted on this growth more than ever before. Wearable sensors can be classified according to low, medium, and high power requirements, type of battery used, and type of functionality, however, they do not impart much inferential information.

Sensing modules are the most important constituent of wearable technology and are dominated by electromagnetic or optical sensing techniques. The development of sensing module is especially important in the healthcare domain because of variation in individual physiological conditions. Physiological monitoring includes biological, physical, or chemical activity in the body, such as cardiac activity, ECG, uric acid, lactic acid, and protein deficiency causing Alzheimer's disease, Parkinson's disease, and so on. While electromagnetic sensors need to deal with data handling aspects, Biochemical sensing modules are still in the developmental phase and need to be miniaturized, flexible, and rapid at the same time, as well as with effective data handling. Early physiological wearable systems were based on discrete sensing module, data storage and computing systems, however, recently, integration of all the components is seen as a more effective way for data handling and power optimization. Sensing and data collection modules can be two types: (a) active sensing – when the user is aware of device functionality, such as spectroscopic sensing and body-attached injectable sensing; (b) passive sensing – when the user is unaware of device functionality, such as sleep and pulse monitoring. Here, we will discuss different technological aspects of sensing modules for diseases and their advantages/disadvantages.

22.2 The Architecture of Wearable Sensing Devices

Typical wearable devices consist of a sensing, data handling, and end-user display module. The most important part of the sensing mechanism of wearable devices is the sensors. Sensors are designed to sense accurate reliable contextual information in a light, compact, and flexible setup to become suitable for wearable devices. A broader understanding of sensing mechanisms can be seen as the period of their development:

(a) Conventional sensing methods: motion sensors, electromagnetic pulse sensing
(b) Non-conventional sensors: strain/stretchable sensors, self-healing, self-powered sensors, photosensors, and humidity sensors
(c) Digital biomarkers

22.2.1 Conventional Sensors

Wearable sensors were introduced in 2000 with the accelerometer and gyroscope-based motion sensors are still used in some devices. Actigraph and SHIMMER are actively used in physiological activity monitoring devices, however, their power requirements, sensitivity, and accuracy pose problems in proper data assessment. Recent progress is focused on the digitalization of data processing and figuring out more informed output. A structured data assessment using Artificial Intelligence/Machine Learning (AI/ML) of sleep and anxiety data can be used for cardiac risk assessment. Gait and motion analysis of limbs and chest can give significant information about Parkinson's or asthma. Such sensing technologies in wearable devices are almost on par with the standard lab setup of photoplethysmographs or polysomnographs and are miniaturized and flexible enough to be compatible with wearable devices. Major digital advancement is required in such wearable devices, including big data analysis of the data provided, communication, and interoperability of multiple systems and devices.

22.2.2 Non-Conventional Sensors

The ever-increasing demand for wearable sensors for a multitude of applications has proportionately increased the requirements and criteria for the implementation and development of various sensing technologies to be used for acquiring and communicating physical data from the body and surroundings. In the last few years, we have seen extensive research in the field of advanced functional metamaterials, resulting in the development of strain-based stretchable sensors, self-healing and self-powered sensors, photosensors, and humidity sensors. These kinds of sensors are well developed and are in use in wearable and flexible devices. However, the potential of these technologies is yet to be utilized. With advancements in nanofabrication and material development, a vast majority of sensing modules can be developed using such sensing techniques. Some of the major perspectives on non-conventional sensing techniques are discussed in the following sections.

22.2.2.1 Non-Conventional Materials

Generally, physical sensors rely on their variation in conductivity, capacitance, piezoelectricity, optical transmittance, pressure, and temperature in comparison with a reference. Flexible and wearable sensors need to be accurate and reliable without hindering comfort and the motion of body parts. To achieve skin-like conformability and flexibility, polymeric materials have been extremely useful like polyurethane (PU), polycarbonate (PC), polyethylene Terephthalate (PET), and so on. Polymeric materials facilitate extremely tailored properties such as transparency, deformability, and ease of fabrication. In recent trends, the most suitable polymeric materials for wearable and flexible sensor developments have proved to be silicone elastomers, especially polydimethylsiloxane (PDMS). Elastomeric polymers such as PDMS provide high deformability, optical transparency, and integration of active sensing elements into the PDMS substrate. Table 22.1 summarizes different classes of sensing materials and manufacturing processes used in the fabrication of smart wearables.

Table 22.1 provides a fair idea of the selection of materials for a particular application. Advancements in the fields of nanomaterials, ionic liquids, and manufacturing processes are ongoing and are continually developing a large variety of materials and integration processes for tailored properties.

22.2.2.2 Non-Conventional Applications

Non-conventional applications of wearable technology can be seen in the clinical and non-clinical domains. Non-clinical wearable items are generally intended for monitoring processes when required, whereas clinical wearables are meant for monitoring human health in real time for extended hours. A schematic representation of emerging wearable technology applications is shown in Figure 22.3.

TABLE 22.1

Summary of Materials and Processes Commonly Used for Flexible and Stretchable Physical Sensing Devices

Class of materials	Flexible template	Young modulus of elasticity (MPa)	Tensile strain	Poisson's ratio	Processing temperature
Polymer	PET	2000 – 4100	<5	0.3 – 0.45	70
	PC	2600 – 3000	<1	0.37	150
	PU	10–50	<100	0.48–0.499	80,120
	PEN	5000–5500	<3	0.3–0.37	270
	PI	2500–10,000	<5	0.34–0.48	70–80
Silicone elastomers	PDMS	0.36–0.87	>200	0.4999	25
	Silbione	0.005	>250	0.4999	25
	Ecoflex	0.02–0.25	>300	0.4999	25
	Active sensing element	**Structure form**		**Size**	**Sheet resistance**
Conductive materials	Metallic nanomaterials (Ag, Au, Cu, Mn, etc.)	Nanoparticles, nanowires nanorods		2–400 nm in dia and 200–1000 nm (length)	0.015–20 Ω sq^{-1}
	Carbon-based nanomaterials	Nanoparticles, nanowires nanorods		10–2000 nm in dia and 500–5000 nm (length)	$5–30 \times 10^6$ Ω sq^{-1}
	Ionic or metallic liquids	Liquid		NA	$2.63 \times 10^{-9} – 0.025 \times \Omega$ cm^{-1}
	Fabrication technique	**Resolution**		**Throughput m/ min**	**Limitation**
Additive process	Gravure printing	50–500		8–100	Limited resolution due to alignment
	Screen printing	30–700		0.6–100	A small selection of inks due to high viscosity
	Inkjet printing	15–100		0.02–5	Not suitable for roll-to-roll production, the coffee ring effect

Adapted with permission [1]. Copyright The Authors, some rights reserved; exclusive licensee Nature. Distributed under a Creative Commons Attribution License 4.0 (CC BY).

22.2.2.2.1 Biomedical

Flexible and wearable devices have been used for biomedical applications since their invention. However, in recent trends, the scope of applications of wearable devices has diversified due to the development of a new class of materials and data processing techniques. Table 22.2 summarizes emerging biomedical applications with the advent of new technologies and materials.

Design and integration of sensing platforms for physiological monitoring and assessment are continually developing to increase spatial resolution, miniaturize device configuration, and increase sensitivity. In the near future, we will see a shift toward digital processing of real-time physiological data acquired through a wearable sensor.

22.2.2.2.2 Biosensing

Physiological monitoring and assessment have opened a new domain of sensing and biomedical applications through wearable devices. Changes in skin resistance, gait motion,

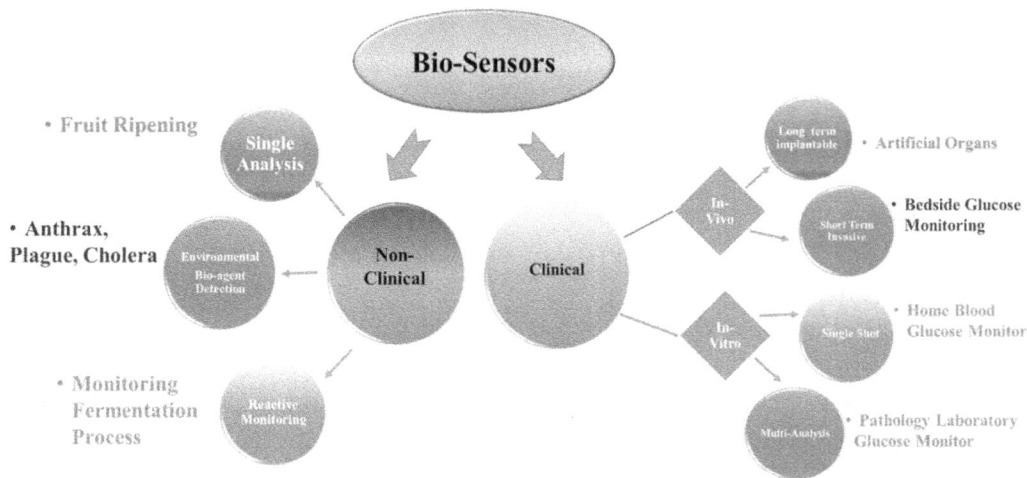

FIGURE 22.3
Types and application domains of biosensors.

TABLE 22.2

Emerging Biomedical Applications of Flexible and Wearable Devices

Application	Material/technology
Artificial electronic skin	Mechano and thermoreceptors-based sensory patches.
Therapeutic and drug delivery platforms	Stretchable and wearable heating element-based sensors.
Physiological monitoring and assessment systems	Spatial-temporal monitoring and assessment of physiological data.

and various body parts in physical activity can impart significant information about diseases like skin diseases, Alzheimer's disease, and Parkinson's disease. Development in big data machine learning and data analysis has also played a key role in the assessment of physiological data to infer significant and useful information.

22.2.2.2.3 Environmental

Environmental wearable sensors are mainly designed for blind persons to sense the ambient within meters and determine local motion and behavior. The visual, thermal, and acoustic environment around the wearable device wearer is an important data need to be monitored in various cases [3]. Environmental data sensing can be critical in an industrial setup, especially where continuous exposure to environmental factors beyond a limit can be detrimental to human health. Environmental sensing technologies are under development, similar to other technologies discussed in this chapter.

22.2.2.3 Non-Conventional Sensing Modules

Non-conventional sensing modules are lab setup detection technologies, miniaturized to be used as wearable devices. Various sensing modules are discussed as follows.

22.2.2.3.1 Photo Sensors

Photosensors are not essentially limited to photon and optical detection techniques. Ongoing developments are focused on photoelectric sensing modules and spectroscopic detection of ultralow concentrations. Photoelectric sensing modules involve paramagnetic, surface plasmon resonance, fluorescence spectroscopy, and so on, whereas spectroscopic techniques involve optical spectroscopy, tomography, and Raman spectroscopy. Primarily these sensing modules are helpful in biological marker detection, which can provide significant information about the disease before it is critical. Pretreatment of disease requires their identification at the origination stage of the disease. Thus, further enhanced optical sensing techniques such as fluorescence and Raman spectroscopy have been explored for the early identification of various problems.

Biomarker detection is another promising aspect of developing the healthcare ecosystem. Disease-specific biomarker detection helps in determining the disease prognosis, taking preventive measures, and monitoring any physio-pathological condition. Optical methods are non-invasive and have considerably high resolution making them suitable for biomedical applications. Table 22.3 lists different biomarker detection techniques and their sensitivity to cardiac problems [4].

Although having several advantages, these setups are costly, sophisticated, bulky and their signal transmission and processing is relatively difficult compared with electromagnetic sensing methods. As discussed previously, wearable devices are first developed for point care real-time detection methods, and further miniaturized on flexible substrates to make them wearable devices. Optical methods are inherently in vitro diagnosis methods, thus transforming these setups into wearable devices is difficult. However, development in microfluidics, nanofabrication, and imaging, optical spectroscopic devices has shown

TABLE 22.3

Different Biomarker Detection Techniques for Cardiac Biomarkers and Their Performance Analysis

Method	Platform	Analytical time	POCT potential	LOD
Horseradish peroxidase (HRP)	Colorimetric	15 min	High	0.027 ng/ml
HRP	Colorimetric	20 min	Low	5.6 ng/ml
Au nanoparticle	Colorimetric	10 min	High	0.01 ng/ml
Fluorescein isothiocyanate (FITC)	Fluorescence	1 min	High	5 pg/ml
Europium (III)	Fluorescence	15 min	High	2 pg/ml
Aptamer	Fluorescence	1 hour	Low	5 ng/ml
Erenna immunoassay	Paramagnetic	15 min	High	0.2 pg/ml
Magnetic nanoparticles	Paramagnetic	5 min	High	0.2 pM
Magnetic nanoparticles	Paramagnetic	4 min	High	0.5 ng/ml
HRP enzyme	Electrochemical	1 hour	Low	1 ng/ml
Quantum dots (QD)	Electrochemical	3.7 hour	Low	4 pg/ml
Au nanoparticle	Electrochemical	1 hour	Low	0.2 ng/ml
Au film dextran layer	SPR	35 min	Low	1.4 ng/ml
Peptide epitope	SPR	15 min	High	0.068 ng/ml
Au nanorod	SPR	5 min	High	10 ng/ml

the potential of having miniaturized and wearable devices for real-time monitoring of biomarkers [5].

22.2.2.3.2 Strain Sensors

Accelerometer and gyroscope-based motion sensors have been developed for a long time and are still used in some devices. Actigraph and SHIMMER are still in use, however, their power requirement, sensitivity, and accuracy pose problems in proper data assessment. Recent progress has focused on the development of pressure/strain-based sensors and electromyography. Bending and unbending of body parts develop strain in the sensing material attached to the joints of the body. Such strain sensors utilize changes in resistance, optical properties, or capacitive properties due to applied pressure or strain in the sensing material. Different kinds of strain sensors used for motion detection are listed in Table 22.4.

Recent developments in high-resolution nanofabrication techniques have allowed researchers to focus on developing a flexible stretchable fabric sensor suitable for wearable devices. Tolvanen et al. [6] fabricated a washable strain-based sensor by coating a thin film of silver ink on a pre-stretched elastomer surface. Buckled film on the flexible substrate can be reversibly changed between buckling and cracking by different levels of stretching, hence varying degrees of electrical, resistive, and optical properties. A versatile degree of equipment can be developed using the microcracking mechanism on the flexible substrates [7, 8]. Wearable devices are supposed to be worn for a long duration and sweating and pollution may significantly deteriorate the device's appearance and performance. Thus, washability is an important parameter that needs to be taken care of while developing such devices. Wu et al. coated a conductive polymer composite layer on the polyurethane yarn for small motion sensing, such as coughing, swallowing, and pulse, with a gauge factor of 39 [9, 10]. Developments in this area are promising because of highly tailored polymer material availability.

There is extensive development ongoing in the preparation of flexible, stretchable, washable, and wearable devices for body motion and vibration sensors, with a vital need for

TABLE 22.4

Different Strain-Based Motion Sensors

Material System	Sensor type	Stretchability (%)	Linearity	Gauge factor	Response time (ms)
rGO/TPU	Resistive	100	Two linear regions	11–79	200
CNT nanopaper/PDMS	Resistive	100	Three linear regions	Up to 2.21	50
CNT/Pre-stretched TPU	Resistive	300	Three linear regions	428.5–83,982.8	70
Parylene/Au film/ pre-stretched elastomeric adhesive	Capacitive	140	Linear	3.05	NA
Ag Nfs/ionic hydrogels	Capacitive	1000	Linear	165	320
Au NP/PDMS	Optical	100	Linear	9.54 db/ε	<12
PDMS fibers	Optical	100	Linear	3.62 db/ε	NA
Graphene PDMS fibers	Optical	100	Linear	13.5 db/ε	NA

algorithms to extract useful information from the motion and gait of the patient. These mechanical stretchable sensors face strong competition with other detection techniques, for example, developments in optical detection techniques such as magnetic resonance imaging (MRI) have been advanced in recent times to detect Parkinson's disease with 85% accuracy in the early stage [11]. Raman spectroscopy methods have also been used for the stratification of the disease [12]. Extracellular vesicles (EV) act as vehicles for Parkinson's disease molecules and thus can be taken as biomarkers. Raman spectroscopy protocols have been developed for the label-free detection of circulating EV as a diagnostic and predictive tool for Parkinson's disease. However, similar to cardiac biomarkers, neural monitoring through optical techniques is costly, bulky, and sophisticated. There is a cost and time advantage associated with digital strain-based sensors, giving them an edge.

22.2.2.3.3 Humidity Sensors

Humidity sensors can be found in a variety of applications in the fields of structural health monitoring, indoor climate control, industrial process control, food processing and storage chains, agriculture, and aerospace. Wearable humidity sensors have found extensive applications in biomedical healthcare monitoring and industrial safety applications. Humidity sensors can be used for biofluid collection in the form of sweat, saliva, blood, and so on. Humidity sensors can be classified into four major classes, namely resistive, capacitive, impedance, and piezoelectric. Table 22.5 summarizes various humidity sensor principles and their sensing materials.

22.2.2.4 Non-Conventional Materials and Integration Methods

Integration of materials in the wearable device is an important aspect of development. The compatibility of materials in terms of reusability, power harvesting, and overall device

TABLE 22.5

Sensing Principles and Materials Used in Wearable and Flexible Humidity Sensors

Sensing technique	Material system used
Capacitive	Hydrophilic polytetrafluoroethylene
	Cellulose acetate butyrate
	Yarn (composition: polyester)
	Nanoscaled polypyrrole
Resistive	Functionalized multiwalled carbon nanotube/hydroxyethyl cellulose
	PEDOT:PSS confined in 1D nanowire
	Polyvinyl alcohol/KOH polymer gel electrolyte (porous ionic membrane)
	Poly(3,4-ethylenedioxythiophene)/reduced graphene oxide/Au nanoparticle
	rGO/WS2
	CuO
	Sulphonate polystyrene poly(3,4-ethylenedioxy thiophene)
Impedance	Biopolymer kappa carrageenan (KC) carbon nanotubes (CNTs)
	Polypyrrole
Piezoelectric	Cadmium doped zinc oxide nanowire nanogenerator (Cd-ZnO nanowire)
	Sodium niobate (NaNbO3 NFs)
Optical	CdTe@Au/NaOH film

performance is an important aspect to be explored. The following integration methods of materials in wearable sensors have been widely explored.

22.2.2.4.1 Self-Healing Materials

Self-healing materials can be broadly defined as materials that do not lose their sensing properties even after being mutilated in the course of time. Self-healing materials can be classified into two major classes: solvent-free or high crosslinked polymers (polyamide, polyurethane) and hydrogels or less crosslinked polymers or structural crosslinking.

Sensors with elastomers and hydrogels as flexible matrices deteriorate in their conductive capability as a result of stretching, which affects the stability and sensitivity of the sensor. General practice is to fill the inter void by conducting material (polyaniline, polypyrrole). Filling solid conducting nanoparticles creates phase separation that reduces the tensile strength, toughness, and even self-healing capability of the polymer matrix. The self-healing mechanism of the polymer materials can be classified into three mechanisms. Table 22.6 lists different types of repair mechanisms, material structures, and physical properties.

22.2.2.4.2 Self-Powered

Flexible and wearable sensors have been developed as an integral part of the personal healthcare assistant measuring physiological parameters passively and continually. Although, researchers have a wide interest in wearable technology for sensing different physiological parameters, a bottleneck of low power output and conversion efficiency always remains. Low power output also hinders the integration into multifunctional devices and gives poor stability and detrimental environmental impacts. There are various mechanisms explored to provide an alternate power source, such as micro-supercapacitors, nanogenerators, piezoelectric energy harvesting, and self-powered smart chemical sensors. The most promising energy harvesting devices under development use the piezoelectric effect through the motion of body junctions to generate energy in a sufficient amount to be used by the wearable device. A recent study uses a polyvinyl difluoride (PVDF) strap to form

TABLE 22.6

Self-Healing Mechanism of Different Materials and Their Physical Properties

Type	Repair mechanism	Tensile strength (MPa)	Repair efficiency (%)	Repair conditions
Dynamic covalent bond (DCB)	Acyhydrazone bond	2×10^{-3}	100	25°
	Disulfide bond	25	86.4	70°
	Diels-Alder bond	22	96	65°
Reversible non-covalent bond (RNB)	Hydrogen bond	0.4	91.5	RT
	Electrostatic interaction	0.23	84	RT
	Metal ligand	0.97	107.17	RT
Multiple healing mechanism (MHM)	Metal ligand/hydrogen bond	10	92.2	RT
	Hydrogen bond/ionic interaction	2.76×10^{-2}	100	RT
	Boroxines bond/hydrogen bond	1.86	91	RT
	Hydrogen bond/covalent bond interpenetrating	3	100	<35°

two helical structures in counter directions in an elastic core [15]. With body movement, the helical harvester was stretched, and torsional and longitudinal stress was experienced. These energy harvesters can be embedded into garments at the limb's junction. With a maximum stretching of 2 cm, an output voltage of 10 V was demonstrated.

There is growing research on the materials and equipment to reduce power requirements, as well as to develop alternative power sources in combination, to either recharge the inbuilt battery or act as a separate power source. Figure 22.4 describes the various challenges and development prospects for wearable self-powered systems.

22.2.2.4.3 *3D Manufacturing*

Three-dimensional technologies have been used for more than three decades and have proved their importance in research and teaching laboratories. With the advancement in miniaturized fabrication, 3D technology has become the next-generation integration and manufacturing tool for wearable devices. 3D manufacturing can be used to fabricate low-cost, lightweight, ultra-thin, and highly flexible and stretchable devices for wearable technology. Emerging additive manufacturing technology in biomedical devices provides on-demand and precise controlled customized fabrication with controllable geometry, using printing, deposition, and so on. There are three major aspects of 3D printing technology:

a) Materials for 3D printing

b) Classification of 3D printing technologies

c) Wearable sensing applications of 3D printing technology

Materials of 3D printing technology are the most important aspect of development; suitable material decides the precision of the desired architecture. Traditionally, a variety of materials have been used for 3D additive manufacturing such as plastic, composites, thermoplastics, resins, ceramics, and metals. There is a sufficient increase in the identification

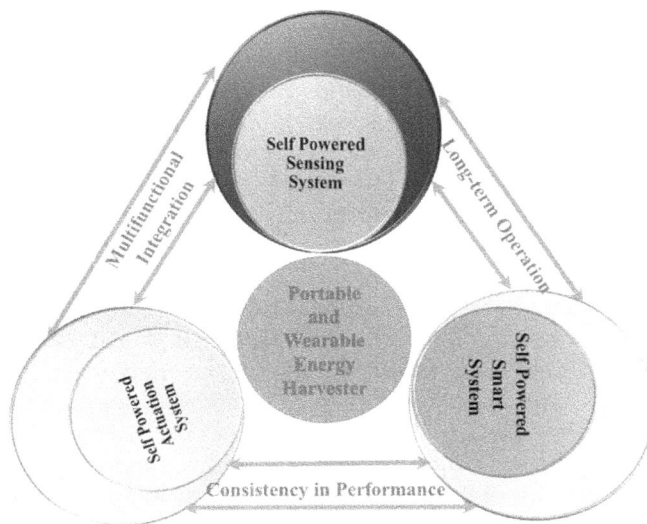

FIGURE 22.4
Various modules of power sources for self-powered wearable sensors.

and development of more compatible and advanced materials to fabricate the required 3D model with the application of heat, light, and/or other energy sources to provide the structure directly from the raw material. Emerging 3D printing technologies and materials have created possibilities to achieve application-specific requirements. For example, polyethylene terephthalate (PETG), thermoplastics (polylactic acid (PLA), acrylonitrile butadiene styrene (ABS), nylon, polyetherimide (PEI), etc.) are generally used for the fabrication of finished parts for end-user application, such as functional prototypes and device integration housing. Moreover, the research community continues to advance 3D printing technologies and has demonstrated a significant improvement in the last decade. In this scenario, 3D printing technology developments can be studied in the following dimensions: (i) development of advanced functional materials and corresponding application-specific printing technology; (ii) improvement in accuracy and printing speed; and (iii) enhancement in fabricated structure resolution from micrometer to nanometer scale. There are several metals with superior electrical and mechanical properties that can be used for fabricating more advanced 3D structures with conductive pathways. For such applications, the 3D printing process and electroplating process have been explored where stainless steel electrodes are commonly used in the 3D printing process and gold, platinum, iridium oxide, bismuth, and nickel are used for electroplating processes.

Polymeric materials are lightweight and their tailoring ability is very impressive, however, materials like polyurethane (TPU), ABS, and PLA were not found suitable for the fabrication of flexible wearable sensors due to their rigid nature, low bending stiffness, bulky interfaces. Mechanical confrontation with the skin, inducing discomfort for wearable devices in long-term wearability. To address such issues, soft and light materials with better stretchability and a lower modulus of elasticity are preferred to make the surfaces of flexible wearable devices show compatibility with human skin. Under such conditions, silicone elastomers are the most favored material for developing wearable sensors because of their low elastic modulus, better flexibility, and high compatibility with the skin. More specifically, polydimethylsiloxane (PDMS, a popular silicone elastomer) has been widely explored to develop lab-on-chip, patient-specific, wearable devices as per the requirement of design models. Also, PDMS-based 3D-printed devices are more reliable because they do not require adhesive to integrate sensing parts in the device at a fixed place in the 3D-printed architecture. Due to the transparency of silicone elastomers (PDMS), they can provide continuous monitoring of the integrative elements in the device architecture during the fabrication process. In many cases, the integration of wearable sensors in the architecture requires conductive stretchable materials, which can be developed by using conductive fillers in the silicone elastomer matrix. Carbon-based conductive nanofiller materials such as carbon nanofiber, graphene, CNT, carbon black, and many other similar developments are under investigation for the development of conductive flexible and stretchable wearable sensors.

An important consideration in 3D manufacturing is creating an effective 3D design. There are many ways to develop a 3D virtual model such as a 3D scanner, digital design software, or a combination of three different images taken from an angle of 120° to create a 3D virtual model [16]. These files are usually converted and stored in "stereolithography" file formats popularly termed STL files. Generation and decoding of these STL files are one of the digital aspects of the development of 3D-printed wearable devices. Almost all the 3D printing software recognizes these STL files, and using an appropriate slicing software (IdeaMaker, Cura, etc.), a G-code file is obtained from the STL files. The G-code file provides the printing instruction details such as feed rate, extrusion rate, printing pathways, and material details [17]. Finally, layer-by-layer deposition of application-specific material by following the design-specific G-code instructions is done to achieve the required 3D device.

The fabrication of 3D architecture and integration of sensing units in the 3D printed architecture to develop a wearable device is a rigorous process. The selection of 3D printing technology is important in determining the ease of printing complex 3D architecture [18]. In the last two decades, increasing growth has been observed in 3D printing technologies, and such technologies have opened various domains of development and fabrication of biomedical wearable device. Some of the major 3D printing technologies include (i) material jetting, (ii) VAT photopolymerization, and (iii) material extrusion.

Three-dimensional printing technology has made the realization of smart wearable biosensor devices such as glucose sensors, pressure sensors, strain sensors, and wearable electronics. Emerging applications of 3D printing technologies are haptic skin, conformal tattoo sensors and artificial skins. Such applications require smart functionalized materials and sensing architecture development (Table 22.7).

22.2.3 Digital Biomarkers (Multifunctional)

The power of wearable devices lies in the passive monitoring of the physiological parameters of the body and the digital processing of the data acquired. A large set of untapped data from active monitoring sensors can be tapped in real time and utilized for healthcare monitoring. Digitization of the sensing modules and smartphone-based applications is an important domain of IoT technologies. IoT devices share the tapped data by connecting to an IoT gateway where data is either analyzed locally or sent to the cloud for analysis. Such technologies have allowed passive tapping of the physiological parameter data for the effective development of the healthcare ecosystem.

Passive monitoring of physiological parameters generates a voluminous amount of data that need to be analyzed. These data can be structured, semi-structured, or unstructured. Big data or data science technologies allow us to figure out useful information in multiple domains of healthcare such as sleep patterns, step counting, stress patterns, and electronic signal processing of cardiac HRV, while pulse monitoring can be termed as a digital biomarker. Digital biomarkers can be defined as contextual information regarding a particular ailment from a large set of physiological activity data. Determination of redundancy in

TABLE 22.7

Some of the Application Materials and 3D Printing Processes in Existing and Emerging Wearable Sensing Devices

Material	Printing technology	Stimulus	Application
PNIPAAM Hydrogel	μSLA	Thermal	Soft robotics, microfluidic devices
Poly(2-vinylpyridine)	FDM	PH	Membrane and photonic gels
Carboxymethyl cellulose	DIW	Moisture	Complex architecture and actuators
Poly(lactic acid) and iron nanoparticles	DIW (UV crosslinked)	Magnetic field	Scaffold for an intravascular stent
PEO/CNT	μSLA	Electric field	Artificial muscles, soft robotics
Acrylic acid and modified thermochromic pigment	SLA	Thermal	Smart temperature sensors, packaging
Commercial resins	Inkjet	Thermal	Aerospace coating protections

the predictive power of HRV and HRD indices [19], identification of new digital biomarkers [20], and patient-centric data pattern generation recognition and prognosis of diseases are important prospects of big data analytical tools. Identification of digital biomarkers is the most important aspect of big data analysis, which can be person-centric or for the general population. Increasing effort by researchers in this domain is required to identify important parameters in terms of external observable symptoms such as skin resistance, motion of limbs, gait, speech, and other physiological parameters. Based on such external physiological parameters, underlying disease identification will be the next achievement in the development of wearable device technology.

22.3 Future Applications

22.3.1 Medical Devices

Wearable devices in combination with cloud storage, machine learning tools, and fast data processing can be used for real-time data monitoring and e-healthcare system development.

22.3.2 Industrial Applications

There are lots of gases produced in industrial production processes. Humidity and electrochemical sensors can be used for safety and monitoring applications of industrial processes.

22.3.3 Sports

Real-time gait and physiological data measurement can be of great help in sports and fitness training. Wearable devices can provide effective solutions in such cases. Such devices will provide a competitive edge in training and monitoring processes.

22.3.4 Entertainment

Wearables in the form of spectacles associated with 3D features have been developed and need to be popularized with extensive features. Such devices can be useful in virtual meetings and learning devices. Development of individual theatres like Google glass have been developed, however, a lot more development is still required in this field.

22.3.5 Consumer Direction Prospects

Until now, the wearable device market has been substantially available with new solutions in the market. In this section, we have identified increasing mass consumer demands. A significant effort in developing wearable devices for these applications is in practice in academia and industries, however, a fully realized device is still a far-reaching objective. The commercial market for wearable devices depends mainly on consumer needs and significant improvements in existing devices. Figure 22.5 depicts various domains of consumer demands in the coming future with increased priority. Individual application domains have been discussed to understand consumer needs and the scope of future developments in flexible wearable devices.

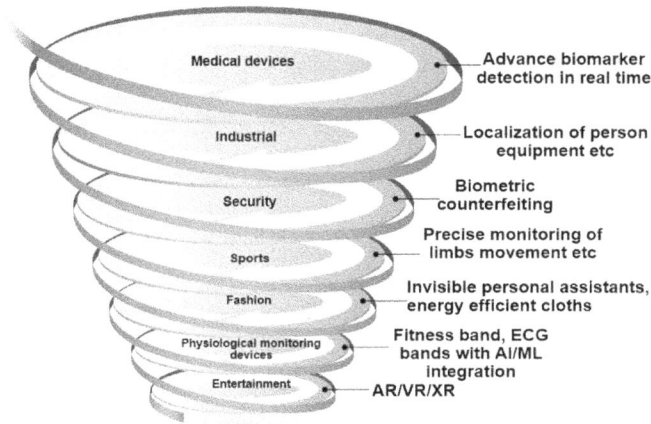

FIGURE 22.5
Future applications of wearable devices.

22.4 Challenges

Digital technology is going to impact our lives in more ways than just personal healthcare and convenience. With such a ubiquitous presence of internet-enabled technologies, there are several concerns raised over the development of wearable devices, which need to be confronted before venturing into a future-ready wearable device. In this section, we will discuss different problems and their possible solution for effective development.

Of prime concern with cloud-enabled wearable devices operating over the internet is the privacy and safety of the data. Privacy of the data available on the cloud is always compromised and can reveal significant health and physical activity-related information that one may not be comfortable with revealing in the public domain. Localized data processing may be one immediate solution for this problem, however, it will increase power requirements and data management issues, and create bulky architecture in the device. Development in blockchain technology offers an effective solution to individual data safety on the cloud. However, data encryption using blockchain technology requires increased data transfer and processing. Blockchain-enabled data safety requires an additional band. The IoT-based wearable device market is increasing at a compounded annual growth rate (CAGR) of 11.8%, and by 2022, the number of devices is expected to be more than two billion. Presently available telecom infrastructure is not sufficient to provide the necessary data transfer protocols and may collapse in a setup where excessive devices are in use. It is expected that 5G telecom technology, to be enabled in the next few years, will be a boost for wearable technology from the data transfer and processing perspective.

Ethical concerns regarding wearable devices are also important. There are many places where electronic devices are not allowed, such as examination halls, electronic device-protected places, or internet jammer-enabled spaces. People supported with electronic wearables such as pacemakers or hearing devices may face difficulties in working in such scenarios. Draining out the power supply battery may pose an additional risk, which is inevitable for power-driven wearables. Local data processing units may provide data safety but requires additional power. Machine learning/AI-based setups will require more power than any other setup, and local processing of data is not possible in such cases. Self-powered devices have offered a promising scope of avoiding device failure due to power shortages. Self-powered devices derive power from body fluids such as the electrolysis of glucose or physiological movements, which can be used for devices like pacemakers; apart from self-powering devices, cloud-based data handling can also significantly reduce the power requirement and risk of device failure. Wireless charging is also an emerging field that needs to be successfully incorporated into wearable devices. Wireless charging will enable externally worn wearables to be charged without hindering individual movements and avoid inconvenience.

Healthcare is a highly regulated sector because health assets are at risk with any new technology introduced. Discretization of the healthcare ecosystem may increase unregulated prescriptions and medications in remote setups. It has been observed that getting real-time data may increase panic among patients and they may seek their own medical prescriptions, which is not recommended. Due to an unusual individual routine, the device may be uncalibrated and could provide impaired results, which will again create risk for the patient. Thus, continuous calibration of the device and keen observation of the regulatory authorities will be required with the development of wearable technology. Digital regulation protocols need to be developed for the regular calibration of a wearable devices and to help regulatory authorities work efficiently. With many advantages provided by wearable devices, it is important for consumers to regulate their health literacy and work in line with the guidelines provided by healthcare regulatory authorities.

References

1. Kenry; J. C. Yeo, C. T. Lim, Emerging Flexible and Wearable Physical Sensing Platforms for Healthcare and Biomedical Applications. *Microsystems and Nanoengineering*, 2, 2016, 16043.
2. F. Salamone, M. Masullo, S. Sibilio, Wearable Devices for Environmental Monitoring in the Built Environment: A Systematic Review. *Sensors*, 21, 2021, 4727.
3. X. Han, S. Li, Z. Peng, A. M. Othman, R. Leblanc, Recent Development of Cardiac Troponin I Detection. *ACS Sensors*, 2, 2016, 106–114.
4. Q. Ma, H. Ma, F. Xu, X. Wang, W. Sun, Micro Fluidics in Cardiovascular Disease Research: State of the Art and Future Outlook. *Microsystems and Nanoengineering*, 7, 2021, 19.
5. H. Souri, H. Banerjee, A. Jusufi, N. Radacsi, A. A. Stokes, I. Park, M. Sitti, M. Amjadi, Wearable and Stretchable Strain Sensors: Materials, Sensing Mechanisms, and Applications. *Advanced Intelligent Systems*, 2, 2020, 2000039.
6. J. Tolvanen, J. Hannu, H. Jantunen, Stretchable and Washable Strain Sensor Based on Cracking Structure for Human Motion Monitoring. *Scientific Reports*, 8, 2018, 13241.
7. J. Zhou, H. Yu, X. Xu, F. Han, G. Lubineau, Ultrasensitive, Stretchable Strain Sensors Based on Fragmented Carbon Nanotube Papers. *ACS Applied Material Interfaces*, 9, 2017, 4835–4842.
8. H. Wu, Q. Liu, W. Du, C. Li, G. Shi, Transparent Polymeric Strain Sensors for Monitoring Vital Signs and Beyond. *ACS Applied Material Interfaces*, 10, 2018, 3895–3901.
9. X. Wu, Y. Han, X. Zhang, C. Lu, Highly Sensitive, Stretchable, and Wash-Durable Strain Sensor Based on Ultrathin Conductive Layer @ Polyurethane Yarn for Tiny Motion Monitoring. *ACS, Applied Material Interfaces*, 8, 2016, 9936–9945.
10. L. Tian, B. Zimmerman, A. Akhtar, K. J. Yu, M. Moore, J. Wu, R. J. Larsen, J. W. Lee, J. Li, Y. Liu, B. Metzger, S. Qu, X. Guo, K. E. Mathewson, J. A. Fan, J. Cornman, M. Fatina, Z Xie, Y. Ma, J. Zhang, Y. Zhang, F. Dolcos, M. Fabiani, G. Gratton, T. Bretl, L. J. Hargrove, P. V. Braun, Y. Huang, J. A. Rogers, Large-Area MRI-Compatible Epidermal Electronic Interfaces for Prosthetic Control and Cognitive Monitoring. *Nature Biomedical Engineering*, 3, 2019, 194–205.

11. A. Gualerzi, S. Picciolini, C. Carlomagno, F. Terenzi, S. Ramat, S. Sorbi, M. Bedoni, Raman Profiling of Circulating Extracellular Vesicles for the Stratification of Parkinson's Patients. *Nanomedicine Nanotechnology, Biology Medicine*, 22, 2019, 102097.
12. T. Delipinar, A. Sha, M. S. Gohar, M. K. Yapici, Fabrication and Materials Integration of Flexible Humidity Sensors for Emerging Applications. *ACS Omega*, 6, 2021, 8744–8753.
13. S. Li, X. Zhou, Y. Dong, J. Li, Flexible Self-Repairing Materials for Wearable Sensing Applications: Elastomers and Hydrogels. *Macromolecular Rapid Communications*, 41, 2020, 2000444.
14. Y. Liu, H. Khanbareh, M. A. Halim, A. Feeney, X. Zhang, H. Heidari, R. Ghannam, Self Powered-Piezoelectric Energy Harvesting for Self-Powered Wearable Upper Limb Applications. *Nano Select*, 2, 2021, 1459–1479.
15. C. L. Ventola, Medical Applications for 3D Printing:Current and Projected Uses. *P&T*, 39, 2019, 704–711.
16. B. C. Gross, J. L. Erkal, S. Y. Lockwood, C. Chen, D. M. Spence, Evaluation of 3D Printing and Its Potential Impact on Biotechnology and the Chemical Sciences. *Analytical Chemistry*, 86, 2014, 3240–3253.
17. X. Tian, J. Jin, S. Yuan, C. K. Chua, S. Tor, K. Zhou, Emerging 3D-Printed Electrochemical Energy Storage Devices: A Critical Review. *Advanced Energy Materials*, 7, 2017, 1700127.
18. K. R. Ryan, M. P. Down, C. E. Banks, Future of Additive Manufacturing: Overview of 4D and 3D Printed Smart and Advanced Materials and Their Applications. *Chemical Engineering Journal*, 403, 2021, 126162.
19. E. Yuda, N. Ueda, M. Kisohara, J. Hayano, Redundancy among Risk Predictors Derived from Heart Rate Variability and Dynamics: ALLSTAR Big Data Analysis. *Annals of Non-Invasive Cardiology*, 26, 2021, 1–7.
20. E. Yuda, M. Shibata, Y. Ogata, N. Ueda, T. Yambe, M. Yoshizawa, Pulse Rate Variability: A New Biomarker, Not a Surrogate for Heart Rate Variability. *Journal of Physiological Anthropology*, 39, 2020, 1–4.

Index

A

Additive manufacturing (AM) 32, 153, 208, 214, 336–40, 336, 337, 342, 342, 344, 346, 361

B

Bands 16, 99

Biomedical 5, 6, 27, 51, 53, 61, 66, 72, 81, 105, 110, 127, 149, 150, 153, 194, 198, 203, 210, 211, 214, 228, 239, 250, 284, 291, 293, 323, 324, 326, 327, 330, 331, 355–57, 359, 361, 363

Biosensor(s) 11, 12, 19, 23, 27, 30–32, 34, 36, 39, 40, 41, 46, 52–54, 56, 59–61, 66, 68, 69, 78, 105–7, 110, 113, 141, 142–44, 148, 216, 226–28, 237, 241, 242, 247, 250–52, 263, 268, 291, 291, 341, 347, 356, 363

C

Capacitance 9, 10, 76, 85, 86, 103, 104, 110, 118, 119, 152, 190, 191, 207, 261, 262, 304, 311–13, 315, 324, 326, 354

Carbon 25, 31, 37, 40, 41, 56, 66, 70, 83, 102, 106, 109, 121, 137, 151, 160, 163, 187, 190, 194, 198, 204, 214, 215, 222, 229, 239, 257, 278, 289, 305, 328, 359

Chalcogenides 150–55, 180

CNTs 25, 29, 37, 40, 41, 46, 87, 91, 187, 191, 198, 204, 211, 214, 289, 323, 341, 344, 359

E

Electromagnetic 166, 290, 295, 336, 353, 357

Electronics 1–4, 6–8, 10, 13–15, 23, 25, 27, 29, 30, 41, 45, 58, 66, 102–5, 108–10, 116, 125, 128, 129, 150, 155, 159, 162, 198, 200, 203, 208, 210, 214, 254–57, 262, 266, 283, 284, 286, 287, 293, 299–302, 308, 313, 316, 322–24, 327, 337, 338, 341, 342, 346, 363

Environment 3, 6, 14, 23, 27, 36, 51, 70, 72, 74, 81, 86, 136, 149, 159, 172, 191, 194, 200, 211, 222, 249, 254, 280, 284, 287, 294, 302, 309, 311, 356

Enzymatic 51, 53, 57, 60, 106, 141–43, 152, 226, 242, 243, 245, 289

F

Field Emission Scanning Electron Microscopy (FESEM) 26, 27

Flexible 2, 3, 6–8, 14, 15, 17, 21–31, 35–45, 52–55, 57, 58, 60, 65–68, 72, 74, 76, 77, 79, 81–92, 99, 101–3, 105–10, 115, 116, 118–21, 124–26, 128, 135, 137–46, 150–56, 159–67, 169, 171–74, 180, 181, 184, 186, 188, 191, 198, 200, 203–5, 207–14, 219–21, 224, 225, 228–32, 237–40, 248, 254–58, 260, 262–66, 271–75, 278–80, 283, 284, 291, 293, 295, 299–301, 305, 314, 322–31, 336–46, 353–62, 364, 367

G

Graphene 7, 25–27, 30, 38–43, 53, 57, 58, 67, 68, 70, 74, 109, 110, 120, 121, 127, 128, 137, 143, 144, 150, 151, 155, 163, 165, 169, 186, 187, 191, 194, 198, 199, 204, 208, 210, 212–14, 222, 224–27, 229–31, 239, 241, 245, 257, 275, 278, 279, 294, 295, 304, 323, 339, 341, 358, 359, 362

H

Humidity 6, 13, 38–42, 52, 57, 68, 74, 76, 77, 81, 86–91, 140, 141, 150, 151, 153–55, 159–61, 169, 171, 172, 174, 194, 200, 213, 263, 274, 275, 280, 287, 288, 294, 295, 304, 308, 309, 315, 316, 321, 329, 331, 353, 354, 359, 364, 367

Hydrogel 40, 41, 43, 68, 70, 82, 83, 85, 86, 88–92, 124, 160, 172, 173, 180–91, 193–99, 220–23, 228–31, 272–74, 278, 312, 313, 340–42, 344, 363

L

Light emitting diode (LED) 104, 141, 167, 169, 170, 292

M

Metal-organic frameworks (MOFs) 43, 52, 53, 55, 56, 59, 169, 257

Metal-oxide-semiconductor 43

For Product Safety Concerns and Information please contact our EU
representative GPSR@taylorandfrancis.com
Taylor & Francis Verlag GmbH, Kaufingerstraße 24, 80331 München, Germany

www.ingramcontent.com/pod-product-compliance
Lightning Source LLC
Chambersburg PA
CBHW080711220326
41598CB00033B/5378

9 781032 289809